VEGETABLES
Characteristics, Production, and Marketing

VEGETABLES
Characteristics, Production, and Marketing

Lincoln C. Peirce
University of New Hampshire

John Wiley and Sons

New York   Chichester   Brisbane   Toronto   Singapore

Library of Congress Cataloging in Publication Data:

Peirce, Lincoln C.
  Vegetable: characteristics, production, and
marketing.

  1. Vegetables. 2. Truck farming. 3. Vegetable
trade. 4. Vegetables—United States. 5. Truck
farming—United States. 6. Vegetable trade—United
States. I. Title.
SB320.9.P45 1987   635   86-32625
ISBN 0-471-85022-5

Printed in the United States of America

10 9 8 7 6 5 4 3 2 1

Preface

Over 85 distinct taxonomic groups constitute the majority of vegetables grown in the United States. On a global basis, the number is much higher, including many plants indigenous to tropical and subtropical regions, but relatively unknown in temperate areas. The focus of this book is on those crops that contribute to the commercial vitality of the vegetable industry in the western hemisphere, particularly in North America. These crops include some of major importance to world food supplies (such as potato, beans, cole crops) and others of relatively local significance (okra, taro).

The vegetable industry is a dynamic one. Cropping practices, regardless of commodity, change continually, incorporating innovations by growers and new technology supplied by biochemists, physiologists, geneticists, engineers, economists, and others at an accelerating rate. It is difficult to partition these technologies within a textbook, for they represent inputs within highly integrated vegetable production systems. A given technology also may have a different purpose for different crops in different production areas.

Students interested in vegetable production or marketing must recognize that techniques and skills alone are insufficient to maintain a progressive business. A broad understanding of basic sciences and of the economics of the marketplace is essential if one is to remain progressive. This text is directed toward the sophomore or junior level student who has acquired an understanding of basic sciences. The first portion of this book is devoted to resources—natural, biological, and economic—that are fundamental to successful vegetable production and to the general vegetable management systems that have evolved to integrate those resources. The remaining chapters include the importance, history, botany, cultural methods, and handling of specific vegetable crops and descriptions and general control methods for the common insect and disease pests.

It is not feasible to describe each variant of

a production system: growers often develop modifications that succeed within their specific environments and market structure. An understanding of the plant and its response to stress and to growing and marketing practices is most important as is a full awareness of the kinds of environmental pressures that may affect growth or market quality. The choice of specific application rate of fertilizers or pesticides or, in many instances, of specific materials will differ according to soil, climate, and legal restrictions within a state or county. Local extension services should be consulted for specific recommendations and for proper application methods within the existing regulatory framework. Specific practices presented in the text are intended as illustrative of systems developed to produce and protect a crop, not as general recommendations to be applied regardless of local circumstances.

Many people have contributed to this book—growers and extension specialists who demonstrate the practicability of different production and marketing systems, scientists who have published their research and otherwise shared their understanding of production problems and opportunities, and those in industry who have created new products that enhance production skills or improve production volume and quality. Particular thanks are extended to Lincoln C. Peirce III and Amy Bartlett Wright for their artwork, to those who offered critique and suggestions for improving each chapter, and to those who allowed their data and photographs to supplement the text.

Lincoln C. Peirce

Contents

VEGETABLES

Characteristics, Production, and Marketing

1

The Vegetable Industry

In the 1940s, American policy was guided by the Jeffersonian ideal that our society would be served best by traditional family-size, owner-operated farms. Substantial change has occurred in the vegetable industry since World War II. Until that time, small family farms did predominate, relying heavily on local labor, suppliers, and consumers to sustain their business. After 1945, the rapid transition to volume marketing systems, the improvements in technology of food handling systems, including refrigeration, developed during wartime, and the changing economic structure of American agriculture, spurred by the federal highway expansion of the 1950s, all favored those growers who could supply the

market with a large volume over a prolonged time period. Small production units were inefficient in that setting and either failed or enlarged to meet the new challenges. They enlarged through purchases of additional land or through production/marketing cooperatives. They maintained competitiveness by adopting new technology and by stressing high quality in vegetable production and handling.

Today's vegetable industry is a highly interdependent network of producers, suppliers, processors, seed growers and dealers, brokers, wholesalers, retailers, and service industries. Income attributed to farm sales is substantial, over $3.9 billion, but the significance of these dollars is the value they generate within sectors of the economy supporting vegetable production and within sectors supported by this production. Because of the efficient food production and distribution system and the competitive nature of the industry, the share of the consumer dollar devoted to food purchases is among the lowest in the world (Figure 1.1).

COMPONENTS OF THE VEGETABLE INDUSTRY

Field Production for Fresh Market

The total vegetable industry includes commercial farms and greenhouses and many small units not included in agricultural marketing statistics. Table 1.1 provides an estimate of the total area in vegetables by state. Both the number of farms and the area cultivated have declined steadily since the 1940s as productivity per farm has increased. The average vegetable farm is small by comparison to grain

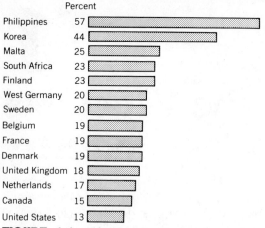

Percent

Philippines	57
Korea	44
Malta	25
South Africa	23
Finland	23
West Germany	20
Sweden	20
Belgium	19
France	19
Denmark	19
United Kingdom	18
Netherlands	17
Canada	15
United States	13

FIGURE 1.1
Share of consumer expenditures for food. 1978 data; UN National Accounts of Statistics and National Sources, 1979. (*Source: Handbook of Agricultural Charts*, USDA Agric. Handb. 609 (1982).)

farms, but the productivity and value per acre are much higher. This high value supports the substantial labor and marketing costs inherent in vegetable production.

In the past, vegetable farming was compartmentalized into distinct kinds of production systems. **Truck farms** were defined as

TABLE 1.1
Distribution of farms and vegetable acreage by state and region, 1954 and 1982[a]

Region and state	Farm (number)		Acreage harvested	
	1954	1982	1954	1982
Northeast				
Connecticut	1,185	511	12,727	8,244
Delaware	1,446	367	40,033	40,421
Maine	2,307	535	18,926	11,278
Maryland	5,345	1,403	94,078	38,331
Massachusetts	2,347	1,011	20,324	15,307
New Hampshire	620	278	3,729	2,974
New Jersey	5,102	1,970	145,879	70,746
Pennsylvania	11,862	3,876	96,154	46,194
Rhode Island	238	120	1,810	1,908
Vermont	577	228	3,804	1,633
West Virginia	1,934	421	3,804	1,288

TABLE 1.1 (Continued)

Region and state	Farm (number)		Acreage harvested	
	1954	1982	1954	1982
Midwest				
Illinois	4,570	1,585	125,437	81,916
Indiana	5,935	1,429	76,415	30,830
Iowa	2,565	429	36,261	8,363
Kansas	852	362	5,330	4,808
Michigan	13,931	3,634	105,358	126,248
Minnesota	7,598	3,152	158,040	177,242
Missouri	2,001	696	12,831	13,797
Nebraska	621	186	2,901	1,677
Ohio	7,897	2,442	63,042	53,594
Wisconsin	21,497	4,262	241,766	280,326
South				
Alabama	11,894	2,341	64,942	24,047
Arkansas	6,397	917	29,220	12,938
Florida	9,766	2,455	323,909	283,780
Georgia	23,384	2,801	167,317	61,973
Kentucky	3,278	1,549	6,882	4,919
Louisiana	6,458	999	34,931	8,597
Mississippi	10,508	1,363	36,025	11,530
North Carolina	22,723	3,938	77,325	54,650
South Carolina	11,125	1,645	86,948	31,538
Tennessee	9,147	2,070	33,203	30,708
Virginia	7,001	1,510	72,716	27,498
Southwest				
Arizona	494	339	69,190	69,864
Colorado	2,139	657	29,640	28,686
New Mexico	1,282	497	9,973	23,502
Oklahoma	2,364	556	26,239	13,987
Texas	21,391	3,434	403,197	210,873
Intermountain				
Idaho	1,635	766	29,237	42,291
Utah	2,429	432	19,488	6,982
West				
California	7,161	4,053	560,116	894,573
Hawaii	—	746	—	4,673
Oregon	2,926	1,554	95,490	134,814
Washington	3,789	2,031	106,787	169,170
Other states[b]	994	431	5,785	3,908

Source: Bureau of the Census, Department of Commerce.

[a] Excludes white potato; vegetable data include all vegetable farms growing for sale.

[b] Alaska, Montana, Nevada, North Dakota, South Dakota, Wyoming.

arising adjacent to transportation systems, dealing with only one or two crops on a substantial acreage for distant marketing. **Market gardens** tended to develop near population centers and supplied a wide array of home-grown produce (typical of today's roadside market operations). The differences among farms today are less distinct than in past years; transportation and market proximity, although important, place few restraints on location of a farm. It is the environment that is the major consideration in locating a successful farm: proximity to water, availability of fertile and well-drained land, length and climatic features of the growing season. These factors have been responsible for development of most major production centers. For small diversified farms, location near a suitable market is still an important advantage.

Fresh market vegetables are produced to some extent in every geographical area of the United States. Of the total acreage and production, the amounts devoted to commercial production of the principal fresh market vegetables by state and their value by year are presented in Tables 1.2 and 1.3. The intensity, size, and objectives of the different production units differ according to climate, demographics (population density, labor supply), and economic pressures resulting from industrial and population growth. Six states, California, Florida, Texas, Arizona, New York, and Michigan, constitute over 70 percent of the harvested acreage and 80 percent of the total U.S. production and value of principal fresh market vegetables. Of these six, the first four produce winter vegetables, thereby inflating percentage domination of the market. California and Florida account for over 50 percent of the nation's fresh vegetable production. During spring, summer, and fall, many areas of the country have substantial acreages in vegetables, both for local and for long-distance marketing.

The trend toward increased farm size, begun after World War II, has persisted. In the major vegetable production areas, corporate purchases of agricultural enterprises, not only of substantial numbers of farming units but also of seed production and distribution enterprises, and other agricultural suppliers and services have provided the capital for rapid development and use of technology. Although bigness has dominated several sectors of the fresh market vegetable industry, the family farm also has thrived, especially in areas with substantial urban populations. Grower-operated roadside markets, pick-your-own systems, and farmer's markets, featuring fresh, high-quality produce, are especially appealing to those unable to enjoy such produce from a home garden. Such operations often are characterized by intensive cropping practices, including intercropping, succession cropping, high plant populations per acre, plastic mulching, and temperature modification to maximize income, particularly within restricted acreages.

Processing Vegetables

Freezing, canning, and dehydration are the major mechanisms by which food is preserved. Growers who supply the raw product to processors do so through contracts that specify, among other conditions of production and sale, some of the production techniques, price per ton at a given level of quality, and standards for acceptance of the harvest (percentage weed seed, foreign matter). The price usually does not approach that of fresh produce in a free market, but the stability of price is attractive to many growers. Production of many processing crops is highly mechanized (Figure 1.2), and cultivars often reflect specific quality components dictated by the type of processing. Processing acreage has become concentrated predominantly in several states. California now accounts for almost 50 percent of the total volume of raw vegetables for processing, and of the four major processing crops, sweet corn, peas, snap beans, and tomatoes, Wisconsin, California, and Minnesota have the largest production (Table 1.2). The value has approximately doubled since 1969 (Table 1.3).

Vegetable Forcing

The greenhouse industry has declined substantially over the past 30 years because of competition from winter production areas and because of high maintenance and energy

TABLE 1.2

Commercial production of principal vegetable crops[a] in the United States by state: Annual average for the period 1981 to 1983

State	Fresh market[b]		Processing[b,c]		Total	
	Acres	Tons	Acres	Tons	Acres	Tons
Alabama	8,250	26,350	1,300	2,690	9,550	29,040
Arizona	45,160	625,470	—	—	45,160	625,470
Arkansas	2,130	10,070	3,670	10,000	5,800	20,070
California	415,500	5,126,470	232,530	5,694,710	648,030	10,821,180
Colorado	16,700	223,250	1,390	15,740	18,090	238,990
Connecticut	3,600	11,750	—	—	3,600	11,750
Delaware	—	—	15,910	48,420	15,910	48,420
Florida	119,930	1,162,250	3,130	23,170	123,060	1,185,420
Georgia	2,100	9,870	930	1,750	3,030	11,620
Hawaii	900	9,400	—	—	990	9,400
Idaho	4,870	124,170	27,000	167,920	31,870	292,090
Illinois	3,730	15,320	58,970	254,270	62,700	269,590
Indiana	1,270	8,050	9,700	131,160	10,970	139,210
Iowa	—	—	4,890	28,850	4,890	28,850
Louisiana	360	1,970	—	—	36	1,970
Maine	—	—	8,540	11,230	8,540	11,230
Maryland	2,600	12,480	18,300	97,140	20,900	109,620
Massachusetts	9,060	43,800	—	—	9,060	43,800
Michigan	34,900	343,980	25,470	210,700	60,370	554,680
Minnesota	2,150	32,300	176,170	687,600	178,320	719,900
Missouri	—	—	610	2,090	610	2,090
New Jersey	19,880	115,020	12,770	109,790	32,650	224,810
New Mexico	8,500	124,150	200	2,310	8,700	126,460
New York	48,010	387,970	70,270	256,220	118,280	644,190
North Carolina	6,700	29,720	2,400	13,160	9,100	42,880
Ohio	19,290	105,600	17,490	343,630	36,780	449,230
Oklahoma	—	—	1,300	4,350	1,300	4,350
Oregon	17,900	323,000	100,100	529,390	118,000	852,390
Pennsylvania	19,470	69,370	16,070	108,660	35,540	178,030
South Carolina	7,300	67,920	300	410	7,600	68,330
Tennessee	4,270	24,570	6,470	15,360	10,740	39,930
Texas	64,100	522,020	4,570	22,660	68,680	544,680
Utah	2,000	34,620	3,240	12,500	5,240	47,120
Virginia	5,130	36,700	2,970	22,760	8,100	59,460
Washington	12,780	203,270	107,130	479,060	119,910	682,330
West Virginia	—	—	27	373	27	373
Wisconsin	6,310	120,170	268,300	907,740	274,610	1,027,910
U.S. total	915,000	9,951,000	1,202,133	10,215,800	2,117,133	20,166,800

Source: Agricultural Statistics (1984).

[a] Including broccoli, carrot, cauliflower, celery, sweet corn, honeydew melon, lettuce, onion, and tomato.

[b] Area for fresh market is area for harvest; area for processing is area harvested.

[c] Data for the following crops in all states: snap bean, sweet corn, green pea, and tomato. Other vegetables processed are included in fresh market data.

TABLE 1.3

Area, production, and value of principal vegetable crops[a] in the United States, 1970 to 1984

	Area[b] (1000 acres)		Production (1000 tons)		Value[c] ($1000)	
	Fresh market	Processing	Fresh market	Processing	Fresh market	Processing
1970	1,674	1,581	11,358	9,297	1,233,222	410,189
1971	1,610	1,558	11,361	9,923	1,438,946	439,980
1972	1,648	1,584	11,578	10,242	1,607,022	466,633
1973	1,637	1,727	11,907	10,662	1,857,859	550,632
1974	1,558	1,776	12,017	11,794	1,885,149	929,785
1975	1,542	1,874	11,994	13,533	2,159,168	1,036,635
1976	1,577	1,625	12,510	11,049	2,260,078	786,606
1977	1,579	1,638	12,741	12,612	2,351,737	945,180
1978	1,644	1,612	13,140	11,323	2,786,530	874,768
1979	1,637	1,652	13,422	12,576	2,919,656	1,030,239
1980	1,610	1,429	13,248	10,807	3,182,975	864,451
1981[d]	865	1,166	9,643	9,222	2,613,119	746,130
1982	922	1,250	10,281	11,180	2,626,319	909,738
1983	928	1,190	9,930	10,246	2,804,157	791,843
1984	1,083	1,371	10,835	11,980	3,089,382	1,007,066

Source: Agricultural Statistics (1985), USDA.

[a] Fresh market data include artichoke, asparagus, lima bean, snap bean, beet, broccoli, brussels sprouts, cabbage, cantaloupe, carrot, cauliflower, celery, sweet corn, cucumber, eggplant, escarole/endive, garlic, honeydew melon, kale, lettuce, onion, green pea, green pepper, shallot, spinach, tomato, and watermelon. Processing data include lima bean, beet, cabbage, sweet corn, cucumber, green pea, spinach, and tomato. Data for other vegetables processed included in fresh market estimates.

[b] Area for fresh market is for harvest; area for processing is area harvested.

[c] Value for all fresh market vegetables (except garlic) on f.o.b. basis; for processing vegetables, value at processing plant door.

[d] Beginning in 1981, statistics discontinued for the following crops: fresh market—artichoke, asparagus, snap bean, brussels sprouts, cabbage, cantaloupe, cucumber, eggplant, escarole/endive, garlic, green pepper, spinach, and watermelon; processing—lima bean, beet, cabbage, cucumber, spinach, and asparagus.

costs. However, greenhouse production has, at times, been a very significant segment of vegetable production, particularly in the eastern United States. It is not a new industry; transparent coverings were used in Roman times for cucumber production. Today, however, it is a highly intensive industry, focused largely on production of tomato, cucumber, and lettuce or on production of transplants for field or home use. Increases in fuel costs accelerated changes in greenhouse production in the 1970s. The glasshouses gradually are being replaced with those constructed of double-layered polyethylene (Figure 1.3) or rigid insulated plexiglass. Many utilize such technologies as nutrient film technique (soilless production using a shallow reservoir of complete nutrient solution) or sand-hydroponic culture and carbon dioxide enrichment to maximize yields per square foot. The quality of the greenhouse product must be high to meet competition from winter production areas and from imported produce and to justify the costs of production. An extension of greenhouse forcing, "factory" production, has developed in which the basic needs for plant growth are supplied automatically in controlled atmospheres under artificial light. Plant growth un-

Components of the Vegetable Industry / 5

der these conditions is rapid and uniform, but the cost of inputs, particularly of energy, can be substantial.

Specialty Crops

In addition to standard production systems, there are specialties that can be profitable for careful and intelligent growers. These specialties include such enterprises as cellar forcing of rhubarb or witloof chicory, or mushroom culture. Proper environmental controls are critical to the success of any of these specialties. The market for such commodities is sufficiently strong to support a high price; however, many of the suppliers are well established, and gaining access to the market is difficult.

FIGURE 1.3
Typical twin-layer plastic greenhouse. Greenhouse vegetable production is confined largely to tomato, cucumber, and lettuce, mainly because of economic feasibility. (Photo courtesy of O. S. Wells, University of New Hampshire.)

FIGURE 1.2
Mechanically harvested tomatoes arriving at a processing plant. Water added to each car flushes the fruit into a sluice that conveys them to the plant. (Photo courtesy of University of New Hampshire.)

TRENDS IN VEGETABLE PRODUCTION AND MARKETING

Only 3.1 percent of the U.S. population is directly responsible for our total agricultural production. This statistic largely reflects the mechanization and efficiency in growing feed and fiber crops and in handling livestock. Relative to these enterprises, vegetable production is substantially more labor intensive; yet the productivity of the average vegetable grower, through assimilation of new cultivars and production and marketing technology, has increased dramatically. Total production of vegetables has remained relatively stable since 1969, whereas acreage devoted to vegetable production has declined. Since the early 1950s, total acreage devoted to vegetable production has declined by 11 percent (Table 1.4). The increased production efficiency in the past 15 years has been focused largely on several major crops, most notably tomato, sweet corn, and lettuce. Prior to 1969, efficiency had been improved for beans, peas, and other easily mechanized crops.

TABLE 1.4
Distribution of U.S. harvested vegetable acreage by crop in 1954 and 1982

Crop	Year 1954 Acres	1954 %	1982 Acres	1982 %
Artichoke	5,601	—[a]	11,204	—
Asparagus	143,227	4	97,202	3
Lima bean (green)	127,042	3	56,113	2
Snap bean	271,937	7	277,538	8
Beet	24,491	1	13,983	—
Broccoli	31,748	1	80,277	2
Brussels sprouts	5,601	—	6,138	—
Cabbage	131,691	4	90,360	3
Cantaloupe	144,070	4	113,981	3
Carrot	80,353	2	83,601	3
Cauliflower	26,821	1	50,168	2
Celery	30,529	1	39,455	1
Cucumber	121,222	3	113,849	3
Eggplant	5,966	—	5,748	—
Endive	5,854	—	2,604	—
Garlic	2,473	—	15,379	—
Honeydew melon	—	—	23,940	1
Lettuce	8,454	—	229,887	7
Dry onion	98,330	3	123,776	4
Green pea	385,507	10	281,350	8
Sweet pepper	73,523	2	70,999	2
Spinach	48,792	1	34,915	1
Sweet corn	608,568	16	642,168	19
Tomato	439,555	12	403,469	12
Watermelon	417,838	11	184,043	6
Miscellaneous vegetables[b]	284,744	8	284,813	9
Total vegetables	3,739,994		3,330,637	

Source: Bureau of the Census, Department of Commerce.

[a] Less than 1 percent.

[b] Includes Chinese cabbage, mustard cabbage, chicory, edible-pod peas, collards, green southern pea, escarole, kale, mustard, green onion, okra, parsley, hot pepper, pimiento, pumpkin, radish, rhubarb, shallot, squash, turnip, watercress, and other vegetables.

The total per capita consumption of fresh vegetables (excluding potatoes) has remained relatively stable since 1962 (Table 1.5) after declining during the decade following World War II. Over the same period, consumption of both canned and frozen vegetables increased by over 12 percent. Because of the modest increase in per capita consumption and the steady population growth (Figure 1.4), market demand for vegetables has, for the most part, remained strong. Eating habits, however, have changed. In recent years, use of fresh and frozen foods has increased, whereas canned food consumption has declined (Table 1.6). Leafy, green and yellow vegetables have become popular, and, in particular, use of such commodities as broccoli, lettuce, and green and yellow sweet peppers has increased substantially. Consumption of starchy foods generally has declined. Among those crops currently leading in fresh sales (Table 1.7), potato, lettuce, and tomato constitute 20 percent of the total.

The dietary contributions of vegetables are discussed in later chapters. Vegetables provide an important source of vitamins, minerals, roughage, and fiber. New cultivars have contributed not only higher productivity but also improved vitamin content, textural quality, and esthetic appeal over the forms used by our forebears.

INTERNATIONAL COMPETITION

Competition in international produce marketing is governed by (1) the ability to preserve quality in transit and (2) the farm gate cost/price relationship (which is affected by cost of credit, production risks, political risks, and monetary exchange rate). The development of international commerce in vegetables has intensified competition within the U.S. market (Figure 1.5, Table 1.8). In many countries, the costs of production, particularly of labor, provide a competitive advantage. In others, subsidies for production costs have enabled their growers to become significant participants in the U.S. produce market. To meet this competition, U.S. growers have made effective use of new technology to minimize their production costs. New marketing technology, however, may not provide a lasting competitive advantage; it may, in effect, increase competition. A

TABLE 1.5
U.S. per capita consumption of fresh vegetables and melons, farm weight in pounds, 1962 to 1983

Year	Aspara-gus	Snap bean	Broc-coli	Cab-bage	Car-rot	Sweet corn	Cucum-ber	Let-tuce[a]	Green pepper	Onion	Melons Water-melon	Melons Canta-loupe	Total vege-tables
1962	0.6	2.3	.3	9.8	7.0	8.3	2.8	20.5	2.3	11.7	14.6	8.5	124.4
1963	0.6	2.2	.4	9.7	7.3	8.2	3.1	21.4	2.5	11.9	15.9	8.7	125.8
1964	0.5	2.1	.3	9.5	6.9	7.8	3.0	21.0	2.3	11.4	14.8	8.2	121.6
1965	0.6	2.0	.3	8.9	7.0	8.1	3.1	21.7	2.3	11.4	15.7	7.9	121.9
1966	0.4	1.9	.3	8.9	6.4	7.4	3.0	21.6	2.4	11.5	14.8	7.3	118.0
1967	0.4	2.0	.3	9.1	6.5	8.0	3.1	22.1	2.6	12.1	14.2	8.1	120.5
1968	0.5	1.9	.4	9.3	7.5	7.8	2.9	22.5	2.8	11.9	14.4	8.6	124.2
1969	0.4	1.8	.4	8.9	6.0	8.0	3.2	22.5	2.6	12.2	13.8	9.1	121.6
1970	0.5	1.7	.5	8.8	5.9	7.9	3.2	23.1	2.4	12.2	14.4	8.8	122.3
1971	0.4	1.6	.7	9.2	6.1	7.5	3.1	23.1	2.5	9.8	14.1	8.5	118.1
1972	0.5	1.6	.7	8.8	6.5	7.8	3.3	23.2	2.7	9.8	13.1	8.7	118.6
1973	0.4	1.5	.8	8.9	6.7	8.0	3.0	23.7	2.8	9.2	13.7	7.9	119.1
1974	0.4	1.5	.8	9.1	6.9	7.6	3.4	24.2	3.0	10.5	11.8	7.0	118.0
1975	0.4	1.6	1.0	9.1	6.4	7.8	3.1	24.2	3.1	9.5	12.1	6.8	116.9
1976	0.4	1.6	1.1	8.8	6.4	8.1	3.6	24.9	2.8	10.0	13.3	6.9	121.0
1977	0.4	1.5	1.2	5.5	5.0	7.7	3.9	26.4	3.4	10.0	13.6	7.5	122.9
1978	0.3	1.4	1.3	8.9	5.9	7.5	4.2	26.2	3.4	10.3	12.8	9.0	125.7
1979	0.3	1.4	1.5	8.5	5.8	7.3	4.3	26.6	3.6	11.2	12.3	8.4	127.0
1980	0.3	1.4	1.8	9.0	6.3	7.2	4.3	27.4	3.6	9.9	11.4	7.2	126.6
1981	0.3	1.4	2.1	8.6	6.5	6.8	4.4	26.2	3.5	9.8	12.3	8.5	125.7
1982	—[b]	—[b]	2.2	—[b]	7.4	6.8	—[b]	25.3	—[b]	11.4	—[c]	—[c]	131.8

Source: Produce Marketing Almanac (1985).

[a] Data not available in 1982.

[b] Reported as miscellaneous vegetables starting in 1982.

[c] Including escarole until 1982; in 1982, lettuce only.

TABLE 1.6
Trends in per capita consumption of fresh, canned, and frozen vegetables, 1972 to 1981

Year	Consumption Fresh (fresh wt) lb	kg	Canned (processed wt) lb	kg	Frozen (processed wt) lb	kg
1972	119	54	56	25	22	10
1973	120	54	57	26	24	11
1974	118	54	56	25	23	10
1975	117	53	55	25	23	10
1976	121	55	55	25	25	11
1977	123	56	55	25	26	12
1978	126	57	54	24	28	13
1979	126	57	57	26	29	13
1980	126	57	51	23	27	12
1981	126	57	48	22	30	14

Source: Produce Marketing Almanac (1985).

TABLE 1.7
Top 10 vegetables in contribution to sales

Rank	Vegetable	Contribution to sales (%)[a]
1	Potato	8.32
2	Lettuce	6.50
3	Tomato	5.29
4	Onion	3.06
5	Mushroom	2.46
6	Celery	2.37
7	Cucumber	2.25
8	Cabbage	2.10
9	Carrot	2.08
10	Broccoli	2.04

Source: Produce Marketing Almanac (1985).

[a] Percentage of total commercial farm sales.

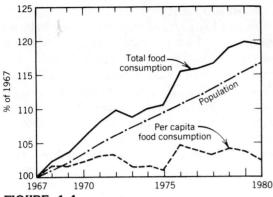

FIGURE 1.4
Growth of population and total food consumption as compared with per capita food consumption, using constant retail prices as weights. (*Source: Handbook of Agricultural Charts*, USDA Agric. Handb. 609 (1982).)

case in point is the development of controlled-atmosphere packaging which promises to extend substantially the longevity of fresh quality of some vegetables. Although this new technology will facilitate orderly marketing of high-quality products, the prolonged shelf life will enable growers worldwide to compete for the same consumer. Those most likely to be affected by such changes are wholesale producers and processors; growers involved in direct retail marketing would be disrupted the least.

TABLE 1.8
Total U.S. imports of selected vegetables, 1982 to 1984

Crop	Shipment (MT)			Increase (%)
	1982	1983	1984	
Tomato	268,812	334,845	373,900	39.0
Pepper	76,571	69,801	98,559	28.7
Cucumber	138,142	177,170	176,125	27.5
Squash	47,269	51,990	60,423	27.8
Cantaloupe	82,774	75,341	111,899	35.2
Onion	74,352	91,821	119,027	60.0

Source: Produce Marketing Almanac (1986), Foreign Agricultural Service data.

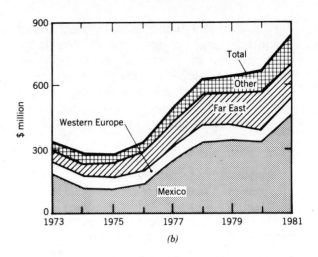

(b)

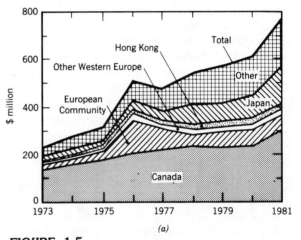

(a)

FIGURE 1.5
Destination of U.S. vegetable exports (*a*) and origin of U.S. vegetable imports (*b*). Excludes melons, dried beans, and dried peas. (*Source: Handbook of Agricultural Charts*, USDA Agric. Handb. 609 (1982).)

EMERGENCE OF SOCIETAL ISSUES IN AGRICULTURE

The transition to large or corporate farms has spawned several important societal issues which may affect local economics, national policy, or, occasionally, consumer confidence in farmers and agricultural scientists. Among these issues are the unionization of farm labor, the displacement of farm labor by mechanized production and harvest systems, the long-term effects of chemical application on groundwater supplies and concerns about spray residues and drift, and taxation policies with respect to industrial or residential development of agricultural lands.

Regulatory Legislation

Prior to the mid 1950s, vegetable producers employed substantial numbers of domestic migrants and imported foreign labor to support needs for production and harvesting. Social activism and the expanding size and complexity of the commercial vegetable farm led to many changes relating to labor. Many of the constraints formerly restricted to manufacturing and business have been extended through federal and state legislation to farming. These constraints include unemployment compensation, workman's compensation, laws governing foreign and domestic labor, occupational safety and health monitoring, and control of product purity and safety through Food and Drug Administration and Environmental Protection Agency regulation. The costs of regulation and the management needed to satisfy federal and state laws have, along with increasing union activity, accelerated management's efforts to reduce dependence on labor, primarily through introduction of mechanization. In some instances, protests by workers through strikes and boycotts have accelerated the kinds of change they have opposed, for costs of farming within an unstable labor environment have discouraged growers, occasionally resulting in liquidation of farms, loss of prime farming land to nonfarm development, and loss of farm employment. A benefit of the public discussion of farm labor problems has been the improvement in compensation and work conditions for farm workers in all sections of the country and an understanding of the problems faced by growers of perishable crops.

Contaminants

Issues involving chemical use normally do not arise until a crisis develops. Neither scientists nor growers can anticipate those problems that accrue slowly over many years; yet, when discovered, the problems require prompt attention. In 1985, selenium levels in parts of the San Joaquin Valley of California reached levels toxic to wildlife and, by inference, to humans. These toxic levels developed because the system built to provide drainage for irrigated fields was incomplete. The failure to provide suitable provisions for drainage of farmland effluent resulted in contamination of a wildlife preserve and eventual removal of thousands of acres of farmland from cultivation. Although this problem is dramatic and perhaps unique, it underscores the need for ecologically sound farming practices.

Land Use

Land use has been an issue for many years. Level farmland, usually well drained, also may be ideal as an industrial or housing site, and in many parts of the United States and Canada, urban development has removed thousands of acres of farmland from production [estimated at 875,000 acres (354,375 ha) in the United States between 1967 and 1975]. This transformation is irreversible, and the rate cannot be sustained while trying to meet world food needs. The pressures that lead to loss of farmland are economic—tax rates that make it impossible to farm at a profit or land values driven upward by competing land use—but these pressures have been the result of industrialization, enhanced by government-sponsored programs, such as federal highway construction, that encourage relocation of American workers to rural sites. Taxation based on current agricultural use, rather than on development potential, has been helpful in efforts to preserve open land near urban areas. In several areas, programs by which the state can purchase development rights from a grower have been enacted, thereby preventing loss of valuable sites to housing or industrialization.

RESOURCE ISSUES IN AGRICULTURE

The expansion of farm size also has crystallized several issues related to preservation of resources that relate to the long-term stability of farming. Two resources, soil and water, historically have been plentiful and inexpensive, and both have been the focus for government-sponsored conservation programs. It is just as important for each grower to practice conservation in managing crop cultural systems.

Soil Management

Soil erosion by wind and water increases as production intensity increases. In Iowa, it was estimated that a loss of just $\frac{1}{3}$ in. (8 mm) of topsoil per year, an amount not visible to the casual observer, would represent the removal of 40 tons/acre (90 MT/ha) of the best soil for production. Furthermore, erosion, by eliminating primarily topsoil, reduces water-holding capacity, removes plant nutrients, and degrades soil structure and uniformity. To compensate for the decline in plant growth because of erosion, growers then must apply additional fertilizer and irrigation, thereby increasing their costs per unit of yield. Modification of tillage practices may be required, and much research information has been gathered with respect to alternative tillage systems (Chapter 5). Some argue for reduced intensity of production to preserve soils, but economic considerations, particularly in highly intensive vegetable production, dictate otherwise. As demand for food increases, new lands, often more prone than current sites to soil erosion, will be cultivated. These sites will incur serious soil losses if conservation practices are neglected.

Water Management

Irrigation has been an indispensable factor in vegetable production, but as water use has increased, its availability has declined. Water tables in some areas have dropped dramatically, and the instances of competing and conflicting water use have increased. The attendant problems of runoff pollution and increasing salinity also have become apparent, particularly in the western states. As competing uses and reduced supplies increase the cost, irrigation systems likely will be modified, concentrating on efficiency of distribution, but the basic need for supplemental water in most vegetable areas will remain.

VEGETABLE GROWING AS A BUSINESS

There is no business more complex than farming. The inputs for manufacturing the final product—temperature, light, water, and nutrients—are impossible to program with the precision of a General Motors; nor can the volume of output be scheduled with as much confidence as in an assembly line. Yet, the product is marketed within the same forces of supply and demand as an automobile, extending far beyond the domestic market. The grower must accommodate the same demands for wages and benefits and meet the same regulatory constraints imposed on most manufacturers. Unlike most manufacturers, however, growers must achieve their productivity within a variable environment, and product quality affects the health of the entire population. Furthermore, the price of their product cannot be adjusted arbitrarily, even when costs for producing a specific crop may erase the possibility of profit. Each grower, therefore, faces many uncertainties and must minimize risks through sound production and marketing decisions.

In managing both the business and production aspects of a farm, detailed records of expenditures and sales are invaluable not only for tax purposes, but also to provide a chronicle of crop production efficiency. An accounting of expenditures must include both fixed and variable costs (Table 1.9). Fixed costs are those incurred whether or not a crop is harvested from the land. They include depreciation, interest on investment, repairs and maintenance, land rent or interest on the land, plus taxes, insurance, licenses, and other mandated costs. Variable costs include those over which a grower has some control—cost of fertilizer, pesticides, utilities, marketing, labor (with fringe benefits), and interest on loans for purchasing inputs.

Beyond the fiscal aspects, however, records should be maintained by yield for each crop, rotations, cultivars planted and dates of planting, soil fertilization, pesticide applications, and other inputs. These records provide invaluable data for planning and occasionally for evaluating a particular production problem.

TABLE 1.9
1983 Celery production costs on a typical 55-acre farm in Michigan

Item	Per acre	Per hectare	Per crate
Yield (crates)	750	1,853	
Price			$6.50
Gross income	$4,875	$12,046	
Growing costs			
Seed	$ 6	$ 15	$0.01
Fertilizer	150	371	0.20
Spray, dust	281	694	0.38
Labor	407	1,006	0.54
Fuel	108	267	0.14
Supplies	17	42	0.02
Utilities	126	311	0.17
Total	1,095	2,881	$1.46
Interest on operating capital (13%, 6 months)	71	27	.09
Total costs and interest	$1,166	$ 2,039	$1.55
Harvest and marketing			
Fuel	$ 11	$ 11	$0.01
Containers, supplies	825	2,039	1.10
Labor	635	1,569	0.85
Cooling	450	1,112	0.60
Marketing	426	1,052	0.57
Total	2,347	5,799	3.13
Fixed costs			
Depreciation	$ 334	$ 825	$0.45
Interest on investment, excluding land	369	912	0.49
Repairs, maintenance	295	729	0.39
Land rent (interest on land and taxes)	340	840	0.45
License, insurance	45	111	0.06
Total fixed cost	1,383	3,417	1.84
Total cost	4,896	12,098	6.53
Net return	−21	−52	−0.03

Source: From Allen E. Shapley, *American Vegetable Grower.* Copyright © 1983 by *American Vegetable Grower.* Reprinted by permission.

SELECTED REFERENCES

Batie, S. S., and R. G. Healy (1983). The future of American agriculture. *Scientific American* **248**:45–53.

Lorenz, O. A., and D. N. Maynard (1980). *Knott's Handbook for Vegetable Growers.* Wiley, New York.

Produce Marketing Association (1985). *Produce Marketing Almanac.* Newark, Del.

Schrader-Franchette, K. (1984). Agriculture, property and procedural justice. *Agriculture and Human Values* **1**:15–28.

USDA (1982). *Handbook of Agricultural Charts,* Agr. Handb. 609. U.S. Govt. Printing Office, Washington, D.C.

USDA (1985). *Agricultural Statistics.* U.S. Govt. Printing Office, Washington, D.C.

STUDY QUESTIONS

1. What factors stimulated the specialization and concentration of vegetable production in the United States?

2. What are the bases for the current position and future potential of the forcing industry in vegetable production?

3. To what extent and in what ways have societal issues changed vegetable production practices?

2
Labor, Mechanization, and Production Efficiency

or to the market price for produce. A number of inflationary pressures then combined to increase all production costs more rapidly than the market price could absorb. These inflationary pressures included, among others, escalating minimum wage and attractiveness of industrial wages, increased land values and taxes, costs of new technology and production supplies, and, in the 1970s, significant increases in energy costs. In addition, legislation affecting availability of foreign labor, working conditions, insurance, and other benefits not only increased cost per worker, but also underscored the need for new skills in personnel management.

Because of the increased costs, growers sought ways of improving efficiency, first by increasing yield per acre and/or expanding the number of acres harvested, later by reducing or eliminating certain production inputs, most notably labor.

Until 1965, production and harvesting of vegetable crops, with a few exceptions, had been accomplished with large and reliable sources of labor, first predominantly domestic resident or migrant labor, later by laborers largely drawn from Mexico, the Philippines, Jamaica, and Puerto Rico. Wages, although a major percentage of total production costs, remained relatively stable and were not unreasonable in relation to other production costs

IMPROVING CROP PRODUCTION

Yields per acre have increased dramatically since World War II for many vegetable crops (Table 2.1). The reasons for these increases are many: concentration of crops within areas having the most suitable environment; increased use of irrigation, improvements in fertilization, and fundamental knowledge of plant nutrition; increases in plant populations per

TABLE 2.1

Fresh market yields per acre of selected crops

Crop	Yield 1950 cwt/A	1950 MT/ha	1955 cwt/A	1955 MT/ha	1960 cwt/A	1960 MT/ha	1965 cwt/A	1965 MT/ha	1970 cwt/A	1970 MT/ha	1975 cwt/A	1975 MT/ha	1980 cwt/A	1980 MT/ha
Snap bean	30	3.4	36	4.0	37	4.1	39	4.4	36	4.0	37	4.1	—	—
Broccoli	47	5.3	53	5.9	57	6.4	59	6.6	77	8.6	81	9.1	87	9.7
Cabbage	170	19.0	154	17.2	176	19.7	177	19.8	193	21.6	219	24.5	—	—
Carrot	172	19.3	186	20.8	219	24.5	219	24.5	234	26.2	277	31.0	270	30.2
Celery	374	41.9	464	52.0	419	46.9	454	50.8	475	53.2	504	56.4	501	56.1
Sweet corn	51	5.7	62	6.9	63	7.1	66	7.4	69	7.7	80	9.0	79	8.8
Cucumber	67	7.5	82	9.2	82	9.2	98	11.0	94	10.5	100	11.2	—	—
Lettuce	127	14.2	156	17.5	174	19.5	190	21.3	197	22.0	230	25.8	268	30.0
Onion	169	18.9	187	20.9	258	28.9	289	32.4	302	33.8	306	34.3	296	33.2
Pepper, green	58	6.5	72	8.1	79	8.8	85	9.5	81	9.1	104	11.6	—	—
Tomato	75	8.4	90	10.1	118	13.2	134	15.0	124	13.9	166	18.6	201	22.5
Cantaloupe	88	9.9	94	10.5	100	11.2	104	11.6	121	13.6	132	14.8	—	—
Watermelon	67	7.5	79	8.8	94	10.5	103	11.5	103	11.5	115	12.9	—	—
Average, all vegetables	99	11.1	112	12.5	130	14.6	142	15.9	151	16.9	154	17.2	—	—

Source: U.S. Fresh Market Statistics (1949–1975). 1980 data from Agricultural Statistics (1984) (USDA).

acre and improved reliability in stand establishment; new and effective methods for controlling weeks, diseases, and insects; and introduction of superior cultivars, including F_1 hybrids, of many crops. Each of these units of technology, discussed in detail in succeeding chapters, usually contributed substantially more in market return than its input cost.

The costs of many inputs did not remain stable, however. Some increased disproportionately, particularly those directly or indirectly dependent on fossil fuel supplies. More importantly, the inflation stimulated by increased fuel prices increased manufacturing and operating costs of equipment, narrowing the economic advantage of extensive mechanization.

MECHANIZATION

Some mechanization is indispensable for efficient production of vegetables. Tractors, tillage and cultivation implements, fertilizer applicators, sprayers/dusters, seeders, and/or transplanters are basic requirements for any competitive vegetable farm. For many growers, such equipment as bed formers and mulch layers also are used regularly within a particular cultivation system. The larger the farm, the greater must be the capacity of the equipment and the greater the inventory of equipment used. At some point, the increasing acreage may favor inclusion of some mechanized harvest equipment (Figure 2.1) as well.

Farmers growing significant acreages of dry beans and peas, potatoes, and some canning crops mechanized their harvests well before economic pressures stimulated extensive mechanization of other vegetables. Equipment evolved slowly to replace labor-intensive harvesting in the high-risk and perishable crops and usually was confined to vegetables for processing. For example, engineering research to develop a cucumber harvester began in the mid-1950s. After repeated failure to secure a design that would remove fruit from the crown of a plant without terminating the plant's productivity, the once-over harvest concept was adopted. By 1963, a machine

FIGURE 2.1
Examples of mechanical harvesting: (*a*) cabbage; (*b*) peas; (*c*) beans; (*d*) sweet corn. Although mechanical harvesting is most suited for processing crops, several fresh market crops are harvested predominantly by machine in some areas. (Photos (*a*) and (*b*) courtesy of R. Becker, N.Y. Agricultural Experiment Station, Geneva. Photos (*c*) and (*d*) courtesy of Pixall Corporation, Clear Lake, Wisc.)

had been developed, and, by 1972, 80 percent of the pickling cucumber acreage in Michigan was harvested mechanically. This equipment, even today, is most suited to harvest of large fruit. Losses of small fruit are substantial. The success in mechanizing fruiting crops, such as cucumber and tomato, stimulated subsequent development for harvesting other crops, including fresh peppers and melons. Lettuce, leafy greens, and other perishable commodities with high labor inputs also received attention.

TABLE 2.2
Mechanical harvest of vegetables: Actual 1980 and projected 1990 labor-hour ratios

Crop	Labor-hour ratio (hand : machine)	Percentage of crop mechanically harvested			
		1980		1990	
		Processing	Fresh	Processing	Fresh
Artichoke	—	0	0	0	0
Asparagus	15:1	1	1	1	1
Bean					
Green lima	100:1	100	0	100	0
Snap	50:1	100	60	100	60
Beet	5:1	100	0	100	0
Broccoli	7:1	0	0	20	0
Brussels sprouts	3:1	0	0	5	0
Cabbage	9:1	80	0	90	0
Cantaloupe	—	0	0	0	0
Carrot	17:1	100	99	100	100
Cauliflower	7:1	0	0	10	0
Celery	1:1	<1	15	<1	15
Corn, sweet	10:1	100	30	100	50
Cucumber	9:1	30	0	40	0
Lettuce	1.2:1	0	0	0	0
Onion	15:1	100	16	100	20
Pea	100:1	100	0	100	0
Pepper, green	—	0	0	0	0
Potato	10:1	100	99	100	99
Spinach	8:1	100	0	100	0
Sweet potato	10:1	75	25	80	30
Tomato					
Fresh	2:1		0		10
Processing	20:1	95		100	
Watermelon	—	0	0	0	0

Source: From S. S. Johnson and M. Zahara (1985) Hort-Science 20:23–28. Reprinted by permission of the authors.

Crops differ in the land labor required to harvest a given volume or weight, and these differences in part may justify whether mechanization can enhance efficiency. The labor-hour ratios of hand versus machine harvest for a series of vegetables have been estimated by Johnson and Zahara (1985) and used to predict future mechanization trends (Table 2.2). These ratios vary widely, and most vegetable crops with ratios greater than 15:1 already are heavily mechanized, particularly for the processed crop. Low labor-hour ratios provide little economic justification for mechanization.

Other factors also influence the adoption of mechanized practices. Certain soils or weather conditions may preclude or discourage use of some equipment. The engineering complexity in developing harvesters for some high-risk and multiple-harvest crops, such as asparagus, increases machinery cost and reduces the likelihood that mechanization will occur, regardless of the labor-hour ratio. Furthermore, the economic significance of the ratio is influenced a great deal by availability and cost of an experienced labor force. Even if mechanized equipment succeeds mechanically in the task for which it was designed, the harvested product must meet the quality standards of the intended consumer.

Where mechanization has been successful, it has increased productivity of labor, facilitated introduction of new technology (chemicals, cultivars, etc.), and stabilized production, and, in some instances, quality. It also has contributed, over a period of time, to the decline of the family farm and to the increase of farm size, and has increased owner debt, changed ownership structure of many farms, and increased the use of nonland resources.

Mechanical Harvesting and Quality

The perishability and fragile nature of vegetables present many opportunities for physical damage within a mechanized harvest system. It is not practical to engineer equipment that will retain original preharvest mature quality. For processing, the consequences may not be limiting: for fresh produce, the grower or handler must determine the level of damage of each crop that can be tolerated before quality losses exceed gains in efficiency.

Harvest damage to tomato fruit, as one example, is related to cultivar, maturity at harvest, time interval between harvest and use, method of picking, and depth of container. Cultivars developed for processing (puree) suffer some damage, but this damage does not reduce product quality. Tomatoes picked mechanically at a mature green stage for fresh market may suffer stem punctures, and abrasion from soil or stones passing through the harvester may be a problem. In several tests, however, the grade of machine-harvested mature greens has been comparable to that of hand-harvested fruit. The use of suitable "jointless" cultivars, those lacking an abscission layer in the fruit stem, reduces stem punctures by eliminating the stem and improves the feasibility of green fruit harvest for fresh market. Efforts to harvest a fresh product at partial or full color, however, result in noticeable increases in damage.

Harvests of cucumber, sweet corn, beans, cabbage, cauliflower, brussels sprouts, green peas, and southern peas present similar results, and most of the mechanically harvested produce therefore is processed. Some leafy and root crops can be harvested successfully for the fresh market. Other vegetables, such as sweet potato, may be extremely susceptible to skinning or bruising which will reduce market acceptance.

Mechanization and Cultural Systems

The introduction of mechanization, regardless of crop, often requires dramatic changes in cultural systems and production philosophy. A case in point is the mechanization of tomato harvest for processing.

Prior to 1962, tomatoes were harvested by hand, several times per week, largely from fields consisting of indeterminate-vined plants grown at densities of 3200 to 4000 plants per acre. Most fields were established with transplants, and weed control was entirely cultural. Conceptually, mechanization was not adapted to this cultural system. In an engineering sense, it was not feasible to harvest the same tomato plant repeatedly; thus, a single destructive harvest (Figure 2.2) became the most effective alternative. To ensure adequate yields for once-over harvest, it was necessary to increase plant population density and to compress the ripening of most fruit on a vine into a relatively short period. Tests of plant populations of 14,000 or more plants per acre not only proved the feasibility of high-density planting, but also indicated that crowding would increase the percentage of total fruit ripening at one time. Genetically superior determinate cultivars (Figure 2.3) also contributed substantially to concentrated ripening and to the resistance of fruit to bruising.

The change to high-density plant populations exposed needs for additional modifications in the production system. Conventional transplanting was not cost effective at high population density. Direct seeding, however, was unreliable in many soils, and its success

FIGURE 2.2
The tomato harvester, introduced in the 1950s, cuts the vines at the soil line, elevating vine and fruit to a shaker that removes fruit. Graders remove trash and unripe fruit not separated by the machine. (Photo courtesy of FMC Corporation, Hoopeston, Ill.)

required improvements in precision seeding that would ensure stand establishment with no thinning. It also was clear that weed control by cultivation was not effective within direct seeded, high-density situations; consequently, several preplant and postplant chemical herbicides were developed that achieved economic levels of control, particularly in the early season when tomato seedlings are vulnerable to weed competition. Fertilization and irrigation rates were increased in the early season to support high plant density but decreased as plants neared maturity to reduce carbon partitioning to new fruit and promote uniform ripening. Increased attention to insect and disease control were required. Finally, growers had to program their planting dates to ensure a steady supply of raw product for the processor. Similar changes in production methods have occurred with mechanization of cucumber harvests.

Mechanical harvesting of processing tomatoes, cucumbers, and other vegetables became successful because different technologies were integrated into a single system. This system's effectiveness was enhanced by culti-

vars developed specifically to meet new cultural needs. Determinate tomato plants bearing firm, bruise-resistant fruit, a high percentage of which ripen at the same time, and gynoecious hybrid cucumbers consistently bearing fruit at each node now constitute most of the processing acreage of these crops. Although some crop harvests have been mechanized successfully, others, such as head lettuce, have not, largely because of the difficulty in ensuring consistently marketable products.

Harvest Aids

Mechanization of any task normally evolves in two phases. The first is the development of modifications that will facilitate hand work. Labor is required, but the equipment may substitute for a percentage of worker motion or for certain tasks (mechanical aids). The second phase is a machine that requires no hand removal of the edible product from the plant or soil.

As harvest equipment evolved, in many instances harvest aids were more efficient than a completely mechanical system. Mobile con-

FIGURE 2.3
Tomato cultivar types for mechanical harvesting. Fruit are firm and high in solids, tolerating mechanical stress. (Photo courtesy of University of New Hampshire.)

FIGURE 2.4
Mobile conveyers are used in many production areas to facilitate hand harvest. Some include all equipment needed to film-wrap and prepare the product for immediate cooling and transport. (Photo courtesy of A. Kader, University of California, Davis.)

veyers, or "mule trains" (Figure 2.4), which were developed to help move the harvested product to shipping boxes, have evolved into integrated and affordable field-packing units. Actual harvest is by hand. After hand trimming, the produce is placed on the mobile conveyer for grading and packing as the machine and crew move slowly through the field. In such crops as cauliflower, normally prepackaged for market, the product may be trimmed and placed in a film bag on the ground. As the mobile unit passes over, the filled bags are picked up, sealed, sized, and crated. Although the labor requirements to operate field-packing units may exceed those of shed-packing facilities, the equipment investment is comparatively low, trimming waste is left in the field, there generally is less damage to the produce, and there is increased efficiency of motion.

ENERGY CONSUMPTION AND PRODUCTION EFFICIENCY

During the early 1970s, world supplies of fossil fuels became restricted. In the previous two to three decades, dependence on fuels had increased dramatically, reflecting mechanization

and other energy-dependent management systems (e.g., irrigation) and increasing use of fertilizer, pesticides, plastics, and other energy-based materials. After World War II, the number of tractors increased by 88 percent, and average horsepower rating rose from 18.0 to 46.6. In the period between 1940 and 1969, consumption of fuel for all agricultural production rose from 3.3 to 7.6 billion gallons (12.5 to 28.8 billion liters). Energy needs for produce handling, storage, transportation, and packaging also increased. It has been estimated that the U.S. food system now expends at least 5 units of fossil energy for each unit of food energy made available to consumers. Of the energy required for the total food system in the United States, agricultural production uses 18 percent, food processing 33 percent, transportation 3 percent, wholesaling and retailing 16 percent, and home preparation 30 percent.

Over 60 percent of the energy used in agricultural production (Figure 2.5) is related to

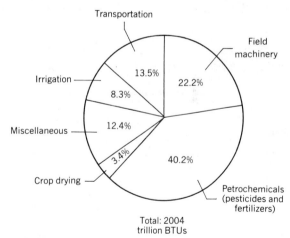

FIGURE 2.5
Energy consumption in total agricultural production. For vegetables, crop drying would not be a major use, and consumption for irrigation, transportation, and petrochemicals would be somewhat greater than the average for all farming shown here. 1980 data. *Source: Handbook of Agricultural Charts USDA Agric. Handb. 609 (1982).)*

operation of field machinery and application of petrochemicals. Although some categories of energy use for vegetable production might vary from these averages, the effects of mechanization and dependency on petrochemicals are clear. Data provided by Lougheed *et al.* (1975) for potato, a carbohydrate crop (Table 2.3), and asparagus, grown for its vitamin rather than energy value (Table 2.4), show over 80 percent of the energy required to produce these vegetables in fuel and fertilizers. Such management techniques as minimum or zero tillage, trickle irrigation, banding rather than broadcasting of fertilizer, high-density planting, and addition of organic matter to preserve nutrient availability may reduce energy costs for some crops. Development of cultivars resistant to diseases or insects or efficient in absorption of nutrients might enable growers to modify some of their management systems toward improved energy efficiency. Information as to soil texture and structure and plant water needs, so that superfluous irrigation can be avoided, and appropriate tractor

TABLE 2.3

Energy relationships in potato production (per hectare)

Input	Time or quantity	Energy equivalent (kcal)	Percentage of total
Labor	108 h	18,900	0.3
Tractor and machinery investment		224,500	3.9
Gasoline	105 liters	862,200	15.0
Irrigation	3 ha-cm	415,300	7.2
Pesticides	15.5 kg	380,400	6.8
Fertilizer (10–10–10)	1335 kg	3,831,600	66.8
Total input		5,732,900	100.0

Output (yield) = 27,800 kg/ha × 767 kcal/kg
= 21,322,600 kcal/ha

Output : input ratio = 3.72 : 1

Source: From E. C. Lougheed *et al.* (1975) *HortScience* **10**:459–462. Reprinted by permission of the authors.

TABLE 2.4

Energy relationships in asparagus production (per hectare)

Input	Time or quantity	Energy equivalent (kcal)	Percentage of total
Labor	230 h	58,168	1.4
Tractor	12 h	247,000	5.9
Gasoline	93 liters	998,000	23.6
Pesticides	3 kg	81,510	1.9
Fertilizer			
N	112 kg	2,074,800	
P	112 kg	375,440	
K	112 kg	259,350	
Herbicide	52 kg	108,680	2.6
Total input		4,192,948	100.0

Output (yield) = 1993 kg/ha × 246 kcal/kg
= 490,278 kcal/ha

Output : input ratio = 0.12 : 1

Source: From E. C. Lougheed *et al.* (1975) *HortScience* **10**:459–462. Reprinted by permission of the authors.

speed and gear ratios to operate equipment efficiently also will reduce energy consumption. Energy costs will fluctuate with world supplies of petroleum. Conservation is important, regardless of current price, because it minimizes vulnerability to cyclic fluctuations.

ENERGY IN GREENHOUSE PRODUCTION

When receiving full sunlight, greenhouses of any type are efficient solar collectors and need little if any supplemental heat. During the night and during cloudy weather, however, greenhouses are inefficient, and heating losses are excessive (75 percent of total thermal loss).

Improvements in greenhouse design, including sidewall insulation, double-layered polyethylene, or cellular rigid plexiglass covering, and modification of exit and entrance doors have decreased thermal loss by 40 to 50 percent. Further measures tested have included (1) foam or styrofoam pellets forced between layers of the double polyethylene cover each night (Figure 2.6), (2) abutting overhead plastic tubes that are inflated at night to form a "blanket," (3) heat exchangers, (4) development of heat sinks (water or rock), and (5) and fabric covers pulled over plants during the night. On existing glasshouses, a polyethylene cover on the glass has provided sufficient dead air space to reduce heat loss (Figure 2.7). Each system reduces dependence on fossil fuel, but each has an initial installation cost and some systems may, if not automated, require additional labor to operate. The efficiency of each system relates not only to its cost relative to energy savings, but also to its reliability.

Vegetables grown in structures engineered to provide optimum levels of nutrients, CO_2, and temperature under artificial lighting substitute electrical energy for heating energy. Although the heating requirement for such structures is small, electrical use is high, and the equipment required to control the environment is costly. The controlled-environment production system has been successful in a

FIGURE 2.7
Existing glasshouses have been equipped to maintain a layer of polyethylene inflated over the glass. The air insulation layer cuts vertical heat loss significantly, but also reduces solar radiation. (Photo courtesy of W. Bauerle, Ohio State University, Columbus.)

technological sense in producing lettuce, a crop successful at relatively low light intensities; however, it may not be sufficiently conservative in energy use to become competitive with winter field production, and its applicability to other crops is in question.

SELECTED REFERENCES

Council for Agricultural Science and Technology (1984). *Energy Use and Production in Agriculture,* Rep. No. 99. Ames, Iowa.

Jensen, M. H. (1977). Energy alternatives and conservation for greenhouses. *HortScience* **12:**14–24.

Johnson, S. S., and M. Zahara (1985). United States fruit and vegetable harvest mechanization and labor use projections to 1990. *HortScience* **20:**23–28.

FIGURE 2.6
Reusable liquid foam system developed at the University of Arizona reduces nighttime heat loss by 70 to 80 percent. Foam is forced between plastic film layers. (Photo courtesy of O. S. Wells, University of New Hampshire.)

Kasmire, R. F. (1983). Influence of mechanical harvesting on quality of non-fruit vegetables. *HortScience* **18**:421–423.

Levin, J. H. (1969). Mechanical harvesting of food. *Science* **166**:968–974.

Lougheed, E. C., J. T. A. Proctor, R. G. Rowberry, H. Tiessen, P. H. Southwell, and J. W. Riekels (1975). Fruit and vegetable production and the energy shortage. *HortScience* **10**:459–462.

Pimentel, D., L. E. Hurd, A. C. Bellotti, M. J. Forster, I. N. Oka, O. D. Sholes, and R. J. Whitman (1973). Food production and the energy crisis. *Science* **182**:443–449.

STUDY QUESTIONS

1. How does mechanization of harvest affect production systems for a specific crop?
2. Define methods by which fossil fuel energy use might be conserved in vegetable production without reducing yield or quality of the product.

3

The Plant and Its Environment

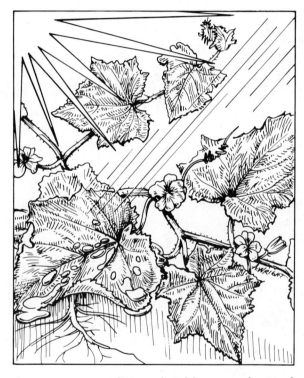

Appearance, quality, and yield are products of the genetic potential of each plant and the environment within which that potential is placed. Environmental constituents—light, temperature, moisture, soil moisture, and other factors essential for plant growth—vary widely. The environmental and genetic factors that control plant growth and development do so largely through specific events within and between cells. Each constituent of a cell assumes a dominant role in plant growth and development which may differ according to the tissue and/or organ within which the cell is

located and to the environmental factors to which a plant is subjected. A familiarity with the composition of a cell, the primary role of each cell constituent, and the integration of cell functions within and between tissues thus is essential in understanding the interaction of the plant or plant parts with components of the environment.

THE PLANT CELL

Cell Wall

Cell walls are rigid, permeable structures enclosing all cell contents (Figure 3.1). The outer **primary wall,** composed largely of cellulose, hemicellulose, pectins, and structural protein, is produced from the cytoplasm of the growing cell. After cell enlargement or differentiation, a **secondary wall,** composed of cellulose, hemicellulose, and lignin, develops on the interior of the primary wall. Neither wall has any role in the transport of materials to and from cells. The cell wall surfaces may provide an area for binding of some ionic materials, and thick-walled cells provide structural rigidity to plant organs.

Membranes

Within the cell, adjacent to the secondary cell wall, is a single **plasma membrane,** consisting of a lipid core with a protein sheath. This membrane is characterized by selective transport of solutes to and from the cell (termed **selective permeability**), perhaps relating to the polarity of lipid molecules and inte-

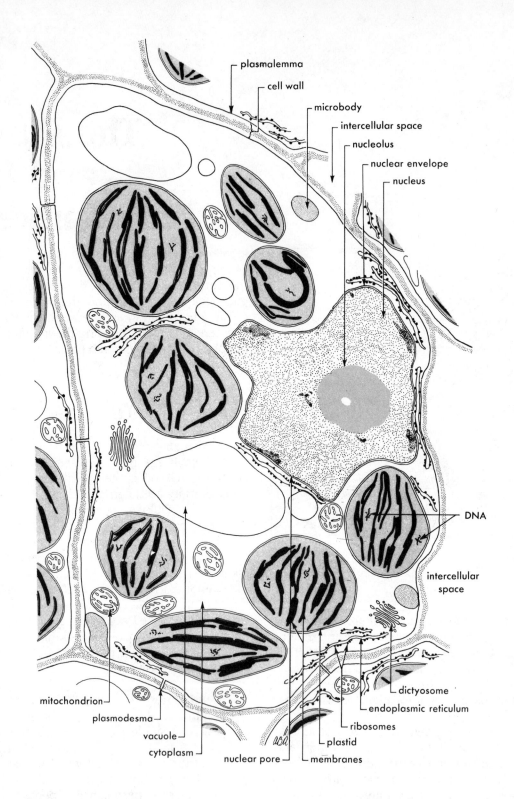

gral (membrane transversing) proteins which facilitate transport of specific ions and small molecules. The plasma membrane also presents a surface for important chemical reactions. Similar membranes, either single or double, also enclose cytoplasmic organelles; and a continuous membranous network throughout the cytoplasm, the **endoplasmic reticulum,** is thought to serve as a transport mechanism and as a surface for chemical reactions and protein synthesis.

Cell Inclusions

Nucleus   The most visible inclusion in most cells is the nucleus. Nuclei occur in all living plant cells, except sieve tubes, and are enclosed by two large-pored membranes. The predominant chemical constituents are protein, lipids, and nucleic acids (RNA and DNA) of which the DNA is the structural component of chromosomes and the site of stored information (genes).

The appearance of a nucleus of dividing cells changes in a cyclic manner. At periods of active division, the chromosomes become thickened and discrete. Following division, the chromosomes relax and appear as a diffuse reticulum of slender strands within the protein matrix. The nucleolus, a prominent, dark-staining body within the nucleus, functions in the storage of RNA and has a role in the synthesis of ribosomal RNA.

Each chromosome is characterized by a double helix of DNA strands, each strand composed of alternating sugar and phosphate groups. The sugar groups of the two strands are connected, as rungs on a ladder (Figure 3.2), by one purine (adenine or guanine) and one pyrimidine (thymine or cytosine). This

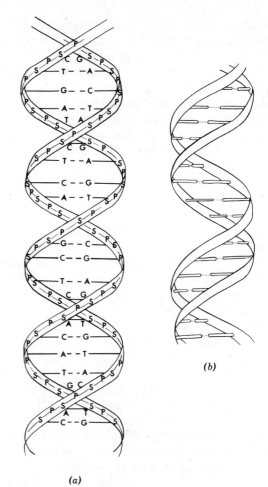

(a)

FIGURE 3.2

Schematic diagram of the double helix in a portion of a DNA molecule. (*a*) One strand is displaced along the axis of the helix for convenience of diagramming; (*b*) diagram of the parallel strands in the helix. Alternating sugar, S (deoxyribose), and phosphate, P, groups make up the backbones of the strands. Attached to each sugar unit is one of the purine or pyrimidine bases: adenine, A; thymine, T; guanine G; cytosine, C. The strands of the helix are held together through hydrogen bonds between the base pairs, adenine to thymine and guanine to cytosine. (From *Botany. An Introduction to Plant Biology,* T. E. Weier, C. R. Stocking, M. G. Barbour, and T. L. Rost. Copyright © 1982 by John Wiley & Sons, Inc., New York, p. 99. Reprinted by permission.)

FIGURE 3.1

Diagrammatic view of a bean leaf cell. (From *Botany. An Introduction to Plant Biology,* T. E. Weier, C. R. Stocking, M. G. Barbour, and T. L. Rost. Copyright © 1982 by John Wiley & Sons, Inc., New York, p. 39. Reprinted by permission.)

cross connection is termed a **base pair.** A **nucleotide** comprises a pyrimidine or a purine attached to the sugar–phosphate. A **nucleotide pair** consists of two nucleotides in which the purine of one is bonded to the pyrimidine of the other by weak hydrogen bonds, a structure that ensures separation and duplication of strands during cell division. A **gene** includes a portion of the double-stranded DNA comprising 300 to 1000 nucleotide pairs. A single **chromosome** contains hundreds of thousands of nucleotide pairs.

Chloroplasts   A single cell within photosynthetic plant organs may contain over 20 chloroplasts. Each chloroplast, as well as other kinds of plastids within the cell, is bound by two differentially permeable membranes. Within the membrane are **grana,** each of which is composed of a stack of "plates," termed **thylakoids,** and is interconnected by intergranal lamellae. The grana are embedded within the **stroma,** a hydrophilic material containing soluble protein, lipid droplets, starch grains, amino acids, sugars, and RNA and DNA. Carbon dioxide assimilation reactions in photosynthesis occur within the stroma, and the phytochemical reactions take place on the lamellar membranes of the thylakoids.

Mitochondria   Numerous mitochondria may be included in a single cell. These organelles are elongate to occasionally spherical bodies enclosed within a double membrane. The outer membrane has pores, and the inner membrane forms flat or mostly tubular extensions, termed **cristae.** The cristae have surface structures containing enzymatic proteins important in electron transport and in oxidative phosphorylation. Within the mitochondria, enzymes catalyze reactions of the citric acid cycle (Krebs cycle) of respiration, producing energy for ATP synthesis. Mitochondria also contain RNA and DNA and appear to synthesize some of their own enzymes.

Microbodies   The predominant microbodies are peroxisomes and glyoxisomes. The former usually are associated with chloroplasts and contain glycolate oxidase. The latter are found in oil-bearing seeds (e.g., cucurbits) and contain enzymes for fatty acid oxidation and for linking fatty acid degradation to sugar synthesis.

Golgi Bodies (Dictyosomes)   The Golgi bodies appear as stacks of units, each enclosed by a single membrane. They contain protein and are believed to function primarily in secretion, including cell wall deposition.

Microtubules   Microtubules are common cytoplasmic inclusions, apparently involved in cell shape and symmetry and in transport within the cell. The **spindle fibers** that develop during cell division are primarily microtubules.

Vacuole   The most conspicuous cytoplasmic inclusion in many cells is the vacuole, enclosed by a single differentially permeable membrane. Vacuolar contents may include inorganic ions, amino acids, organic acids, sugars, flavonoids, phenolics, alkaloids, and other compounds. Collectively, the vacuolar contents are termed cell sap. The function of the vacuole is varied and complex. The turgor of a cell is determined by the water content of a vacuole. The vacuole also serves as a waste disposal site for the cell.

Cell Function and Organization

Although cells contain the same structures, they are highly specialized according to function and location within the plant. Functionally, cells have been categorized as ground cells, protective cells, supporting cells, and conducting (vascular) cells.

Ground cells, or **parenchyma,** are thin-walled, elastic cells with large vacuoles, often with visible starch grains. They are living cells at maturity and function in photosynthesis and storage. Parenchymal cells with chloroplasts are termed **chlorenchyma.**

Protective cells include epidermal and cork or periderm cells. Epidermal cells are alive at maturity and may contain anthocyanins, but seldom contain chlorophyll. Cork cells are not alive at maturity. Protective cells normally have waxy deposits on exposed surfaces, cutin on epidermal cells and suberin on cork cells.

Supporting cells include **sclerenchyma** and **collenchyma.** Sclerenchyma provide substantial structural support and these cells are not alive at maturity. They include **fibers,** elongate cells with thick secondary walls, and **sclereids** (stone cells), which occur singly or in groups. Collenchymal cells are similar to fibers in shape, but are alive at maturity and have a

primary cell wall. Celery "strings" include collenchyma and associated vascular tissue.

Conducting cells are subdivided into those of the **xylem** (tracheids and vessels) which transport water and largely inorganic ions and those of the **phloem** (sieve cells and sieve tube cells) which conduct food. Tracheids are elongate with thickened secondary walls. The side walls have thin-walled areas (pits) which allow intercell solute/water flow. Tracheids have high tensile strength and contribute to the structural support of the plant. **Vessel elements** have perforated end walls, and when

TABLE 3.1
Classification and characteristics of plant tissues

Permanent	
Simple	
Epidermis	Single or multilayered cells which may include trichomes and guard cells. The exposed cutinized surface affords protection to underlying tissues.
Parenchyma	Tissues of thin-walled cells. In the stem, the inner parenchyma is the pith, the outer parenchyma the cortex. In leaves, it includes the spongy parenchyma and the palisade layer. These cells are active in photosynthesis, respiration, and other metabolic functions.
Sclerenchyma	Fibers located in stems and other organs and stone cells of some fruits. The fibers afford structural support.
Collenchyma	Bundles of collenchymal cells, often in stems and petioles, also providing some structural support.
Complex	
Cork	Protective tissue of secondary plant tissues with suberized exposed surface. Cells are loosely arranged. At intervals, unsuberized cells form the lenticels, which function in gas exchange. The cork consists of three cell types: cork cambium, phellem (produced inward from the cork cambium), and phelloderm (produced outward from cork cambium).
Conducting (vascular)	
Xylem	The primary tissue arises from the procambium; secondary xylem arises from the vascular cambium. The tissue consists of fibers, parenchyma, tracheids, and vessel elements.
Phloem	Primary and secondary phloem arise in the same manner as primary and secondary xylem. The tissue consists of parenchyma, fibers, sieve tubes, and companion cells.
Meristematic	Small thin-walled cells with capacity for division. They are characterized by dense cytoplasm and few vacuoles.
Apical	Located at plant apices, gives rise to primary plant growth.
Lateral	
Vascular cambium	Gives rise to secondary xylem and phloem.
Cork cambium	Gives rise to phellem and phelloderm.
Intercalary	Specialized apical meristems common in some monocots (corn) such that growth occurs not at the plant apex, but within the stem.

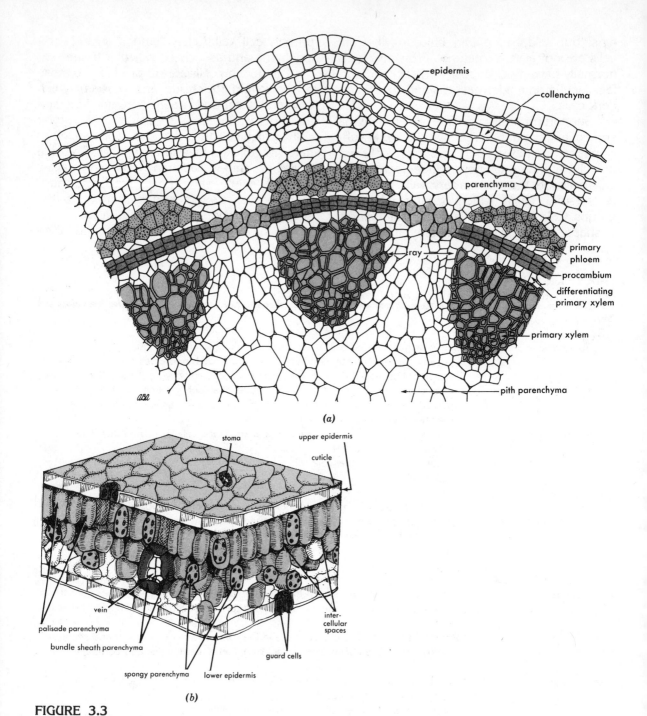

FIGURE 3.3
Section of dicotyledon stem (*a*), leaf (*b*), and root
(*c*). (From *Botany. An Introduction to Plant Biology*, T. E. Weier, C. R. Stocking, M. G. Barbour, and T. L. Rost. Copyright © 1982 by John Wiley & Sons, Inc., New York, pp. 130, 171, 183. Reprinted by permission.)

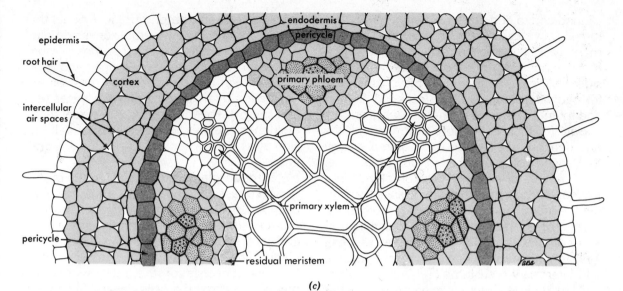

(c)

FIGURE 3.3 (*Continued*)

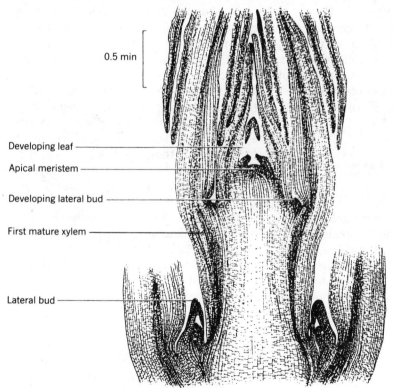

FIGURE 3.4
Longitudinal section of a shoot tip. (From *The Living Plant*, 2nd ed., by Peter Martin Ray. Copyright © 1963, 1972 by Holt, Rinehart & Winston, Inc. Reprinted by permission of CBS College Publishing.)

aligned end to end, they form long conductive tubes. **Sieve cells** are elongate cells, alive at maturity but containing no nucleus when functioning in food conduct. Both end and side walls have perforations. **Sieve tubes** are formed when the sieve tube cells are aligned end to end. The sieve elements always are associated with one or more **companion cells** and depend on them for the required energy for transport of solutes and for other functions.

The several cell types are organized into **permanent** and **meristematic** tissues. Permanent tissues include simple tissues (epidermis, parenchyma, sclerenchyma, collenchyma) and complex tissues (cork, vascular tissues). Meristematic tissues include apical, lateral, and intercalary meristems (Table 3.1).

The various tissues are organized further to form specific plant organs. Each plant organ is anatomically and morphologically distinctive. Further, plants of different species often show unique developmental or morphological features. Monocots, especially, are structurally different from dicots. The general anatomical features of representative plant organs for dicots are summarized in Figures 3.3 and 3.4.

GROWTH AND DEVELOPMENT

Cell Division
Growth occurs through an increase in both number and size of cells. The increase in number occurs within apical, lateral, and intercalary meristems through cell division (**mitosis**). Mitosis is an equational division: daughter cells duplicate the previous cell in chromosome number and genetic constitution. Subsequently, cell appearance and function will change as new tissues form.

Reproduction
Reproduction in plants occurs vegetatively (asexual propagation) and sexually (seed propagation). Vegetative reproduction through rhizomes, tubers, bulbs, fleshy roots, or corms is more common among tropical plants than those grown in the temperate zones and often occurs within polyploid species. Most temperate zone vegetables are propagated from seeds, which arise from either cross- or self-fertilization by male and female gametes (egg and sperm). The focal process in sexual reproduction is **meiosis,** the sequence of cell division by which the haploid egg and pollen grain arise.

All seed plants are characterized by two growth phases differing in chromosome number. One, the **sporophytic** generation, is of the normal somatic chromosome number and constitutes most of the growth phase of seed plants. The **gametophytic** generation is haploid and consists of the microspores and megaspores and the cells derived from them (including pollen grains and egg cells, respectively). The sequential appearance of these two growth phases is termed **alternation of generations.** It is in the transition from sporophytic to gametophytic generations that meiosis occurs.

Meiosis takes place in specialized spore mother cells (megaspore mother cell and microspore mother cell for female and male gamete formation, respectively), giving rise to **microspores** and **megaspores** (male and female spores, respectively), each of which first appears as a tetrad, or group of four (Figure 3.5).

Of the four megaspores, only one normally develops, and it undergoes three successive mitotic divisions within the embryo sac, giving rise to the **egg nucleus,** two **polar (endosperm) nuclei,** three **antipodal cells,** and two **synergid cells.** Only the egg nucleus and the polar nuclei are important in subsequent fertilization. On occasion, however, some embryo sac cells or occasionally cells outside the embryo sac may be stimulated to develop maternal embryos without fertilization (**apomixis**). Apomixis is relatively rare among the vegetable crops.

The four haploid microspores of the tetrad each develop a thickened wall characteristic of

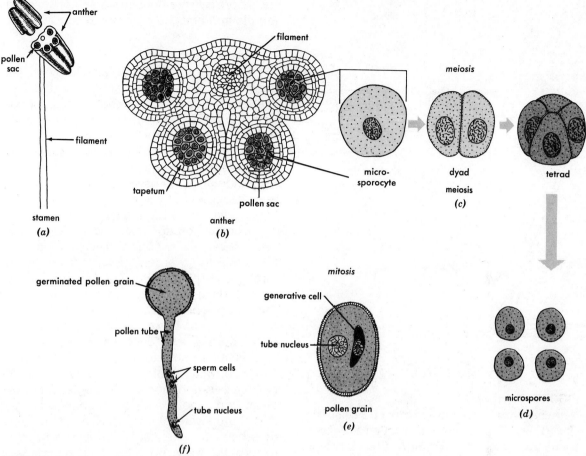

FIGURE 3.5
Development of pollen from a microsporocyte to the pollen grain: (*a*) stamen; (*b*) cross section of anther; (*c*) development of tetrad of cells from the microsporocyte by meiosis; (*d*) four microspores; (*e*) pollen grain; (*f*) germination of pollen grain.

(From *Botany. An Introduction to Plant Biology*, T. E. Weier, C. R. Stocking, M. G. Barbour, and T. L. Rost. Copyright © 1982 by John Wiley & Sons, Inc., New York, p. 286. Reprinted by permission.)

a pollen grain. Within the pollen grain (Figures 3.5 and 3.6), a mitotic division occurs, forming one **tube nucleus** and one **sperm cell.** The latter divides again, either in the pollen grain or in the pollen tube following pollen germination, providing two **sperm nuclei.**

Fertilization terminates the gametophytic generation and restores the somatic chromosome number. The union of egg with one sperm nucleus is followed by mitotic divisions and by cell differentiation, eventually forming the **embryo.** The fusion of the second sperm

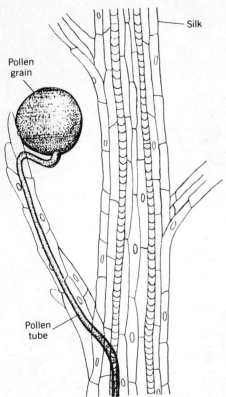

Pollen grain

Silk

Pollen tube

FIGURE 3.6
Germination of pollen grain of corn on a hair of the silk. (From *Botany*, 5th ed., by Carl L. Wilson, Walter E. Loomis, and Taylor A. Steeves. Copyright © 1952, 1957, 1962, 1967, 1971 by Holt, Rinehart & Winston, Inc. Reprinted by permission of CBS College Publishing.)

nucleus with polar nuclei also is followed by mitosis and by storage of carbohydrates, eventually forming the seed cotyledons or **endosperm.**

It should be evident in this reproductive process that self-pollinated plants will not express the genetic variability within offspring shown by cross-pollinated plants. It also should be recognized that genes affecting the endosperm (such as genes controlling sugar or starch content in sweet corn) may have an immediate effect on that constituent in the mature seed of a cross. Genes affecting other traits normally are not apparent until the seeds are planted and the offspring are grown.

Water Transport

Most plant tissues contain 70 to 95 percent water. Water is the solvent and the cell medium within which reactions occur, a reactant in important metabolic processes, and it forms part of the structure of protoplasm. Under water stress, nutrient uptake slows, and cells plasmolyze and lose turgor, affecting gas exchange and photosynthesis. The availability and movement of water, therefore, are essential conditions for plant survival.

Water absorption takes place near the root tips, where cells are thin-walled and root hairs are present. Movement of water from the soil through the plant is governed by the free energy gradient within the soil–plant–air continuum. Substances move from a system of high free energy to one of low free energy until an equilibrium is reached. Free energy of a system is expressed as chemical potential of a mole of water. Water potential (ψ) is the difference between the chemical potential of water at some point within the continuum and that of pure water at a pressure of 1 atm. Since the chemical potential is, by definition, equal to zero, the water potential generally will be a negative value, normally expressed in bars or pascals (1 bar $= 10^5$ Pa).

The water potential at any point of the continuum is a reflection of one or more forces: hydrostatic potential (ψ_p), osmotic potential (ψ_π), absorptive (matric) potential (ψ_m), and occasionally others. In plant cells, osmotic potential is the major force. Matric potential is the predominant force in soils.

Osmotic pressure is the pressure necessary to prevent water flow from a region of high concentration to one of low concentration across a selectively permeable membrane. This water flow is governed by differences in water potential which are related to solute concentrations on both sides of the membrane. The higher the concentration of salts and sugars in a cell, the lower the free energy of the cell sap; thus, water flow across the cell membrane will be toward the cell. As this migration occurs and increases the water content of the cell, turgor pressure is increased until equilibrium is reached with osmotic pressure. The osmotic potential is the osmotic pressure at this equilibrium, expressed as a negative value. The hydrostatic or pressure potential normally is a positive value and equals osmotic pressure at equilibrium.

Matric potential is governed by adsorption of water to surfaces. Adsorption reduces free energy of water, and the matric potential values therefore are negative.

The water potential of the atmosphere is much less than that of any part of the plant, and the water potential within any part of the plant normally is much less than that of the soil. Therefore, there is a strong gradient from the soil through the plant to the atmosphere, driving the movement of water. Affecting the gradient are such factors as relative humidity (vapor pressure) of the atmosphere, air temperature, wind velocity, soil texture and salinity, and surface area of the plant canopy. The lower the vapor pressure of the atmosphere, or the greater its temperature, the greater the rate of evaporation of water from leaf surfaces (the lower the free energy of water) and the greater the water potential gradient. Since matric potential reflects adherence of water to soil particle surfaces, it follows that soils with a large fraction of clay, characterized by a large surface area, would have a more negative matric potential than soils composed largely of sand. Soil matric potential also becomes more negative as the soil solution becomes saline, thereby reducing the gradient and consequent water flow.

The same forces driving water flow through a plant to the air also drive the flow between tissues and organs within the plant. The flow occurs spontaneously between the source (region of high free energy) and the sink (region of low free energy) and is termed **diffusion.** Diffusion gradients are affected primarily by osmotic potential or relative concentrations of solutions in cells. The distribution of water within and among tissues is a very dynamic process, optimizing cell turgor and metabolic activity within a particular plant environment.

Photosynthesis and Respiration

The total yield of a plant, both roots and top growth, is a reflection of **net photosynthesis,** the amount by which the rate of photosynthesis exceeds that of respiration on a cumulative basis. Both photosynthesis and respiration are fundamental to all metabolic activity in the plant, providing, on the one hand, the raw product (carbohydrate) for plant metabolism and, on the other, the transformation and conservation of solar energy necessary to support metabolic processes.

Photosynthesis   Photosynthesis is the only significant biological event in which inorganic compounds from nonfood sources (CO_2, water) are converted to food supporting plant and animal life. Photosynthesis is the process by which radiant energy is trapped by pigments (mostly chlorophyll) of green plants and then is used to reduce atmospheric carbon dioxide. This process is composed of two basic reactions, the **light reaction** and the **dark reaction.** The light reaction is the photochemical interception and transmission of light energy. The dark reaction is the reduction of CO_2 to form carbohydrates and other compounds.

The radiation reaching a plant leaf is reflected, transmitted through the leaf, or absorbed. Radiation affects plants only if absorbed. Further, it is known that only light of wavelengths between 400 and 700 nm is active photosynthetically. Of the direct solar radiation, approximately one-third is photosynthetically active, whereas two-thirds of the diffuse or reflected light may be photosynthetically active. Short wavelengths (blue light) within the visible spectrum are absorbed by carotenoids and chlorophyll, long wavelengths (red light) by chlorophyll. Chlorophyll occurs within chloroplasts predominantly as chlorophyll a and b. Chlorophyll a is the primary absorbing pigment and occurs in forms that differ slightly in peak absorption wavelength and therefore exploit a greater portion of the visible light spectrum. Other chlorophyll forms apparently serve as "antennae," intercepting light and transferring the intercepted energy to primary absorbing or trapping sites of the chloroplast where the photochemical reaction takes place.

Light Reaction In the initial phase of the light reaction, a photon of light is absorbed by chlorophyll, exciting the pigment molecule. The energy from this excitation may be lost by radiation (as heat or light) or may be transferred to another molecule and conserved. The overall reaction by which energy is conserved is expressed as

$$\text{light} + 4H_2O + 2NADP + 2ADP + 2P_i \rightarrow$$
$$O_2 + 2H_2O + 2NADPH + 2ATP + 2H^+$$

Two photosystems, referred to as PS I and PS II and differing slightly in the absorption spectrum of chlorophyll, are linked by the electron transfer process. The excitation of chlorophyll occurs in PS I, and the electron produced is passed through enzymes until it binds hydrogen ions, produced in PS II, with NADP (nicotinamide adenine dinucleotide phosphate). The electron void in the chlorophyll molecule resulting from this transfer is filled with an electron from water, a result of reactions within PS II. As electrons are passed from PS II to PS I, ATP (adenosine triphosphate) is formed.

Dark Reaction (Carbon Fixation) The hydrogen and associated electrons carried by NADP (designated as NADPH$_2$) and ATP provide the active ingredients for the dark reaction, in which CO_2 is reduced (combined with hydrogen) to form sugars (Figure 3.7). The overall reaction may be expressed as

$$3CO_2 + 9ATP + 6NADPH + 6H^+ \rightarrow$$
$$\text{triose phosphate} + 6NADP$$
$$+ 9ADP + 8P_i + 3H_2O$$

This reaction, unlike the light reaction, is temperature dependent and also is affected by CO_2 supply.

The initial product of the dark reaction is phosphoglyceric acid, a three-carbon compound, predominant in most plants ("C$_3$" plants). Some plants, most notably corn and some related genera, form malate (or aspartate), a four-carbon acid, as an initial product. These "C$_4$" plants also are anatomically distinctive, showing two types of cells containing chloroplasts: bundle sheath cells, which surround the vascular tissue, and mesophyll cells, which radiate outward from the bundle sheath cells. The chloroplasts in the bundle sheath cells lack the typical grana of mesophyll cells. Further, there is a partitioning of the carbon-fixing pathway between the two cell types. The outer mesophyll forms malate which then is transported to the bundle sheath cells where reactions forming sugar are completed.

Respiration Respiration is an oxidative degradation of organic compounds, the major product of which is usable energy. Functionally, respiration and photosynthesis are opposite reactions. However, they are closely interrelated. Respiration is totally dependent on photosynthesis; without the products formed by photosynthesis, respiration would lack the substances necessary for energy conversion.

Energy is essential for the many metabolic events within cells and, ultimately, for supporting life. Energy is conserved by the photosynthetic process; it is converted to usable form by respiration. This conversion is an oxidative process for which the carbohydrates (CH_2O) combine with oxygen, forming water and energy. This energy may be in the form of heat released from the system, or it may be stored in specialized molecules, most commonly ATP. ATP is formed from adenosine diphosphate (ADP) by adding one atom of phos-

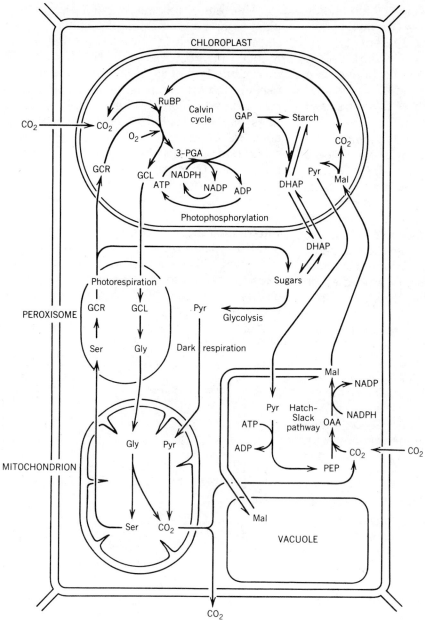

FIGURE 3.7
Diagram of photosynthesis and respiration pathways in plants, showing activity associated with cell organelles. DHAP, dihydroxyacetone phosphate; GAP, glyceraldehyde phosphate; GCL, glycollate; GCR, glycerate; Gly, glycine, Mal, malate; OAA, oxaloacetate; PEP, phosphoenolpyruvate; 3-PGA, 3-phosphoglyceraldehyde; Pyr, pyruvate; RuBP, ribose 1,5-biphosphate. (From *Introduction to Crop Physiology*, F. L. Milthorpe and J. Moorby. Copyright © 1979 by Cambridge University Press, New York, p. 101. Reprinted by permission.)

phorus, an addition involving a high-energy bond. This bond is unique, since, when hydrolyzed, it releases substantial free energy. This energy may be transferred to specific bonds in molecules of ATP, or it may be bound to a hydrogen acceptor, such as nicotinamide adenine dinucleotide (NAD), where it may be used by the cell or used to form ATP.

The carbohydrates normally stored within plant cells are complex and must be digested to simple sugars (i.e., glucose, $C_6H_{12}O_6$) before oxidation occurs. Two sets of reactions are required for complete respiration of glucose. The first, glycolysis, takes place in the cytoplasm and produces two molecules of pyruvate. Of the energy in glucose, little is transferred to ATP in this first reaction. However, two molecules of $NADH_2$ are produced which have the potential to produce four ATP molecules.

The two molecules of pyruvate then are oxidized in a series of steps, known as the Krebs cycle (or the citric acid or tricarboxylic acid cycle), to form carbon dioxide and water. The energy associated with bonding of the pyruvate molecules is released and stored in ATP in a process (oxidative phosphorylation) mediated by enzymes located in the mitochondria. Fats and amino acids also may be respired in the Krebs cycle.

The energy released in oxidative phosphorylation is substantial. For each mole of glucose, 36 moles of ATP is formed (Figure 3.8). Each mole of ATP transfers 7 kcal of energy. The total energy transferred thus is 252 kcal (7 × 36), and when compared to the total release of energy from a fully oxidized mole of glucose (673 kcal), the transfer efficiency is 37 percent. The remaining energy is released as heat or as other forms of energy.

The conversion of pyruvate to carbon dioxide and water is blocked in the absence of an oxygen supply. When oxygen is unavailable, the respiration pathway is altered, and fermentation occurs, producing either alcohol or lactic acid.

The rate of respiration differs in different plant organs. It is almost undetectable in seeds but may be high in flowers. The rate also is reduced in most matured plant tissues (but the mature fruit of some species show a steep rise, termed the climacteric): meristematic regions have a high respiration rate, whereas rates of differentiated or fully developed tissues are perhaps half those of the meristematic rate. The inherent respiration rate of a plant tissue or organ becomes a factor in determining production and handling systems for each crop.

Photorespiration   The respiration as described typifies most of the temperate zone crop plants, those classified as C_3 plants. Because it occurs in the absence of light (but also in the presence of light), it is termed dark respiration. Photorespiration also occurs in C_3 plants, largely within the glycolate pathway in

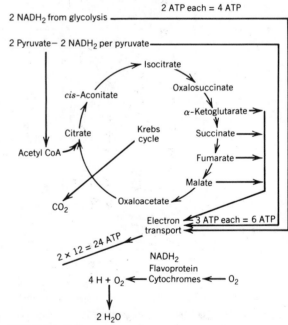

FIGURE 3.8
Krebs cycle, showing electron transport. (From *Crop Production* by S. R. Chapman and L. P. Carter. W. H. Freeman and Company. Copyright © 1976.)

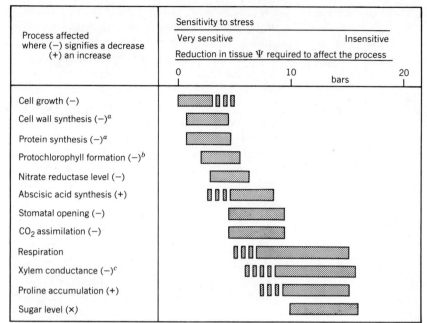

Process affected where (−) signifies a decrease (+) an increase	Sensitivity to stress		
	Very sensitive		Insensitive
	Reduction in tissue Ψ required to affect the process		
	0	10 bars	20
Cell growth (−)			
Cell wall synthesis (−)[a]			
Protein synthesis (−)[a]			
Protochlorophyll formation (−)[b]			
Nitrate reductase level (−)			
Abscisic acid synthesis (+)			
Stomatal opening (−)			
CO_2 assimilation (−)			
Respiration			
Xylem conductance (−)[c]			
Proline accumulation (+)			
Sugar level (×)			

FIGURE 3.9

Influence of water stress on function and activity of cells of mesophytic plants. Broken bars indicate effects not firmly established.[a] Rapidly growing tissue.[b] Etiolated leaves.[c] Should depend on xylem dimensions. (From T. C. Hsiao et al. (1976) *Philosophical Transactions, Royal Society of London Series B 273:479–500, as modified by A. H. Fitter and R. K. M. Hay (1981) Environmental Physiology of Plants.* Academic Press, New York. Reprinted by permission.)

peroxisomes and mitochondria (Figure 3.7). Photorespiration results in substantial CO_2 loss, thereby reducing net photosynthesis (dry matter accumulation and crop yield). Furthermore, there is no energy conservation in this process; hence, photorespiration is considered to be wasteful. Within C_4 plants, photorespiration does not occur, and such plants therefore are more efficient than C_3 plants in accumulating dry matter.

THE INFLUENCE OF ENVIRONMENT

Plants may react to stress imposed by one or several environmental factors by showing reduced productivity, poor quality, poor flowering or fruit set, or tissue necrosis and senescence. These symptoms of stress relate to changes brought about at the cellular or tissue level (Figure 3.9). The stress factor may affect the rhizosphere or the plant canopy.

Root Environment

Soils   Soil is not required for plant growth. However, it provides an efficient system for anchoring the plant and for maintaining the balance of ions essential for plant development. The physical properties of a soil, related to size, arrangement, and characteristics of individual soil particles and aggregates, in large measure determine its suitability for vegetable production. Although a wide array of soil types can accommodate successful production, the characteristics of each type may dictate choice of specific crop, production objectives, and management systems.

Soil **texture** is described by the proportions, by volume, of sand, silt, and clay (Figure 3.10). The extent to which soil particles form small aggregates defines the soil **structure.** In most production systems, texture is constant; the proportion of constituents is unaffected by crop management. Structure, however, can be altered drastically by tillage systems or cultural practices that increase or decrease aggregation or by modification of soil chemistry by fertilizer and irrigation water. Compression caused by heavy equipment, failure to maintain organic matter content, and tillage of wet soils all tend to destroy aggregates. Excess salts accumulated through fertilization and irrigation affect tilth and flocculation.

The basic constituents of a mineral soil differ in particle size and shape. Sand particles are relatively large (0.05 to 2.0 mm) and irregular in shape; silt and clay particles are small (0.05 to 0.002 mm and <0.002 mm, respectively). Stones or gravel (>2.0 mm), although constituents of many soils, do not contribute significantly to their properties. Sand has relatively little surface area per unit mass. Silt particles, although small with greater surface area than sand, behave more as sand particles than as clay. Clay particles are platelike in configuration, and their large surface areas maintain a negative electrical charge to which cations in the soil solution are attracted. The attracted ions are in equilibrium with those in the soil solution and can be exchanged from the soil particle to the solution in response to changes in equilibrium (e.g., resulting from depletion of soil solution by plant uptake). The amount of clay in a soil therefore contributes to its **cation exchange capacity** (CEC), defined as the amount (in milliequivalents) of exchangeable cation per 100 g of a dry soil. Cation exchange is important in plant nutrition, acting as a buffer against radical changes in nutrient status or balance. Sandy soils have a low CEC; clay soils a high CEC.

Soils have three physically distinct phases: solid, liquid, and gas. The solid soil particles occupy approximately 50 percent of a loam soil volume. The remaining pore space is filled with water and dissociated ions, and air. The **bulk density** of a soil reflects total pore space and is calculated as weight per volume of dry soil. It indicates water capacity of a soil and also gives an indication of compaction. A normal tilled loam soil will have a bulk density of 1.1 to 1.4 g/cm^3. Low bulk density would characterize sands and well-aggregated soils; heavy clays or soils with poor aggregate structure would have high values. For satisfactory production of vegetables, bulk density should be below 1.4 g/cm^3 for clays and below 1.6 g/cm^3 for sands.

Movement of water and air differs within soils differing in bulk density. The rate of water movement through a tube is proportional to the square of the tube diameter; thus, water flow through a soil is related largely to the pore size. In addition to pore size, the nature of soil particles, their compaction, and the degree of

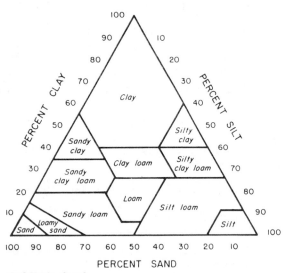

FIGURE 3.10
The texture triangle, showing relative percentages of sand, silt, and clay in each textural class. (*Source*: Soil Survey Manual, USDA Agric. Handb. 18 (1951). Reprinted with permission.)

saturation also affect water movement. Because clay particles are small and have a high surface-to-mass ratio, they have a strong affinity for water, resulting in layers of oriented water molecules that provide the characteristic cohesion, shrinkage and swelling, and absorptive capacity in clayey soils. Slow penetration of water through saturated clayey soils reflects both small pore spaces and the surface attraction afforded by clay particles to water molecules. Soil aggregation modifies effects of soil texture increasingly as the clay fraction increases. Aggregation adds pore space, thereby improving both water and air movement. Compaction will destroy aggregates and cement together fine soil particles, particularly clay, such that both water percolation and aeration are stopped or retarded. Poor water movement and the poor aeration that may accompany it result in decreased oxygen and increased carbon dioxide in the root zone. The lack of oxygen slows the conversion of insoluble plant nutrients to soluble forms, the uptake of nutrients, and the development of a normal root system. The major vegetable crops are very sensitive to poor soil aeration.

Organic Matter   Although organic matter is not essential for plant growth, the organic fraction contributes to the dynamic properties of a soil. The colloidal particles of the partially decomposed organic fraction (humus) which, like clay particles, are negatively charged, provide far more cation exchange capacity than clay surfaces. Some essential elements are bound by organic particles, reducing losses by leaching. The natural decomposition process then releases these elements slowly, along with nitrogen, phosphorus, sulfur, and carbon dioxide. Decomposition of organic matter also contributes to soil structure, providing, among other products, polysaccharides that are the primary agents for binding together soil particles into aggregates.

Associated with organic matter, sustained by it, and speeding its decay are a myriad of microorganisms, including viruses, myxomycetes, protozoa, algae, yeasts, fungi, actinomycetes, and bacteria, which take part in many reactions in the soil. Some are involved in a specific reaction; others are active in many reactions. Some may be responsible for plant disease; others enhance plant growth through nitrogen fixation. Still others contribute to nutrient loss. Within a normal cycle of plant growth and death, organic matter is formed continually and continually is lost by decomposition. The decaying plant residue provides carbon as a source of food and energy for soil microorganisms; the nitrogen released during decay, then, is utilized, at least in part, in building microbial protein. The demand for nitrogen by saprophytic microorganisms is related directly to the carbon fraction of the plant residue. Organic matter with high carbon relative to its nitrogen (high $C:N$ ratio) provides the food source to support an increase in microbial activity, but the nitrogen released in decay of such organic matter will be insufficient for their metabolic needs. Therefore, decomposition will proceed slowly unless nitrogen fertilizer is added. As decomposition proceeds, carbon is lost as carbon dioxide, whereas the nitrogen is retained by the microbes, and the $C:N$ ratio gradually narrows. Mature plant residue usually has a $C:N$ ratio of $30:1$ to $50:1$, and all nitrogen released from its decay will be utilized by microorganisms. As the ratio decreases below $20:1$, the nitrogen released increasingly will support both microorganisms and plant growth.

The microbial population is in a state of equilibrium, reflecting associations, interaction with environmental factors, competition for the same food source (antagonism), and possible antibiosis among different microbes. The level of microbial activity differs in different soils and fluctuates as food source, temperature, pH, and soil moisture change. In general, however, a certain soil type has a characteristic microbial composition. Of the total population of microbes, bacteria are the most abundant, followed by actinomycetes (predominantly *Streptomyces*) and fungi.

Nitrogen Cycle   Nitrogen is a plentiful element, but only a small percentage is available for plant growth. Plants acquire nitrogen as NH_4^+ and NO_3^- from decomposition of organic matter, from fertilizers, from transformation of atmospheric nitrogen by electrical storms, or by biological fixation of free nitrogen (Figure 3.11). Within unfertilized areas, the primary source of nitrogen is decomposition of organic matter. During decomposition, nitrogen first is released as NH_4^+ (ammonification). The ammonium is oxidized stepwise to nitrite (NO_2^-) by action of bacteria of the genus *Nitrosomonas* and then to nitrate by action of *Nitrobacter*.

Nitrogen fixation also can be a major source of soil nitrogen. Up to 50 to 280 kg/ha of nitrogen per year is fixed by rhizobia in sym-

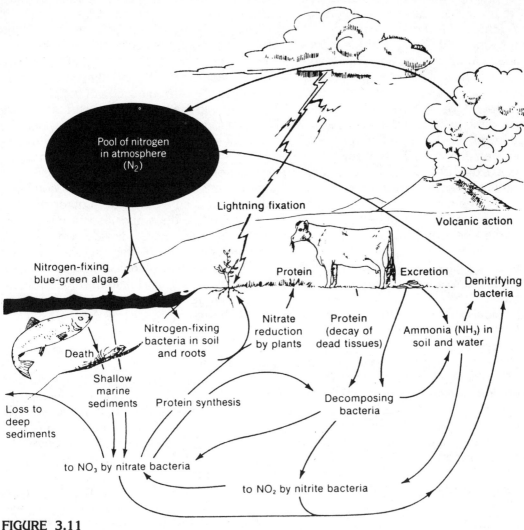

FIGURE 3.11
The nitrogen cycle. (From *Population, Resources, Environment,* 2nd ed., by Paul R. Ehrlich and Anne H. Ehrlich. W. H. Freeman & Company. Copyright © 1972.)

TABLE 3.2
A key to nutrient disorders of vegetable plants

1. Symptoms appear on leaves and stems of petioles.	2
Flowering or fruiting is affected.	13
Storage organs are affected.	14
Plant growth throughout the field is variable. Some plants appear normal, some show severe marginal leaf necrosis, and others are stunted. Determine soil pH.	Acid or alkaline soil complex
2. Youngest leaves are affected first.	3
Entire plant is affected or oldest leaves are affected first.	9
3. Chlorosis appears on youngest leaves.	4
Chlorosis is not a dominant symptom. Growing points eventually die, and storage organs are affected.	8
4. Leaves are uniformly light green and then yellow; poor, spindly growth follows. Most common in areas with acidic, highly leached sandy soils low in organic matter.	Sulfur
Uniform chlorosis does not occur.	5
5. Leaves wilt, become chlorotic, then necrotic. Onion bulbs are undersized; outer scales are thin and lightly colored. May occur on acidic soils, on soils high in organic matter, or on alkaline soils.	Copper
Wilting and necrosis are not dominant symptoms.	6
6. Distinct yellow or white areas appear between veins, and veins eventually become chlorotic. Symptoms are rare on mature leaves. Necrosis is usually absent. Most common on calcareous soils ("lime-induced chlorosis").	Iron
Yellow/white areas are not as distinct; veins remain green.	7
7. Chlorosis is less marked near veins. Some mottling occurs in interveinal areas. Chlorotic areas eventually become brown, transparent, or necrotic. Symptoms may appear later on older leaves. In peas and beans, the radicle and central tissue of cotyledons of ungerminated seeds become brown ("marsh spot"). Most common on soils with pH over 6.8.	Manganese
Leaves may be abnormally small and necrotic. Internodes are shortened. Beans, sweet corn ("white bud" of maize), and lima beans are most affected; potatoes, tomatoes, and onions are somewhat affected; uncommon in peas, asparagus, and carrots. Reduced availability in acidic, highly leached, sandy soils, in alkaline soils, and in organic soils.	Zinc
8. Tissues are brittle. Young, expanding leaves may be necrotic or distorted followed by death of growing points. Internodes may be short, especially at shoot terminals. Stems may be rough, cracked, or split along the vascular bundles (hollow stem of crucifers, cracked stem of celery). Most likely on leached, acidic soils, in alkaline soils, and in organic soils.	Boron
Brittle tissues are not a dominant symptom. Growing points are usually damaged or dead ("dieback"). Margins of leaves developing from the growing point are first to turn brown or necrotic, expanding corn leaf margins are gelatinous and necrotic, expanding cruciferous seedling leaves are cupped and have necrotic margins; old leaves remain green. Common on acidic, highly leached, sandy soils. May result from excess Na, K, or Mg from irrigation water, fertilizer, or dolomitic limestone (celery blackheart, brown heart of escarole, lettuce tipburn, internal tipburn of cabbage, internal browning of brussels sprouts, hypocotyl necrosis of snapbeans).	Calcium
9. Plant exhibits chlorosis.	10
Chlorosis is not a dominant symptom.	12
10. Chlorosis is interveinal or marginal.	11
General chlorosis progresses from light green to yellow. Entire plant becomes yellow under prolonged stress. Growth is immediately restricted, and plants soon become spindly and drop older leaves. Most common on highly leached soils or with high-organic-matter soils at low temperatures. Soil applications show dramatic improvements.	Nitrogen
11. Chlorosis is marginal or appears as blotches that later merge. Leaves show yellow chlorotic interveinal tissue on some species, reddish purple progressing to necrosis on others. Youn-	Magnesium

TABLE 3.2 (*Continued*)

ger leaves are affected with continued stress. Chlorotic areas may become necrotic and brittle and curl upward. Symptoms usually occur late in the growing season. Most common on acidic highly leached, sandy soils or on soils with high K or high Ca.	
Chlorosis is interveinal, with early symptoms resembling N deficiency (Mo is required for nitrate reduction); older leaves are chlorotic or blotched, with veins remaining pale green. Leaf margins become necrotic and may roll or curl. Symptoms appear on younger leaves as deficiency progresses. In brassicas, leaf margins become necrotic and disintegrate, leaving behind a thin strip of leaf ("whiptail," especially of cauliflower). Common on acidic soils or highly leached alkaline soils.	Molybdenum
12. Leaf margins are tanned or scorched or have necrotic spots (may be small black dots which later coalesce). Margins become brown and cup downward. Growth is restricted, and dieback may occur. Mild symptoms appear first on recently matured leaves, then become pronounced on older leaves and finally on young leaves. Symptoms may be more common late in the growing season due to translocation of K to developing storage organs. Most common on highly leached, acidic soils and on organic soils due to fixation.	Potassium
Leaves appear dull, dark green, blue-green, or red-purple, especially on the underside and at the midrib and veins. Petioles also may exhibit purpling. Restriction in growth may be noticed. Availability reduced in acidic and alkaline soils and in cold or organic soils.	Phosphorus
Terminal leaflets wilt with slight water stress. Wilted areas later become bronzed and finally necrotic. Very infrequently observed.	Chlorine
13. Fruit appears rough, cracked, or spotted. Flowering is greatly reduced. Tomato fruit show open locule, internal browning, blotchy ripening, or stem end russeting. Occurs on acidic soils, on organic soils with free lime, and on highly leached soils.	Boron
Cracking and roughness are not dominant symptoms. Fruit exhibit water-soaked lesions at bottom end, later becoming sunken, dark, or leathery (blossom end rot of tomato, pepper, and watermelon). Common on acidic highly leached soils.	Calcium
14. Internal or external necrotic or water-soaked areas of irregular shape (hollow stem of crucifers, internal browning of turnip and rutabaga, canker or blackheart of beet, water core of turnip) appear. May occur on acidic soils or alkaline soils with free lime, or on highly leached soils.	Boron
Cavities develop in the root phloem, followed by collapse of the epidermis, causing pitted lesions (cavity spot of carrots or parsnips). Common on acidic highly leached soils.	Calcium

Source: From Jean E. English and Donald N. Maynard (1978)
HortScience **13**(1). Reprinted by permission of the authors.

biotic association with leguminous plants. Although few vegetables are efficient in that respect, legumes such as clover may be used in rotation to enhance soil nitrogen. The rhizobia reside within root nodules, deriving their energy from the host plant. In turn, they reduce free nitrogen (N_2) gas from the atmosphere to NH_4^+, which then is used for growth of the host. As the plant dies, the residue decomposes, releasing nitrogen through ammonification. Nonsymbiotic fixation by some bacteria and blue-green algae also may occur, but the amount fixed is one-third or less that associated with legumes.

At the same time that nitrogen is added to the soil in available form, it may be lost by **immobilization** (used to form new organic materials) or by **denitrification.** The latter, in which bacteria utilize the oxygen of the nitrate ion, resulting in volatilization of free N_2 or N_2O, can be extensive. It is particularly serious in acidic waterlogged soils and increases as soil temperature increases.

Essential Elements   Sixteen elements are required for plant growth. Those used in largest amounts (macronutrients) include nitrogen, carbon, hydrogen, oxygen, phosphorus,

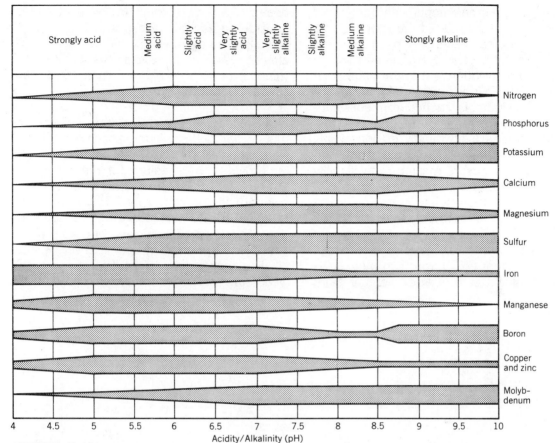

FIGURE 3.12
Relationship between soil pH and availability of elements essential for plant growth. The thickness of each horizontal band indicates relative availability. (From *Chemical Fertilizers* by Christopher J. Pratt. Copyright © 1965 by Scientific American, Inc. All rights reserved.)

potassium, sulfur, calcium, iron, and magnesium. Microelements, used in trace amounts by the plant, include manganese, zinc, boron, molybdenum, chlorine, and copper. Of the macroelements, nitrogen is derived ultimately from air, hydrogen and oxygen are provided by water, carbon is provided by carbon dioxide; and the remaining macro- and microelements are derived from mineral or organic particles in the soil. Each element contributes a specific function in plant growth and development, and deficiencies normally produce visible aberrations in plant appearance. The symptoms described in Table 3.2 frequently affect appearance and marketability as much as yield. Deficiencies occur in soils that lack an inherent source of an element but also may occur in soils of high or low pH (Figure 3.12) or in those in which nutrient levels may be unbalanced. Excessive calcium, for example, may depress the availability of potassium.

Of the macroelements, nitrogen and phosphorus are considered the most critical. Both are major constituents of plant cells and are essential for protein synthesis and energy transfer. Both are abundant in the plant environment, but not in forms available to plants. Nitrogen, as nitrate, is leached from soils subjected to frequent irrigation or rainfall. Phosphorus becomes fixed rapidly in unavailable forms and must be applied to many soils, often in bands, to ensure maximum availability to the plant roots. Neither is taken up in adequate amounts at low soil temperature, a frequent problem in northern soils at spring planting.

Excesses or imbalances of several micronutrients (manganese, boron) can be toxic to plants, causing necrosis and eventual plant death. Elements present in the soil but not considered essential for growth (e.g., silicon) are taken up by the plant with no deleterious effects. Others may be toxic to the plant (aluminum) or to animals and man (lead, cadmium, selenium).

Soil Solution   The soil solution is defined as that portion of the soil pore space occupied by water and various dissolved soluble salts. Water provides the medium for dissociation of ions and for movement of ions to and from the soil particle and plant root (Figure 3.13). The nutrient salts contained in a normal soil solution generally are insufficient for crop production and are particularly sparse in very acid or alkaline soils. In an undisturbed system, weathering, organic decay, and cation ex-

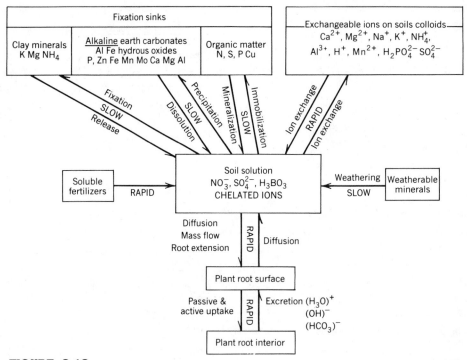

FIGURE 3.13
Equilibria involved in movement of nutrients to and from roots. (From M. E. Summer, Alleviating Nutrient Stress, in *Modifying the Root Environment to Reduce Crop Stress*, G. F. Arkin and H. M. Taylor, eds. Copyright © 1981 by American Society of Agricultural Engineers.)

change continually renew the soil solution; plant roots and leaching continually deplete it.

The predominant exchangeable cations in solution are H^+, Ca^{2+}, Mg^{2+}, K^+, and Na^+. Others, including NH_3^+, Mn^{2+}, Zn^{2+}, Cu^{2+}, and Al^{3+}, are present in lesser amounts. The interception of cations and anions by roots may occur through diffusion or mass flow of ions in the soil solution and through root extension as plants develop. Uptake into the plant is either passive or active, the latter requiring expenditure of energy by the plant. Stresses that slow plant growth and metabolism will reduce uptake of nutrients.

Soil pH   The pH of a laboratory solution is defined as $-\log$ (base 10) H^+ concentration, and the scale is based on the dissociation of water:

$$HOH \rightleftharpoons H^+ + OH^-$$

Since the ionization constant of water $(K_w) = 1 \times 10^{-14}$, the concentrations of H^+ and OH^- may be expressed as

$$\log 1/K_w = \log 1/[H^+] + \log 1/[OH^-] = 14$$

where $\log 1/H^+$ and $\log 1/OH^- = pH$ and pOH, respectively

When concentrations of H^+ and OH^- are equal in solution, the reaction is neutral. Each pH increment (n) from neutrality increases acidity or alkalinity by a factor of 10^n.

Within a soil, unlike a laboratory solution, the pH may not reflect total acidity, for clay and organic particles act as a surface for cation exchange, thereby contributing a potential source of H^+. In general, the larger the fraction of clay : organic particles, the greater the disparity between pH (active acidity) and total soil acidity. Therefore, soils identical in pH but different in texture will require vastly different treatments to effect a pH change.

Mycorrhizae   Symbiotic mycorrhizae have been known for many years to colonize roots of plants. In vegetables, the common class of mycorrhizae is the vesicular–arbuscular (VAM) group, so named because the fungi form vesicles, structures thought to have storage capacity, and arbuscules, possible sites of nutrient transfer.

Mycorrhizae are nonpathogenic, obligate symbionts with a very wide geographical range. The VAM hyphae penetrate plant roots and frequently extend outward from the root beyond the root depletion zone. There is evidence that VAM colonization enhances uptake of nutrients and water in most vegetable plants and *Rhizobium* fixation of nitrogen in legumes.

The response to VAM colonization differs among species. Those within the Brassicaceae and Chenopodiaceae are not colonized. Species that are colonized may show little apparent response, whereas others show consistently enhanced growth, especially in unfertilized soils, a response related to a dramatic increase in phosphorus uptake. Only one vegetable, onion, has shown a dependency on VAM colonization; its performance is improved in soils colonized by VAM regardless of phosphorus levels.

Soil Temperature   Soil temperature is a major determinant of microbial activity and organic matter decay, of germinability of seeds, and of root uptake efficiency. The amount of heat absorbed as radiant solar energy varies. Dark-colored soils have a low reflectance, which enhances absorption of solar energy. Energy absorbed then may be lost to the atmosphere by reradiation or convection or may be transferred to deeper soil layers. Fluctuations in soil temperature, which occur in response to season and daily heat influx and loss, relate to the solid, liquid, and gas proportions of the soil and to atmospheric conditions. Thermal conductivity of water is greater than that of air; thus, heat flow is increased (surface heat is dissipated rapidly) in heavy or wet soils relative to those well aerated. The lower the air temperature, the more rapid the loss. Thus, light-colored sandy soils, in comparison to dark silty clay loams, may absorb less solar energy, but less is lost by transfer or reradiation because of low water-holding capacity of sands. Peat soils, with their high water-holding capacity, not only lose heat by reradiation, but also prevent downward penetration of solar energy.

Soil temperature affects root dry weight, with root development declining above and below the optimum for the species. It also affects germination and uptake of water and nutrients and can affect the size, quality, and shape of storage roots.

Soils for Vegetable Production   Soil often determines the most suitable vegetables to be grown and/or the market to which production should be directed. Plant features, including fibrous or tap root system, density and depth of root growth, and crop maturity date may be optimized more within some soil types and sites than others.

Mineral Soils   Of the range of soil textures, all but those with excessive amounts of clay are useful in vegetable production. Light soils, those with a significant sand content, are termed early soils. They are frost free early in the spring and attain warmer average temperatures during the season than do heavy soils. They are relatively infertile, with low cation exchange and low water-holding capacity, and therefore require careful management. Such soils tend to be uniform, however, and especially suitable for root crops, salad crops, greens, vine crops, sweet potatoes, and other warm season crops. For early market production, they are superior to heavy mineral soils or organic soils.

Silty clay loams and similar loamy soils, in which the clay content is a significant portion of the total volume, drain more slowly than sands and, in northern areas, often cannot be tilled until well after spring thaw. They are inherently fertile, however, with relatively high cation exchange, and are among our most productive vegetable soils. Such soils are used frequently for growing processing vegetables and those for late storage or distant marketing late in the season.

Organic Soils   Soils are organic if the content of organic matter is, proportionately, 20 to 30 percent or greater where the clay fraction is 0 to 50 percent. Most organic soils have been reclaimed as drained swamps or shallow lakes. The organic deposits may be 6 to over 25 ft (2 to 8 m) deep and normally are of uniform tilth and fertility. With respect to management, they present several important characteristics:

1. Nitrogen is released slowly during the growing season, reducing the need for supplemental nitrogen fertilization.

2. Potassium and several micronutrients tend to be deficient, requiring corrective fertilization.

3. Oxidation of organic particles eliminates up to 1 in. (2.5 cm) of the muck per year under high solar radiation and extended production seasons, less in short, cool season areas.

4. Organic particles insulate and, as a consequence, the soils become frost free very late in northern latitudes and solar radiation does not penetrate rapidly or deeply.

5. Because mucks arose from swampland, an effective drainage system usually must be developed to maintain soil productivity.

6. The surface properties of colloidal organic particles attract not only cations and water, but also active ingredients of several herbicides and thereby may reduce herbicidal effectiveness.

7. The abundant organic matter provides an open, loose structure not easily compacted by equipment, a structure particularly suitable for uniform shape of root and bulb crops.

Although careful management is required, organic soils support very uniform quality and high yields of root crops, salad crops, sweet corn, bulb crops, and potatoes. Warm season crops are not usually well suited to organic soils because of soil temperature and excessive nitrogen. Such plants often "go to vine" or show poor fruit set.

Environment of the Plant Canopy

The plant canopy provides the surface area for light interception and gas exchange. The characteristics of this canopy as well as the environmental constituents are therefore major determinants of eventual crop yield. The characteristics of the plant canopy relate to plant architecture (leaf area, leaf and branch angle) and crop architecture (plant population density and spatial arrangement). The environmental constituents are light (intensity, duration, quality), temperature, humidity, atmospheric gases, and air movement.

Plant and Crop Architecture   Interception of light energy is related to the leaf surface exposed to the sun. Although leaf size is a factor in light interception, photosynthesis will increase little once expanding leaf size begins to shade the lower leaves. It is for this reason that plants with slightly erect leaf angle may be efficient, since light penetration to lower leaves is enhanced. Similarly, increasing plant density will increase photosynthetic capacity on an area basis, but as plants become crowded, shading reduces photosynthetic contributions of individual plants. The effects of cropping density are discussed with respect to competition and yield in Chapter 5.

Environmental Factors

Light   Light affects plants primarily through processes mediated by absorbing pigments, chlorophyll and phytochrome. Of these processes, photosynthesis and photoperiod are central to successful production.

The net photosynthesis (amount by which the photosynthetic rate exceeds respiration rate) is directly related to duration of light, to light intensity (heat), and to carbon dioxide content of the surrounding atmosphere. The rate of photosynthesis is controlled by stomates through control of water flow and gas exchange. As temperature and light increase, the leaf may become stressed, and stomates will close, restricting intake of CO_2. Under conditions of rapid photosynthesis, carbon dioxide, present at approximately 300 ppm in normal atmosphere, may become depleted within a plant canopy in still air. Under low irradiance, plant growth becomes etiolated. The photosynthetically active radiation is below the compensation point (where respiration rate equals rate of photosynthesis), and plants eventually die. Excessive irradiance may damage the photoreceptor system, inactivating key enzymes, pigments, and other metabolites.

Garner and Allard (1920) demonstrated that stimulation of reproductive growth was related to duration of light. Their work and research by others identified three categories of plants:

Short-day plants	Flower only if light period is less than a specific threshold
Long-day plants	Flower only if light period exceeds a specific threshold
Day-neutral plants	Flowering not related to duration of light exposure

Subsequent study has shown variation in photoperiodic responses. Some plants are qualitative in response, and flowering is induced when a specific daylength threshold has been passed. Others respond quantitatively, and the threshold may increase or decrease in response to high or low temperature. In reality, it is the night duration that is important, as shown when long-day plants growing in short days are induced to flower by interrupting the long night with a brief light exposure. The photoperiodic response has been found to be related to the two forms of the pigment phytochrome. One form, P_r, is receptive to orange-red light (600 to 680 nm) and inhibits flowering. The other, P_{fr}, is receptive to far-red light (700 to 760 nm) and induces flowering. P_r is the first form of the pigment to be synthesized, but is converted rapidly to P_{fr} when exposed to red light. P_{fr} is unstable and reverts to P_r in the absence of light or when exposed to far-red light. Thus, the length of the dark period influences the relative amount of the flower inhibitor, P_r.

Length of day also affects plant responses other than flowering. Bulbing in each cultivar of onion is dependent on a specific length of day. Potato tuberization is triggered in part by declining daylength. The phytochrome system also has been implicated in many plant phenomena not daylength related, including seed germination, phototropism, and leaf initiation (Table 3.3).

Temperature   Temperature affects rate of plant growth, mineral and water uptake by roots, flowering and pollen viability, blossom fertilization and fruit set, carbohydrate and growth regulator balance, rates of maturation and senescence, and quality, yield, and shelf life of the edible product. As outlined in the following list, some of these components of

plant or product performance are altered because temperature changes rates of metabolic activity; others reflect physical or structural impairment of function or of tissues; still others relate to differentiation of reproductive growth.

1. **Respiration** During daylight, photosynthetic activity normally exceeds that of respiration, resulting in a net accumulation of carbohydrates. In the absence of light, however, respiration utilizes these reserves. The higher the night temperature, the greater the loss of reserves (Figure 3.14). High night temperatures therefore limit both yield and quality of such vegetables as root crops and potatoes.

    Respiration also is a major factor in fruit ripening and in postharvest quality. All plant parts respire, and the degree to which carbohydrate losses can be minimized during and following harvest has a major influence on product quality. Fruit of some species (e.g., tomato) show an accelerated respiration rate (climacteric) coinciding with peak maturation, after which there is a noticeable decline. Just before or accompanying the climacteric is a marked increase in ethylene activity, and rapid changes in pigmentation, flavor, and texture occur during this period. Conditions that reduce the respiration rate will delay all of these processes and the onset of senescence.

2. **Vernalization** Most of the biennial vegetables initiate reproductive growth after extended (several weeks or months) exposure to low temperature and/or a specific daylength. This stimulus is more effective as plants reach adult growth; seedlings and transplants are minimally sensitive. Induction of flowering in sensitive plants is a quantitative response to low temperature; the duration of exposure needed to initiate reproductive growth declines as the temperature declines. It is also possible to devernalize—reverse the direction of differentiation—by

TABLE 3.3
Some phytochrome-controlled responses in angiosperms

Seed germination	Photoperiodism
Hypocotyl hook opening	Flower induction
Internode extension	Expansion of cotyledons
Root primordia initiation	Succulency
Leaf initiation and expansion	Epinasty
Leaflet movement	Leaf abscission
Electric potentials	Tuberization
Membrane permeability	Bud dormancy
Phototropic sensitivity	Sex expression
Geotropic sensitivity	Unfolding of monocot leaves
Anthocyanin synthesis	Rhythmic phenomena
Plastid formation	

Source: Reprinted with permission from P. F. Wareing and I. D. J. Phillips. *The Control of Growth and Differentiation in Plants.* Copyright © 1978. Pergamon Press.

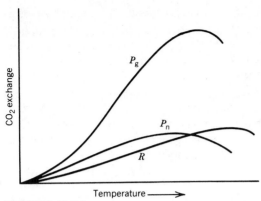

FIGURE 3.14
Effect of temperature on gross photosynthesis, P_g, net photosynthesis, P_n, and respiration, R. (From I. P. Ting, *Plant Physiology.* Copyright © 1982, Benjamin Cummings Publishing Company, Menlo Park, Calif., p. 449, Fig. 16.19. Reprinted with permission.)

exposing partially vernalized plants to high temperature.

The premature appearance of a flower stem in vegetables, termed **bolting,** can cause substantial loss, particularly in crops requiring relatively little cold exposure (e.g., celery). It is a problem that can occur in winter production areas and occasionally in the north in early summer.

3. **Chilling and freezing** Most plants tolerate temperatures 2 to 4°F (1 to 2°C) below freezing without physical damage. The freezing points of most cells are depressed because of dissolved solutes, and most cell solutions tend to supercool 2 to 4°F (1 to 2°C) more. Supercooled cell sap will form ice in response to enucleation by cell inclusions or by external factors such as bacteria. Ice formation within a cell generally is lethal. Plants that tolerate freezing do so because they tolerate extracellular ice formation. Tender plants that have been preconditioned (hardened) also tolerate extracellular freezing, but species vary in the degree to which they can be preconditioned.

Chilling damage occurs at temperatures above freezing and primarily affects plants of tropical or subtropical origin. Only corn, among the warm season crops, is resistant to chilling effects. Chilling results in a number of changes in plant function. Movement of water and solute into roots is slowed, and permeability of cellular membranes may be altered after plants have been returned to normal growing temperatures. Visible effects of chilling on plant organs vary, from surface lesions to abnormal pigmentation. Stored produce may be predisposed to pathogen infection by field chilling prior to storage.

4. **Fertilization and fruit set** Excessively high or low temperatures may reduce pollen viability or germinability on the stigma, or eventual fertilization. Poor fertilization usually is characterized by abortion of the flower or premature abscission of the fruit. Tomato fruit set declines as temperatures exceed 84°F (29°C) and is enhanced with a 10°F (5.6°C) diurnal variation [75°F (24°C) average day, 65°F (18°C) night]. Corn ears often show blank tips where pollination has failed because of hot, dry conditions.

5. **Quality** Temperature affects vegetable quality, not only through respiration, but also because of tissue damage or altered morphology following exposure to excessive heat or chilling and because of changes in relative amounts of sugar and starch in the edible product. Warm season vegetables—tomato, pepper, and squash—are especially sensitive to chilling, normally at temperatures below 45 to 50°F (7 to 10°C). The consequences of chilling are surface pitting or spotting and poor color development. High temperatures in these and other vegetables also cause physical damage (sunburn) or may affect product color or shape. Temperature stress at the time of fruit set or during late fruit development may cause defects (blossom scar, catfacing, puffy fruit in tomato) that render the product unmarketable.

Crops for which flavor is highly correlated with sugar or starch content include asparagus, peas, sweet corn, sweet potato, potato, and several of the root crops. Several of these lose sugar rapidly under high temperature, reflecting high respiration rate. Carrots, parsnips, beets, kale, and several other crops increase noticeably in sugar content under cold [32°F (0°C)] storage. Potatoes also gain sugar at storage temperatures below 50°F (10°C), but this accumulation is a deleterious factor for tubers intended for chipping. Sweet potato sweetness is enhanced by high-temperature curing [85°F (29°C), 85 percent relative humidity].

High temperature near harvest time accelerates maturation and narrows the time interval during which harvest can occur. This problem is particularly serious for peas and corn, melons, and several other vegetables, but affects a majority of crops to some degree.

Moisture   Water affects plant growth primarily through the soil solution. The relative humidity within a plant canopy, however, does influence transpiration rate and turgor of leaves and the intensity of related physiological disorders, such as tipburn. Humidity also affects plant growth indirectly by supporting development of disease and insect infestations. In some species (e.g., beans), low humidity seems to increase flower abortion, and pollen viability is decreased in many plants.

Atmospheric Gases   The role of CO_2 has been cited with respect to photosynthesis; other normal constituents of the atmosphere (nitrogen and oxygen) also are fundamental for plant growth. Increasingly, however, atmospheric pollutants, such as ozone, sulfur dioxide, and other products of industrialized societies, have become atmospheric constituents with serious effects on plant growth and yield. The most common and destructive pollutant is ozone, a natural consequence of electrical

TABLE 3.4
Variation in sensitivity of selected tomato and radish cultivars exposed to ozone concentrations of 40 and 35 pphm, respectively, for $1\frac{1}{2}$ h

Rank	Tomato	Percentage injury[a]	Radish	Percentage injury
1	Roma VF	61	Cherry Belle	35
2	Red Cherry	47	Crimson Giant	34
3	VF145B-7879	36	Comet	32
4	Pearson	35	Champion	31
5	Marglobe	33	Red Boy	25
6	Ohio WR-25	30	Calvalrondo	24
7	Heinz 1350	26	E. Scarlet Globe	24
8	Ohio WR-7	24	French Breakfast	23
9	Manapal	23	Icicle	17
10	VF13L	17		
11	VF145B	14		
12	Heinz 1439	10		

Source: From R. A. Reinert (1975) *HortScience* **10**:497. Reprinted by permission of the author.

[a] Injury of three most severely injured leaves, based on an average of 27 plants.

storms but also a product of combustion. Up to 90 percent of all pollution damage is caused by this photochemical oxidant. Its effects first were identified in 1958 in Los Angeles, California, and subsequent research has revealed sensitivity among many vegetable crops. The severity of damage is related to light, relative humidity, temperature, and other environmental factors and to cultivar (Table 3.4). Symptoms of toxic levels include pigmented lesions, surface bleaching, and necrosis or chlorosis. Leaves between 65 and 95 percent of full development have been reported as most susceptible. Spinach, radish, potato, tomato, onion, bean, sweet corn, celery, and melon are among the crops sensitive to ozone. Environmental factors that suppress plant growth tend to reduce sensitivity to ozone.

SO_2 is prevalent where coal or oil is burned or near volcanic activity. Significant crop losses have been identified, largely because of loss of photosynthetic tissue. Absorbed into mesophyll cells, SO_2 reacts with water to form sulfite ions which then are converted slowly to sulfate. Excesses of either ionic form are toxic, but sulfite is particularly so. Cells affected by toxic levels may collapse or may remain intact but with a bleaching of chlorophyll that reduces photosynthesis. A number of vegetables are sensitive.

Other air pollutants include fluorides, nitrogen oxides, ethylene, chlorine, and solid particulates, often heavy metals. These pollutants tend to be localized problems near the source of industrial or automobile emissions.

SELECTED REFERENCES

Arkin, G. F., and H. M. Taylor (eds.) (1981). *Modifying the Root Environment to Reduce Crop Stress,* Monogr. 4. ASAE, St. Joseph, Mich.

Baird, L. A. M., and B. D. Webster (1978). Relative humidity as a factor in the structure and histo-chemistry of plants. *HortScience* **13**:556–558.

Bleasdale, J. K. A. (1974). *Plant Physiology in Relation to Horticulture.* MacMillan & Co., London/Basingstoke.

Donahue, R. L., R. W. Miller, and J. C. Shickuna (1977). Soils. *An Introduction to Soils and Plant Growth.* Prentice-Hall, Englewood Cliffs, N.J.

Fitter, A. H., and R. K. M. Hay (1981). *Environmental Physiology of Plants.* Academic Press, New York.

Garner, W. W., and H. A. Allard (1920). Effect of the relative length of day on growth and reproduction in plants. *Journal of Agricultural Research* **18**:533–606.

Jacobson, J. S., and A. C. Hill (eds.) (1970). *Recognition of Air Pollution Injury to Vegetation: A Pictorial Atlas,* Rep. 1. Air Pollution Control Assoc., Pittsburgh, Pa.

Johnson, C. B. (ed.) (1981). *Physiological Processes Limiting Plant Productivity.* Butterworths, London.

Milthorpe, F. L., and J. Moorby (1979). *An Introduction to Crop Physiology.* Cambridge Univ. Press, New York.

Reinert, R. A. (1975). Monitoring, detecting, and effects of air pollutants on horticultural crops: Sensitivity of genera and species. *HortScience* **10**:495–500.

Salisbury, F. B. (1979). Photoperiodism. *Horticultural Reviews* **4**:66–105.

Sprague, H. B. (ed.) (1964). *Hunger Signs in Crops.* David McKay, New York.

Taylor, O. C. (1975). Air pollutant injury to plant processes. *HortScience* **10**:501–504.

Ting, I. P. (1982). *Plant Physiology.* Addison–Wesley, Reading, Mass.

USDA (1957). *Soils,* Yearbook of Agriculture. U.S. Govt. Printing Office, Washington, D.C.

Winchester, A. M. (1977). *Genetics.* Houghton Mifflin, Boston, Mass.

STUDY QUESTIONS

1. How does respiration rate affect location of production and timing or harvest of some crops?

2. What attributes are provided by soil organic matter and how do they affect growth of the plant?

3. In what ways does low temperature affect productivity and quality of warm season vegetables?

4

Modifying Plant Environment

PRODUCTION SITE

The siting of a production field determines both soil and canopy environment to some extent. A southern slope allows greater soil warming [1 to 5°F (1 to 3°C)] and generally earlier maturity than a field sloped to the north. Farms located within the influence of large bodies of water often escape low-temperature damage. Air drainage afforded by adequate slope also is important in avoiding frost.

REGULATING SOIL MOISTURE

Of the moisture reaching the soil, some is lost by runoff, some by evaporation, and some by percolation to the water table. The largest portion, however, is transpired by plants, the rate of loss fluctuating with duration and intensity of solar radiation, wind speed, and relative humidity. Plants wilt under weather conditions favoring rapid transpiration, even at a soil water potential of −5 to −30 kPa (field capacity). Although this transient wilting (incipient wilting) is not indicative of soil moisture deficit, it does indicate plant stress. Such stress increases in frequency and intensity as soils dry. At the permanent wilting point, when water uptake ceases, plant tissues and cells are irrevocably damaged.

Excessive soil moisture (0 to −5 kPa) also stresses plants by limiting the oxygen available for root development, and saturated soils increase the activity of soilborne pathogens and reduce available nitrogen (denitrification). It is

Crop management consists of a series of techniques applied to modify the crop microenvironment, thereby enhancing plant productivity and quality. To a considerable extent, these techniques are directed toward the root environment: soil moisture, temperature, pH, fertility, and organic matter content. However, the canopy environment may be modified to reduce frost and freezing injury or to enhance earliness, fruit set, and quality of vegetables.

therefore important to maintain a balance of soil moisture, including systems for drainage as well as for irrigation.

Irrigation

Nearly one-half of the water used in the United States is applied to crops as irrigation. Of the total agricultural cropland, over 61 million acres (24,700,000 ha) are irrigated. Over 740,000 acres (300,000 ha) are estimated to be trickle irrigated, most of this area in vegetable and fruit crops. Supplemental water is especially important in vegetable production, for periods of water stress induced by soil water deficits will decrease yield and impair quality of the edible product. Irrigation, therefore, has become a major production input for vegetable growers.

Natural rainfall seldom maintains adequate soil moisture throughout a season. Crops suffer perceptibly if water is not applied at regular intervals of 1 to 3 days (sandy soils) to 7 to 10 days (clay loams). The growth stage at which a plant suffers most from a soil water deficit varies among vegetable crops. Fruit-bearing vegetables generally are most susceptible at flowering and at fruit set, but also require moisture during fruit enlargement. For nonfruiting crops, the period of enlargement of head, root, tuber, or stem is most critical (Table 4.1).

In general, plants should be irrigated when 40 to 50 percent of the available soil water has been used, although it does depend on soil type and the crop being grown. The amount of water held by a clay is much greater than that held by sand (Table 4.2); therefore, losses by transpiration and evaporation will affect sandy soils more quickly than heavy soils. Crop wilting as an indicator of water deficit must be interpreted for each soil texture and set of weather conditions. Furthermore, wilting may occur after stress has impaired quality or growth. Tensiometers, neutron probes, and resistometers measure soil moisture status with greater accuracy than is afforded by observing plant appearance and may be used to study

patterns of water use and drainage during the growing season. Evaporation pans, although not measuring soil moisture, are useful in estimating irrigation needs.

Resistometers, used by scientists to monitor profiles of water usage, measure electrical current passed through porous gypsum blocks placed in the soil near the root zone. The greater the soil moisture, the less the resistance to electrical flow. These devices generally are not used to program irrigation frequency, since they are not sensitive at high moisture levels. Neutron probes are highly accurate instruments but generally are restricted

TABLE 4.1

Approximate susceptibility to water deficits at several stages of crop development

	Reduction in yield (%)			
Crop	Vegetative	Flowering and fruit set	Enlargement	Ripening
Corn	25	60	32	12
Bean	12	52	44	12
Pea	14	46	40	—
Tomato	25	52	43	25
Watermelon	35	45	45	20
Cabbage	13	—	28	36
Onion	30	—	45	21
Potato	30	—	41	13

Source: Modified from R. Morse (1983). *Vegetable Grower's News,* Virginia Polytechnic Institute & State University. Data used by permission of the author.

TABLE 4.2

Typical moisture percentages in several soil types

Soil type	Field capacity (dry wt basis)	Permanent wilting (%)	Available water (%)
Sand	6	3	3
Sandy loam	10	5	5
Silt loam	20	10	10
Clay loam	35	19	16
Clay	40	20	20

Source: Department of Irrigation, University of California Davis.

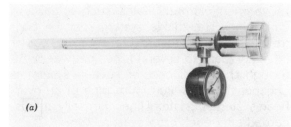

(a)

(b)

FIGURE 4.1

Tensiometer for measuring matric potential of soil. (*a*) Design showing enclosed tube, gauge, and tip through which water moves, creating a vacuum within the tube. (*b*) Placement of tensiometers within a row at two different depths. (Photos courtesy of Irrometer Company, Riverside, Calif.)

to scientific use because of high cost. The most common instruments used to monitor soil moisture are tensiometers. A tensiometer, which measures soil matric potential, consists of a water-filled tube, sealed at one end and with a porous tip at the other. When the porous tip is placed in a dry soil, capillary forces attract water from the tube, creating a vacuum (Figure 4.1). This vacuum is measured by a gauge calibrated in units of pressure (bars or kilopascals). These instruments generally are used in pairs at several field locations. Of each pair, one measures the shallow root zone, the other the deep zone. Tensiometers are most

useful at high moisture levels and become increasingly inaccurate as soil moisture decreases.

Evaporation pans also may be used to estimate evaporation and transpiration water loss. Evapotranspiration (the loss from a given area by soil evaporation and plant transpiration over a specified time) will be 70 to 80 percent of pan evaporation, 80 to 100 percent in soils of high water content or where crop foliage covers the ground. For crops in which foliage is not dense, evapotranspiration will be reduced in relation to evaporation. Evaporation is determined by using four or five 1-liter or 1-quart cans full of water, placed near the crop plants. Water level in the cans is measured daily to determine evaporation loss. In early summer in the northeast, evaporation may be $\frac{1}{4}$ in. (0.5 to 0.6 cm) per day or more. These evaporation data for an average midsummer day may be used to compute water requirements for row crops by the formula

$$Q = 50 \times E_p \times S$$

where Q = gallons required/100 ft of row
   E_p = average daily pan evaporation in inches
   S = row spacing in feet

Types of Irrigation Systems   Irrigation is applied with overhead, surface, and subsurface systems. The applicability of a particular system must be determined by site, crop grown, soil texture, and water cost and availability.

Subsurface irrigation may involve an underground solid-set system or may rely on a natural shallow water table that allows adjustment of the water level as needed. The latter system is utilized in Florida, where a shallow hardpan facilitates delivery of subsurface moisture and drainage of excess moisture. Such circumstances are somewhat unique, and although this system is efficient, growers must use caution in applying chemical treatments (herbicides, systemic insecticides) that

FIGURE 4.2
Trickle irrigation system showing uniform distribution of water within each row. (Photo courtesy of Chapin Watermatics, Inc., Watertown, N.Y.)

might migrate to nontarget areas or affect groundwater quality.

Surface systems include **furrow** and **trickle** irrigation. Of the two, trickle or drip irrigation is the most versatile. Trickle irrigation (Figure 4.2) utilizes plastic twin-wall lateral lines placed in the crop row, with emitters spaced to provide proper placement of the water. The emitters for row crops are line-source types, consisting of equally spaced holes along a single double-wall tube. The discharge pressure, normally less than 15 psi (1.05 kg/cm^2), will deliver from 0.1 to 1.0 gal/min (0.02 to 0.2 m^3/h) per 100 ft (30 m) of line. The tubing is installed in lengths of up to 300 ft (90 m), depending on slope. In areas of excessive slope, the discharge rate will be uneven along the delivery tube, and pressure-compensating emitters may be preferred. These emitters control water pressure within the emitter, improving distribution uniformity.

Water delivery to the emitter tubes (lateral lines) is by a large main line and smaller submain pipes (Figure 4.3). The water pressure is controlled with valves, and water must be filtered to prevent clogging of emitters. Some water sources may require chemical treatment to avoid buildup of algae or bacteria or to re-

move chemical precipitates, such as calcium carbonate. The trickle system may be controlled by timers to provide water at regular intervals (daily to several times per week). Although the initial cost of such a system is cheaper than the aluminum pipe of an overhead system, the lateral lines generally are discarded after each crop season.

Trickle irrigation systems offer several advantages. Amounts of water are reduced because placement is near the root zone, although Bernstein and Francois (1973) found this to be an advantage primarily when plants are young. Limited water sources and low pumping pressures can be utilized, and control of water placement and amount is relatively precise. Labor requirements are reduced, and some field operations can continue during irrigation. Fertilizer can be delivered through the irrigation system to the site of maximum utilization. Disadvantages include plugging of the emitters with soil particles or algae, uneven moisture distribution between plants (which might affect plants with spreading root systems), and the high level of management required for best results, even though the system is amenable to automation. In addition, the fine root structure tends to concentrate within the top 1 in. (2.5 cm) of soil, which, in some circumstances, can exacerbate damage from soil salinity. Trickle irrigation is not well suited for frost control.

Furrow irrigation (Figure 4.4) draws water gravitationally from a header ditch by siphons or other systems at a rate that ensures migration along the furrow between plant beds and drainage from the far end of the field. Fields must be surveyed and planed (using a laser-equipped land plane) to a 0.2 to 0.5 percent slope prior to shaping beds. Water use in such a system relates to soil texture and length of the furrow. The lighter the soil, the more rapid the infiltration rate. The length of furrow should be shortened and the slope increased for light soils to avoid excessive gravitational water loss.

The relative water distribution and use effi-

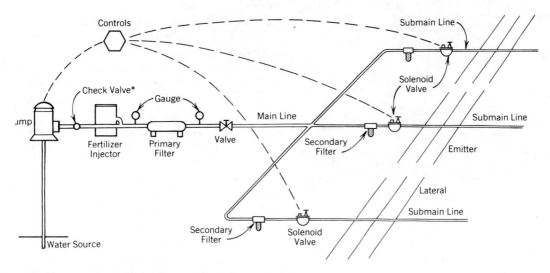

Controls

Submain Line

Check Valve*

Gauge

Solenoid
Valve

ump

Main Line

Submain Line

Fertilizer
Injector

Primary
Filter

Valve

Secondary
Filter

Emitter

Secondary
Filter

Solenoid
Valve

Lateral

Submain Line

Water Source

*A backflow preventer or vacuum breaker is required in some areas.

FIGURE 4.3
Components of a trickle irrigation system. (From Publ. NRAES-4, 1981. Northeast Regional Agricultural Engineering Service, Cornell University, Riley-Robb Hall, Ithaca, N.Y. 14853.)

ciencies of each surface irrigation system and the crop yield responses to each system differ for each vegetable and each soil type. Trickle irrigation is a very efficient supplier of moisture, but coverage is confined to a small area. It has been most efficient in twin-row production systems, reducing water applied per plant substantially. The efficiency of furrow irrigation increases as furrow length decreases. This system provides excellent water penetration, but distribution is uneven and there is substantial loss by evaporation and runoff. The effluent from furrow irrigation also must be discharged without impairing the quality of surface or domestic water supplies, particularly with respect to toxic residues.

Overhead irrigation systems (Figure 4.5) are applicable to any terrain and to porous or nonporous soil textures, where evaporative rates are not excessively high. Both solid-set and moving systems are used. Solid-set irrigation involves several lateral pipes, each with evenly spaced risers fitted with nozzles engi-

FIGURE 4.4
Furrow irrigation of staked tomatoes. (Photo courtesy of Chapin Watermatics, Inc., Watertown, N.Y.)

FIGURE 4.5
Overhead irrigation using movable aluminum pipe. (Photo courtesy of University of New Hampshire.)

neered to apply uniform amounts of water in an overlapping circular pattern. These laterals are supplied by a mainline pipe with water pumped from a surface or, more commonly, a well-water supply. Labor is frequently required to move the pipe, unless permanent installations cover the entire area, and water pumping costs can be high. To reduce labor needs, several mobile systems have been developed, including self-propelled center-pivot, side-roll or wheel pipe, and traveling guns (long-trajectory nozzles). In principle, each is similar to a solid-set system, but each is designed to maximize area of coverage with minimum setup time.

Overhead systems should be designed to provide a rate of application less than the infiltration capacity of a soil. Application rates (Table 4.3) therefore may be higher for light than for heavy soils. Low precipitation rates reduce soil puddling and structural damage of fine-textured soils and minimize runoff and erosion. Low-precipitation systems also cost less than high-volume systems.

The design of an appropriate overhead irrigation system must consider area of coverage, rate of application, distance from source to delivery point, and vertical lift. These factors determine power needs (water horsepower) according to the formula

$$whp = (gpm \times tdh)/3960$$

where gpm = gallons per minute
tdh = total dynamic head ($= h_{st} + h_p + h_f$, where h_{st} = static head or vertical lift pressure, h_p = pressure needed to operate the most distant sprinkler, and h_f = pressure needed to overcome friction)

An increase in the size of the system, the distance of the field from the water source, or the extent of vertical lift increases the power needs. The greater the power need, the higher the cost to irrigate. Although the initial cost of aluminum pipe systems is high, the cost is amortized over many years of use.

Delivery of water by sprinkler systems can be computed as

$$precipitation = \frac{gpm \times 96.3}{C \times S}$$

where gpm = gallons per minute
96.3 = the difference between water pumped and that received by the soil
C = distance between adjacent lines
S = spacing between emitter heads on a line

The gpm may vary from 2.5 for low-pressure, small-emitter heads to over 200 for the big-gun nozzles.

Salinity   Irrigation water contains varying amounts of dissolved salts, and soil fertilization also contributes salts to the soil solution. As plants draw water and as moisture evaporates from the soil surface, dissolved salts accumulate in the root zone, especially in soils in which drainage is restricted by high water table or hardpan. Excessive salt levels reduce the free energy gradient in the soil–plant–air continuum, retarding uptake of water by plants, reducing yield, and eventually leading to leaf

TABLE 4.3
Precipitation rates for various nozzle sizes, pressures, and spacings

Nozzle size (in.)	Pressure (psi)	Discharge (gpm)	Diameter of spray (ft)[a]	Precipitation rate (in./h)[b]		
				30 × 40	30 × 45	40 × 40
$\frac{1}{16}$	50	0.80	61–73	0.064		
$\frac{1}{16}$	55	0.85	62–74	0.068		
$\frac{1}{16}$	60	0.88	63–75	0.071		
$\frac{1}{16}$	65	0.93	64–76	0.075		
$\frac{5}{64}$	50	1.25	62–72	0.100	0.089	
$\frac{5}{64}$	55	1.30	64–74	0.104	0.094	0.079
$\frac{5}{64}$	60	1.36	67–76	0.110	0.097	0.082
$\frac{5}{64}$	65	1.45	68–77	0.116	0.103	0.087
$\frac{3}{32}$	50	1.80	69–77	0.145	0.128	0.108
$\frac{3}{32}$	55	1.88	70–78	0.151	0.134	0.113
$\frac{3}{32}$	60	1.98	71–79	0.159	0.141	0.119
$\frac{3}{32}$	65	2.08	72–80	0.167	0.148	0.125
$\frac{7}{64}$	50	2.44	72–80	0.196	0.174	0.147
$\frac{7}{64}$	55	2.56	74–81	0.205	0.182	0.154
$\frac{7}{64}$	60	2.69	76–82	0.216	0.192	0.161
$\frac{7}{64}$	65	2.79	77–83	0.224	0.199	0.168
$\frac{1}{8}$	50	3.22	78–82		0.230	0.193
$\frac{1}{8}$	55	3.39	79–83		0.242	0.204
$\frac{1}{8}$	60	3.55	80–84		0.253	0.213
$\frac{1}{8}$	65	3.70	81–85			0.222

Source: University of California Coop. Ext. Leaflet 2265.

[a] Shows range of diameters of spray reflecting different models of sprinklers.

[b] Precipitation rate at spacings indicated in feet. The three-digit numbers are shown to indicate progression as nozzle size and pressure increase. Equipment seldom performs with this precision.

scorch, wilting, and death (Figures 4.6 and 4.7). High salinity also may alter soil structure, reducing flocculation and tilth and increasing soil crusting. More salt accumulates within furrow irrigation systems than in drip or overhead systems.

Salinity is measured as electrical conductivity (ECe) in millimhos per centimeter at 25°C. Crops showing yield losses at ECe values of 2 to 4 are considered to be salt sensitive (bean, onion, carrot, lettuce). At an ECe between 4 and 8, many crops are damaged. A few vegetables (e.g., beet) can grow at ECe values of 8 or above (Table 4.4).

Salt levels are reduced in the root zone by deep plowing or by leaching. Deep plowing places the region of highest salt accumulation below the root zone and is most effective when upward capillary water movement is minimal. The leaching requirement (volume of water required to lower ECe to a safe level) relates to the quality of irrigation water and to the salt sensitivity of the crop. Assuming a salt-free water source, 6 in. (15 cm) of water will leach 1 ft (30 cm) of soil of 50 percent of the accumulated salts.

Relative Humidity Irrigation, particularly when applied overhead, adds temporarily to the relative humidity within the plant canopy.

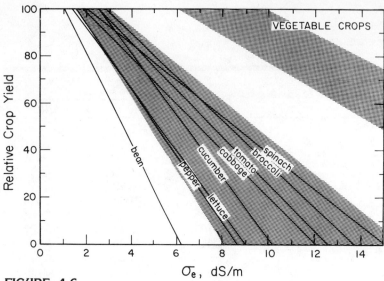

FIGURE 4.6

Effect of soil salinity on relative yield of eight vegetable crops. Bean is the most sensitive, spinach and broccoli the least. (Photo courtesy of U.S. Salinity Lab, Riverside, Calif.)

TABLE 4.4
Tolerance of certain vegetables to salinity

Vegetables (in order of decreasing tolerance)	ECe (mmho/cm at 25°C) at which yields are decreased by		
	10%	25%	50%
Beet	8	10	12
Spinach	5.5	7	9
Tomato	4	6	8
Broccoli	4	6	8
Cabbage	3	4	7
Cucumber	3	4	6
Muskmelon	3	4	6
Potato	3	4	6
Corn	2.5	4	6
Sweet potato	2.5	4	6
Lettuce	2	3	5
Pepper	2	3	5
Radish	2	3	5
Onion	2	3	4
Carrot	2	3	4
Bean	1.5	2	4

Source: Adapted from L. Bernstein (1970). *Salt Tolerance of Plants,* USDA Agric. Inf. Bull. 283. In O. A. Lorenz and D. N. Maynard, *Knott's Handbook for Vegetable Growers.* Copyright © 1980 by John Wiley & Sons, Inc., New York.

Within canopies of high-density plantings, this humidity persists and may enhance fruit set and leaf turgor but also may encourage activity of plant pathogens, particularly bacterial pathogens. The greater the wind circulation, the shorter the duration of high humidity and the smaller its effect on plants. Rain or irrigation water may destroy market quality of some commodities. Plastic "awnings" have been used over tomatoes in some locations and seasons to reduce fruit cracking that results from exposure to frequent rainfall.

Drainage   Soils must be well drained to be productive. When natural drainage is insufficient, tile systems are installed, or surface shaping and ditching is necessary. Increasingly, growers are utilizing raised beds, once developed for surface irrigation, but also effective in maintaining proper aeration of soils of low to medium infiltration rate, particularly in the plant root zone. Root and tuber crops grown in raised beds often are improved in quality and appearance.

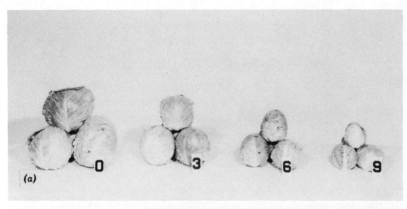

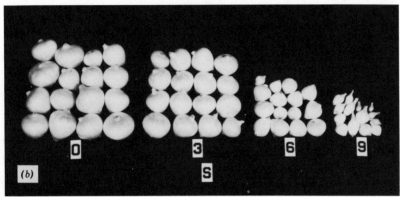

FIGURE 4.7
Reduction in size of cabbage heads (*a*) and onion
bulbs (*b*) in soils of increasing salinity. (Photo
courtesy of U.S. Salinity Lab, Riverside, Calif.)

Drainage of organic soils presents unique problems, for the height of the water table directly affects the loss of soil due to oxidation and decomposition. Most fields of organic soils are ditched to provide drainage, the depth of the ditch providing control of maximum water table. Tile systems may be placed 3 to 5 ft (1 to 1.5 m) deep in such soils, depending on the characteristics of the soil and subsoil, at intervals ranging from 50 ft (15 m) to well over 5000 ft (1500 m). The desired water table for most crops would be approximately 2 ft (0.6 to 0.7 m).

Subirrigation and drainage may be provided by the same system, using corrugated flexible perforated plastic pipe placed at a soil depth of approximately 20 in. (50 cm). Water

and nutrients are pumped in to irrigate, and each pipe has an outlet to allow drainage. This system can be operated as a hydroponic system, automating to feed the plants through the subirrigation water. Such systems have been used effectively in both sands and organic soils.

MODIFYING TEMPERATURE

Each growth stage, from germination to maturation, is affected by temperature. Each stage is characterized by an optimum range which may differ among species and among cultivars within a species. Temperatures above or below the optimum range generally diminish productivity, either per plant or per acre. Low temperature slows germination and seedling growth, affects plant stands, intensifies root rot diseases, restricts water and nutrient uptake, and may cause plant mortality, flower or fruit abortion, or abnormal ripening. High temperatures induce dormancy of seeds of several salad and root vegetables and may reduce yield and/or quality of most cool season crops.

Forcing Structures

Greenhouses provide a means of temperature control and are invaluable for northern transplant production. Most are constructed of double-layered polyethylene separated by air pressure; this air layer reduces heat loss by approximately 40 percent. Fuel requirements for growing transplants are low in most areas; however, the capital investment can be justified only by use of the structure for an extended season.

Hotbeds represent a minimal investment but require a subsurface heat supply. A typical structure is shown in Figure 4.8. The heat source may be evenly spaced electrical heating elements placed over an insulation board. Flue heat, supplied by a furnace or by waste heat from industries, also is effective, although the electrical systems provide superior uniformity.

Coldframes are identical in structure to hotbeds but lack supplemental heat and therefore are used primarily for growing cool

FIGURE 4.9
Seeding in plastic mulch underlaid with drip irrigation. (Photo courtesy of Chapin Watermatics, Inc., Watertown, N.Y.)

season crops or to harden plants prior to field transplanting. They are inexpensive to construct and operate.

Both hotbeds and coldframes should be sited to maximize southern exposure and should be near a water source. Plants growing within the environment of a forcing structure require careful, constant management. Drastic and rapid temperature and moisture fluctuations can occur, producing erratic growth and poor-quality plants.

Mulching

Early field planting of many plants increases both early and total yields. Yet, in the spring, cold soils of northern areas will limit the number of days that planting can be advanced. Transplants produced in forcing structures extend the effective growing season, but some transplants also do not tolerate cold soils.

Polyethylene mulch (Figure 4.9) has provided substantial gains in earliness and in total and marketable yields in some years (Table 4.5). Plastic provides a greenhouse effect during the day, warming the soil considerably rel-

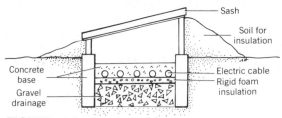

FIGURE 4.8
General structure of a hotbed, side view. Heat can be provided by electric cable (as shown) or by flue-distributed hot air. Sides and bases are pressure-treated wood or concrete.

TABLE 4.5
Comparison of performance of vegetables mulched with black polyethylene and those cultivated on unmulched soil

Crop and yield measurement	Unmulched	Black polyethylene
Broccoli		
Number of heads	26.0	50.5
Total weight (g)	840.6	1589.0[a]
Lettuce		
Number of heads	10.3	19.3[a]
Total weight (g)	1795.0	5994.5[a]
Cucumber		
Number of fruit	25.0	87.5[a]
Total weight (g)	1058.0	3396.0[a]

Source: H. J. Hopen and N. F. Oebker (1975) *HortScience* **10**:160. Reprinted by permission of the authors.

[a] Significant increase, $P = .05$.

ative to unmulched soil. However, during the night, plastic does little to slow radiational heat loss.

Modification of soil temperature is only one benefit of plastic mulch. Plastic also modifies fluctuations in soil moisture, partially by eliminating evaporation, and minimizes crusting of soil from hard rains and leaching of nutrients. Black plastic is preferred for its weed suppression, a property that also substantially reduces transpirational water loss from soils. In addition, carbon dioxide levels have been reported to be increased under black plastic. Although these effects all contribute to improved plant performance, experiments comparing black plastic mulch with bare soil and with mulch plus row cover (in which soil temperature would be increased) indicate temperature modification to be the major factor contributing to plant growth and early maturity (Table 4.6).

Organic mulches do not increase soil temperature; they tend to act as insulators between air and soil temperatures. Organic mulching materials do suppress weeds, reduce crusting, and preserve soil moisture to some degree. Gradual decomposition also adds to soil organic matter. Decomposition of organic mulches may reduce nitrogen availability at the soil–mulch interface, requiring additional fertilization.

Row Covers
Plants covered with translucent paper, fabric, or plastic always show growth superior to that of uncovered plants when air temperatures are cool and the plants are exposed to full sun. In cloudy weather, the benefits are not realized (Tables 4.7 and 4.8). Individual hot-caps or tents (Figure 4.10) must be applied to each plant and vented as growth fills the tent volume and as temperature increases. Loy and Wells (1982, 1983) have demonstrated that row covers—either a polyethylene strip wide enough to cover one or several rows, slitted to provide ventilation, and placed on evenly spaced wire hoops, or a translucent synthetic fabric that requires no support or ventilation ("floating row covers")—have enabled northern growers to extend the season by approximately 2 weeks. Normally, row covers are placed over black plastic mulch to eliminate weed problems, and trickle irrigation placed under the plastic will ensure uniformity of soil moisture. If black plastic mulch is not

TABLE 4.6
Yields of pepper, tomato, and muskmelon as affected by mulching and/or row cover treatment

	Early marketable yield		
Treatment	Pepper (bu/acre)	Tomato (bu/acre)	Muskmelon (tons/acre)
Bare soil	60 a[a]	54 a	—
Black plastic mulch	219 b	64 a	1.5
Row cover[b]	479 c	73 ab	9.3 b
Row cover[c]	512 c	130 c	9.5 b

Source: J. B. Loy and O. S. Wells (1983) *Proceedings, 17th National Agricultural Plastics Congress,* pp. 58–59. Reprinted by permission of the authors.

[a] Yields separated by Duncans multiple range, $p = .05$.

[b] Spunbonded polyester.

[c] Slitted clear polyethylene.

TABLE 4.7

Transmittance of photosynthetically active radiation through two row cover materials

| | Photosynthetically active radiation | | | |
| | Sunny | | Cloudy | |
Row cover	μmol m^{-2} s^{-1}	% of control	μmol m^{-2} s^{-1}	% of control
None (control)	1345	—	105	—
Slitted polyethylene	1160	86	60	57
Polyester	1040	77	69	66

Source: J. B. Loy and O. S. Wells (1982) *HortScience* **17**:406. Reprinted by permission of the authors.

used, an herbicide would be required to prevent weed competition. The row covers are removed when daytime temperatures exceed 84 to 89°F (29 to 32°C).

Row covers, particularly those of synthetic (polyester or polypropylene) fabric (Figure 4.11), have been applied successfully over direct seeded early vegetables, enhancing germination and emergence, and have been especially valuable in accelerating early growth of vine crops. Vegetables susceptible to flower abortion at high temperature (e.g., tomato) have been inconsistent in performance.

Row covers are not recommended for frost protection, although they may reduce the damage from light frost. They have been ef-

fective in excluding early season infestations of several insect pests, including flea beetles, tarnished plant bug, aphids, and cucumber beetles.

Frost Protection

Early vegetables in northern states and winter vegetables in southern areas are exposed to occasional freezing temperatures. Physiological hardening of tissues may offer protection of 1 to 5°F (1 to 3°C), and hotcaps or row covers provide not much more. Application of foam provides up to 18°F (10°C) pro-

TABLE 4.8

Mean air and soil temperature[a] during June 29 to July 6, 1981 (Durham, N.H.), with row cover and mulch treatments

| | Air temperature (°C) | | Soil temperature (°C) | |
Treatment	Day	Night	Day	Night
Slitted polyethylene + mulch	30.1	18.6	28.5	25.9
Polyester cover + mulch	27.9	18.4	26.4	24.7
Black polyethylene mulch	—	—	25.8	24.5
Bare soil (control)	22.9	17.0	23.2	22.0

Source: J. B. Loy and O. S. Wells (1982) *HortScience* **17**:406. Reprinted by permission of the authors.

[a] At depth of 7.5 cm.

FIGURE 4.10

Row covers laid over young seedlings for increasing daytime temperature and crop earliness. The covers shown are spunbonded polyester and do not require support over the plants. (Photo courtesy of O. S. Wells, University of New Hampshire.)

FIGURE 4.11
Individual hills of squash protected by weather-resistant paper tents. As temperatures rise and plants develop, the tents must be vented. (Photo courtesy of University of New Hampshire.)

tection, but foam stability and application over large areas have detracted from its usefulness. The most effective technique for protecting plants is mist irrigation, which can prevent damage at temperatures as low as 19 to 21°F (−6 to −7°C).

Irrigation for frost control, using microsprinklers or spray jets, is begun at 34°F (1°C) or higher if the dewpoint is below 27°F (−3°C). As the air temperature drops below freezing, ice forms, releasing heat (heat of fusion). At the same time, plants also are radiating heat to the atmosphere and losing heat by evaporation. The heat generated by ice fusion must exceed radiant energy and evaporative cooling losses to prevent tissue damage. Therefore, as temperature declines or as wind speed increases, the amount of irrigation water required for protection must increase. Excessive water application is not desirable, especially on soils with a low infiltration rate. In most instances, 0.11 in. (0.3 cm)/h will protect to 25°F (−4°C) in calm air; up to 0.35 in. (1 cm)/h is required for temperatures of 21°F (−6°C) and a wind of 5 to 6 mph (2.5 to 3 m/s) (Table 4.9). As air temperature rises, irrigation must continue until the dewpoint exceeds the freezing point. If terminated sooner, the heat loss by evaporative cooling will depress temperature of plant tissue, resulting in freeze

damage that may be more severe than might occur with no ice protection.

Icing by irrigation has protected young tomato and pepper plants, celery, potato, artichoke, and other row crops. However, maturing fruit of tomato and pepper, even though protected from freeze, may suffer chilling damage with consequent loss in marketability.

Solarization

In areas of intense solar radiation, clear plastic applied to moist, fallowed soil can be used to increase soil temperature to a level lethal to some soilborne pathogens and some weed seeds. This procedure, termed solarization, requires 4 to 8 weeks and will increase soil temperature in the top 1 in. (2.5 cm) to 135°F (57°C). Solarization of individual flat or raised beds is a cost-effective alternative to chemical fumigation. Solarizing an entire area with a solid plastic sheet may be too expensive to justify.

Soil and Plant Cooling

Alleviation of heat stress, where required, is achieved by shading or by irrigation. Cooling by irrigation requires application rates of 0.03 to 0.17 in./h (0.76 to 4.32 mm/h) and gener-

TABLE 4.9
Estimated sprinkling rate required for cold protection

Tempera-ture (°C)	Wind speed (m/s)					
	0.5	1.0	1.5	2.0	2.5	3.0
−2	0.14[a]	0.19	0.23	0.27	0.30	0.32
−3	0.21	0.29	0.35	0.40	0.44	0.48
−4	0.29	0.39	0.47	0.53	0.59	0.64
−5	0.36	0.49	0.58	0.67	0.74	0.80
−6	0.43	0.58	0.70	0.80	0.89	0.97
−7	0.50	0.68	0.82	0.93	1.03	1.13
−8	0.57	0.78	0.93	1.07	1.18	1.29

Source: From J. F. Gerber and J. D. Martsolf, in *Modification of the Aerial Environment of Plants* (ASAE Monograph), p. 329. Copyright © 1979 by American Society of Agricultural Engineers. Reprinted by permission.

[a] Data in centimeters of water per hour.

ally is begun at 81°F (27°C) for cool season crops and at 93°F (34°C) for warm season crops. Air temperature can be cooled 5 to 12°F (3 to 7°C), and the evaporative cooling from foliage is substantial. Field establishment of lettuce and other crops with thermodormant seeds has been enhanced by sprinkler irrigation, and potato tubers have developed higher levels of solids when cooled to maintain an optimum environment. Shading systems are not applicable to large field production of vegetables, but may be feasible for establishing transplants. Celery, for example, often is seeded in beds shaded with a mesh net during germination and early growth. Shading reduces temperatures at the soil surface by 3 to 7°F (1.5 to 4°C) and increases relative humidity slightly.

Wind Protection

Many sites offer little natural protection from wind, and some form of windbreak may be required. Windbreaks include trees or hedgerows, tall annual plants (corn, sorghum, wheatgrass, ryegrass, and others), snowfence, reed mats, or plastic screens. In some regions, winter windbreaks enhance snow collection, providing improved spring moisture. In most vegetable sites, windbreaks are used to preserve plant stands and enhance early growth.

Young plants are especially susceptible to abrasion by soil particles moved by wind or by erosion. Abrasion may arrest growth of parts of the plant or it may be severe enough to destroy the plant entirely. Abrasion is one cause of curled asparagus spears, damaged spinach leaves, or poor stands of cucurbits and beans. In high winds, entire plants may be uprooted because of erosion. Damage from erosion and abrasion is most severe in dry soils, and abrasive effects of sandy soils exceed those of loams. Young seedlings in dry organic soils are especially prone to uprooting ("blowout") by wind erosion.

Windbreaks also can change crop microclimate. The reduction in effective wind speed reduces evaporation, increasing air tempera-

ture within a plant canopy. Humidity also increases, reducing moisture stress and evapotranspiration, and there may be some effect on carbon dioxide levels. Improvements in plant growth and yield have been measured, indicating that net assimilation of carbon dioxide must increase.

MODIFYING ORGANIC CONTENT OF SOILS

Soil organic matter must be maintained by adding farm manures, peat, or sewage sludge and/or by using a green manure or cover crop. Crop production, in itself, through decay of roots and crop debris, contributes constantly to organic matter content. The more mature the material, the more slowly it decomposes. Green manure crops seeded in the fall or between production cycles usually are rather immature plants when turned under, and decomposition occurs rapidly. Of the crops used as green manure, the legumes decompose more rapidly than grasses. In many production areas, cover crops and green manures are the only means available for maintaining some organic matter in soils, although these crops provide relatively little to the soil humus.

Animal manures are preferred and once constituted the primary method by which both organic matter and soil fertility were maintained. Animal manures consist not only of animal waste but also of an absorptive litter—generally straw—providing a material that decomposes slowly, adding substantially to soil humus. The nitrogen, phosphorus (expressed as phosphate), and potassium (expressed as potash) contents of an average farm manure are not high, 0.5, 0.25, and 0.5 lb/ton (5, 2.5, and 5 kg/MT), respectively, with some variation reflecting animal source. Poultry manure, for example, tends to be higher in nitrogen than other sources. All nutrients, however, can be supplied most efficiently with commercial fertilizers; nevertheless, the indirect benefits of organic matter on soil microorganisms and on cation exchange capacity account for substantial gains in soil productivity.

Fresh animal manures evolve considerable ammonia; well-rotted manure does not. Fresh manure may be applied well in advance of planting to allow some dissipation of ammonia, or small amounts of fresh manure may be incorporated at planting along with supplemental phosphate and potash. Well-rotted manure is preferred and may be applied at anytime. For small-seeded crops, the decay process is essential to ensure a soil free of clods.

Composts used predominantly by home gardeners generally consist of plant debris. The solid fraction of sewage waste also can be composted to provide a safe organic amendment. Composting is an aerobic microbiological process during which the organic materials are decomposed, giving off substantial heat which further accelerates the microbial activity. About one-half of the total dry matter is decomposed in compost that is ready for use. Composts are used in the same way as animal manures and contribute the same value to crop yield.

MODIFYING SOIL pH AND FERTILITY

Soil pH should be within the range 6.0 to 6.8 for most vegetable crops. Exceptions reflect a need to suppress pathogenicity of soilborne organisms, not an inherent adaptation to high or low pH. Soils of the eastern United States tend to be acidic; those of the Great Plains and the west are predominantly neutral to alkaline.

Acidic soils must be limed regularly to maintain an optimum pH. The predominant liming materials are ground limestone (calcium carbonate) and dolomitic lime (magnesium carbonate and calcium carbonate). Each consists of ground rock particles, the sizes of which determine the rate at which soil pH is neutralized. Marl and chalk, basic slag, ground oyster shells, and hydrated lime have been used in the past but seldom are applied in

TABLE 4.10
Limestone needed to change soil reaction

Change in pH desired in plow layer	Limestone					
	Sand	Sandy loam	Loam	Silt loam	Clay loam	Muck
lb/acre						
4.0–6.5	2,600	5,000	7,000	8,400	10,000	19,000
4.5–6.5	2,200	4,200	5,800	7,000	8,400	16,200
5.0–6.5	1,800	3,400	4,600	5,600	6,600	12,600
5.5–6.5	1,200	2,600	3,400	4,000	4,600	8,600
6.0–6.5	600	1,400	1,800	2,200	2,400	4,400
MT/ha						
4.0–6.5	5.8	11.2	15.7	18.8	22.4	43.0
4.5–6.5	4.9	9.4	13.0	15.7	18.8	36.3
5.0–6.5	4.0	7.6	10.3	12.5	14.8	28.2
5.5–6.5	2.7	5.8	7.6	9.0	10.3	19.3
6.0–6.5	1.3	3.1	4.0	4.9	5.4	9.9

Source: From O. A. Lorenz and D. N. Maynard, *Knott's Handbook for Vegetable Growers.* Copyright © 1980 by John Wiley & Sons, Inc., New York, p. 88. Reprinted by permission.

volume because of expense or unavailability.

Because H^+ ions are attracted to clay and organic particles, mucks and clay loams require larger amounts of lime to change pH than do sandy soils (Table 4.10). However, heavy soils may require application less frequently than sands, perhaps only once each 5 to 10 years. At an optimum level, about 80 to 90 percent of the soil cation exchange capacity involves calcium and magnesium, 2 to 5 percent potassium, and the rest hydrogen. Calcium ions usually predominate, with levels 5 to 10 times the magnesium ion content.

As lime is added, it dissolves to form calcium bicarbonate or magnesium bicarbonate. The calcium and magnesium ions displace H^+ on the soil particles, CO_2 is released by unstable bicarbonate ions, and the H^+ then combines with OH^- to effect neutralization. As pH increases, phosphorus and some of the micronutrients become solublized and available for plant growth. Others, such as manganese, occasionally at toxic levels in acidic soils, become less available as pH increases, and iron, boron, and phosphorus also become increasingly unavailable.

Application of lime should precede cropping by approximately 6 months to allow neutralization to take place. A limed soil should be tilled thoroughly throughout the plow layer, since movement of lime in soil is extremely slow. Highly acidic soils seldom are adjusted by a single lime application; repeated treatments are recommended in effecting a substantive pH change, particularly in well-buffered soils.

In many western states, soil pH is high, with consequent high salt levels. Application of sulfur or sulfate forms of fertilizer will acidify alkaline soils to some extent, releasing bound micronutrients.

The major elements applied as fertilizer include nitrogen, phosphorus, and potassium (Figure 4.12). Micronutrients may be added occasionally to soils or may be applied to foliage to correct deficiencies. The ammonium and nitrate ions are the two nitrogen forms readily available to plants. Of the two, ammonium is not leached as rapidly as nitrate but, if present in excess, can be toxic to several vegetable crops, largely by limiting potassium availability. Most nitrogen is applied as ammonium nitrate (33.5 percent nitrogen), but anhydrous ammonia (80 percent nitrogen) injected into the soil or irrigation water and urea (45 percent nitrogen) also have been used widely. Urea is relatively inexpensive but must be incorporated thoroughly to prevent substantial gaseous nitrogen losses. Ammonium sulfate,

SOURCE	PROCESS	PRODUCT
PHOSPHATE ROCK	GRIND TO ABOUT 0.1 mm ⟶	GROUND PHOSPHATE ROCK (0–35–0)
SULFUR AIR WATER	CATALYTIC OXIDATION AND HYDRATION ⟶	SULFURIC ACID

FIGURE 4.12
Processes used in manufacturing fertilizers, and resulting elemental ratios. (From *Chemical Fertil-* **izers by Christopher J. Pratt. Copyright © 1965 by Scientific American, Inc. All rights reserved.)**

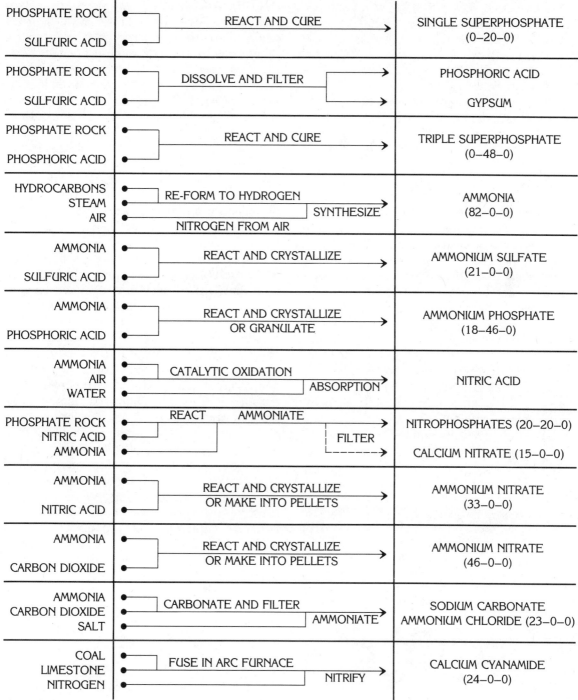

FIGURE 4.12 (*Continued*)

sodium nitrate, and ammonium phosphate are sources of nitrogen used occasionally. Of the other major elements, phosphorus is applied primarily as water-soluble enriched superphosphate (triple superphosphate), some as ammonium phosphate. Potassium usually is applied as potassium chloride, occasionally as potassium sulfate.

Micronutrients once were maintained in soil largely through applications of animal manure. Commercial fertilizers also contained some impurities that contributed to soil levels. Animal manures no longer are as available, and applications of micronutrients have become important components of soil management programs. Boron is added as borax; copper, zinc, manganese, and iron normally are applied as sulfates or as chelated compounds. Each normally is applied only if tissue tests reveal levels approaching a deficiency, and excessive application rates must be avoided (Figure 4.13).

The product of nitrification, NO_3^-, is weakly attracted to exchange sites on clay and organic particles and therefore is susceptible to leaching. Nitrogen stabilizers, such as nitropyrin, have been effective under some soil and management conditions in reducing leaching losses. Applied with nitrogen fertilizer to loam soils, the availability of NO_3^- persists substantially longer than in soil receiving fertilizer alone, often promoting improved growth or reducing the amount of fertilizer needed. However, nitropyrin itself is attracted to organic particles and is not as effective in organic soils; longevity also is reduced in sandy soils.

Fertilizers are applied most frequently as granular materials, although liquid applications formulated to reflect specific crop and soil macro- and micronutrient needs are popular in some regions. The uptake of nutrients required to support a given level of crop growth provides one basis for determining fertilizer needs (see Chapters 10 to 21). In addition, soil texture and natural fertility, cropping practices (particularly plant population density and length of growing season), and leachability of the fertilizer elements must be considered in determining a fertilization schedule for each site. On most soils, from one to several sidedressings of fertilizer are required, especially for crops requiring a 3- to 4-month growing season. These sidedressing applications normally are of ammonium nitrate, since nitrogen is leached rapidly and demands for nitrogen increase as plants develop. In general, sands require frequent applications, particularly of nitrogen; soils with a high cation exchange capacity sustain optimum fertility levels for a greater portion of the season.

Organic soils also may require nitrogen, most frequently during cool weather. Little nitrification takes place at temperatures below 41°F (7°C), but it increases rapidly as temperature increases. Therefore, most northern growers add nitrogen to muck at planting time to ensure adequate levels for early growth. Organic soils also tend to be low in potassium and in micronutrients and may require substantial corrective applications when first brought into production.

Where irrigation is applied regularly, fertilizer may be applied through the irrigation system (chemigation). Of the major nutrients, nitrogen is the most practical to apply because of its solubility. If applied frequently in amounts matched to crop needs, the total ap-

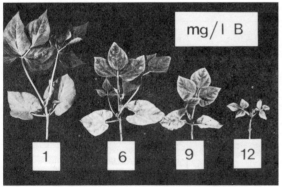

FIGURE 4.13
Reduction of stem and leaf growth of snap bean by excessive boron. (Photo courtesy of U.S. Salinity Lab, Riverside, Calif.)

TABLE 4.11
Yields of four vegetables with and without starter solution in the transplant water

| Fertilizer treatment[a] | Crop yield | | | |
	Tomato (bu/acre)	Pepper (bu/acre)	Cucumber (bu/acre)	Muskmelon (fruit/acre)
None (control)	176 b[b]	204 b	194 b	5,142 b
Starter solution				
12–62–0	329 a	494 a	529 a	9,317 a
20–20–20	316 a	464 a	473 a	8,591 a
16–32–16	313 a	494 a	505 a	10,043 a
10–52–17	300 a	504 a	543 a	—

Source: Data courtesy of O. S. Wells, University of New Hampshire.

[a] The field had received complete fertilizer prior to disking.

[b] Means separated by Duncan's multiple range test, $P = .05$.

plication may be less than that applied in a standard broadcast or band system. Potassium and, especially, phosphorus are not applied as efficiently as nitrogen through irrigation. Phosphorus becomes fixed (insoluble) when applied to the soil surface and may penetrate only 1 to 2 in. (2 to 5 cm).

Research has shown dramatic responses in yields of transplanted crops when a high analysis starter solution is added to the transplant water (Table 4.11). These fertilizers have analyses of 8 to 20 percent nitrogen, 21 to 52 percent P_2O_5, and 0 to 17 percent K_2O. The high P_2O_5 enhances early rooting and plant establishment. Excessive nitrogen should be avoided at transplanting, since it promotes topgrowth at the expense of root development. When combined with other intensive production practices, the yield responses for some crops are magnified further (Table 4.12).

Most row crop fertilizer applications are, at least in part, banded, usually 2 to 4 in. (5 to 10 cm) to the side and below placement of seeds. Banding concentrates available nutrients in the root zone and minimizes leaching losses, particularly of nitrogen. Phosphorus, because of rapid fixation in complexes unavailable to plants, must be delivered to the root zone to ensure maximum benefit. In areas of excessive moisture or in porous soils, plastic may be placed over the row to reduce leaching.

TABLE 4.12
Effects of starter solution applied in transplant water, as affected by plastic mulch

| Soil treatment | Yield (bu/acre) | | | |
| | Tomato | | Pepper | |
	Mulch	No mulch	Mulch	No mulch
Starter solution	342	287	575	401
No starter solution	173	181	285	126

Source: Data courtesy of O. S. Wells, University of New Hampshire.

SELECTED REFERENCES

Arkin, G. F., and H. M. Taylor (eds.) (1981). *Modifying the Root Environment to Reduce Crop Stress,* Monogr. 4. ASAE, St. Joseph, Mich.

Barfield, B. J., and J. F. Gerber (1979). *Modification of the Aerial Environment of Plants,* Monogr. 2. ASAE, St. Joseph, Mich.

Bernstein, L., and L. E. Francois (1973). Comparisons of drip, furrow, and sprinkler irrigation. *Soil Science* **115**:73–86.

Elfving, D. C. (1979). Crop response to trickle irrigation. *Horticultural Reviews* **4**:1–48.

Hopen, H. J., and N. F. Oebker (1975). Mulch effects on ambient carbon dioxide levels and growth of several vegetables. *HortScience* **10**:159–161.

Knavel, D. E., and H. C. Mohr (1967). Distribution of roots of four different vegetables under paper and polyethylene mulches. *Proceedings, American Society for Horticultural Science* **91**:589–597.

Loy, J. B., and O. S. Wells (1982). A comparison of slitted polyethylene and spunbonded polyester for plant row covers. *HortScience* **17**:405–407.

Loy, J. B., and O. S. Wells (1983). Use of spunbonded polyester as a plant row cover over vegetables. In *Proceedings, 17th National Agricultural Plastics Congress,* pp. 54–62.

Marsh, A. W., R. L. Branson, S. Davis, C. D. Gustafson, and F. K. Aljibury (1979). *Drip Irrigation,* University of California Coop. Ext. Leaflet 2740.

Marsh, A. W., H. Johnson, Jr., F. E. Robinson, N. McRae, K. Maybery, and D. Ririe (1977). *Solid Set Sprinklers for Starting Vegetable Crops,* University of California Coop. Ext. Leaflet 2265.

Oebker, N. F., and H. J. Hopen (1974). Microclimate modification and the vegetable crop ecosystem. *HortScience* **9**:564–568.

Pratt, C. J. (1965). Chemical fertilizers. *Scientific American* **212**:62–72.

Ross, D. S., R. A. Parsons, W. R. Detar, H. H. Fries, D. D. Davis, C. W. Reynolds, H. E. Carpenter, and E. D. Markwardt (1981). *Trickle Irrigation in the Eastern United States,* Northeast Regional Agric. Eng. Serv. Publ. NRAES-4. Cornell University, Ithaca, N.Y.

STUDY QUESTIONS

1. How is the choice of irrigation system affected by slope, soil type, salinity, and water availability?

2. By what systems can the growing season be extended in northern climates? Give the reasons for the relative effectiveness of each system.

3. What are the effects, in terms of plant growth, of the interrelationships among soil texture, structure, salinity, and availability of nutrients from the soil solution?

5

Tillage and Crop Establishment

Iowa. Modifications of conventional tillage and the use of herbicides that reduce the amount of soil disturbance therefore may be more appropriate for certain sites and crops than conventional primary and secondary tillage.

Conventional Tillage

Field Preparation In soils with optimum levels of nutrients and water, root growth will occur primarily in zones of low mechanical resistance (low impedance). Tillage promotes root development by decreasing bulk density, thereby reducing soil impedance, and by creating mixtures of coarse and find soil aggregates that improve air and water movement. The structural enhancement, in turn, increases activity of beneficial soil microorganisms, improves nutrient availability, and modifies soil temperature.

Plowing and harrowing (Figures 5.1 and 5.2) and subsequent cultivation constitute conventional tillage. Primary tillage provides initial soil movement that eliminates surface vegetation and deepens the root zone. The moldboard plow has been used most commonly, cutting to a depth of 5 to 8 in. (12 to 20 cm) or up to 10 to 12 in. (25 to 30 cm) where the crop or soil conditions dictate. For most effective results, soils should be plowed at relatively low soil moisture. Operators should avoid plowing with one wheel in the plow furrow (to minimize compaction), and plow angle must be adjusted to ensure complete coverage. For soils that are poorly drained or high in silt and/or clay, fall plowing is an alternative

TILLAGE

Tillage has four objectives: preparation of an effective seedbed (incorporating plant residues), moisture infiltration, moisture control and aeration, and weed control. While meeting these objectives, soils must be protected against loss through erosion, a concern that dictates the kind of tillage system and the time at which it is utilized. It has been estimated that as much as 13 tons/acre (30 MT/ha) of soil per year is lost from contour-tilled corn land in

FIGURE 5.1
Standard moldboard plow (10-bottom) for primary tillage. (Photo courtesy of Deere & Compnay, Moline, Ill.)

not place surface vegetation as deep in the root zone as moldboard plows. Chiseling does not contribute to compaction below the tilled zone, is more energy efficient than moldboard plowing (Figure 5.3), and requires almost 50 percent less time than a conventional plowing system.

Rototillers are common tillage units in home gardens, but seldom are used for tilling large acreages. Repeated rotary tilling can compact the soil at the base of the tilled layer, and excessive beating action of the tines can destroy soil aggregation.

Secondary tillage is designed to break into relatively small soil aggregates the plow ridge formed by moldboard plows, smoothing the field in the process. Normally, disk harrows are employed, occasionally followed by spring tooth harrows or cultipackers. Relatively shallow disking reduces the chance of compaction within the root zone, especially in heavy soils. To minimize the number of passes over a field, a "float" or drag cultivator and/or other units may be attached in tandem behind a disk.

Soils newly plowed under favorable conditions will have a bulk density less than 1.0. Each subsequent harrowing will increase bulk density: in four or five passes, cultivation will

within areas exposed to freezing winter temperatures, as long as the field topography precludes erosion. The freezing action on soil clods is effective in forming small soil aggregates.

Depth of plowing should be related to soil profile and depth of rooting of the crop to be grown. Excessively deep plowing is wasteful of power, and the fine soil particles, which are valuable constituents in the seed zone (surface layer), may settle at the bottom of the plow layer, contributing to compaction potential. Deep plowing may be justified when eliminating pan layers, adjusting soil salinity, or preparing a field for root crop culture.

Primary tillage also can be accomplished with chisels. Chisels break up the soil but do

FIGURE 5.2
Disk can be used for primary or secondary tillage. Canted alignment of disks pulverizes and smooths furrows. (Photo courtesy of Deere & Company, Moline, Ill.)

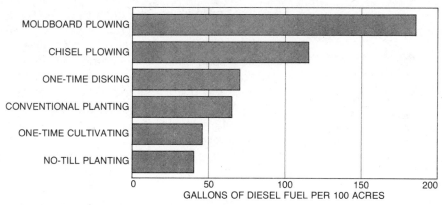

Bar chart with categories on the vertical axis: MOLDBOARD PLOWING, CHISEL PLOWING, ONE-TIME DISKING, CONVENTIONAL PLANTING, ONE-TIME CULTIVATING, NO-TILL PLANTING. Horizontal axis labeled GALLONS OF DIESEL FUEL PER 100 ACRES, scaled 0, 50, 100, 150, 200.

FIGURE 5.3
Energy costs of no-till versus conventional types of tillage, in terms of fuel used by a 100-hp diesel tractor (includes plowing, disking, planting, and cultivation costs). (From *Agriculture Without Tillage* by G. B. Triplett, Jr., and D. M. VanDoren, Jr. Copyright © 1977 by Scientific American, Inc. All rights reserved.)

increase bulk density to 1.4. A range of 1.1 to 1.4 is considered optimum for most crop plants. At a density of 1.6, water movement is curtailed.

At optimum bulk density, fine granules should be prevalent in the seed zone, coarse material in the root zone. Crop performance is favored by granules $\frac{1}{8}$ to $\frac{1}{4}$ in. (0.3 to 0.6 cm) in size and is depressed by granules smaller than $\frac{1}{16}$ in. (0.15 cm). Large granules or soil lumps are not deleterious if they do not impede water contact with seeds. Granules normally persist for 30 days, gradually disintegrating under forces of natural weathering and rainfall. In well-tilled and well-managed clay loams, the granules may persist throughout the season.

Raised Beds   Many row crops have been grown on raised beds (Figure 5.4) to accommodate furrow irrigation. In some areas, ridges or raised beds are used to improve drainage. Raised beds also increase depth of the root zone, creating a tilled soil layer substantially greater than a normal plow layer. Extending the root zone is particularly useful in growing

carrots, parsnips, potatoes, sweet potatoes, and similar crops. Beds are formed easily in well-prepared rock-free soil with tractor-mounted shovels and shapers.

FIGURE 5.4
Raised bed system of culture. This system accommodates surface irrigation and also improves drainage from the plant row. (Photo courtesy of University of New Hampshire.)

TABLE 5.1

Effect of compaction on soil density and pickling cucumber dry weight

Depth (cm)	Soil density (g/cm³)		Root dry wt (mg/dm³)	
	Noncom-pacted	Compacted	Noncom-pacted	Compacted
0–8	1.54 a[a]	1.52 a	336 a	271 ab
8–15	1.56 a	1.67 b	203 bc	206 bc
15–23	1.58 a	1.72 b	154 cd	6 e
23–21	1.67 b	1.72 b	83 de	0 e

Source: From D. A. Smittle and R. E. Williamson (1977) *Journal of the American Society for Horticultural Science* **102**:823. Reprinted by permission of the authors.

[a] Mean separation, within soil density and root dry weight, by Duncan's multiple range test, *P* = .05.

Soil Compaction   Many soils present physical deficiencies that may be alleviated by proper tillage. Most soils are damaged by poor or untimely tillage. High impedance usually is caused by (1) natural clay or hardpan layers; (2) compaction from forces (weight) exerted during plowing, cultivation, or other field operations; (3) loss of soil particle aggregation, which may reflect low organic matter status, or excessive tillage or irrigation; and (4) loss of soil water.

Pan layers occur naturally in many soils, occasionally within a rooting zone. Leveling of soils also may create such impervious layers. Pan layers impede not only root penetration, but water drainage as well. Smittle and Williamson (1977) found that root systems encountering such pan layers become altered (Table 5.1), restricting penetration or, in the case of root crops, impairing plant quality (Figure 5.5). Yields also can be reduced, perhaps because of reduction in nutrient uptake (Table 5.2).

Soils lacking a natural pan layer can develop one through compaction. Soils with high clay fractions are especially prone to compaction, but compression of sands may persist over a longer duration. The pressure exerted by plow and disk passage may contribute to gradual development of a dense layer just below plow depth. Tillage of moist soil acceler-

ates this compaction, and increased mechanization has increased the frequency of compaction-related problems. Prevention is the most inexpensive management approach. The number of machine operations should be minimized, and soil should be tilled only when soil moisture is low. Irrigation frequency and rate should match soil texture and infiltration rate. Where compaction layers have formed, deep chiseling or plowing will shatter the hardpan if done when the soil moisture is relatively low. Clay hardpan layers, however, must be mixed with other soil particles; otherwise, the clay particles reseal very quickly. Natural pan layers normally reform 1 to 3 years after tillage.

Conservation Tillage

One projection estimates that corn yields are reduced 131 lb/acre per year (147 kg/ha) for every centimeter of soil lost by erosion from cultivated fields. Contour planting, strip cropping, terracing, no-till planting, and strip tillage are methods designed to minimize soil erosion losses and maintain long-term pro-

FIGURE 5.5
Shape of carrot roots encountering a pan layer created by heavy implements used on moist soil. (Photo courtesy of University of New Hampshire.)

TABLE 5.2

Effect of compaction on pickling cucumber fruit shape (length/diameter ratio) and tissue and soil nitrate concentration

Treatment	Yield		Fruit shape (length/ diameter)	Nitrate (ppm)[a]	
	MT/ha	$/ha		Petiole	Soil
Compacted	12.2 b[b]	648 b	2.22 b	1910 b	22 a
Noncom- pacted	20.1 a	860 a	2.42 a	3708 a	21 a

Source: From D. A. Smittle and R. E. Williamson (1977) *Journal of the American Society for Horticultural Science* **102**:823. Reprinted by permission of the authors.

[a] Nitrate values are averages of samples collected 27, 34, and 31 days from planting.

[b] Mean separations, within columns, by Duncan's multiple range test, $P = .05$.

ductivity. Contour planting, strip cropping, and terracing are used for cultivating areas with excessive slope; no-till and strip tillage are forms of "mulch tillage" in which the previous crop residue or cover crop largely remains in place throughout the subsequent cropping season. The crop debris in a no-till system captures surface water, and the decaying roots provide channels for water penetration. The bulk density of soils managed by a no-till or minimum-till system is generally higher than that of conventionally tilled soils, and water infiltration is less. Nitrate content of no-till soils also has been shown to be less than that of plowed soils. However, with appropriate adjustments in management, no-till and strip-till systems have provided a feasible alternative to conventional tillage and have proven effective in reducing erosion losses.

No-Till Planting   The first significant modification of the conventional planting system to evolve was termed zero-till or no-till. Rather than preceding seeding by conventional field preparation, seeds are sown with a slot-type seeder through a stubble left from a previous crop or provided by herbicide treatment of a sod. Increases in residual organic matter and

soil moisture have been reported, but, in some locations, reduced soil temperatures in no-till fields have delayed germination and/or maturity, and the crop debris has increased disease and insect problems. Decreases in soil pH also seem to follow repeated no-till cropping, and surface-applied lime is relatively ineffective in neutralizing within the root zone. Yield increases for corn have been rather consistent; yields for other vegetables have in some instances equaled those of conventional tilled soils, but results have been inconsistent. Among the vegetables tested, sweet corn, squash (Figure 5.6), cucumber, beans, peas, cabbage, and several others have been produced successfully with no preparatory tillage. Typical of the problems encountered in no-till production are those shown in Table 5.3 from work at Virginia. The poor performance of no-till in 1982 was related to poor weed control. The superior fall no-till performance in 1983 was due to increased soil moisture in no-till plantings during a drought period. In other tests, no-till has been successful on soils that tend to crust and where weed growth is suppressed throughout the growing season.

FIGURE 5.6

No-till production of squash in a wheat stubble. (Photo courtesy of D. E. Knavel, University of Kentucky, Lexington.)

TABLE 5.3

Influence of tillage on total yields of cabbage in 1982 and 1983

| | Yield (1000 kg/ha) | | |
| | Spring | 1983 | |
Tillage system	1982	Spring	Fall
Conventional	26.6 a[a]	21.9 a	18.0 b
No-till	0.0	23.6 a	33.5 a

Source: Vegetable Grower's News **38**(6), Virginia Polytechnic Institute and State University (1984). Reprinted by permission of the authors.

[a] Means in columns followed by the same letter do not differ significantly, $P = .05$.

Strip Tillage   Several modifications of minimum-tillage systems have been described as strip tillage. Within a killed or overwintered crop stubble, a 6-in. (15-cm) band may be tilled to improve soil structure in the seed zone. The remaining stubble provides the same benefits as on no-till fields. A variation of strip tillage involves wide strips, 42 in. (1.1 m) or wider, cut in a living sod. The tilled strips accommodate multiple rows of cash crops, whereas the interrow sod provides superior erosion control. The composition of the living sod has little effect on crop growth until late in the season, when legumes may sustain soil nitrogen for an extended time. A late maturing crop, such as winter squash (Table 5.4), thus may respond to clover; crops that mature early in the season will not. Subsequent tests of several vegetables have revealed differences in crop tolerance to competition by interrow sod, but slight declines in performance of individual plants can be minimized by irrigation and yields per acre can be improved by increasing plant density.

Living Mulch   Within established crop plantings, a cover crop may be seeded to provide erosion control, some weed suppression, and, eventually, a source of organic matter. The species used as a "living" mulch should be relatively nonaggressive and low in water use. In tests of cabbage and sweet corn (Nicholson

and Wien, 1983), turfgrasses of relatively low vigor (Chewing's fescue, Kentucky bluegrass) were superior, reflecting the significant relationship between crop yield and dry matter production of the living mulch. The time of seeding, the relative depths of rooting of crop and mulch, and the characteristics of the two root systems also may determine suitability of a plant as a living mulch.

Intercropping   In the major food-producing regions, crops are grown in monoculture, an efficient production system, but one occasionally vulnerable to sudden changes in environment or in populations of insect and disease pests. In many areas of the world, intercropping is an accepted practice, using compatible species to maximize use of space. Certain intercrops also may reduce erosive action of water and wind. Competition (Table 5.5) determines yield potential of mixed cropping systems, similar to plant response to increasing population density (see section on stand establishment) or to strip tillage within an established sod. Interaction between rates of root and canopy growth and between relative plant sizes of interplanted crops influences the extent of competition and the degree to which it affects final yields. Short season crops—peas, radish, lettuce, kohlrabi—planted between full season vegetables or species chosen for complementary depth and spread of

TABLE 5.4

Fresh weight and percentage dry matter of bush winter squash as influenced by interstrip sod composition

Sod composition	Fresh weight (MT/ha)	Dry matter (%)
Grass	36.8	9.6
Grass/clover	43.4[a]	10.0

Source: S. Loy, M.S. thesis, University of New Hampshire (1984).

[a] Significant, $P = .05$.

root systems may preclude serious competition. Planting dates also may be staggered to avoid competition, planting a second crop when the first is well established or near maturity. In general, competition late in plant development does not depress yield significantly. The earlier the competition, the greater the yield loss. The greatest loss normally will occur within the least aggressive crop.

Cultivation

The most important effect of cultivation is weed control. By eliminating the transpiring leaf surface of weeds, soil moisture is conserved and nutrients are retained for crop growth needs. Repeated cultivation, however, can increase soil bulk density through compression caused by tractor weight, particularly on moist soil. Cultivating equipment includes duckfoot (flared) shovels or sweeps, chisel tines, rotary hoes (Figure 5.7), rototillers, and similar devices. Most are effective for young weeds, when disturbance of the weed root system is most easily accomplished.

FIGURE 5.7
Cultivation of young corn with flanged disks that uproot small weeds. (Photo courtesy of Deere & Company, Moline, Ill.)

CROP ESTABLISHMENT

Of the major vegetable crops, most are seed propagated. Those with large seeds and many with small seeds are established successfully by field seeding. Direct field seeding is the most economical crop establishment system; however, seeds of some crops are sensitive to temperature or other stress factors and may germinate poorly in certain soils. Others, because of small size or irregular shape, are difficult to sow at a controlled rate.

Seeds and Seeding

Seed Quality Inferior seed quality inevitably leads to variable rates of emergence, poor stands, and consequent lack of uniformity in the harvested product. Seed quality is the single most important determinant of stand. **Quality** of seed refers to its purity (trueness to cultivar type), germinability, and vigor. **Germinability** is defined as the capacity for emergence under favorable environmental conditions. **Vigor** describes the capacity for emergence under unfavorable environments. Germinability can be certified easily by test,

TABLE 5.5
Yields of four vegetables as influenced by cropping system

| Year and system | Marketable yield/ha | | Total yield (MT/ha) | |
	Muskmelon (1000)	Collards[a] (1000 kg)	Cabbage	Tomato
1981				
Intercrop	10.01 a[b]	9.59 a	23 a	87 b
Monocrop	13.29 a	11.13 a	23 a	125 a
1982				
Intercrop	16.58 a	15.97 a	37 a	91 a
Monocrop	24.74 b	20.67 a	42 a	122 a

Source: From J. E. Brown, W. E. Splittstoesser, and J. M. Gerber (1985) *Journal of the American Society for Horticultural Science* 110:351, 352. Reprinted by permission of the authors.

[a] Collards were harvested once in 1981, twice in 1982.

[b] Mean separation in columns by Duncan's new multiple range test, P = .05.

but only certain aspects of seed vigor can be tested reliably.

Seed vigor has been defined as

the sum total of those properties of seed which determines the potential level of activity and performance of the seed or seed lot during germination and seedling emergence.[1]

Seed performance in this definition includes biochemical activity during germination (enzymatic and respiratory activity) and uniformity of seed germination and seedling growth in a wide array of favorable and unfavorable environments. Factors affecting seed vigor include those controlled by the seed grower or processor (nutrition of the parent plants; stage of maturity at harvest; freedom from mechanical damage, environment, or weathering), seed weight or specific gravity, deterioration and aging, genotype, and pathogens.

Seed Production Methods Seed vigor reflects the growth of the plant from which the seeds were harvested. Proper management of seed production fields, including appropriate spacing, nutrition, irrigation, and pest control, contributes not only to the volume of seed harvested, but also to its germinability (both percent and rate). Seeds harvested before or after physiological maturity usually show reduced germinability. Those allowed to remain on a plant may show decreased germinability because of weathering (rainfall, temperature). Weathering effects may vary among different crops according to the physical or physiological features of the seed that contribute to the degree of dormancy. Once harvested, care in handling the seed also is critical in maintaining seed vigor.

Seed Size There is substantial evidence of a relationship between seed size and seed vigor within a cultivar. Large seed often re-

[1] Approved by International Seed Testing Association, Madrid, Spain, 1977.

flects the conditions under which the seed was produced. Hanumaiah and Andrews (1973) showed that the vigor of seeds larger than the average for a cultivar is expressed not only by increased percentage germination, but also by improved rate of emergence and subsequent seedling growth (Table 5.6).

Effect of Aging Seed longevity differs among species and among cultivars within species. Such differences, seldom important unless seeds must be stored for a prolonged period, become magnified under suboptimal storage environments. Seeds of most species, when stored under low humidity (50 percent relative humidity) and moderate to low temperature, will survive with relatively little deterioration over several years. Seeds not stored properly, or lacking full vigor when harvested and stored, will lose germinability rapidly. The loss of germinability is irreversible.

Seed Processing Seeds threshed at excessive speed or otherwise damaged during postharvest handling may fail to germinate; others may germinate but show effects of seed damage. For example, snap and lima bean seed may incur hypocotyl fractures during han-

TABLE 5.6

Fresh and dry weights of 36-day-old seedlings of 'Purple Top' turnip and 'Wisconsin All Season' cabbage from different-age seeds

Crop	Seed size	Fresh weight of 10 seedlings (g)	Dry weight of 10 seedlings (g)
Turnip	Large	35.4 a[a]	4.3 a
	Medium–large	21.0 b	2.7 b
	Medium	15.5 c	2.0 c
	Small	15.1 c	1.8 c
Cabbage	Large	13.0 a	1.8 a
	Medium	7.6 b	1.1 b
	Small	5.7 c	0.7 c

Source: From L. Hanumaiah and H. Andrews (1973) *Proceedings, Association of Official Seed Analysts* **63**:122, 124. Reprinted by permission of the authors.

[a] Mean separation within columns and for each crop by Duncan's multiple range test, $P = .05$.

dling, and emerging seedlings of such seeds will lack a growing point (termed **baldheads**).

Pathogens   Of all factors determining seed reliability, the most important in many species is freedom from disease. Serious field diseases carried in or on seeds affect beans, celery, lettuce, tomato, asparagus, and cole crops, among others. In some instances, the geographic site of seed production provides insurance against infection or infestation of seed lots. Bean seed production, for example, was relocated to arid western states where seedborne diseases, favored by humidity and frequent rainfall, do not pose a problem. Certification programs have provided insurance against some diseases. Other than certain seed crops (e.g., dry bean), however, seeds of vegetables are not certified; however, certification of vegetative seed, such as potato tubers, has been very effective. Seed cleaning and treatment are used to eradicate some seedborne pests.

Source of Seed   Because of the physiological complexity of seed and the interaction of factors that determine quality, growers should use fresh seed purchased from a reputable seed company. Carryover seed, unless held in controlled, dry storage, seldom has the vigor of fresh seed obtained from a reliable seedsman. Home-grown seed presents a high risk of disease and loss of cultivar purity. Hybrid cultivars cannot be propagated without access to the original inbred parents.

All purchased seeds must be labeled by cultivar name, year of seed production, purity (for some seeds), germination test results (and year of test), and seed treatment used for insect or disease control. Large seed units also list a stock number, which identifies the source (grower or area) of the seed and other information required by the grower.

Cultivars

Sexually Propagated Plants   Following the rediscovery of Mendel's laws in 1900, signifi-cant improvements of many food and feed crops have occurred through organized plant breeding programs. Prior to this modern era of plant improvement, substantial advances had been made, primarily through selection of adapted forms as crops were introduced into new areas, but these advances took place over an extended time. Modern technology has made it possible to develop improved cultivars with 5 to 10 years, an important advantage when an agricultural crop is threatened by disease or when changes in technology make existing cultivars obsolete. Furthermore, genetic resistance to insects, disease, drought, cold temperatures, excessive heat, and environmentally induced defects has contributed toward the relative stability of crop production in many parts of the world.

Hybrid Cultivars   Of the many scientific advances supporting cultivar development since the discovery of Mendel's work, the understanding of hybrid vigor has been the most significant. The introduction of hybrids revolutionized the corn industry, and the use of hybrids since has extended to many other vegetable crops, including spinach, onions, cole crops, tomato, pepper, eggplant, cucumber, muskmelon, and others.

The term **hybrid** may be used to describe any of several kinds of parental combinations suited to specific situations. The F_1 hybrid is the direct product of a cross of two *inbred* parents (Figure 5.8). Because the two parents each show no genetic plant-to-plant variation, the hybrid also will be very uniform. For many crops, the hybrid also shows greatly enhanced vigor, which may appear as increased vegetative growth or reliable plant establishment, increased number of fruit per plant, early maturity, or superior yield. The extent of this hybrid advantage is related directly to the parents, to their genetic diversity (generally, the more distant genetically, the greater the hybrid vigor), to their normal breeding behavior (naturally self-pollinated plants generally show less hybrid vigor than those naturally cross-pollinated), and to the degree of homozygosity

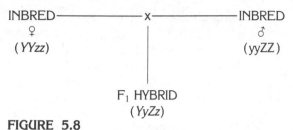

INBRED—————x—————INBRED
♀   ♂
(*YYzz*)   (*yyZZ*)

F$_1$ HYBRID
(*YyZz*)

FIGURE 5.8
Schematic of single-cross hybrid development, listing the genetic expectation for a simple two-gene model. Self-fertilization of the F$_1$ hybrid would result in extreme variation among offspring.

(genetic uniformity) within each parent. Enhanced yield has been the most obvious advantage of hybrids, especially in corn and tomato. For many vegetables, however, uniformity, not yield, is the primary value of an F$_1$ hybrid. Uniformity is especially important where a crop is harvested mechanically, and it also contributes toward improved quality of processed vegetables and uniform appearance of fresh market commodities. In producing F$_1$ hybrids of corn, the inbreds often are rather unproductive small plants. Consequently, seed yields can be low and seed cost correspondingly high. While high seed cost is not a concern for growers of high-return crops, it is unacceptable to those growing grain or silage corn. The three-way and double-cross concepts were introduced to offer hybrid seed economically to the grower. A double cross is a cross of two F$_1$ hybrids [(A × B) × (C × D)], and the three-way cross uses the F$_1$ hybrid as the female parent in a cross with an inbred [(A × B) × C]. Using the productive F$_1$ as the seed parent enhances seed productivity per acre severalfold. For vegetable growers, such hybrids are not suitable in most instances, since neither is as uniform as a single cross.

Unique Hybrids Occasionally, genetic mutations have resulted in significant developmental changes in plants, and scientists have capitalized on these changes to produce rather unique hybrids. In asparagus, the isolation of perfect-flowered plants led to the development of "supermales" (homozygous for the male gene). When supermales are used as a parent with a normal female, all-male progenies are produced, with consequent superior yields and uniformity. While not strictly F$_1$ hybrids, they are crosses that represent significant improvements over standard cultivars.

Gynoecious (all female flowered) cucumbers, isolated with normal monoecious types, result in gynoecious hybrids. These hybrids are very productive for commercial fresh or processed markets, especially where the crop is harvested mechanically, since fruit are produced uniformly at each node (Figure 5.9). This trait is especially useful in parthenocarpic greenhouse cucumbers, since, unlike field-grown plants, a pollen source is not required to stimulate fruit development.

Seedless watermelons are hybrids resulting from the cross of a diploid (2*N*) male parent with a tetraploid (4*N*) female. The resulting hybrid is triploid (3*N*) and highly sterile. When pollinated with standard diploid pollen from any cultivar, the ovaries of blossoms on these triploid plants enlarge, but fertilization and subsequent seed development will not occur. Although the hand labor requirement and, consequently, the cost to produce such cultivars are high, the superior quality and uniqueness of the seedless watermelons command a high market price.

Standard Cultivars For many crops, non-hybrid cultivars are as productive as hybrids; in others, hybrids may not be available. Standard cultivars predominate among such self-pollinated crops as beans and peas and still constitute a significant share of cultivars available in tomato, pepper, and eggplant. F$_1$ hybrids have not been developed for most salad crops and many root crops.

FIGURE 5.9
Gynoecious cucumber provides only female flowers at each node. This trait simplifies hybrid development and also contributes toward yield. (Photo courtesy of University of New Hampshire.)

For a few crops, there is increasing interest in multiline cultivars, essentially composites of closely related selections from a breeding program. The genetic heterogeneity offered by multilines has been effective in several agronomic crops and in green peas in stabilizing yield within erratic environments and also in reducing the vulnerability to sudden changes in disease pressure.

Asexually Propagated Cultivars   Several vegetables are not propagated by seed. These include, among others, rhubarb, globe arti-choke, Jerusalem artichoke, white potato, sweet potato, horseradish, and garlic. From some of these crops, seeds seldom form; from others, the seeds constitute a level of variability unacceptable to commercial growers.

Cultivars of sexually propagated species represent clones of one or more superior stock plants. Therefore, the plants maintain a high level of uniformity over successive years. Variation, if it occurs, is through mutation (sports and chimeras) or inadvertent mixtures. Mutations occasionally become valuable sources of improved cultivars, perhaps involving changes in color (with possible effect on vitamin content), shape, or other qualitative traits.

Quality of asexually propagative stock is maintained through continued selection of superior plants, through a rigid certification system, or through tissue culture. The last system has been used successfully in producing transplants of asparagus, potato, and sweet potato.

STAND ESTABLISHMENT

Competition, Population Density, and Planting Patterns

Competition of adjacent plants for a common pool of moisture and sunlight occurs in most cropping situations. As plant population density increases, competition increases which will change rate and magnitude of growth unless nutrients, moisture, or other limiting factors are increased. In monoculture, the major factors determining interplant competition are plant spacing, weed growth, and time of emergence of one plant in relation to its neighbor. Assuming adequate weed control, the ultimate performance of the crop can be determined to a large extent by in-row and between-row spacing, by planting arrangements, and by those cultural techniques that ensure uniform seed emergence and stands.

In normal crop spacings, little or no competition occurs among adjacent plants until foliage growth overlaps. Early growth is not restricted by competition, and the yield potential of a plant usually is established before competition affects growth.

As population density increases, competition may occur among relatively young plants, well before an adequate photosynthetic area has been established. Since yield potential for most crops relates to the extent of early vegetative growth, high populations inevitably will modify performance of constituent plants, usually reducing size and/or number of harvested units. Fery and Janick (1971) described the competition effects on an individual tomato plant, noting a reduction in numbers of flower clusters, flowers per cluster, and clusters with marketable fruit (Figure 5.10). On a per acre basis, yields improve as population density increases, but the rate of improvement decreases as competition inten-

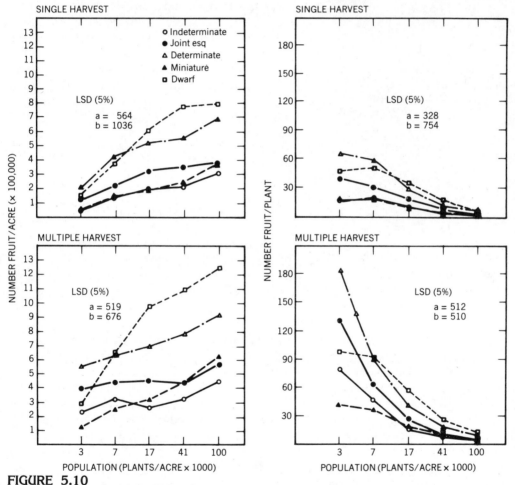

FIGURE 5.10
Effect of plant population on number of fruit per plant (right) and per acre (left) from single (top) and multiple (bottom) harvests. LSD values a = between populations within entries; b = between entries within populations. (From R. L. Fery and J. Janick (1971) *Journal of the American Society for Horticultural Science* 96:172–176. Reprinted by permission of the authors.)

sifies. Eventually, the cost of additional inputs is not justified by the slight gains in total performance. For some crops, per acre and per plant yields may decline substantially under high populations (e.g., when heading crops fail to produce marketable heads).

Within the same population density, the arrangement of plants can alter productivity. Loy (Table 5.7) found twin rows of pepper [24 in. (0.6 m) between plants, 18 in. (0.4 m) between rows] superior to single rows in which plants were spaced 12 in. (30 cm) apart. Presumably, such effects reflect improved distribution of root systems, reducing competition, particularly in the early season among young plants. Mack (1972) reported improved yield per hectare for sweet corn as populations increased to 16,000 to 22,000 plants. The most effective planting arrangement was triangular, affording each plant maximum root zone. Close spacing also was shown to reduce weed competition in corn. Similar responses have been reported for beans, watermelon, and other species. In some instances, high-density plant populations can lead to improved quality. For example, peppers have shown reduced sunscald because of the increased shading in dense plantings.

The degree of competition is perhaps most influenced by the supply of soil moisture and nutrients. The effects of high-density populations in crop production can be minimized or delayed by increasing irrigation and fertilizer applications. Such increases, however, need not be applied at a constant rate per plant. A 100 percent increase in plant population per acre may require an increase of only 50 percent or less in fertilizer or soil moisture, depending on soil type and crop.

Seeding Systems   Direct field sowing of most large-seeded vegetables will result in successful stands if soil moisture and temperature are within the optimum range for the specific crop and if seed quality is high. Frequently, however, crops must be established when environmental factors are not optimum, and emergence often is delayed. The most common environmental problems are high moisture and cold soils, which, when combined, may retard rate of germination sufficiently to predispose the crop to root rot diseases or damping-off. The grower may adjust rate and depth of seeding to improve germination/emergence in problem soils somewhat, but these adjustments seldom ensure a complete stand. In dry soils, deep seeding may promote contact of the seed with soil moisture, thereby accelerating germination. Shallow seeding [$\frac{1}{2}$ in. (1.5 cm)], which exposes the seeds to warmer temperatures than would be found at 1-in. (2.5-cm) depths, may be preferred in early spring, when temperatures are erratic. In any unfavorable environment, increased seeding rates may be used to compensate for a reduction in seed germination brought about by suboptimal conditions. Seed priming or conditioning by soaking seeds for up to 2 weeks in an osmoticum, such as polyethylene glycol, and then drying, has been employed to enhance uniformity of germination and consequent stand. The pretreatment encourages pregermination but inhibits radicle emergence until the seeds have been placed in the field. This system has been effective in improving the stands of carrots and other crops in which germinability is reduced by thermodormancy.

Standard seeding equipment consists of single- or multiple-row planters or drills, with

TABLE 5.7
Effect of plant pattern on yield of peppers

Yield response	Single row	Twin row
Marketable fruit		
Number/ha	164,081	202,755[a]
MT/ha	21.1	26.7[a]
Total fruit		
Number/ha	198,893	244,898[a]
MT/ha	23.4	29.9[a]

Source: S. Loy, M.S. thesis, University of New Hampshire (1984).

[a] Significantly higher than single row value at $P = .05$.

seeding rate controlled by any of several mechanical systems geared to tractor speed. Small-seeded crops traditionally have been sown thickly, followed by hand thinning; however, thinning is very labor intensive. To reduce dependence on extensive hand thinning, several seeding concepts have been introduced, including block or clump seeding, seed tapes, precision seeding using a vacuum system with coated seeds, and gel seeding or fluid drilling.

In block seeding, a cluster of seeds is placed at spaced intervals along a row. One plant in each block, perhaps the earliest to germinate, may become dominant, with seedlings emerging late or those plants not vigorous generally posing no competitive threat. Where several plants in a clump develop to maturity, each contributes reduced yield per plant. For some crops, the total productivity of the clump, however, would be at least equal to that of a single plant. For crops that produce heads or in which competition might reduce the size of the edible part, some thinning may be required. An alternative to block seeding that gives identical results is the mechanical removal of plants emerging in a solid row seeding, leaving clumps at spaced intervals.

For small-seeded vegetables, spacing of seeds affixed to a biodegradable tape that is planted directly in furrows provides a precision seeding method that is easily mechanized. However, seed viability seems to decline in these tapes, particularly if stored for any length of time. The cost to the grower also is a negative factor. Seed tapes are appropriate for home gardeners but have not been feasible for large acreages.

Precision seeding, utilizing Stanhay or other equipment (Figure 5.11), involves placement of individual seeds at a preset rate, generally closer than the final stand density, but substantially farther apart than conventional seedings. Because the shape, size, and surface characteristics of some seeds are not amenable to precision seeding, coatings have been developed to homogenize the shape, size,

FIGURE 5.11
Four-row precision bed seeder. This seeder allows adjustment for seed size and spacing, multiple rows, and seed depth. (Photo courtesy of Triangle M Tractors, Morocco, Ind.)

and flowability of seeds, thereby enhancing the efficiency of precision seeding equipment. The seed coatings may be vermiculite, clay, diatomaceous earth, or charcoal with a binder of methyl ethyl cellulose, gum arabic, polyvinyl alcohol, or sugar. The hygroscopic properties of these coatings enhance imbibition of water by the seeds, thereby improving emergence, although germination inhibitors released from seed coats of some species of seeds may be held hygroscopically to the coating material, depressing emergence. There is evidence that activated charcoal placed within the seed coating will reduce inhibition. While perfect stands cannot be assured, precision seeding has reduced labor costs by minimizing thinning, although some is still necessary, and the disruption of root systems of surviving plants is reduced where thinning is required.

Gel seeding, or fluid drilling, is a sowing system involving treatment and pregermination of seeds, separation of germinated from ungerminated seeds, storage of germinated seeds, and preparation and application of the gel–seed mix. The system has been demonstrated successfully for a range of crops in England (Gray, 1981), and several crops have been gel seeded on a small scale in the United

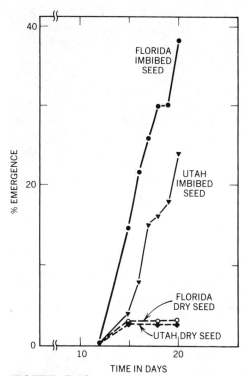

FIGURE 5.12
Percentage emergence of dry and pregerminated (imbibed) seeds of 'Florida 683' and 'Tall Utah 52–70' celery. (From N. L. Biddington, T. H. Thomas, and A. J. Whitlock (1975) *HortScience* 10:620–621. Reprinted by permission of the authors.)

States. Seeding rate is controlled by the number of seed per volume of gel [generally sodium alginate, synthetic clay (Laponite), or other substance]. Prior to incorporation into the gel, the seeds are preconditioned or treated to remove germination inhibitors and then pregerminated. Once pregermination has begun, the seeds with emerging radicles [shorter than $\frac{1}{16}$ in. (1 mm)] must be separated from ungerminated seed before mixing with gel in order to ensure final stand uniformity. Some storage of pregerminated seed may be necessary before the field can be seeded. The fluid drill expresses the seed mix by piston action, either as a continuous pressure feed or as a surge at regular intervals (peristaltic action). In dry soils or at soil temperatures triggering thermodormancy in seeds, gel seeding is very effective (Figure 5.12). In optimum environments, the advantages as yet do not justify the expense.

Another system of precision seeding has used scatter plates to develop broad bands, rather than rows, from the same volume of seed. This system has been effective for root crops in sand and peat soils.

Anticrustants   In some areas, rainfall or irrigation may occur before seedling emergence. The beating action of raindrops produces a hard crust on soils containing a significant clay fraction, which impedes emergence, particularly of such small-seeded crops as carrot, lettuce, onion, and brassicas. Anticrustants, of which vermiculite and phosphoric acid are the most effective, may be used to cover the seed placed in the seeding furrow. Phosphoric acid (H_3PO_4) applied over the seeded row helps to maintain soil aggregation. Test results indicate that vermiculite improves greatly the stands in soils prone to crusting, resulting in moderate to substantial yield increases.

Transplanting

Transplanting is the only feasible establishment system for some crops (sweet potato, celery). For most crops, however, it is an option used to grow vegetables otherwise restricted by a short growing season, to enhance potential yields, to enable growers to improve early yield (and market prices), and to ensure uniform stands and spacing.

Transplants are grown in forcing structures or outdoor beds. The applicability of the latter has expanded with the use of plastic row covers; however, heated and well-ventilated plastic greenhouses or hotbeds are preferred in northern areas, particularly for warm season crops.

A successful plant production system requires suitable materials (medium and container) and careful management (light, water, ventilation, nutrition, and proper timing). The medium should be sterile, light textured, and friable, with adequate water-holding capacity. Although composted soils can be used, most plants are grown in a soilless mix consisting of peat and vermiculite, amended with nutrients and lime as needed (Table 5.8). A mix of soilless medium with a seeding gel has been tested with some success. Containers favored by growers include peat pots, Speedling trays, or plug trays (Figures 5.13 to 5.15), all of which reduce transplanting shock by minimizing root breakage. All are convenient to use. The plug tray is part of an integrated growing system involving a 1- to 2-in. (2.5- to 5-cm)-deep tray or board in which holes $\frac{1}{2}$ in. (1.2 cm) or larger in diameter are formed. A 12-in. (30 cm) square tray may contain up to 400 such holes. A special plug mix of peat with a binder maintains the integrity of the root ball throughout the automatic transplanting. Size of container influences transplant size, but the effect on ultimate yield is variable. With proper handling, even plants grown in small plugs or pots develop rapidly in the field and produce normal yields.

The filled containers may be seeded directly, or young seedlings (one true leaf) may be transferred from seeding trays. Flats or pots must have complete drainage and adequate ventilation. Speedlings, in particular, require air circulation under the flats to ensure air pruning of roots protruding through the drainage holes.

FIGURE 5.13
Automatic seeder (*a*) and transplanter (*b*) for use with plug trays. (Photos courtesy of Blackmore Transplanter Company, Ypsilanti, Mich.)

Planted flats should be placed in full sunlight, but in moderate temperatures. During germination and early emergence, water is applied frequently; however, as plants develop, the quantity of water is decreased to slow the rate of plant growth. As the plant size increases per unit volume of medium, water management becomes increasingly critical.

Restricting plant growth under forcing conditions is termed hardening and is practiced extensively in the north to minimize transplant shock or chilling damage of vegeta-

TABLE 5.8
Typical soilless medium for transplant production

Quantity	Ingredient
1 bu	Sphagnum peat moss
1 bu	Horticultural vermiculite
1 lb	Dolomitic lime (finely ground)
4 oz	Superphosphate (0–20–0)
2 oz	Calcium nitrate
7 oz	Calcium sulfate (gypsum)
8 gr	Fritted trace elements

FIGURE 5.14
Speedling system showing tapered cell that encourages downward orientation of fine root system. (Photo courtesy of Speedling, Inc., Sun City, Fla.)

FIGURE 5.15
Speedling trays placed on T-bars. Exposed base maintains root growth within each cell. (Photo courtesy of Speedling, Inc., Sun City, Fla.)

bles, particularly of the warm season annuals. Reducing water supply to the point of incipient wilting, maintaining cool temperatures, or withholding fertilizer constitute effective hardening measures as long as the plants are exposed to full sunlight. Photosynthate then is not utilized in rapid growth but is stored, contributing to thickened cell walls and increased osmotic concentration (decreased freezing point) of the cell sap. Well-hardened seedlings may develop increased waxy coating on the leaves (as in crucifers), and foliage of all hardened seedlings is noticeably darker than those not hardened. Some plants, such as pepper, can be overhardened and may not recover to produce optimal growth. Hardening normally takes 7 to 10 days, and constant attention is required.

Plant size for field transplanting varies among species. Fruiting vegetables, such as

FIGURE 5.16
Mechanical transplanting system. Uniform placement of plant and water is controlled by gear linkage to weight-bearing wheels. (Photo courtesy of Mechanical Transplanter Company, Holland, Mich.)

tomato or melon, must not show reproductive growth prior to transplanting, since early fruit set tends to divert photosynthates from vegetative growth before the plant has developed to a size adequate to support maximum productivity. Just prior to transplanting, the root medium must be moistened thoroughly, and a high-phosphorus starter solution is recommended to encourage early root development.

Field planting is predominantly by machine (Figure 5.16). Plants grown in plug trays or other containers developed as part of an integrated system are handled with minimum hand labor. Bare-root plants or those grown in traditional containers are planted by machines requiring manual placement of plants in a spacing wheel or in the furrow.

SELECTED REFERENCES

American Society of Agronomy (1976). *Multiple Cropping,* Spec. Publ. 27. Madison, Wisc.

Arkin, G. F., and H. M. Taylor (1981). *Modifying the Root Environment to Reduce Crop Stress,* ASAE Monogr. 4. ASAE, St. Joseph, Mich.

Biddington, N. L., T. H. Thomas, and A. J. Whitlock (1975). Celery yield increased by sowing germinated seeds. *HortScience* **10:**620–621.

Delouche, J. C. (1980). Environmental effects on seed development and seed quality. *HortScience* **15:**775–784.

Fery, R. L., and J. Janick (1971). Effect of time of harvest on the response of tomato to population pressure. *Journal of the American Society for Horticultural Science* **96:**172–176.

Gray, D. (1981). Fluid drilling of vegetable seeds. *Horticultural Reviews* **3:**1–27.

Hanumaiah, L., and H. Andrews (1973). Effects of seed size in cabbage and turnips on performance of seeds, seedlings and plants. *Proceedings, Association of Official Seed Analysts* **63:**117–125.

Hemphill, D. D. (1982). Effects of transplanting, imbibition and gel on stands and harvest variability of lettuce. *HortScience* **17:**256–257.

Mack, H. J. (1972). Effects of population density, plant arrangement and fertilizers on yield of sweet corn. *Journal of the American Society for Horticultural Science* **97:**757–760.

Nicholson, A. G., and H. C. Wien (1983). Screening of turfgrasses and clovers for use as living mulches in sweet corn and cabbage. *Journal of the American Society for Horticultural Science* **108:**1071–1076.

North, C. (1979). *Plant Breeding and Genetics in Horticulture.* Halsted Press (Wiley), New York.

Robinson, R. W. (1954). Seed germination problems in the Umbelliferae. *Botanical Review* **20:**531–550.

Smittle, D. A., and R. E. Williamson (1977). Effect of soil compaction on nitrogen and water use efficiency, root growth, yield and fruit shape of pickling cucumbers. *Journal of the American Society for Horticultural Science* **102:**822–825.

Thompson, J. R. (1979). *An Introduction to Seed Technology.* Halsted (Wiley), New York.

Triplett, G. B., Jr., and D. M. VanDoren, Jr. (1977). Agriculture without tillage. *Scientific American* **236:**28–33.

USDA (1961). *Seeds,* Yearbook of Agriculture. U.S. Govt. Printing Office, Washington, D.C.

Wilson, H. M., and C. S. Winkelbleck (1973). *Tillage: Basic Principles and Techniques,* New York State College of Agriculture and Life Sciences, Cornell University, Ext. Bull. 1176.

STUDY QUESTIONS

1. What are the various forces that contribute to soil compaction, and how can the problem be alleviated?

2. What factors contribute to high germination in seed? Can the inherent percentage germination of seed be increased?

3. Describe "hardening" of young seedlings in terms of the plant's physiological processes.

creasing share of total production costs. Over a 9-year period in Florida, overall control expenditures increased by 6.6 percent (Table 6.1). Differences among specific crop trends reflect not only amounts of chemical applied, but also escalating costs of regulation, introduction of new materials, and, in some instances, reduction of costs due to improved resistance or improved control procedures.

CHARACTERISTICS OF PLANT PESTS

Disease Agents

Plant disease may be parasitic or nonparasitic. Examples of the latter are the necrosis, mottling, and stunting caused by poor plant nutrition, air or water pollution, or weather-related factors. Defects caused by such agents, extremely important in determining the yield and quality of many vegetable crops, are discussed for specific crops in later chapters.

Parasitic diseases are caused by fungi, bacteria, viruses, and mycoplasmas. All of these pathogens compete with crop plants by using metabolites produced by the host or those stored in harvested produce, and most reduce the physiological efficiency of the plant by reducing photosynthetic capacity and/or water and nutrient uptake or by disrupting metabolic processes at the cellular level. All diseases, regardless of causal organism, develop in the same general way. **Inoculation,** the movement of the organism to the plant surface, precedes **penetration** into plant cells through wounds or stomates. Following penetration, the pathogen **incubates,** then **infects**

It is difficult to measure the losses that occur from plant diseases, insects, nematodes, weeds, and birds and other animals. Assuming complete control of crop pests, yield increases of 30 percent have been estimated. Losses to a grower, however, must include not only actual reduction in yield or quality, but also costs of applying the various systems of pest control, whether direct costs of spray or cultural control methods or indirect costs hidden in taxes (to pay for administering legal control systems) or seed prices (reflecting development of new cultivars).

The dollars spent for pest control constitute a substantial and, in some instances, in-

TABLE 6.1

Comparison of spray and dust costs as a percentage of total growing cost for Florida commodities, 1967 to 1968 and 1976 to 1977

Commodity	Percentage of growing cost	
	1967–1968	1976–1977
Snap bean	6.1	11.3
Cabbage	8.1	9.6
Celery	18.8	29.2
Sweet corn	32.2	22.1
Cucumber	16.0	10.5
Pepper	13.2	14.8
Potato	12.2	11.8
Radish	7.4	14.3
Squash	14.2	11.0
Tomato	15.9	16.1

Source: From S. L. Poe (1981) *HortScience* **16**:503. Reprinted by permission of the author.

the tissue. Infection occurs when the pathogen actually becomes parasitic. A period of several days to 6 to 8 weeks may elapse before symptoms of disease appear. At this time, abnormalities in plant growth or actual tissue death become evident. Once the pathogen has become established in the infection stage, it produces secondary inoculum that may infect adjacent tissue or plants or may provide overwintering potential.

Fungi comprise the largest group of disease organisms and abound in most environments. Some are specific to few plant species; others have a wide host range (Table 6.2). Some are characterized by many *forma speciales*—identical in activity and appearance but biochemically different and attacking different host genera. Sporulation in fungi may

TABLE 6.2

Common pathogens with a broad host range

Pathogen	Class	Disease caused
—	Mycoplasma	Aster yellows (tomato, carrot, lettuce, and many other plants)
—	Virus	Cucumber mosaic (celery, cucumber, melon, lettuce, tomato, pepper)
—	Virus	Potato X and Y (potato, pepper, tomato, eggplant)
—	Virus	Curly top (tomato, beet, potato)
Botrytis cineraria	Fungus	Gray mold of bean; storage rot of carrot; gray mold of lettuce; leaf blast of onion, neck rot of onion
Erwinia carotovora	Bacterium	Storage rot of carrot; soft rot of celery, cabbage, cauliflower, onion, pepper, tomato, and other crops
Fusarium oxysporum	Fungus	Yellows or wilt: formae speciales on many crops (pea, muskmelon, tomato, asparagus, watermelon, celery, cabbage, sweet potato, and other crops)
Fusarium roseum	Fungus	Dry rot of carrot; fruit rot of muskmelon; basal rot of onion
Fusarium solani	Fungus	Root rot of bean, basal rot of onion, root rot of pea
Verticillium spp.	Fungus	Wilt of tomato, pepper, eggplant, potato, and other crops
Phytophthora spp.	Fungus	Damping-off of seedlings
Pythium spp.	Fungus	Damping-off of seedlings; root rots
Pellicularia filamentosa	Fungus	Lower stem rot of bean; stalk rot of celery; wirestem and head rot of lettuce; wirestem of crucifers; lower stem rot of pea; soil rot of tomato
Sclerotinia sclerotiorum	Fungus	White mold of bean; cottony soft rot of carrot; pink rot of celery; white rot of cabbage; sclerotinia drop of lettuce; stem rot of tomato
Streptomyces scabies	Bacterium/fungus[a]	Scab on potato and beet
Thielaviopsis basicola	Fungus	Root rots of bean, pea, cucurbits, and solanaceous plants

[a] *Streptomyces* is classified as a "higher bacterium" or a bacteria-like fungus.

be sexual, asexual, or both, and spores may be transmitted by wind, rain, seed, soil, or animals, including man. Some fungi (e.g., *Fusarium oxysporum*) persist in soils indefinitely; other species remain only as long as there is host tissue.

Life cycles of fungi are variable. The simplest cycle requires a single host in which the entire process—infection, sporulation, and overwintering—involves a single spore type. In contrast, some rust fungi may produce several successive spore types during the cycle. In some instances, only one or two spore types may be parasitic on the primary (crop) host, the others infecting an alternate host which thus becomes essential for completion of the life cycle.

Bacteria cause fewer diseases than do fungi and viruses. The disease organisms comprise five distinct groups: *Agrobacterium, Corynebacerium, Erwinia, Pseudomonas,* and *Xanthomonas.* Bacterial increase occurs at an enormous rate under optimum conditions and is encouraged particularly by surface moisture from rainfall, irrigation, or high humidity and by moderately warm temperature. Bacteria may be transmitted by seeds or vegetative propagation, by insects (Table 6.3), by tillage and planting equipment, or by wind-blown aerosols following rain or irrigation.

Viruses cannot function outside a host, but most have a wide host range and more than one vector. Aphids, whitefly, and leafhoppers are the most common vectors, but various

TABLE 6.3
Diseases of vegetable crops transmitted by insects

Disease	Plant	Causal organism	Vector
Downy mildew	Lima bean	*Phytophthora phaseoli*	Honey bee (*Apis mellifera*)
Blackleg	Cabbage	*Phoma lingam*	Cabbage maggot (*Hylemya brassicae*)
Potato scab	Potato, beet	*Streptomyces scabies*	Flea beetle (*Epitrix cucumeris*)
Bacterial wilt	Cucurbits	*Erwinia tracheiphila*	Cucumber beetles (*Acalymma vittata, Diabrotica undecimpunctata*)
Stewart's wilt	Corn	*Xanthomonas stewartii*	Flea beetle (*Chaetocnema pulicaria*), corn rootworms (*Diabrotica longicornis, D. undecimpunctata*), and seedcorn maggot (*Hylemya cilicrura*)
Soft rot	Many crops	*Erwinia carotovora*	Maggots (*Hylemya brassicae, H. cilicrura*)
Curly top	Many crops	Virus	Beet leafhopper (*Cirdulifer tenellus*)
Aster yellows	Many crops	Mycoplasma	Leafhopper (*Macrosteles divisus*)
Corn streak	Corn	Virus	Leafhopper (*Cicadulina mobila, C. zeae, C. storeyi*)
Spindle tuber	Potato	Virus	Flea beetles, aphids, grasshoppers, tarnished plant bug (*Lygus oblineatus*), Colorado potato beetle (*Leptinotarsa decimlineata*)
Leaf roll	Potato	Virus	Green peach aphid (*Myzus persicae*)
Bean mosaic	Bean	Virus	Green peach aphid and other aphids
Cabbage mosaic	Cole crops	Virus	Aphids (*Myzus persicae, Brevicoryne brassicae*)
Cucumber mosaic	Cucurbits and other crops	Virus	Melon aphid (*Aphis gossypii*) and other aphids, Cucumber beetles
Spotted wilt	Tomato	Virus	Thrips (*Thrips tabaci* and others)
Yellow dwarf	Onion	Virus	Aphids (up to 50 species)

Source: From *Destructive and Useful Insects,* C. L. Metcalf, W. P. Flint, and R. L. Metcalf. Copyright © 1951 by McGraw–Hill Book Company, Inc., New York, pp. 14–16. Reprinted by permission.

beetles, grasshoppers, earwigs, thrips, mealybugs, and free-living and cyst-forming nematodes also may transmit virus diseases. Viruses overwinter in insects, in crop debris, or in many weed species.

Mycoplasmas are the smallest independent organisms and also must live close to plant cells. These pathogens once were classified as viruses and cause similar symptoms. The most common disease resulting from mycoplasmal infection is aster yellows, which affects many vegetable, ornamental, and agronomic crops as well as weeds. Leafhoppers serve as vectors.

Diagnosis   Diagnosis and verification of the causal organism of a disease involve an established protocol based on Koch's postulates (proposed in 1881 by Robert Koch, a German microbiologist). The conditions for verification are as follows:

1. The pathogen always must be associated with a specific set of symptoms.
2. The organism must be isolated and grown in pure culture.
3. A healthy plant must be inoculated with this pure culture and must develop the same symptoms originally associated with the presence of the pathogen.
4. The organism finally must be isolated from the inoculated plant and compared with and found identical to those in pure culture.

Insects

Plant pests collectively called insects include the true insects (class Insecta) and species of several other classes: mites (Arachnida), symphylans (Chilopoda), slugs (Gastropoda), and nematodes (eelworms). Each insect pest has a characteristic life cycle, categorized as **simple** or **complete** metamorphosis (Figure 6.1). The latter includes egg, larva, pupa, and adult stages (moths, beetles). Insects with gradual metamorphosis may develop directly from egg to adult with no intervening stages (leafhopper, aphids).

Insects are described according to feeding habits as **chewing, sucking,** or **rasping.** Chewing insects consume the plant roots, leaves, or stems as adults and/or larvae. Evidence of feeding usually is very apparent. Sucking insects, including aphids, whitefly, tarnished plant bug, and others, derive sustenance by puncturing plant cells and then sucking plant sap. Some sucking insects also inject toxin into the plant cells that may, in sufficiently high insect populations, cause distorted plant growth. Except for toxin effects, feeding by sucking insects is not always evident. The rasping insects, primarily thrips, cut plant cells and feed on the exposed plant sap. The injury is apparent as small patches of killed tissue, often not penetrating through the leaf.

Weeds

Weed pests are classified as grasses, sedges, and broadleafs and may be annual, biennial, or perennial. Propagation over long distances is by seed; both seed and vegetative reproduction may be responsible for distribution within a local area. The number of seeds disseminated by a single weed plant is sufficient to maintain a weed stand over a large area for a long time. Curly dock, for example, may produce up to 40,000 seeds per plant. Of the weeds listed in Table 6.4, the most common broadleaf pests are pigweed, thistles, ragweed, lamb's-quarter, and chickweed. Crabgrass, quack grass, foxtails, nutsedge, and Johnson grass are the most serious grass and sedge weeds.

Weeds compete directly with crops for nutrients, water, and light. This competition is particularly damaging during early growth of crop plants. Potato yield declines of 12 percent have been measured for each 10 percent dry weight increase in weeds. In some instances, weeds also affect quality. For example, the buds of field daisies and Canada thistle, which are the same size and weight as mature green peas, are difficult to separate by mechanical viners or strippers. As a consequence, some may become mixed in the canned product, thereby reducing quality and consumer ap-

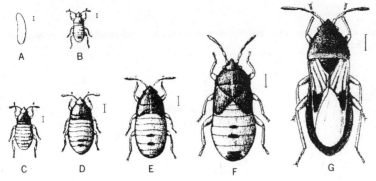

Stages in the development of a bug (simple metamorphosis).
A, egg; B-F, nymphal instars; G, adult.

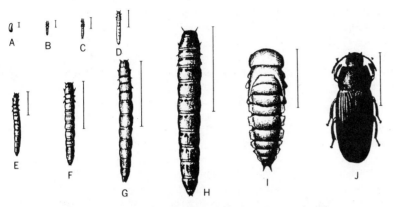

Stages in the development of a beetle (complete metamorphosis).
A, egg; B-H, larval instars; I, pupa; J, adult.

FIGURE 6.1
Simple and complete metamorphosis in insects.
(From *A Field Guide to the Insects,* by Donald J.
Borror and Richard E. White. Copyright © 1970
by Donald J. Borror and Richard E. White. Reprinted by permission of Houghton Mifflin Company.)

TABLE 6.4
Weeds of cultivated areas classified by family

Family	Genus and species	Common name	Life cycle
Amaranthaceae	*Amaranthus blitoides*	Prostrate pigweed	Annual
	Amaranthus retroflexus	Redroot pigweed	Annual
Apiaceae (Umbelliferae)	*Daucus carota*	Wild carrot	Biennial
Ascelepiadaceae	*Asclepias syriaca*	Milkweed	Perennial
Asteraceae (Compositae)	*Ambrosia artemisiifolia*	Common ragweed	Annual
	Ambrosia trifida	Giant ragweed	Annual
	Chrysanthemum leucanthemum	Field daisy	Perennial
	Cirsium arvense	Canada thistle	Perennial

TABLE 6.4 (*Continued*)
Weeds of cultivated areas classified by family

Family	Genus and species	Common name	Life cycle
	Galinsoga parviflora	Galinsoga	Annual
	Helianthus annuus	Sunflower	Annual
	Taraxacum officinale	Dandelion	Perennial
Brassicaceae	*Barbarea vulgaris*	Yellow rocket	Perennial
	Brassica kaber	Wild mustard	Annual
	Brassica nigra	Black mustard	Annual
	Capsella bursa-pastoris	Shepherd's purse	Annual
	Lepidium campestre	Pepper weed	Annual
Caryophyllaceae	*Stellaria media*	Chickweed	Annual
Chenopodiaceae	*Chenopodium album*	Lamb's-quarter	Annual
Convolvulaceae	*Convolvulus arvensis*	Field bindweed	Perennial
	Ipomoea purpurea	Morning glory	Annual
Cyperaceae	*Cyperus esculentus*	Yellow nutsedge	Perennial
	Cyperus rotundas	Purple nutsedge	Perennial
	Eichornia crassipes	Water hyacinth	Perennial
Euphorbiaceae	*Euphorbia corollata*	Flowering spurge	Perennial
	Euphorbia esula	Leafy spurge	Perennial
	Euphorbia masculata	Spotted spurge	Annual
	Euphorbia supina	Prostrate spurge	Annual
Gramineae	*Agropyron lepens*	Quack grass	Perennial
	Cenchrus incertus	Field sandbur	Annual
	Cenchrus longispinus	Longspine sandbur	Annual
	Cynodon dactylon	Bermuda grass	Perennial
	Digiaria sanguinalis	Large crabgrass	Annual
	Echinochloa crus-galli	Barnyard grass	Annual
	Panicum dichotomiflorum	Fall panicum	Annual
	Setaria faberi	Giant foxtail	Annual
	Setaria glauca	Yellow foxtail	Annual
	Setaria viridis	Green foxtail	Annual
	Sorghum halepense	Johnson grass	Perennial
Labiateae	*Lamium amplexicaule*	Henbit	Biennial/annual
Malvaceae	*Abutilon theophrasti*	Velvetleaf	Annual
	Malva neglecta	Mallow	Annual/biennial
Polygonaceae	*Polygonum aviculare*	Prostrate knotweed	Annual
	Polygonum pensylvanicum	Pennsylvania smartweed	Annual
	Rumex acetosella	Red sorrel	Perennial
	Rumex crispus	Curly dock	Perennial
Portulacaceae	*Portulaca oleracea*	Purslane	Annual
Solanaceae	*Datura stromonium*	Jimsonweed	Annual
	Physalis heterophylla	Ground cherry	Perennial
	Solanum carolinense	Horse nettle	Perennial
	Solanum nigrum	Nightshade	Annual
Tetragoniaceae	*Mollugo verticillata*	Carpetweed	Annual

peal. Similarly, leafy weeds may contaminate mechanically harvested spinach or other leafy crops destined for processing.

Weeds also are hosts for many virus diseases and provide effective overwintering sites for both insects and diseases. They obstruct hand harvests and interfere with efficient farm management.

GENERAL PEST CONTROL SYSTEMS

Disease and Insect Pests

Extensive genetic variation within many pest species may support continual population adjustment to a single control system. Chemicals, for example, applied on a regular basis provide an effective screen against susceptible individuals. Surviving individuals within a single species may interbreed, and successive populations often show a gradual increase in tolerance. A variety of strategies is therefore important for effective pest control.

The bases for controlling insect, mite, nematode, and disease pests are **legal, cultural** (physical), **biological, chemical,** and **integrated** systems utilizing a series of complementary techniques.

Legal Methods   Attempts to prevent introduction of pest problems to an area have achieved some success through state and federal legislation establishing seed certification, plant quarantines, and plant inspection services. Seed certification has been particularly effective in reducing disease problems in vegetatively propagated vegetables (such as potato and sweet potato). Seed may be certified only if plants from which it was obtained had been inspected and found within established tolerances for specified diseases. Tolerances range from zero for virulent pests with high reproductive rates to levels of other organisms that will not impair subsequent yield and quality.

Plant quarantines can be effective if the movement of the pathogen or insect can be controlled by man. It is most effective where isolation is afforded by a natural barrier— ocean or mountain range. Once introduced into an area, control by quarantine becomes difficult, particularly of pests carried by wind and as contaminants of produce or plant material freely exchanged among different areas. Quarantines must have legal sanction, must be biologically attainable, and must be administered strictly, with effective inspection, usually at point or origin (farm field) or at point of entry.

Legislation also may govern the kinds of seed that may be sown and/or the time it may be planted. For example, it is unlawful in Monterey County, California, to plant lettuce seed that has not been indexed for virus. Also in California, western celery mosaic has been minimized by defining a vulnerable period during which planting of the crop is prohibited, thereby breaking the annual disease cycle.

Cultural Methods   Cultural practices seldom eliminate disease or insect pests. The objectives should be to minimize inoculum or infestation levels and to avoid practices that would encourage dispersal or increase reproductive rate or growth of the pest.

Seed Treatment   Prior to planting, treatments often are applied to seed or vegetative propagules to eradicate disease pests; many seeds are treated with hot water, and heat treatment of potato tubers has in some instances eliminated leafroll virus. Such treatments must be done carefully, since there is a narrow range between lethal effects on the pest and on the host. Fermentation of tomato seed is effective against the bacterium causing canker. A recent approach toward eradicating disease from vegetatively propagated crops, meristem tissue culture, has been effective in eliminating several viruses.

Modifying Environment The environment of a crop may be managed by altering cultural techniques. Temperature may be adjusted by changing the time of planting, a technique that also may avoid certain insect pests. Aluminum mulches have been used to discourage migration of aphids to some crop plants. White and black polyethylene also discourage aphid populations, but to a lesser degree. The reflective surface of these materials repels insect visitation, an effect apparently related to light/dark perception by affected insects. Mulching and/or row covers also will modify soil temperature, thereby affecting aggressiveness of root rot fungi and other soil inhabitants, and row covers have been effective as a physical barrier eliminating early season insect damage from flea beetles, aphids, Colorado potato beetles, and others.

Excessive moisture increases frequency and intensity of serious disease problems. Soil moisture is managed by drainage, irrigation, and site selection. Moisture on leaf surfaces can be minimized by surface irrigation. Fertilization with nitrogen, combined with excessive moisture, promotes soft plant growth that is more prone to infection than is hardened tissue, even by organisms to which it is ordinarily resistant. Soil pH can be manipulated to control only a few soilborne pathogens (those causing clubroot of crucifers and potato scab).

Cropping Practices Manipulation of the crop also can modify pest levels. While a dense plant canopy precludes aggressive weed growth it tends to increase disease and insect intensity. Plant spacing may be adjusted to provide aeration through the plant canopy, effecting a measure of control of such fungi as *Sclerotinia sclerotiorum*.

Rotation is recommended for improving soils and for conservation, but its greatest benefit may be a reduction in pest levels. Many foliar disease organisms persist as long as foliage remains in the soil. Two to four years may be required to decompose plant litter completely. As a consequence, a 3- to 4-year rotation is recommended for control of several diseases (bean blights, cabbage blackleg and black rot, anthracnose diseases, and others). Rotation also may reduce insect populations, and, by alternating sod with row crops, weed propagation may be reduced. Rotation should take into consideration the host range of the pest, avoiding related species or those vulnerable to the same organism.

Rotation with sod crops is difficult for vegetable growers, for a concern for lack of income from sod fields often overrides the long-term ecological benefits. Yet, periodic sod cover contributes to the stability of soil organic matter, and soil structure is enhanced through the action of fibrous root systems. Although difficult to measure directly, rotation, including an occasional sod year, even may be cost effective over a prolonged period, if reduced levels of pests require fewer chemical inputs.

Sanitation Sanitation often is overlooked in pest control. Cleaning tillage equipment before moving to a new location reduces the risk that a soilborne problem will be transferred throughout an entire farm. Failure to remove plant trash piles and other sites of disease and insect overwintering will increase the inoculum supply or insect numbers in successive years.

Other cultural practices that may reduce pest levels include elimination of alternate hosts, control of weeds around the perimeter of a field as well as those directly competing with crop growth, and burning or killing of crop debris or tops to lower infestation levels or to prevent contamination of the harvested product.

Biological Methods The most effective form of biological control is **genetic resistance** (Figure 6.2). Single-gene immunity or resistance to some pathogens has been very successful; for other diseases, genetic mutation of the organism or unreliable resistance within the host species has minimized effectiveness of resistant cultivars or discouraged their development. Cultivars described as tolerant

FIGURE 6.2
Genetic resistance to *Verticillium* wilt in tomato. Two plants on the left are infected, showing reduced growth and yellow and dying leaves; the plant on the right carries single-gene resistance. (Photo courtesy of University of New Hampshire.)

may show some disease, but the organism is slowed in its development and reproduction, and a satisfactory crop may be produced. Tolerance normally is based upon multiple-gene action and tends to be stable over varied environments. When available, genetic resistance and tolerance constitute the least expensive and most effective control and the only effective approach for preventing infection by many soilborne organisms. The mechanisms of resistance or tolerance are not understood in most cases but probably have more than one basis. A number of **phytoalexins** have been isolated from plants, including phaseolin (beans), pisitin (peas), and ipomoeamarone (sweet potato). These substances are fungistatic and appear to be natural plant defense mechanisms activated when plant cells become infected.

Insect control through resistance has not advanced to the level of usefulness characterizing disease control. There is considerable evidence that insects prefer one cultivar over another, but feeding is not deterred by growing only a nonpreferred cultivar. Resistance often seems related more to the morphology of the plant or plant organ. Development of corn with tight ear husks, for example, may reduce eventual earworm infestation but does not prevent it. Evidence of repulsion has been described for corn (DIMBOA on corn borer) and cucurbits (cucurbitacin on beetles).

Nematode resistance has been introduced into cultivars of several important vegetable crops. Resistant cultivars of tomato, sweet potato, and potato have minimized losses at far less cost to the grower than that of fumigation.

Biological control also may utilize a form of **antagonism** of one organism against another, which may be based on (1) competition for essential resources, (2) antibiosis, (3) predation or parasitism, or (4) alteration of reproductive capacity. Of these possible mechanisms, the most useful have been predation or parasitism and alteration of reproductive capacity.

Bacillus thuriengensis has been very successful in controlling soft-bodied larvae. Applied as a spray, the bacterium parasitizes such larvae as cabbage worm and looper, tomato hornworm, and fruit worm, and presents no ecological hazard. Its activity is sufficiently rapid to ensure minimal crop feeding damage.

Alteration of reproductive capacity of insects has been achieved by sterilization, chemically or by gamma irradiation. Females mate with the sterilized male insects, but the eggs produced do not hatch. Propagation of an infestation is thereby eliminated or at least slowed while other eradication methods are applied. Previously, this technique had been successful in eradicating oriental fruit fly from Guam, and Mexican fruit fly from the southern border area of California. Recently, outbreaks of mediterranean fruit fly in California and Florida have been attacked with this and other techniques.

Organic gardeners often utilize predator insects to control such pests as aphids (Figure 6.3). Lady beetles and praying mantids are two most frequently used. In general, the effective-

FIGURE 6.3
Biological control: lady bug predation on aphids. (Photo courtesy of A. H. Retan, Washington State University, Pullman.)

ness of predator insects decreases as the density of the pest problem increases.

Chemical Methods Agricultural chemicals—fungicides, bactericides, miticides, and nematicides—constitute the single most important approach for controlling disease and insect pests. Yet, chemical application is no panacea: the materials are toxic to humans, in many instances extremely so, and must be applied in strict compliance with label instructions and with a great deal of respect for their levels of toxicity. Legality of use may vary from state to state.

Chemicals are applied as **protectants** or **eradicants** to seeds and other propagules, to the plant canopy, to soils, and to storage facilities and stored produce.

Seed Treatment Seed treatments (dusts or slurries) are used to protect germinating seed and emerging plants against soilborne pathogens and insects. Such treatments are important when seeding in cool soils, since delayed germination increases the risk of damping-off or of insect feeding. For maggot control in crucifers and onion, an insecticide (diazinon) applied in the seed furrow or transplant hole

acts in the same way as seed treatment, but provides an expanded zone of protection.

Fumigation Fumigants are designed to eradicate soilborne disease organisms, insects and nematodes and some weed seeds, or to eradicate pests from produce and from storage facilities. For soil treatment, the volatile materials are injected into a friable soil free from undecomposed plant litter and then sealed with a plastic cover or, in some instances, by $\frac{1}{2}$ in. (1.2 cm) of irrigation water. A field may require treatment once each second or third year to maintain pest-free conditions. Fumigation of plant produce or of storage facilities has been used to control insects, particularly weevils infesting dried beans or peas; recently, however, some fumigants have been eliminated because of carcinogenic properties.

Sprays Sprays (Figure 6.4) provide the most effective vehicle for transporting chemical to a plant surface. Equipment varies, but boom sprayers with nozzles spaced and directed to provide complete coverage of both leaf surfaces are most often used for row crops. Fan-blast units, similar to orchard

FIGURE 6.4
Simple spray boom rig, used here to apply pre-emergence herbicide. (Photo courtesy of Deere & Company, Moline, Ill.)

sprayers, produce a very fine mist extending some distance from the sprayer. Coverage may vary, however, over this distance. The small droplet size allows the material to penetrate the canopy, achieving maximum coverage. Materials may be applied as normal water dilutions or, for other than boom sprayers, as concentrates. Application by aircraft is effective over large acreages. Regardless of equipment, spraying must be done under calm conditions to avoid drift.

Sprays for Disease Control   The most widely used fungicides/bactericides are carbamates (dithiocarbamates), fixed coppers and other copper compounds, and sulfur compounds. **Carbamates,** including compounds of iron, manganese, zinc, or sodium, are effective against foliar fungal organisms and fruit rots. Most are persistent over a period of 7 to 10 days and require a commensurate waiting period before produce may be consumed. Each of these materials has a slight to moderate mammalian oral toxicity. **Fixed coppers** have a wide spectrum of control, including both bacterial and fungal foliage diseases. Cupric hydroxide (kocide) also is effective against fungal diseases, including some downy mildew strains. **Sulfur** is an effective eradicant of powdery mildew, and sulfides are used in controlling damping-off and several other fungal diseases. Other fungicides/bactericides vary in active ingredient.

Sprays for Insect Control   Insecticides are ingested or are absorbed through the skin or inhaled by the insect. The mode of action of most materials currently in use is as a poison to the nervous system. During the insect life cycle, one or more stages may be more vulnerable than others to a chemical control. Adults may not succomb to the same dosage level that is effective on larvae. For the most part, egg masses are resistant to chemical action but can be destroyed by cultural techniques.

Of the insecticides, most now applied are termed second generation pesticides and fall into any of four categories. **Chlorinated hydrocarbons,** including DDT, chlordane, dieldrin, and others, once provided most chemical insect control. These substances, however, are relatively nonbiodegradable and persist in the food chain for very long periods of time. Furthermore, laboratory tests implicated DDT as a causal agent in tumor formation in mice and as the cause of abnormalities in reproduction of birds. Therefore, most chlorinated hydrocarbons have been banned from crop use. **Organic phosphates** comprise a more recent group of chemicals which include parathion, malathion, and methyl parathion. These materials degrade rather rapidly, leaving little or no residue. However, the materials are extremely toxic to humans, and extreme caution must be exercised during application. **Carbamates** are relatively new insecticides which decompose rapidly and, in general, present less mammalian toxicity than organic phosphates. Such materials as carbaryl and lannate (high toxicity) are widely used in vegetable production. **Pyrethroids** constitute the newest class of insecticides, comprising synthetic analogs of the natural plant constituents pyrethrin I and pyrethrin II, discovered in 1924. Pyrethrum derived from *Chrysanthemum cinerariafolium* and other chrysanthemum species has been used over two centuries as an insecticide. Development of the most recent materials proceeded only after the less expensive chlorinated hydrocarbons became unavailable. The synthetic pyrethroids, like the chlorinated hydrocarbons, are lipophilic compounds, almost completely insoluble in water. Unlike the chlorinated hydrocarbons, however, they are degraded by metabolic action of organisms and thus do not persist in the food chain. Their water insolubility gives them excellent weatherability, and they are absorbed readily on leaves, increasing their longevity. Many of the synthetic pyrethroids are used commercially with the addition of a synergent (commonly piperonyl butoxide) which inhibits the oxidases, produced in some insect tissues, that detoxify the insecticide. The pyrethroids have been effective against Colorado potato beetle, whitefly, corn borer, and other insects normally difficult to control. These chemicals should not be overused, however; insect populations develop resistance to each formulation rather quickly.

A number of insecticides (for example, carbaryl) are extremely toxic to bees and should not be used during periods of bee visitation. The active ingredient may be transported by the bees to the hive, resulting in complete kill.

Rates of Application   Spray materials are sold as emulsifiable concentrates or as wettable powders. Each contains a specified percentage of active ingredient or acid equivalent. This percentage must be used in calculating the proper dilution of the spray mix. Furthermore, it is essential that the spray equipment be calibrated carefully to deliver the proper dose to the plant or soil surface. To calibrate a sprayer (in U.S. equivalents):

1. Clean the sprayer and all nozzles; adjust nozzles to proper height.
2. Fill the clean spray tank with water, and mark the top level.
3. Using label information, spray a measured area (width of coverage × distance traveled, in square feet) at a constant speed and pressure.
4. Measure the water required to refill the tank to its previous level (in gallons). As an alternative, containers may be suspended under each nozzle to collect and measure the water used.
5. Multiply the gallons used by the number of square feet in a acre (43,560) and divide by the area sprayed (in square feet) to derive gallons per acre.
6. Add the correct amount of active ingredient to the tank to meet the recommended rate per acre.

Effectiveness of sprays is improved with **spray adjuvants:** stickers, spreaders, or ex-

tenders. Stickers and spreaders reduce surface tension, promoting coverage on the leaf surface, or bind the chemical to the leaf, increasing weatherability. Extenders form an elastic film that prolongs chemical activity, serving the objectives of both stickers and spreaders. All adjuvants are not compatible with all spray materials; technical information must be consulted before mixing.

Dusts and granules are applied more for controlling insects than other pests. Dusts should be applied in still air, when hygroscopic moisture on leaf surfaces will attract and bind dust particles. Granular pesticides utilize clay particles or vermiculite as the carrier for an adhered chemical. Granules are particularly effective for delivering herbicides but have been used to control soil insects and diseases.

Integrated Pest Management

Integrated pest management (IPM) is defined as

> *the selection, integration and implementation of pest control based on predicted economic, ecological and sociological consequences. IPM seeks maximum use of naturally occurring pest controls, including weather, disease agents, predators and parasites. In addition, IPM utilizes various biological, physical and chemical control and habitat modification techniques. Artificial controls are imposed only as required to keep a pest from surpassing intolerable population levels predetermined from accurate assessments of the pest damage potential and the ecological, sociological and economic costs of the control measures.*[1]

IPM thus is a system in which scientifically obtained data provide the basis, and well-trained personnel monitor arrival and densities of individual plant pests (Table 6.5). The data base then is used to justify application of a

[1] D. R. Bottress (1979) *Integrated Pest Management.* Council on Environmental Quality, U.S. Govt. Printing Office, Washington, D.C.

TABLE 6.5

Systematic levels of operation for interdisciplinary crop pest management

Level of program operation	Example and nature of activity
Basic research	Obtain fundamental data from research on crop growth, pest biology, development of new techniques, methods of sampling and assessment, new tactics, model equations of rates, economic damage thresholds, description of processes.
Syntheses	Conceptualize system components and interactions from research data base, economic analysis of costs, effects of weather, pest prediction, choice of control tactics, monitoring, simulation from the models.
Demonstration	Validate reliability of tactics, effects of integration of tactics, yield potential, system response, use of natural enemies, validate cost effectiveness of strategy, pilot tests, decision making, management.
Training	Scout instruction, consultant and grower education, specialists, short courses and workshops on concept and techniques, curriculum development.
Implementation	Delivery for wide-scale use, scouting, decision making, management tactics.

Source: From S. L. Poe (1981) *HortScience* **16**:504. Reprinted by permission of the author.

specific control system, which may or may not involve agricultural chemicals.

The simplest system is one in which the appearance of a pest, for example, late blight (*Phytophthora infestans*) of potato, can be predicted by certain atmospheric conditions. Late blight sporulation will occur following several days of very high (near 100 percent) humidity and temperatures at or near 68 to 70°F (20 to 21°C). Such conditions can be monitored by instruments interfaced with computers accessible to growers. Provided with such a warning system, the frequency of spray application may be adjusted according to risk. In most years, the number of sprays and consequent spray costs have been reduced (Table 6.6).

For insect migration, trained scouts, using insect traps to provide counts, supply a similar kind of data base. An understanding of insect life cycles and how they are affected by weather, crop growth, weeds, and other facets of crop production then can be integrated to develop the most economically effective and efficient control.

IPM must be developed for each region. Environmental conditions vary widely as do intensities of pest occurrence. Growers should work with their county extension specialists if attempting to incorporate IPM into their crop management system.

Weed Control Systems

To some extent, the general methods described for controlling disease and insect pests also apply to weed control. Weed control practices are predominantly cultural and chemical, although a few effective biological approaches have been developed. Weed control also is an integral part of IPM programs.

Cultural Cultural control methods include use of weed-free seed, cultivation, mulching, steam sterilization of greenhouse benches, rotation, adjustment of population density, and no-till culture. Cultivation is an important means for suppressing weeds, particularly during early development of crop plants, and can reduce the weed seed levels in soils dramatically over a period of several years (Figure 6.5). Persistent cultivation of perennial weeds gradually will deplete stored carbohydrates, eventually eliminating them or, at the least, diminishing competition with crop plants.

For many crops, cultivation must terminate when crop growth restricts tractor movement (lay-by). While weeds developing after lay-by may or may not seriously impair crop yields, they can impede spray and harvest operations. Mulching with black plastic has been an effective alternative, particularly for northern growers who benefit also from the increased soil temperature under mulch. Burning or flaming has been used to a limited extent but must be done before the weeds exceed 2 in. (5 cm) in height. The flame causes the cells to rupture with eventual death of the foliage.

Biological The most successful biological control of a weed occurred following introduction of a leaf-feeding beetle, *Chrysolina quadrigemina,* in 1945/1946, for the purpose of eradicating the Klamath weed in western states. Over a period of several years, up to 99 percent control was achieved. Limited success has been reported with fungi pathogenic only to a specific weed. Such specificity in insects or fungi is rare.

The most interesting biological approach to weed control is based on allelopathy. Allelopathy has been suggested as a possible basis for weed suppression, although the level of suppression attainable may be insufficient to avoid competition. Allelopathy refers to the detrimental effects of higher plants of one species on germination, growth, or development of another; the effects are exerted by a chemical released by the "aggressor" plant. In most instances, the chemical release occurs during decay of litter or roots and may affect crop or weed growth. Allelopathy has been observed with certain combinations of no-till crop and cover, and suppression of weed populations of up to 80 percent by certain lines of cucumber has been reported (Putnam and Duke, 1978). Allelopathy also may be the basis for success of certain species in companion plantings and interplantings. Whether chemically active principles can be isolated and used remains for future research to determine.

A related phenomenon, autotoxicity, has been reported for a number of perennial plant species, including asparagus. Materials from decaying asparagus roots reduce germinability and growth of asparagus and may account, at least in part, for the decline of some production areas.

TABLE 6.6
Mean plot marketable yields, insect damage, cost of pesticides, and cost of monitoring under an integrated crop management (ICM) system for celery in comparison to standard commercial management

Year	Treatment	Trimmed weight (kg)	Worm damage[a] (%)	VLM damage[b] (leaves/plant)	Cost ($/ha)	
					Pesticides	Scouting
1974	ICM[c]	47.5	27	9	94.13	10.69
	Commercial	40.6	19	11	650.00	—
1975	ICM	28.3	10	9	150.67	17.29
	Commercial	40.9	1	4	496.29	—
1976	ICM	30.1	4	9	182.92	15.80
	Commercial	35.3	7	7	1058.34	—
1977	ICM	50.5	7	10	205.42	20.74
	Commercial	50.1	4	8	571.23	7.41

Source: V. L. Guzman, W. G. Genung, D. D. Gull, M. J. Janes, and T. A. Zitter (1979) *Proceedings Florida State Horticultural Society* **92**:88–93. Reprinted by permission of the authors.

[a] Before stripping.

[b] Number of leaves after stripping, but before trimming.

[c] Direct seeded in field: 5-month growing period.

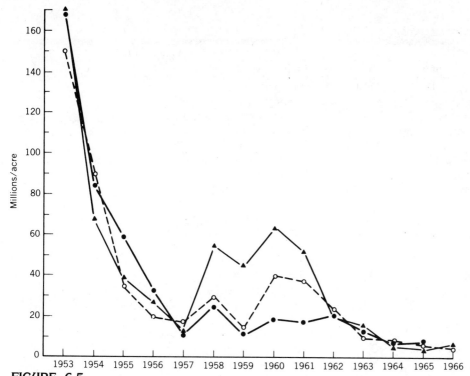

FIGURE 6.5
Mean numbers of viable weed seeds in the top 6
in. of soil, cultivated from 1953 to 1966. The
three lines represent different series of plots.
(From The Changing Populations of Viable Weed
Seeds in an Arable Soil, by H. A. Roberts. *Weed
Research* **8:253–256.** Copyright © **1968** by
Blackwell Scientific Publications, LTD., p. 255.
Reprinted by permission.)

Chemical   Weeds are controlled chemically
by managers of most commercial vegetable
acreages, usually in combination with cultivation. Agricultural herbicides had been used to
some extent since 1900, but the most important advance was the introduction in 1944 of
2,4-D[(2,4-dichlorophenoxy)acetic acid], a selective auxinlike material effective on broadleaf
weeds. This material was not used widely by
vegetable growers because of extreme sensitivity of many horticultural crops. It was primarily a corn herbicide, but its success spur-

red efforts to develop materials effective on a
wide range of crops against a broad spectrum
of weeds. Today, the range of herbicides from
which a grower may select is extensive (Table
6.7).

Herbicides are applied to weed foliage
or, more commonly, to the soil. The successful elimination of weeds without incurring crop
damage is based upon (1) inherent selectivity
of the herbicide; (2) a selectivity imposed
by distinguishing features of crop or weed
plants; or (3) selectivity imposed by applica-

TABLE 6.7
Herbicides appropriate for vegetable production

Herbicide group	Generic name	Target weeds[a]	Crop on which used
Acids	Sulfuric acid	G/BL	Muckland onion
Aliphatics	Dalapon	G	Asparagus, bean, pea, potato
	Glyphosate	NS	Preplant, general nonselective
Amides	Alachlor	G/BL	Potato, sweet corn
	Propachlor	G/BL	Pea, sweet corn
	Diphenamid	G/BL	Okra, potato, sweet potato, tomato, pepper
	Pronamide	G/BL	Endive, lettuce, potato
Benzoics	Chloramben	G/BL	Asparagus, bean, tomato, pepper, squash, sweet potato, pumpkin
Bipyridylines	Paraquat	NS	Asparagus, tomato, potato; preharvest desiccant on potato
Carbamates	Chlorpropham	G+/BL−	Bean, lettuce, onion, pea, squash, southern pea, spinach, tomato
Dinitroanilines	Trifluralin	G/BL	Most vegetable crops
	Benefin	G/BL	Lettuce
	Butralin	G/BL−	Cucumber, watermelon
	Profluralin	G/BL−	Okra, pea, chickpea, southern pea, bean, parsley
Diphenyl ethers	Nitrofen	G/BL	Cole crops, carrot, celery, bulb onion, parsley
Petroleum oils	Stoddard solvent	G/BL	Carrot, parsley, parsnip, celery
Phenoxys	2,4-D[b]	BL	Sweet corn
Salts	Calcium cyanimid	G/BL	Asparagus
Triazines	Atrazine	G/BL	Sweet corn
	Prometryn	G/BL	Celery (seedbeds)
	Simazine	G/BL	Artichoke, asparagus, corn
Ureas/uracils	Diuron	G/BL	Artichoke, asparagus
	Linuron	G/BL	Asparagus, carrot, celery, sweet corn
	Monuron	G/BL	Asparagus
	Chloroxuron	G/BL	Carrot, mustard greens
Unclassified	DCPA[c]	G/BL	Most vegetable crops
	Dinoseb	NS	Bean, sweet corn, cucumber, potato
	Pyrazon	BL	Beet
	Bensulide	G/BL	Wide spectrum of crops

[a] G = grasses; BL = broadleaf; NS = nonselective. + and − indicate wide and narrow spectrum, respectively.

[b] (2,4-Dichlorophenoxy)acetic acid.

[c] 2,3,5,6-Tetrachloro-1,4-benzenedicarboxylic acid dimethyl ester.

tion technique or timing. Herbicides affect sensitive plants in several ways: as mitotic poisons, by hormonal action, by interfering with chlorophyll formation, by interfering with photosynthesis or respiration, by affecting enzyme activity, or by altering nitrogen metabolism.

Inherent selectivity is provided by the specific chemical structure of an herbicide. Slight changes in this structure may alter the level of toxicity toward a specific target weed or host plant. Combinations of herbicides or of an herbicide with an insecticide or fungicide also may change toxicity differentially.

Selectivity may relate to morphological or physiological differences between crop and weed species. Morphological differences include characteristics of the leaf surface (hairiness, corrugation, waxiness); angle of the leaf; location and natural protection of buds; and nature of the root system or of underground propagative tissue (rhizomes, etc.). Morphological features that cause droplets of chemical to roll off leaf surfaces (as from pea or cabbage leaves) or fail to reach a growing point (as in some grasses) ensure that such plants will escape phytotoxicity from general foliar sprays. Physiological differences between crop and weed may involve differential absorption or translocation rates of the chemical (the higher the rate, the greater the phytotoxic effect); reactions with cell constituents resulting in complexes with the herbicide that immobilize it; metabolic detoxification of the chemical by the crop while it remains toxic to the weed; or the reverse—toxification of otherwise nontoxic chemicals by a weed plant.

Selectivity imposed by application system includes timing (preemergent, preplant); placement or shielding (sprays directed to avoid contact with crop plants); varying dosage (some crop species may be unaffected by slightly reduced concentration); and formulation (use of granular material rather than spray).

An example of selectivity by timing is the "stale seedbed system," in which the seedbed is tilled and prepared, but maintained crop free until the weed seeds germinate. When the emerging weeds have produced true leaves, the area is sprayed with a nonselective herbicide, generally a contact material. Shortly thereafter, the crop is seeded, taking care to disturb the soil as little as possible. The stale seedbed system is effective because the weed seeds lying deeper than 1 to 2 in. (2.5 to 5 cm) in the soil are not likely to germinate unless brought to the surface by cultivation.

Herbicides are applied **preplant, preemergence** (after seeding, but before emergence), **at emergence,** and **postemergence** with re-

spect to the crop. The material may be most effective as a contact (post weed emergence) or as a barrier preventing weed seed emergence (pre weed emergence). Those applied to the soil to prevent emergence interact with soil particles, affecting herbicidal activity. In general, activity is reduced in dry soils, in soils with a high humus fraction, and in acidic soils. Certain kinds of clay particles also affect herbicidal efficiency. Therefore, certain soils especially useful for vegetable production, particularly organic soils, are difficult to keep weed free. Recommendations for application rates are specific for each soil texture, and many herbicides are not recommended for organic soils. Most soil applications must be incorporated and/or sealed with approximately $\frac{1}{2}$ in. (1.2 cm) of rainfall or irrigation water for maximum weed control.

Fate of Herbicides Many herbicides are designed to persist through a major portion of the growing season. In most instances, the material should dissipate by the time a new crop is planted. Excessive persistence of herbicides may restrict severely successive crop growth or result in a complete crop failure. Corn land treated with atrazine, for example, may not be suitable for sensitive vegetables for up to 2 years after treatment.

The processes by which herbicides are lost from a soil may be physical, chemical, or microbial (Figure 6.6). Physical processes include volatility, the rate of which is affected by temperature, air movement, and attachment to soil particles; leaching, a term indicating movement of the chemical in any direction by water flow; and soil erosion. Excessive leaching and soil erosion not only reduce herbicidal effectiveness, but also may contaminate groundwater or surface water. Chemical degradation processes include photochemical decomposition, adsorption to soil particles (the most significant factor restricting chemical uptake by plants and microorganisms), chemical reactions with soil constituents, and uptake by plants and microorganisms. Microbial

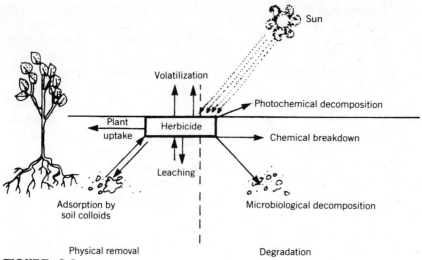

FIGURE 6.6
Processes by which soil-applied herbicides may be inactivated. The herbicide may be lost from the soil by physical removal of the herbicide molecule (left) or by degradation (right). (From *Weed Control Handbook*, Vol. I, *Principles*, J.D. Fryer and S.A. Evans, eds. Copyright © 1968 by Blackwell Scientific Publications, LTD., p. 152. Reprinted by permission.)

decomposition may be the most important process by which the molecules of organic herbicides are broken down. Those factors that favor soil microbiological activity thus will favor breakdown of the chemical. In contrast, poor soil aeration or cold soils may result in retention of herbicide toxicity longer than desired.

GENERAL PLANT PESTS

The most frequently encountered pest problems for specific vegetables are discussed in Chapters 10 through 21. In addition, there are several with wide host ranges that occur in most production areas.

Diseases
Damping-off is a seedling disease complex caused by species of *Pythium*, *Phytophthora*, *Fusarium*, *Pellicularia*, *Sclerotinia*,

Phoma, or *Aphanomyces.* Most soils contain one or more of the causal organisms. Stem tissue collapses or rots at the soil line, causing the seedling to fall over or to become girdled, wilt, and die. The disease is most prevalent under humid greenhouse conditions but can occur in the field. Chemical treatments are relatively ineffective under conditions that favor damping-off. Preventative measures include soil sterilization, air movement around emerging seedlings, careful watering, sanitation, and full exposure of the plants to sun. The use of well-drained soilless medium has decreased the incidence of damping-off significantly. Many of the root rot diseases of developing field plants also are caused by the same soilborne fungi (Figure 6.7).

Insects
Cutworms (Figure 6.8) are the larvae of any of several species of noctuid moths. They feed on roots of many plants, often emerging

FIGURE 6.7
Damping-off in a young seedling. (From *Diseases of Plants*, slide set 37, Diseases of Carrots, Eggplant, Peas, Peppers, Sweet Corn. Reproduced by permission of the American Phytopathological Society.)

at night to feed on shoots. The damage characteristic of this insect is a cut stem at or slightly above the soil line on young seedlings. Damage can be severe, particularly in crops following a sod rotation. Control seldom is complete, but diazinon disked into a friable soil before planting reduces the larval population.

Japanese beetle (*Popillia japonica*) adults (Figure 6.8) feed on most vegetables in addition to many fruit, ornamental, and grass plants. The beetle is metallic green or bronze, $\frac{1}{3}$ to $\frac{5}{8}$ in. (8 to 15 mm) long, and feeds on the surfaces of fruit and leaves, skeletonizing the latter. The insect overwinters as a grub, emerging in June or July. Feeding is especially noticeable on bright sunny days. Control can be improved by controlling grassy areas, which provide food for the grubs, and by chemical applications. Milky disease spores also have been used to treat grassy areas for grub control.

Tarnished plant bugs (*Lygus oblineatus*) are small [$\frac{1}{4}$ in. (6 mm)] insects, coppery brown with brown and yellow flecks (Figure 6.8). They overwinter as adults on plant matter, particularly perennial crops and weeds.

During warm spring days, they feed on plant buds, injecting a toxin that often causes dieback of the affected shoot. Plants affected include cole crops, potato, bean, celery, beet and chard, cucumber, turnip, and most fruit crops. Their life cycle is completed in 3 to 4 weeks; three to five generations may occur within one season. Chemical control is difficult, and the presence of the insect is obscured by their color and their tendency to hide within the foliage.

Grasshoppers (many species within the family Locustidae) attack almost all cultivated plants but are especially damaging to corn. The insect has a worldwide distribution and has been responsible for major famines in years past. Most grasshoppers overwinter as eggs in masses $\frac{1}{2}$ to 2 in. (1.2 to 5 cm) below the surface of pastures, field margins, and roadsides. Hatching occurs in mid- to late spring, and feeding persists until frost. Within cultivated fields, numbers are reduced because the habitat for egg laying is disturbed. Damage still can be substantial, however, if numbers in adjacent pastureland become large. Control may be achieved by general insecticide applications to row crops.

Slugs

Within a moist plant canopy, or in no-till plantings in which the plant stubble may provide moisture and protection, garden slugs may cause extensive damage. Slugs (Figure 6.8) are legless, slimy, soft-bodied pests that feed on plant foliage, causing large holes in leaves, or completely destroying small seedlings. They are particularly troublesome in greenhouse production, favored by the humid conditions. Control in the field is best achieved by cultural practices that provide aeration and soil drying. In greenhouses, poisoned baits are effective.

Garden Symphylan

In some production areas, the garden symphylan or symphylid (*Scutigerella immaculata*) is a serious pest, attacking germinating seeds, plant roots, and parts of the plant can-

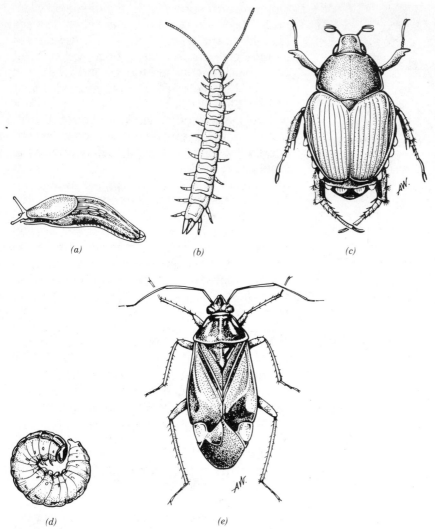

FIGURE 6.8
General crop pests: slugs (*a*); garden symphylan
(*b*); Japanese beetle (*c*); cutworm (*d*); tarnished
plant bug (*e*).

opy in contact with soil. Symphylans (Figure
6.8) resemble centipedes and are not true insects. They predominate in the top 6 in. (15
cm) of soil but migrate deeper in dry or in
cool, wet seasons. Up to two broods can occur
in a season. Thorough tillage and soil treatments with insecticides such as diazinon will
retard symphylan activity sufficiently to pro-

duce a crop. Soil fumigation may provide
three or more years freedom from damage.

Spider Mite
There are several spider mites that damage plants, of which the most significant for
vegetable production is the two-spotted mite
(*Tetranychus bimaculatus*). This pest builds

on upper and lower leaf surfaces during hot dry weather, affecting bean, corn, tomato, eggplant, celery, onion, and other plants. The first evidence may be a pale yellow or browned leaf. The under surface will appear to be powdery, and close examination will reveal tiny white eggs suspended in a very fine web. Under the web are many adult mites which feed on plant sap. Control is achieved by repeated application of a miticide for 7 to 10 days to eradicate not only active adults, but also those newly emerging.

Nematodes

Nematodes are neither pathogens nor insects. They are microscopic eelworms (0.7 to 1.7 mm in length), many of which are harmless, living on dead tissue. Several types, however, produce disease symptoms not unlike those caused by soilborne wilt fungi. Others may provide an avenue for disease infection. The only effective control in most regions of the country is fumigation.

Sting nematodes (*Belonolaimus* spp.) cause extensive root damage. New roots fail to develop, and coarse stubby roots that develop are characteristic of infestation. Plants become stunted and yellow, and no resistance is known. This nematode primarily affects southeastern states north to New Jersey. Control is through fumigation, rotation, and cultivation.

Root knot nematodes (*Meloidogyne* spp.) first feed on the root epidermis, penetrating with their stylet and sucking plant sap. Larvae eventually penetrate the root, taking up a permanent site and feeding on surrounding tissues. Galls develop in affected sites (Figure 6.9), apparently in response to growth substances. The northern root knot nematode can withstand soil freezing; other species cannot. Plants appear stunted and wilted, and the most effective control is resistance.

Lesion nematodes (*Pratylenchus* spp.) enter the root by forcing cells apart or by creating lesions through which soilborne pathogens also may enter. The pathogens then often become the major problem. Lesion

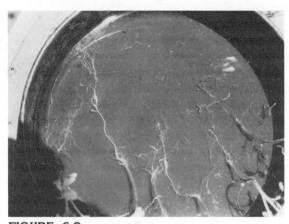

FIGURE 6.9
Root knots of young tomato seedlings caused by nematodes that restrict the function of the root system, causing stunting and yellowing of the plant. (Photo courtesy University of New Hampshire.)

nematodes affect a wide range of plants and must be controlled with soil fumigants.

Cyst nematodes (*Heterodera* spp.) include the golden nematode, serious in potato fields on Long Island. The cyst is the oxidized cuticle of a female nematode which contains up to 500 viable eggs. The nematodes penetrate the root, and larvae become relatively sedentary. Symptoms of cyst nematode do not differ from other nematode problems—stunting, chlorosis, and wilt. Resistance has been incorporated into some potato cultivars.

Weeds

The 10 broadleaf weed and grass species dominating cultivated fields in a range of areas and environments are pictorially described in Figure 6.10 and 6.11. Some are more widely disseminated than others, but all are intense competitors with cultivated crops. Of those pictured, all but quack grass and Johnson grass are annuals. Other prominent weed pests include yellow nutsedge and several species of mustard.

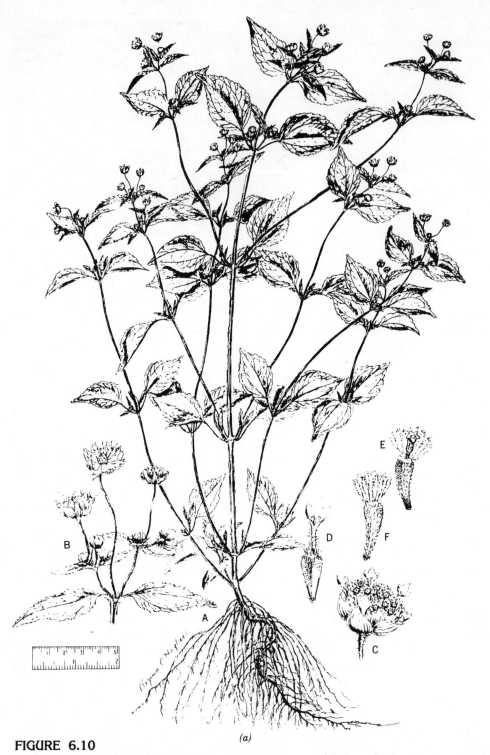

(a)

FIGURE 6.10

Common garden weeds. (a) Galinsoga: A, habit; B, enlarged flowering branch; C, flower head; D, ray flower; E, disk flower; F, achene with pappus. (b) Field bindweed: A, habit; B, rootstock; C, leaf variation; D, flower showing five stamens of unequal length; E, capsule; F, seeds. (c) Common purslane: A, habit; B, flower and capsules; C, flower open; D, seeds. (d) Redroot pigweed: A,

(b)

FIGURE 6.10 (Continued)
habit; B, pistillate spikelet; C, utricle; D, seeds. (e)
Lamb's-quarters: A, habit, small plant; B, floral
spike; C, flower; D, utricle; E, seed. (From *Se-*

lected Weeds of the United States, USDA Agric.
Handb. 366 (1970), pp. 133, 147, 153, 291,
411. Reprinted by permission.)

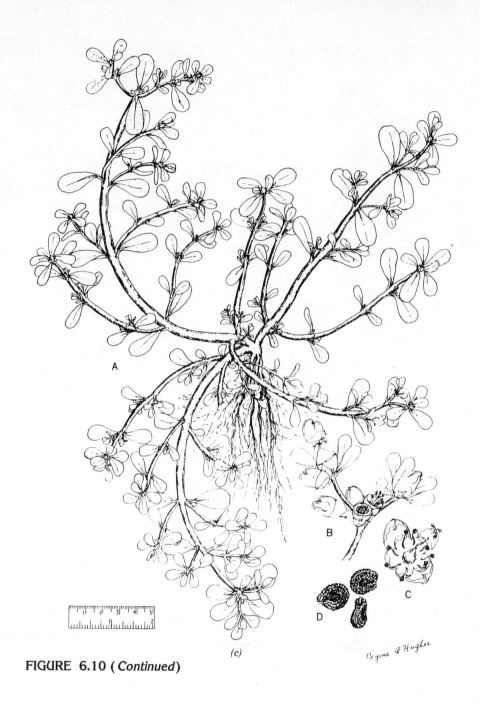

A

B

C

D

(c)

FIGURE 6.10 (*Continued*)

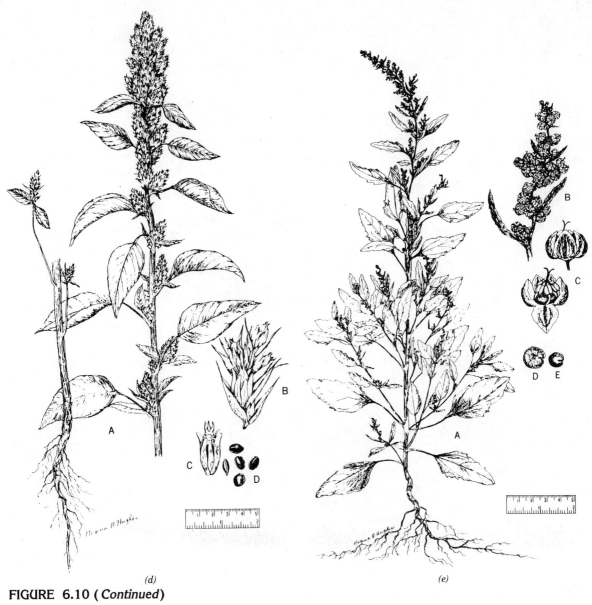

(d)

(e)

FIGURE 6.10 (*Continued*)

(a)

FIGURE 6.11

Common garden weeds. (*a*) Curly dock: A, habit;
B, fruit a, surrounded by persistent calyx; b, showing three valves); C, achene. (*b*) Pennsylvania
smartweed: A, habit; B, spike; C, achenes. (*c*)
Johnson grass: A, habit; B, spikelet; C, ligule; D,
florets; E, caryopsis. (*d*) Foxtail: A, giant foxtail; B,

(b)

FIGURE 6.11 (*Continued*)
green foxtail; C, yellow foxtail (a, habit; b, spikelet; c, ligule; d, caryopsis). (*e*) Quackgrass: A, habit; B, spikelet; C, ligule; D, florets. (From *Selected* *Weeds of the United States*, USDA Agric. Handb. 366 (1970), pp. 35, 83, 87, 125, 131. Reprinted by permission.)

(c)

FIGURE 6.11 (*Continued*)

FIGURE 6.11 (*Continued*)

Regina O Hughes

(e)

FIGURE 6.11 (*Continued*)

Other Pests

In some geographical areas, birds, raccoons, groundhogs, rabbits, deer, and other animals become serious pests, but the methods for control are restricted by lack of selectivity. Poisoned baits generally are not acceptable by the general public. In most instances, physical or chemical repellents are used. Bird repellents include carbide guns or other noise-makers, devices that resemble predator birds, and a chemical repellent (measural) which may be applied to the seed. Chemical treatments to wide crop areas are not labeled nor are they economically feasible. Animals are controlled largely through fencing, where practical, but most frequently through traps, fumigants, or use of firearms.

SELECTED REFERENCES

Anderson, W. P. (1977). *Weed Science: Principles.* West, St. Paul, Minn.

Bottrell, D. R. (1979). *Integrated Pest Management.* Council on Environmental Quality, U.S. Govt. Printing Office, Washington, D.C.

Dixon, G. R. (1981). *Vegetable Crop Diseases.* AVI, Westport, Conn.

Horsfall, J. G., and E. B. Cowley (eds.) (1977). *Plant Disease: An Advanced Treatise.* Academic Press, New York.

Klingman, G. C., F. M. Ashton, and L. J. Noordhoff (1982). *Weed Science: Principles and Practices.* Wiley, New York.

Lockerman, R. H., and A. R. Putnam (1979). Evaluation of allelopathic cucumbers (*Cucumis sativus*) as an aid to weed control. *Weed Science* **27:**54–57.

Metcalf, C. L., W. P. Flint, and R. L. Metcalf (1951). *Destructive and Useful Insects.* McGraw–Hill, New York.

Poe, S. L. (1981). An overview of integrated pest management. *HortScience* **16:**501–507.

Putnam, A. R., and W. B. Duke (1978). Allelopathy in agroecosystems. *Annual Review of Phytopathology* **16:**431–451.

Roberts, H. A. (1968). The changing population of viable weed seeds in a arable soil. *Weed Research* **8:**253–256.

Woodford, E. K., and S. A. Evans (eds.) (1965). *Weed Control Handbook,* Vol. 1. Blackwell, Oxford.

STUDY QUESTIONS

1. Considering all possible costs, which of the general methods of disease control is the least expensive?

2. Why do such cultural methods as rotation, trickle irrigation, high-density cropping, and cultivation influence the effectiveness of chemical pest control?

3. Describe the role of synergists and adjuvants in chemical pest control.

4. Are allelopathic interactions in weed–host populations most damaging to the crop or to the vigor of the weeds? Cite examples of both.

5. Is it possible to use nonselective herbicides in actively growing crops? Explain.

7

Alternative Production Systems

The purpose of this chapter is to describe two very different production approaches—organic and hydroponic—and, in so doing, underscore the range of possible techniques by which vegetables can be produced. In market volume, neither of these systems competes with conventional field production. However, there are circumstances within which each system fulfills a need, with economic gain to the producer.

ORGANIC CULTURE

In a natural balanced ecology, diverse arrays of plant and animal species coexist, each represented by substantial genetic diversity. Within such a mixed population, failure of one ecotype because of an environmental change or susceptibility to a disease or insect pest seldom was critical, for individuals relatively unaffected by these factors increased, providing population resiliency.

As competing plant species were eliminated by crop management systems and production moved toward monoculture, insects and pathogenic fungi and bacteria, freed of some of their natural competitors and predators and provided with a concentrated food source, increased to become serious plant pests. To decrease losses within monocultures threatened by increasing disease or insect pressure, materials were developed that eliminated, in a rapid and cost-effective manner, many major crop pests.

Most of this evolution has occurred since the early 1940s, accelerated by a transition from small, diversified crop–livestock farms to highly specialized endeavors concentrated within those geographical areas providing the best environment or the most favorable economics for profitable production. Crop production in most instances became separated from the livestock industry.

These agricultural changes accelerated development of today's highly mechanized monoculture and increased grower reliance on agrochemicals to support that system. Today's agriculture is, without question, efficient, but it does not lack criticism. The criticism, of which the following statements are examples, has some merit, but the arguments, pro and con, occasionally lack objectivity or scientific evidence.

1. Soils are not being managed for long-term productivity; organic matter levels therefore have declined, increasing the need for agrochemicals.
2. Surface runoff from areas receiving commercial fertilizer contributes to reduced quality of surface and groundwater supplies.
3. A variety of pesticides are used on plants and soils without, in some instances, full understanding of the long-term effects on the soil and on other components of the environment.
4. Continued heavy use of pesticides does not eliminate a problem; it provides an artificial screen by which highly resistant pest forms succeed and increase.
5. Materials applied to plants and produce are not necessarily without health consequences.

The demand for low-cost food, the cost of alternative methods of production on a large scale, and the expectation by consumers that produce will be free of pest damage are not unimportant factors that will encourage conventional production systems.

Although most food crops will continue to be produced conventionally, some consumers and producers are attracted philosophically to natural systems (organic farming or gardening) for crop production and protection. The techniques used to produce a crop organically are drawn not only from biological relationships that support ecological stability within an area, but also from the cultural techniques contributing to successful monoculture.

The kinds of organic production systems range from those utilizing no synthesized product (pesticides, fertilizers, plastics) to those in which pesticides of low toxicity and supplements of commercial fertilizer occasionally may be applied.

Soil Management

The fundamental focus of organic crop production is the improvement of soil structure, fertility, and productivity through incorporation of organic matter. Most soil currently in conventional production is lower in organic matter than desirable. To bring these soils into organic production, substantial applications of organic matter over a period of 4 to 5 years are required. The establishment of an optimum organic level is considered critical for ensuring vigorous plant growth and, because of that vigor, tolerance to disease and insect pests.

Organic matter may be supplied as animal manure, compost, peat, sewage sludge, and/ or green manure. Animal manures and compost provide the greatest fraction of slowly decomposing lignins and, as a consequence, the greatest contribution to soil humus.

Manure normally is applied at 10 to 20 tons/acre (22 to 40 MT/ha) each year, then tilled in to speed decomposition. Rotted manure is preferred for spring applications; fresh manure should be applied in the fall to allow time for breakdown and reduction of ammonia content.

Compost is used by many organic gardeners as the only available organic matter source. Compost may include grass, leaves, straw, stems or other plant residue, and fruit or vegetable peelings, but should not have grease, meat scraps, or plant material infected with diseases that might infect crops eventually grown on a treated site. The materials to be composted are layered in a pile or trench with lime or other calcium source (wood ashes or sea shells) and a nitrogen source (manure or fertilizer) and kept moist (Figure 7.1). Early decomposition will involve mesophyllic forms of aerobic bacteria and fungi [optimum temperature for activity, 77 to 86°F (25 to 30°C)]. As decay proceeds, heat is evolved, and the temperature within the compost will increase

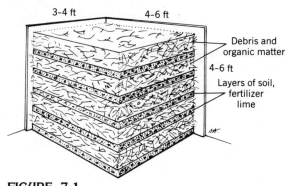

FIGURE 7.1
Diagram of compost pile construction showing layering of masses. In addition to organic matter, nitrogen and calcium (energy) sources are needed for microorganism activity.

to 140 to 148°F (60 to 65°C), optimum for thermophilic bacteria, fungi, and actinomycetes. During this period, most pathogenic bacteria and some fungi and viruses are killed. Since the maximum temperature occurs in the center of the pile, and aeration is necessary for activity of these microorganisms, the pile should be restacked periodically, turning the outer surface of the pile inward and eliminating the matting and compression of organic material. Failure to provide aeration will result in fermentation, recognizable by objectionable odors and incomplete decomposition.

Peat, because of its expense, is not used frequently for field-grown crops. It is useful in greenhouse production, but peat from some sources may be highly acidic, requiring careful liming.

Sewage sludge constitutes only 0.5 percent of the total organic waste produced in the United States (compared to approximately 22 percent for animal manure). The source of sludge is an important factor in considering possible agricultural use. Industrial wastes often include heavy metals which can be taken up by some plants. Sludge has been used in digested form (given primary treatment by anaerobic digestion but not heat treated), acti-

vated form (treated, inoculated with microorganisms, aerated, and heat treated), or as a compost. When mixed with a bulking material—wood chips, corn cobs, bark, etc.—and provided with forced aeration, sludge will compost within several weeks. The resulting material has been found to be an excellent soil amendment, although its use on food crops is restricted by federal regulation.

Manure, peat, compost, and composted sewage sludge are considered to be soil amendments, not fertilizers, although each provides small amounts of nitrogen, phosphorus, and potassium (Table 7.1). Heavy applications of manures or other organic amendments may provide sufficient levels of nitrogen to sustain plant growth, but phosphorus and potassium levels from this source normally are insufficient. The availability of each constituent is governed by the rate of decomposition of organic particles, and, in cold soils, the rate slows. Further, as discussed in Chapter 3, decomposition temporarily increases demand for nitrogen by soil microorganisms until the carbon:nitrogen ratio decreases to approximately 15. Other fertilizer sources often must be used to supplement nutrients released by manures and composts and to ensure sufficient nitrogen during decomposition. A few of these sources, such as dried blood, provide nitrogen more quickly than decomposing manure (Table 7.1).

Compared to commercial fertilizer ratios, the nutrient constituents of most organic sources or natural deposits are low and slowly available. For example, 5 to 10 tons (4.5 to 9 MT) of granite dust would be required to obtain the same plant response provided by 300 lb (136 kg) of muriate of potash (60 percent K_2O). Over several years, as rock sources of phosphorus and potassium are added to the soil and as the cation exchange offered by organic particles increases, exchangeable levels of these elements may improve somewhat.

Soil pH can be adjusted using ground limestone (calcium or magnesium) or other calcium sources (Table 7.2). Ground lime-

TABLE 7.1
Constituents[a] of natural deposits and organic by-products used for crop production

Source and primary use	Nutrient (%)			Availability[b]
	N	P_2O_5	K_2O	
Soil amendments				
Animal manure[c]				
Cattle	0.25	0.15	0.25	+
Horse	0.30	0.15	0.25	+
Sheep	0.60	0.33	0.75	+
Swine	0.30	0.30	0.30	+
Poultry				
75% water	1.50	1.00	0.50	+ +
50% water	2.00	2.00	1.00	+ +
30% water	3.00	2.50	1.50	+ +
15% water	6.00	4.00	3.00	+ +
Compost, garden	1.50–3.50	0.50–1.00	1.00–2.00	– –
Compost, mushroom	0.40–0.70	57.00–62.00	0.50–1.50	– –
Kelp[d]	0.90	0.50	4.00–13.00	– –
Peat	1.50–3.00	0.25–0.50	0.50–1.00	– – –
Sewage sludge				
Activated, dry	2.00–6.00	3.00–7.00	0.00–1.00	+
Digested	1.00–3.00	0.50–4.00	0.00–0.50	– –
Fertilizers				
Animal tankage, dry	7.00	10.00	0.50	+
Bone meal, raw	2.00–6.00	15.00–27.00	0.00	– –
Bone meal, steamed	0.70–4.00	18.00–34.00	0.00	–
Corn cob ash	0.00	—	4.00–8.00	+ + +
Cottonseed meal	6.00	2.50	1.70	–
Dried blood	12.00	1.50	0.57	+ +
Fish meal	10.00	4.00	0.00	– –
Fish scrap	3.50–12.00	1.00–12.00	0.08–1.60	– –
Granite dust	0.00	0.00	3.00–5.00	– – –
Greensand	0.00	1.35	4.00–9.50	– – –
Guano, bat	5.70	8.60	2.00	+
Guano, Peru	12.50	11.20	2.40	+
Milorganite	5.00	2.00–5.00	2.00	+
Rock phosphate	0.00	20.00–32.00	0.00	– – –
Sodium nitrate	16.00	0.00	0.00	+ + +
Soybean meal	6.70	1.60	2.30	–
Tobacco stems	2.00	0.70	6.00	– –
Composts, mulch				
Alfalfa hay	2.45	0.50	2.10	– – –
Alfalfa straw	1.50	0.30	1.50	– – –
Grass hay	1.20	0.35	1.75	– – –
Clover hay	2.10	0.50	2.00	– – –
Pea vines (green)	2.08	0.60	2.00	– – –
Oats (green)	1.50	0.65	2.20	– – –

128

TABLE 7.1 (*Continued*)
Constituents[a] of natural deposits and organic by-products used for crop production

Source and primary use	Nutrient (%)			Availability[b]
	N	P₂O₅	K₂O	

Source and primary use	N	P$_2$O$_5$	K$_2$O	Availability[b]
Rye (green)	2.00	0.80	2.80	- - -
Sawdust	4.00	2.00	4.00	- - -
Wheat (green)	2.14	0.20	2.48	- - -
Wheat straw	0.50	0.15	0.60	- - -

Source: Extension Agent's Guide to Organic Gardening, Culture and Soil Management, Pennsylvania State University, University Park (1972); and *Organic Gardening,* Washington State University Coop. Ext. Bull. 0648 (1982).

[a] Constituents are variable, depending on source.

[b] +, ++, +++: moderately available, rapidly available, very rapidly available; −, −−, −−−: moderately to slowly available, slowly available, very slowly available.

[c] Variable, depending on amount of litter, leaching, and storage.

[d] Contains chlorides, carbonates, and sulfates of sodium.

stone and shells are relatively slow acting; wood ashes provide a rapid pH adjustment. Repeated use or excessive application rates of wood ashes have increased pH levels in home gardens from 5.2 to 8.0 within several years. Soils should be tested prior to lime or ash application.

TABLE 7.2
Materials utilized in neutralizing acidic soils

Source	Calcium as CaO (%)
Clam shells	50
Ground shell marl	35–42
Oyster shells	43–50
Wood ashes[a]	32
Limestone	
Dolomitic	35 (+ 15% or more MgO)
Calcitic	45–50

Source: From Extension Agent's Guide to Organic Gardening, Culture and Soil Management, Pennsylvania State University, University Park (1972). Reprinted by permission of the authors.

[a] Depends on tree species.

Where land is available, most organic rotations involve a sod, usually legume based, and green manures or cover crops are integral to the maintenance of soil fertility and physical condition. Following application of organic matter, the cover sod is tilled in, usually with shallow rototilling or disking, seldom turning the soil with a moldboard plow. By avoiding placement of organic matter deep in the soil, decomposition is speeded, and nutrients become available more rapidly for plant uptake. Because of the increased cation exchange capacity afforded by the organic matter, the nutrients contained in shallow soil layers are lost by leaching more slowly than those at the base of a plow layer.

Earthworm populations are enhanced in soils enriched with organic matter and can accomplish significant tillage action by building permanent channels and transporting organic matter throughout a variable depth of soil. The night crawler, the manure worm, and the common field worm are the principal species found in most soils, particularly those that receive regular treatments of organic matter or are mulched with organic materials. The night crawler is the most effective of the species in enhancing water penetration in heavy soils, particularly if pH is maintained above 5.5. Most species prefer loamy soils of moderate pH.

Mulching

Surface mulches for organic crop production may be organic or manufactured of paper, plastic, or aluminum. Organic materials provide some nutrient value as they decompose (Table 7.1) and may be incorporated into the soil following harvest. Most organic mulches consist of rather coarse stemmy material, high in lignins and cellulose, low in nitrogen. As decomposition at the mulch–soil interface occurs, the demand for nitrogen increases, often resulting in nitrogen deficiency symptoms in mulched crops. Supplemental nitrogen may be required when organic mulch is used.

Plastic and paper mulches are used primarily for weed control and for modifying soil temperature and moisture. Some plastics are photobiodegradable, breaking apart after several weeks or months of solar exposure. Buried edges of such materials, however, do not decompose.

Pest Control

The most troublesome aspect of organic farming or gardening is control of disease and insect pests. Without recourse to synthesized chemicals, cultural techniques and biological control systems must be applied. Such control measures often are variable in effectiveness, depending on environment, pest population, and site.

Cultural methods include sanitation, rotation, cultivation, adjusting planting date, use of clean seed and sterile seeding and transplanting medium, application of mulch and row covers, spacing and vertical training to increase aeration, and avoidance of cultural operations while foliage is moist. All of these measures are equally applicable to conventional farming. In addition, mixed plantings (intercropping) may reduce disease and insect intensity by reducing the concentration of host tissue.

The most important approach for enhancing disease and insect control is through soil building. It is known that plants stressed by poor soil conditions are affected more by insects and diseases than those growing under optimum conditions. Under heavy pressure from insects or diseases, however, even healthy plants will be damaged.

Biological methods constitute an important approach toward pest control for organic producers, but one for which the limits are not always appreciated. The techniques are based on genetic resistance, allelopathy, and other chemically based interactions among plants, and on color perception, biological attractants, and predation among insects.

Genetic resistance is as useful to the organic farmer as to the large commercial grower. Many cultivars that include resistance to one or more diseases or nematodes or show reduced insect feeding are suitable for either commercial or home garden purposes. Resistance is the least expensive of all control measures.

Allelopathic inhibition of weed seed germination has been discussed. In addition, some plants are known to repel certain pests and attract others (Table 7.3). The isothiocyanate from cruciferous crops has been reported to attract diamondback moth (*Plutella xylostella*) but to reduce emergence of cyst nematodes (*Heterodera* spp.) from infested roots. Re-

TABLE 7.3
Several companion plantings and target pests

Companion planting	Target insect
Beans and rosemary	Mexican bean beetle (*Epilachna varivestis*)
Cabbage and thyme	Imported cabbage worm (*Pieris rapae*)
Carrot and bulb crops	Carrot rust fly (*Psila rosae*)
Cucumber and radish	Cucumber beetle (*Acalymma vittatum, Diabrotica undecimpunctata*)
Cole crops and mint	General insect control
Eggplant and catnip	Flea beetle (*Epitrix fuscula*)
Tomato and basil	Tomato hornworm (*Manduca quinquemaculata*)
Bean and French marigold	Mexican bean beetle (*Epilachna varivestis*)

peated cropping of potato in one area enhances soil populations of *Bacillus subtilis,* a bacterium antagonistic to *Streptomyces scabies,* the causal organism of potato scab. Companion plantings of marigold and beans have reduced populations of Mexican bean beetle (*Epilachna varivestis*) but also have reduced bean growth, because of either allelopathy or competition. In some instances, extracts of plants have been sprayed on crops to deter insect feeding (thyme or sage to deter leafhopper and cabbage worm infestations on beans and crucifers, respectively).

The effectiveness of trap crops is based on their chemical constituents. Black nightshade (*Solanum nigrum*) and eggplant (*Solanum melongena*), both very attractive for Colorado potato beetle oviposition, may be used to deflect initial insect infestations from potato or tomato. On nightshade, neither beetle larvae nor adults feed on foliage. Without an alternate food source nearby, the insects die. Eggplant foliage is consumed voraciously, and once these plants are infested with the beetle, they and the insects should be destroyed. Otherwise, the lack of food will force insect migration to the main crop.

Color perception is the basis for success of certain reflective mulches as deterrents to aphid feeding. Several insects also are attracted by one of several yellow hues. Traps made of yellow plastic and coated with tanglefoot can be suspended throughout the garden to reduce populations of aphids and other insects.

Biological attractants, usually sex attractant pheromones, are effective in luring insects of the species from which the attractant was obtained. Other insects are unaffected. Such traps are more useful in determining the extent of an insect infestation than in providing control.

Insect predators, primarily lady beetle and praying mantids, are used for controlling aphids and several other insect pests. The lady beetle, released within a field, will migrate to the best source of food. If aphid infestations within that field are light, beetle migration will deplete the introduced population, and aphids will not be controlled to any extent. Under heavy infestations, the beetles will provide some control. However, many of the aphids transmit disease as they first penetrate the plant tissue, and control by predation seldom can prevent disease transmission.

Mantids are introduced as egg masses. As the adults emerge, they cannibalize their own species unless there is an abundant food source nearby. Thus, the mantid population generally is reduced substantially soon after emergence.

Chemical Controls

Several natural plant products have been developed as insecticides; none has proven effective against disease pathogens, except in controlling those insects that transmit disease.

Nicotine sulfate is a botanical insecticide, moderately toxic with a wide range of activity. It degrades rapidly after application and offers no residual problem on sprayed produce.

Pyrethrin (refer to Chapter 6) is a botanical insecticide derived from several *Chrysanthemum* species. Pyrethrins are contact materials that degrade rapidly after application and have been common ingredients in household insecticides due to their low mammalian toxicity.

Rotenone is a moderately toxic botanical insecticide and extremely toxic to fish. It is derived from roots of *Derris* spp. and *Lonchocarpus* spp. and is very effective against a range of beetles.

Ryania is a botanical derived from roots and stems of *Ryania speciosa.* The active ingredient, an alkaloid, is effective against some caterpillars. Although remaining active longer than other botanicals, it does not seem to affect predator insects. Its use on vegetables has not been labeled.

Sabadilla, derived from seeds of species of *Schoenocaulon,* a lily of Central and South America, lacks persistence and appears to have some toxicity to bees. Although of low mammalian toxicity, it is irritating to eyes and throat.

Few diseases can be controlled with natural materials. Both sulfur and lime have been used in certain situations to control fungal diseases. Sulfur is particularly effective against powdery mildew in some plant species, but it also may be phytotoxic.

COMPARISONS OF ORGANIC AND CONVENTIONAL SYSTEMS

It is difficult, if not impossible, to compare directly conventional agriculture with organic production methods. A soil farmed conventionally for a period of time must be restored organically for 4 or 5 years before it is representative of an organic system. Furthermore, pest control in adjacent organic and conventional test plots presents a dilemma. Most studies comparing the two systems in one location probably are flawed; claims of superiority of one system over another based on results from different tests or areas may reflect far more than cultural system differences.

In general, the following statements are supportable.

1. Fertilizers influence mineral composition of a plant (Table 7.4); however, there are no totally objective data indicating that organic metabolites synthesized within the plant differ because of mineral composition or source of the minerals.

2. Equivalent yields can be attained by either system assuming equivalent nutrient input and availability. Declines in organically produced yields occur immediately after transition to an organic system, but yields recover in time. The average yield obtainable in an organic system, however, does not approach the average for conventional production.

3. Because of increased use of rotations and cover crops, erosion control probably is superior within an organic system.

TABLE 7.4
Yield and composition of two spinach cultivars fertilized with organic and inorganic fertilizer

Fertilizer	Fresh weight (g/pot)		Nitrate-N in leaf (%)	
	Hybrid 424	America	Hybrid 425	America
None	13 a[a]	18 a	0.03 a	0.04 a
Dried blood	31 cd	26 a	0.27 a	0.47 c
Castor pomace	34 cd	22 a	0.36 a	0.51 c
Cottonseed meal	37 d	26 a	—	—
Sewage sludge	32 cd	22 a	0.28 a	0.58 c
Cow manure (dry)	21 ab	18 a	0.15 a	0.19 b
Ammonium nitrate	26 b	21 a	0.20 a	0.62 c

Source: From A. V. Barker (1975) *HortScience* **10**:51. Reprinted by permission of the author.

[a] Means followed by the same letter do not differ. Mean separation by Duncan's multiple range test, $P = .01$.

4. Labor intensity is greater on organic farms than on conventional farms.

5. Costs for fertilizers and pesticides, which constitute about 33 percent of the total energy costs for conventional farming, can be eliminated in an organic system. However, these costs may be offset by increased labor and mechanical equipment use. The net returns have been reported to be the same for some crops, mostly agronomic.

6. Superior quality attributes have been reported for organically grown produce; however, such claims have not been substantiated. Some tests have shown superiority in organic culture only for one cultivar, while another is unaffected (Table 7.4). The structural superiority of an organic soil conceivably could enhance textural quality of root and stem crops by maintaining uniform soil moisture and nutrient levels. However, the frequency of surface blemishes from insect feeding and/or pathogen infection also will increase, reducing marketable yield or acceptable quality.

7. Crops sprayed with agrochemicals are more

likely to have chemical residues than those grown organically. The tolerances imposed by federal and state authorities and the registration procedures for existing and new chemicals provide a substantial safety margin for human consumption, and there is no evidence that suggests any relationship between allowable spray residues and human health.

HYDROPONIC SYSTEMS

At the opposite end of the spectrum are systems that provide essential elements, water, and oxygen to plants growing in the absence of soil (hydroponics). Hydroponic systems have been classified as liquid (plants growing without a supporting medium) and aggregate (roots anchored by an inert medium, generally gravel, sand, vermiculite, rockwool, or other materials). Within each category, the system may be open, in which the nutrient solution is not recirculated, or closed, reusing water and nutrients through a drainage–recirculation mechanism. Although not integral to hydroponic production, controlled environments normally are used.

Prior to the use of plastics, hydroponic systems generally were not competitive because of engineering inadequacies and costs and because of the difficulty in maintaining uniformity of the root environment throughout the life of the plants. Although costs remain high, and only a few vegetable crops can be produced profitably within hydroponic systems, the capabilities for controlling plant growth through environmental management have improved greatly.

Liquid Systems

In recent years, the nutrient film technique (NFT) has been adopted successfully for producing several vegetable and ornamental crops. The basic NFT system involves recirculating a shallow film of nutrient solution (Table 7.5) over bare roots underlaid with plastic. Movement of the nutrient solution is provided

TABLE 7.5

Nutrient concentrations and chemicals for tomatoes in NFT

Element	Desirable concentration (ppm)	Chemical
Nitrate nitrogen	150–200	KNO_3, NH_4NO_3, $Ca(NO_3)_2$
Ammonium nitrogen	0–20	NH_4NO_3, $(NH_4)_2SO_4$
Potassium	300–500	KNO_3, K_2SO_4, KH_2PO_4
Phosphorus	50	KH_2PO_4, NaH_2PO_4, $CaHPO_4$
Calcium	150–300	$Ca(NO_3)_2$, $CaSO_4$, $CaHPO_4$
Magnesium	50	$MgSO_4$, $Mg(NO_3)_2$
Iron	3[a]	FeEDTA, FeEDDHA
Manganese	1	$MnSO_4$
Copper	0.1	$CuSO_4$
Zinc	0.1	$ZnSO_4$
Boron	0.3–0.5	H_3BO_3
Molybdenum	0.05	$(NH_4)_6MO_7O_{24}$
Sodium	250[b]	
Chlorine	200[b]	

Source: From C. J. Graves, The Nutrient Film Technique. *Horticultural Reviews* **5**:1–44. Copyright © 1983 by the AVI Publishing Company, Westport, Conn. 06881.

[a] 5 to 10 ppm is preferable during the early stages of growth before the start of fruit picking.
[b] Maximum amount.

by gravity flow through sloping gullies within which the plant lower stem and roots are placed.

The basic system is diagrammed in Figure 7.2. Gullies, preferably no longer than 65 ft (20 m), may be formed from soil graded to a 1.5 to 2 percent slope, constructed of concrete (which seems to contribute temperature stability and minimize contamination) or suspended from an elevated superstructure. The gullies are lined with a heavy-mil plastic, usually black with a white upper surface (for solar reflection).

Young plants contained in inert, sterile cubes are placed in the gullies on capillary mats to ensure contact with the stream of nutrient solution. As the roots emerge from the cube, they form a solid mat, only partially sub-

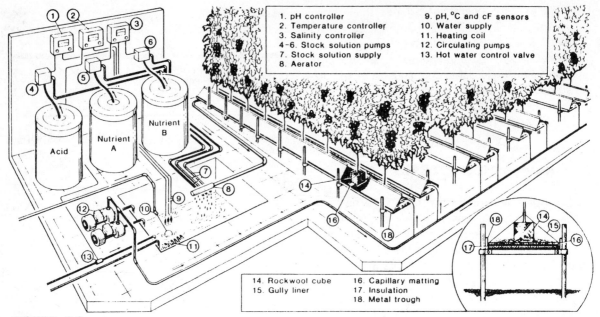

FIGURE 7.2
Features of a nutrient film system. (From C. J. Graves, The Nutrient Film Technique. *Horticultural Reviews* 5:1–44. Copyright © 1983 by the AVI Publishing Company, Westport, Conn. 06881.)

1. pH controller
2. Temperature controller
3. Salinity controller
4–6. Stock solution pumps
7. Stock solution supply
8. Aerator
9. pH, °C and cF sensors
10. Water supply
11. Heating coil
12. Circulating pumps
13. Hot water control valve
14. Rockwool cube
15. Gully liner
16. Capillary matting
17. Insulation
18. Metal trough

merged. Generally, twin rows are placed between walkways (Figure 7.3).

There are several variations of NFT, including the use of flat plastic tubes in place of gullies, solid pipes to carry solution and plants, or moving belts. A floating system (Figure 7.4), by which the plants are placed in holes of polystyrene sheeting floating on a nutrient solution, and aeroponics (Figure 7.5), using the nutrient solution to mist roots suspended in a darkened tube or channel, have been shown to be technologically feasible.

The advantages of NFT are careful control of root environment, simplicity of watering, uniformity of fertilization which can be adjusted as the plants develop, possibility of metering in chemicals for controlling pathogens, and rapid crop turnaround. Disadvantages include need for a high level of technical skill and understanding of chemistry, high capital cost, and risk of imbalance of nutrients and consequent crop growth problems.

FIGURE 7.3
Tomato cropping using nutrient film technique. (Photo courtesy of Merle Jensen, University of Arizona, Tucson.)

FIGURE 7.4
Floating hydroponic system: long lines of floats with lettuce plants set in holes are moved easily on nutrient pools. (Photo courtesy of Merle Jensen, University of Arizona, Tucson)

Aggregate Systems

A gravel or sand (Figure 7.6) rooting medium may be a relatively inexpensive alternative to NFT. The nutrient solution is metered to the plants by trickle irrigation and the effluent is collected by way of sloping drainage tubes underlaid with plastic beneath the sand and returned to a reservoir for recycling. The aggregates provide anchoring and aeration for roots, and the engineering requirements are minimal. Popular alternatives are rockwool cubes (Figure 7.7) and bag culture, an open system in which nutrient solution is applied to the peat-lite mixture contained in plastic bags.

Management

The nutrient solution in any hydroponic system, particularly a closed system, must be monitored constantly. Because the aqueous solution is poorly buffered, pH may change rapidly, usually increasing as plants remove nutrients. As the pH changes, nutrient uptake by plants can be affected. Below pH 5, cation uptake is suppressed; above pH 7, anion uptake is reduced. The balance of nutrients also will change. In general, levels of the macronutrients should be tested every 2 to 3 weeks,

microelements each 5 to 6 weeks. The salinity should be maintained at 2500 to 3500 μmho; however, for crops such as tomato, the conductivity may be increased to 9000 μmho at time of first flower to promote fruit production, later reducing the salts to the recommended optimum.

Oxygen is essential for proper root development and plant growth. The nutrient solution itself does not have sufficient oxygen to support the density of plants normally grown in one trough. However, the partially submerged roots and the solution film derive sufficient oxygen from the atmosphere to sustain plant growth.

FIGURE 7.5
Aeroponic system: plants suspended with roots maintained in high humidity, receiving nutrients through misting. (Photo courtesy of Merle Jensen, University of Arizona, Tucson.)

FIGURE 7.6
Development of a sand culture system. Perforated drainage pipes over plastic are covered with sand to a depth of 1 ft (30 cm). Nutrient solution applied by a trickle system is returned through the drainage pipe for recycling. (Photo courtesy of Merle Jensen, University of Arizona, Tucson.)

Applications

The major applications for NFT have been in tomato and lettuce production. In The Netherlands, lettuce has been intercropped in the walkway between twin rows of tomato. Before tomatoes begin to ripen, the lettuce has been harvested.

FIGURE 7.7
Plants in small rockwool cubes set on rockwool slabs in holes cut in plastic. Each plant receives drip irrigation with nutrient solution. (Photo courtesy of Merle Jensen, University of Arizona, Tucson.)

NFT has been economically feasible when used within a greenhouse, so that supplemental lighting is not required. In other facilities in which lighting is required, the cost for the electricity generally approaches 50 percent of the total operating budget.

SELECTED REFERENCES

Antonelli, A., R. Byther, A. Halvorson, R. Thornton, and B. Barritt (1982). *Organic Gardening,* Washington State University Coop. Ext. Bull. 0648.

Dindal, D. L. (1981). *Ecology of Compost,* University of California Coop. Ext. Leaflet 21200.

Fletcher, R. F., P. A. Ferretti, R. W. Hepler, A. A. MacNab, and S. G. Gesell (1972). *Extension agent's guide to Organic Gardening, Culture and Soil Management.* Pennsylvania State University, University Park.

Graves, C. J. (1983). The nutrient film technique. *Horticultural Reviews* **5:**1–44.

Jensen, M. H., and W. L. Collins (1985). Hydroponic vegetable production. *Horticultural Reviews* **7:**483–558.

Parr, J. F., and G. B. Wilson (1980). Recycling organic wastes to improve soil productivity. *HortScience* **15:**162–166.

Topoleski, L. D. (1981). *Growing Vegetables Organically,* New York State College of Agriculture and Life Sciences, Cornell University, Coop. Ext. Inf. Bull. 39.

William, R. D. (1981). Complementary interactions between weeds, weed control practices and pests in horticultural cropping systems. *HortScience* **16:**508–512.

STUDY QUESTIONS

1. Devise several alternative schemes by which diseases and insects could be controlled organically.

2. Given an adequate reservoir of plant nutrients in the soil, why is it unlikely that the source—organic or inorganic—does not affect plant constituents or crop quality?

3. What are the major differences in root environment between NFT- and soil-cultured plants and how do these differences affect management?

8

Quality Control and Marketing

Quality is a nebulous term; it implies different dominant features in different vegetables, and many of the separate attributes that contribute to quality are perceived quite differently by different people.

Kramer and Twigg (1966) have classified those attributes that contribute to food quality as follows[1]:

Sensory Attributes
Sight—appearance of the
  product
    Color
    Gloss
    Viscosity
    Slze and shape
    Defects
Touch—kinesthetic (texture)
    Hand or finger feel
    Mouth feel
Smell and taste—flavor
    Odor
    Taste

Hidden Attributes
Nutritive value
Harmless adulterants
Toxicity

QUALITY CONTROL

Constituents of Quality

The delivery of high-quality fresh or processed vegetables to the consumer is the ultimate goal in marketing. Quality of the raw product is attained largely during plant growth, the product of cultivar, environment, and management practices. The extent to which this quality is preserved for the consumer relates to techniques of postharvest handling and processing.

Both **color** and **gloss** are products of light. When light strikes an object, some is reflected, some absorbed, some transmitted, and some refracted. When all light is reflected, the object appears white; total absorption results in black

[1] From A. A. Kramer and B. A. Twigg, *Fundamentals of Quality Control for the Food Industry.* Copyright © 1966 by The AVI Publishing Company Westport, Conn. 06881. Reprinted by permission.

color. When light is absorbed partially at all wavelengths, the object appears gray. When light is absorbed more at one wavelength than at others, a color (hue) other than black or white will be perceived. The plant pigments controlling base color of vegetables or their products include flavonoids (red anthocyanins and yellow flavones), chlorophylls (greens), and carotenoids (yellow to orange). Each specific pigment absorbs light at a different wavelength, thereby contributing characteristic color.

The amount of reflection at a given wavelength determines the intensity of color (chroma); a glossy appearance results from directional reflection of light, whereas a dull finish results from reflection of light evenly at all angles.

Both color and gloss, although inherent properties of a vegetable, also can be affected by the cleanliness of the product (dirt, spray residues) and by surface moisture. A film of moisture on a green leaf reflects light directionally, giving the object a bright, glossy appearance.

Viscosity, a property normally associated with appearance of prepared foods, such as catsup or pureed baby foods, is perceived by both sight and feel. It is a measure of resistance to flow, the opposite of fluidity. Such terms as consistency, thickness, and wateriness may refer to the apparent viscosity of a product but occasionally are used also to describe texture.

Size and **shape** are attributes important in marketing because they elicit from the consumer early impressions of quality. Size is a standard criterion for grading and packaging produce. Uniformity of size and shape contributes to attractiveness and sales appeal. Shape, for the most part, should be symmetrical and characteristic of the species and cultivar.

Defects may be genetic or physiological, or due to insects or disease, mechanical damage, or extraneous material. Defects inherent in a crop or due to interaction with specific environmental factors may relate to structure (fasciation of asparagus spears, blossom scar or lobing of tomato), color (appearance of yellow kernels in white corn or chimeras on leaves) or degree of development of plant tissues or organs. The latter might include such defects as blotchy ripening in tomato or fibrous flesh in sweet potatoes. Surface blemishes, insect stings, necrotic areas, and tissue deformed by insect feeding or virus infection account for a major portion of all defects. Mechanical defects include cuts, bruises, and discoloration caused during harvest and produce handling. Extraneous material can be a quality problem in both raw product and processed containers. For example, the appearance of a crate of snap beans is diminished by inclusion of stem and leaf fragments, and similar materials may reduce the price paid for raw products for processing.

Texture is described by the interaction of

that group of physical characteristics that are sensed by the feeling of touch, are related to the deformation, disintegration and flow of the food under application of a force, and are measured objectively by functions of force, time and distance.[2]

Texture can be sensed by finger or hand pressure, which helps to form an opinion of a product's firmness, softness, and/or juiciness. Of more concern as a quality constituent is the texture sensed by the mouth—chewiness, fibrousness, grittiness, mealiness, stickiness, oiliness, and dryness. The structural factors which ultimately affect texture include those related to cell constituents—starch, enzymes, and phytin—and those associated with the cell wall—polysaccharides and lignin. The swelling properties of starch granules, for example, contribute to flouriness or mealiness in potato. Lignin contributes to fibrousness or chewiness in asparagus, celery, and other crops. Calcium pectate, formed by the activity

[2] M. C. Bourne (1980) *HortScience* **15**:51–57.

TABLE 8.1
Volatile compounds of celery

Formaldehyde	Carvone
Acetaldehyde	Diacetyl
Propionaldehyde	
Hexanol	Ethyl isovalerate
Heptanol	cis-3-Hexen-1-yl pyruvate
Octanol	Decyl acetate
Undecanal	Linalyl acetate
Dodecanal	Terpinyl acetate
Neral	Geranyl acetate
Citronellal	Citronellal acetate
	Neryl acetate
Isoamyl alcohol	Carvyl acetate
Hexanol	Terpinyl acetate
Heptanol	Geranyl butyrate
	Benzoyl benzoate
n-Valeric acid	
Isobutyric acid	D-Limonene
Pyruvic acid	Myrcene
3-Isobutylidene-3a,4-dihydrophthalide	
3-Isovalidene-3a,4-dihydrophthalide	
3-Isobutylidene phthalide	
3-Isovalidene phthalide	
Sedanonic anhydride	

Source: From H. J. Gold and C. W. Wilson III, *Journal of Food Science* **28**(4): 487. Copyright © 1963 by Institute of Food Technologists.

of pectinmethylesterase, contributes firmness to tomato fruit.

Texture not only contributes to one's impression of quality, but also may be correlated with some flavor constituents. For example, texture as measured by a tenderometer is highly correlated with content of alcohol-insoluble solids (sugars) in peas and sweet corn, and the starchiness of potato is highly correlated with a mealy texture.

Flavor is one of the most important constituents of quality but has little effect on initial sale of a product. It is sensed differently by different individuals, and impressions often are affected by other nonflavor quality attributes (color, texture) and by other flavors sampled at the same time.

Flavor is a composite of taste and odor. Taste is four dimensional, distinguishing sweet, sour, salty, and bitter principles. Odor is more important, a major contributor of flavor for many vegetables. Odor relates to a myriad of volatile compounds (Tables 8.1 and 8.2) specific for each crop. Some of the compounds, as in onion, develop enzymatically only after tissues are ruptured. Others may develop only during heat processing.

Hidden attributes cannot be perceived in most instances, except by inference from sensory attributes. They are, however, important to health. Insofar as color (yellow and green vegetables) is an indicator of vitamin A content, consumers may choose the most intense color and, by association, increase food value. For the most part, however, there is little purposeful discrimination among vegetables for their food value. Toxicants occur in certain crops, and adulterants may be used in preparation foods. Natural toxicants (Table 8.3) seldom pose health problems, and known hazards are minimized by federal restrictions

TABLE 8.2
Volatile compounds of canned snap beans

Ethanol	3-Pentanone
cis-Hex-3-en-1-ol	Diacetyl
n-Hexanol	2-Heptanone
2-Methyl-2-hexanol	3-Octanone
Oct-1-en-3-ol	Ethyl acetate
Furfurol	Hex-3-en-1-yl acetate
Benzyl alcohol	Ethyl phenyl ether
	Furfuryl methyl ether
Acetaldehyde	Methyl benzyl ether
2-Methylpropanal	Veratrole
3-Methylbutanol	2-Methoxymethyl benzyl ether
Methylthiothanal	2-Butoxytoluene
n-Hexanal	2-(2-Methoxyethyl) methoxybenzene
trans-Hex-2-en-1-al	Phenyl ether
Methional	Aryl-methoxy phenol
Furfural	Biphenyl
5-Methylfurfural	
2-Methoxyfurfural	Pulegone
2-Methyltetrahydrofuran	Linalool
Pyridine	α-Terpineol
	α-Phellandrene

Source: From M. A. Stevens, R. C. Lindsay, L. M. Libbey, and W. A. Frazier (1970) *Proceedings, American Society for Horticultural Science* **91:**839. Reprinted by permission.

TABLE 8.3
Naturally occurring toxins in vegetables[a]

Class and constituent	Vegetable
Alteration of hormonal action	
Goitrogens (thioglycosides)	Brassicas and other crucifers
Amino acid inhibitors	
Trypsin inhibitor	Bean, potato
Antivitamins	
A	Soybean
D	Soybean
E	Pea, bean
Photocarcinogens	
Furocoumarins	Parsnip
Enzyme inhibitors	
Protease inhibitors	Lima bean, soybean, fava bean
Cyanogenetic glycosides	Lima bean, green bean
Glucose-6-phosphate dehydrogenase inhibitor	Fava bean
Cholinesterase inhibitor	Solanaceous crops, squash, pumpkin
Alkaloids	Potato, other solanaceous plants
Amylase inhibitors	Taro, dry bean
Invertase inhibitors	Potato
Physiological disorganizers	
Hemagglutinins	Legumes
Nitrate and nitrite	Spinach, other leafy greens
Oxalates	Rhubarb,[b] spinach, chard, New Zealand spinach
Allergens	Many plants

Source: National Academy of Science (1973); Yamaguchi (1983); Ivie *et al.* (1981).

[a] Constituents used primarily as medicines not included.

[b] The poisonous reputation of rhubarb leaves is not due to oxalates alone; other poisonous constituents are primarily responsible.

applicable to release or importation of new cultivars. Adulterants must be approved for use in or on foods by the Food and Drug Administration.

Development of Quality

Development, prematuration, maturation, ripening, and senescence are terms used to describe the stages through which a vegetable passes during production and handling (Figure 8.1). Development begins with the formation of the edible part and ends when there is a change in growth pattern or when natural en-

largement ceases. This stage occurs mostly before harvest and includes prematuration and at least part of maturation. Prematuration starts with development and extends to the first edible stage. Maturation involves development of size and quality of the edible part and is the period at which commercial crops are harvested. During maturation, the appropriate harvest duration may be very brief (as in sweet corn or pea) or relatively long (dry bean or winter squash). Senescence applies to normal physiological changes in flavor, composition, and other attributes and cessation of growth. It is not as inclusive as the term deterioration, which includes not only senescence but also effects of disorders, disease, and mechanical damage. Deterioration may occur at any time;

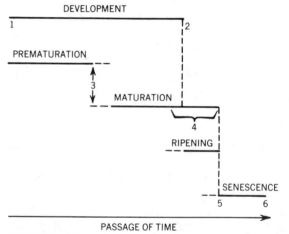

FIGURE 8.1
Stages during the lifespan of fresh vegetables and fruit: (1) initiation of edible part; (2) termination of natural or desirable growth in size or type; (3) start of period of usefulness, but too immature for most users; (4) period of maximum usefulness; (5) degradative changes become predominant; (6) end of usefulness for human consumption. (From A. L. Ryall and W. J. Lipton, *Handling, Transportation and Storage of Fruits and Vegetables,* Vol. I, *Vegetables and Melons.* Copyright © 1972 by the AVI Publishing Company, Westport, Conn. 06881.)

senescence is a chronological event in the life of a plant. Senescence may occur both before and after harvest. After harvest, it is characterized by yellowing of leaves and deterioration of tissues. Senescence before harvest may be indicated by such phenomena as seedstalk formation, growth cracks, or lignification of tissues.

Ripening applies to fruiting crops and may involve increases in respiration, ethylene activity, and water loss and consequent changes in color, flavor, and texture. In fruit with a climacteric, the changes associated with ripening and the onset of senescence occur rapidly, and deterioration soon follows if environmental conditions are not modified to slow the rates of respiration and water loss.

Each attribute of quality is affected to some extent by the environment within which the crop is grown, including the management practices applied during the development phase. In most instances, quality is not improved once a crop is harvested, except for crops in which sugar–starch conversion is important.

Soil Fertility   There are many conflicting reports of effects of fertilization practices on attributes of quality. In general, the major effect of fertilization is to correct nutrient deficiencies which do cause serious quality defects (Table 8.4). Increasing soil fertility beyond this point may increase edible product size in some instances (lettuce, greens), but not in others (tomato). Moisture probably is more important than fertility level in controlling product size.

Shape is related to nutrition in several crops. Cucumber fruit develop "bottlenecks," or constricted stem ends when grown in low-potassium regimes. Potassium also has been reported to improve sweet potato shape.

Texture is affected greatly by nitrogen in such crops as spinach, lettuce, and cabbage, and increased tenderometer values have been observed for peas receiving increased nitrogen. Specific gravity of both potato and sweet potato is reduced by increases in potassium

fertilizer (high specific gravity is considered as high quality). Insufficient fertilization has been reported to increase stringiness in celery, probably indirectly through increased growth stress.

Color may be affected directly or indirectly by fertilization. Nitrogen is necessary for the dark green color of leafy crops, and phos-

TABLE 8.4
Quality defects caused by deficiencies or toxicities of nutrient elements

Vegetable	Nutritional status[a]	Effects
Lettuce	B−	Distortion of terminal bud, cupped leaves
	Mn−	Mottle, interveinal chlorosis
	Cu−	Cupped leaves, loose heads
	Mo−	Poor heart formation
Cole crops	B−	Browning, hollow stem
	K−	Loose heads, poorly developed
	Mo−	Whiptail (cauliflower)
	Ca−	Tipburn
Greens	Mn−	Chlorosis, mottling
	B−	Hollow stem, yellow leaves
	K−	Leaf scorch
Solanaceous	Ca−	Blossom end rot
	N+	Green gel in tomato fruit
Root crops	B−	Split or hollow roots, internal spotting
	K−	Misshapen roots
	Ca−	Misshapen roots, cavities on roots
Potato	B−	Tubers small and rough
	P−	Internal lesions on tuber
Salad crops	Ca−	Tipburn, blackheart
	Mg−	Mottling (celery)
	B−	Cracked stem (celery)
	K−	Leaf scorch, short petioles (celery)
Legumes	Mn−	Spotting on seed cotyledons
	K−	Poor pod fill
	Ca−	Poor seed development
Vine crops	K−	Blossom end cracking, poor fruit shape (cucumber)
	P−	Poor color (cucumber)
Bulb crops	N+	Doubles, splits (bulbing onion)
	Cu−	Thin, pale scales
Sweet corn	P−	Poor earfill

Source: From F. C. Olday (1979) *Proceedings, 1979 Illinois Vegetable Grower's School.* University of Illinois Coop. Ext. Serv. Dept. Hort. Ser. 12. Reprinted by permission of the author.

[a] −, Deficiency; +, excess.

phorus has been reported to improve the red color of beet roots. Indirectly, high fertilization rates increase vine growth in tomato, providing foliage cover for developing fruit. Fruit shielded from direct radiation by foliage develop higher lycopene:β-carotene ratios (deeper red color).

Flavor is difficult to assess, but excessive nitrogen has been reported to impair flavor of celery, tomato, cucumber, and pepper and result in strong objectional flavor in brassicas. In melon, increasing fertility level may enhance photosynthetic area, increasing percentage sugar in harvested fruit. Nutritional changes due to fertilization have been variable. Although vitamin losses from vegetables with excess or deficient levels of certain elements have been reported, these data have not been duplicated consistently and may relate more to other environmental conditions than to soil nutrients. There is some evidence that some tissue mineral levels may be affected by soil fertility.

Soil Moisture   Soil moisture, in addition to its role in nutrient uptake, will influence product texture and color and frequency of defects. Moisture stress increases thickness of cell walls and may increase lignification, resulting in a fibrous or woody texture. Fibrousness in root crops, celery petioles, and chewy leaf tissue of greens and salad crops often develop because of erratic water regimes which result in intermittent slowing of growth. Water stress also has been observed to increase flavor strength of many crops (carrot, cabbage, onion), whereas excess water generally reduces sugars and quality of muskmelon, carrot, and other vegetables. Erratic water supply also leads to serious market defects, including tipburn of lettuce, cracking and blossom end rot of tomato, cracking of melons and cabbage heads, second growth (knobbiness) of potato tubers, and internal disorders of potato tubers.

Temperature   Hot growing temperatures increase respiration and depress storage of

carbohydrates in plants and could be expected to affect product size and quality, particularly in root and tuber crops. In addition, extreme soil and/or air temperatures affect color, shape, size, texture, and flavor of many commodities. Carrots produced in low temperature tend to be a lighter orange than those grown under moderate conditions, and roots also become long and tapered. At high temperature, root length is reduced. Tomatoes ripening under high temperature take on an orange hue because of an increase in β-carotene relative to lycopene, and sweet corn earfill often is poor in hot weather. Chilling temperatures impair quality of several crops, causing surface pitting on eggplant, peppers, cucumbers, and several other warm season crops, russetting of snap beans, and decay of melons. Extreme temperatures affect the texture of several crops by slowing the rate of growth (often a water stress), and the sugar–starch relationship in root and tuber crops is altered greatly by temperature extremes.

Disease   Postharvest losses have been estimated as high as 10 to 30 percent of the total harvested yield, highest in countries lacking facilities for handling and transit of produce. A significant portion of this loss is attributable to biological agents—fungi and bacteria especially—that may penetrate the vegetable before or during harvest (Table 8.5). Bruises and cuts incurred during this process contribute to eventual spread through storages or crates.

Management practices affect not only the texture, appearance, and flavor constituents of produce, but also the susceptibility to disease. Frequency of irrigation, especially near harvest, can increase incidence of decay in tomato, perhaps because of increased cracking or decreased solids. Proper drying of onion bulbs prior to harvest decreases susceptibility to neck not. Pest control measures must be applied diligently, and careful handling throughout production and marketing is essential. The percentages of loss and their causes are listed in Table 8.5.

TABLE 8.5
Retail and consumer losses in selected vegetables in the New York market[a]

Crop	Retail			Consumer			
	Disease	Disorder	Injury	Disease	Disorder	Injury	Total
Bean, snap	0.6	6.3	2.6	1.0	2.1	7.2	19.8
Cucumber	2.4	1.4	1.2	0.9	1.2	0.8	7.9
Lettuce							
Summer	1.5	0.4	2.7	1.2	2.8	3.1	11.7
Winter-wrapped	2.4	1.2	0.6	0.8	4.0	6.4	15.4
Winter-unwrapped	3.1	5.5	6.0	0.4	4.1	6.6	25.7
Onion							
Grano-Granex	1.7	0.6	0.7	3.5	0.2	0.5	7.2
Spanish	6.4	0.2	1.4	4.2	0.1	2.1	14.4
Early Yellow Globe	0.1	0.1	0.1	2.7	0.1	0.9	3.9
Pepper, bell	3.2	3.5	2.5	0.8	0.2	0.4	10.6
Potato							
Katahdin	—[b]	—	—	1.1	1.0	1.5	3.6
White Rose	—	—	—	2.1	0.4	0.7	3.2
Sweet potato							
Dry	3.0	2.4	0.5	9.5	0.8	0.4	16.6
Moist	1.6	3.1	0.7	4.2	3.4	0.4	13.4
Tomato	4.2	0.6	1.5	6.5	0.3	1.1	14.2

Loss (% of total shipped)

Source: From R. A. Cappellini and M. J. Ceponis, in *Postharvest Pathology of Fruits and Vegetables: Postharvest Losses in Perishable Crops,* H. E. Moline, ed., Northeast Regional Res. Publ. NE-87.

[a] Averages for 3-year marketing season.

[b] Losses too small to measure.

Application of Pest Control Although timely applications of pesticides are important in reducing eventual produce loss and in preserving quality, some pest control materials may affect appearance and taste of some vegetables. Apart from the unsightly appearance of a spray residue, some chemicals may alter color or cause russetting or leaf burning. Stoddard solvent herbicide applied to carrots or celery at a stage of growth later than recommended will impart an unpleasant flavor to the mature crop. Other pesticides, such as synthetic pyrethroids, must be applied only at specific temperatures; otherwise, plant leaves can become chlorotic. Most of the deleterious effects of pesticides on quality are a consequence of application at the wrong time or at the wrong temperature or, because of miscalculation or poor equipment calibration, at an excessive dose.

Growth Regulators Field treatment with growth regulators has altered quality of some vegetables. Maleic hydrazide applied to onion or potato several weeks prior to harvest prolongs shelf life by suppressing sprouting. Application of ethrel on tomato or muskmelon to concentrate or accelerate ripening may shorten shelf life, but will enhance uniformity.

Maintenance of Quality

While the edible part of a plant is developing on the plant, it derives its constituent qualities from transpiration and nutrient uptake, photosynthesis, respiration, and metabolic activity of the entire plant. Once the crop is harvested, each leaf or stem or fruit becomes an independent living entity in which respiration and transpiration become the major concerns.

Respiration   Initial respiration rates of different vegetables vary widely (Table 8.6) and increase markedly in response to increasing temperature. High rates characterize young tissue, including meristematic tissue (asparagus spears) and developing seed (green peas). Low rates are the rule for roots, mature fruit, or dry seeds. During respiration, the substrate

and oxygen are consumed, releasing carbon dioxide, water, and energy. The substrate loss in respiration is considered minor, only 2 to 3 percent, for example, for potato and cabbage stored for 6 to 8 months at proper temperatures. Of most concern in postharvest handling is the release of carbon dioxide and heat energy. Accumulation of carbon dioxide over 1 percent of the storage or transit atmosphere can destroy produce, and accumulation of heat accelerates the respiration rate and consequent release of heat.

Transpiration   At harvest, the water source for each commodity is severed. For those vegetables that form an abscission layer or are physiologically conditioned prior to harvest, the change is not abrupt. For others, continued transpiration through stomata on surfaces of detached plant parts can lead to major water losses, and some water also is lost through bruises, surface cracks, and stem scars. The rate of water loss increases with increasing temperature and/or decreasing relative humidity and atmospheric pressure. The rate of loss also is increased from produce with a large transpirational area per unit weight. The loss will occur more rapidly from loose-leaf vegetables than from heading forms, and leafy crops in general transpire more rapidly than stem or fruit crops. For most vegetables, a water loss of 5 to 10 percent causes noticeable deterioration.

Other Changes   Respiration and transpiration clearly are the major concerns in maintaining vegetable quality. Changes in enzyme activity, carbohydrate metabolism, organic acid composition, and other plant constituents also affect quality. Of these changes, the release of ethylene may be the most critical event. Ethylene is involved in the ripening of fruiting vegetables and is added to the atmosphere to ripen tomatoes artificially. For leafy crops, however, ethylene, even in very small quantities, accelerates senescence. Concurrent storage of leafy crops and ethylene-evolv-

TABLE 8.6
Respirational heat loss of selected vegetables

Crop	Loss (BTU/ton-day)		
	32°F	50°F	70°F
Low rates			
Beets	1,600	3,000	—
Cabbage	—	4,000	9,500
Garlic	660	2,000	2,200
Onion, dry	—	1,600	3,700
Pepper, sweet	—	3,200	9,600
Potato	—	2,200	3,500
Tomato	—	3,300	7,600
Turnip	1,300	4,300	7,000
Watermelon	—	1,650	4,700
Moderate rates			
Bean, snap	4,400	12,800	28,600
Brussels sprouts	5,300	18,600	—
Carrot	—	6,900	15,500
Cauliflower	—	7,400	17,600
Head lettuce	3,700	8,800	13,200
Summer squash	2,800	—	21,400
High rates			
Asparagus	17,600	67,000	110,000
Broccoli	5,800	20,300	—
Mushroom	9,600	—	69,600
Okra	—	19,900	57,400
Peas			
Unshelled	8,500	25,700	65,000
Shelled	16,600	—	87,000
Spinach	4,700	16,500	—
Sweet corn	—	24,600	63,400
Watercress	5,800	20,300	—

Source: From A. L. Ryall and W. J. Lipton, *Handling, Transportation and Storage of Fruits and Vegetables*, Vol. I, *Vegetables and Melons.* Copyright © 1972 by the AVI Publishing Company, Westport, Conn. 06881.

ing fruit within the same chamber should be avoided.

Systems for Maintaining Quality

Systems for handling and maintaining quality of fresh and processed vegetables are designed to reduce deterioration due to wilting, decay, overmaturity, chilling, or regrowth and sprouting. Respiration, transpiration, pathogen infection, and mechanical damage all accelerate deterioration and require control of temperature, humidity, and air circulation, attention to sanitation, and care in handling the produce. Although each component of the postharvest environment contributes in a major way to quality control, temperature is the most important, for it modifies the effects of all other factors.

Cooling All cool season vegetables and several crops adapted to warm seasons should be cooled as rapidly as possible after harvest. The field temperature within some produce during a sunny day may exceed 70 to 90°F (21 to 32°C). Vacuum cooling, hydrocooling, top-icing, and forced-air cooling are utilized to remove field heat. In general, cool season crops (and sweet corn) may be cooled to 32 to 40°F (0 to 4.4°C); warm season crops normally are injured by temperatures below 42°F (5.6°C).

Vacuum cooling is based on rapid evaporation from the surface of a vegetable. Evaporation of 1 lb (0.45 kg) of water from 100 lb (45 kg) of vegetables will reduce their temperature by approximately 10°F (5.6°C). For those crops with high water content and large surface area, evaporative cooling is the most rapid means for eliminating heat (Table 8.7). Produce to be cooled is placed in a sealed chamber (Figure 8.2) and atmospheric pressure is reduced to 4.0 to 4.6 mm Hg. At 4.6 mm pressure, water boils at 32°F (0°C). Vaporization requires energy which is drawn (at a rate of 1000 BTU/lb of water) from the object being cooled. Within 28 to 30 min, the interior temperature of the object will be reduced to 32 to 40°F (0 to 4.4°C), depending on initial field temperature and surface area. Prewrapped vegetables also may be vacuum cooled, but films must be perforated to allow free evaporation. Vacuum systems are used primarily for cooling salad vegetables but have been used, although not as efficiently, for cole

TABLE 8.7
Effectiveness of vacuum cooling vegetables with large surface-to-volume ratios

| | | Duration of vacuum cycle (min) | Temperature of commodity (°F) | |
			Beginning of cycle	End of cycle
Commodity	Container			
Brussels sprouts	Quart cup	20	68	38
Cabbage	None	20	68	40
Endive	Crate	20	68	36
Belgian endive	Bundle	14	67	40
Escarole	Crate	20	68	36
Lettuce	Cellophane wrap	10	75	34
Parsley	Crate	20	68	34
Spinach	Cellophane bag	20	65	34
Spinach	Bushel basket	10	67	37

Source: B. A. Friedman and W. R. Radspinner (1956) USDA Agric. Marketing Serv. Rep. 107.

FIGURE 8.2
Hydrovac (Salinas, Calif.) cooler for vegetables. (Photo courtesy of A. Kader, University of California, Davis.)

crops, artichoke, asparagus, sweet corn, and others (Table 8.8). Prewetting the produce will enhance cooling somewhat.

Many vegetables are cooled by immersion (Figure 8.3) in cold flowing water (**hydrocooling**). The rate of cooling depends on heat transfer from the interior of plant tissues to their surface, which is a function of product size and surface area exposed. The rate of heat loss will be increased by decreasing water temperature and by increasing water flow rate. The temperatures of produce and water can become a factor in disease transmission; chlorine often is added to the cooling bath to ensure against inoculation during hydrocooling or washing.

Top-icing (or slurry icing) normally is used to preserve freshness of broccoli and other vegetables in transit. For several commodities, however, cooling after harvest has been achieved by blowing snow ice throughout the load (contact icing), allowing the ice to melt quickly to cool the produce. Approximately 1

TABLE 8.8
Effectiveness of vacuum cooling vegetables with medium surface area-to-volume ratios

Commodity	Container	Duration of vacuum cycle (min)	Temperature of commodity (°F)	
			Beginning cycle	End of cycle
Artichoke	Crate	16	66	50
Asparagus	Bunch	10	64	36
Bean, snap	Hamper	20	80	60
Bean, snap	Hamper	12	69	45
Bean, snap	Cellophane bag	14	70	43
Broccoli	Wirebound crate	20	65	45
Broccoli	Crate	13	67	44
Cauliflower	Crate	20	76	44
Cauliflower	Crate	13	62	46
Celery	Cellophane bag	20	66	47
Sweet corn				
Husked	Cellophane tray	10	59	36
Husked	Cellophane tray	10	75	39
Unhusked	Crate	20	83	43
Leek	Wirebound crate	20	68	36–40
Mushroom	Basket (9 lb)	20	67	39
Mushroom	Prepackaged	15	68	43

Source: B. A. Friedman and W. A. Radspinner (1956) USDA Agric. Marketing Serv. Rep. 107.

Vacuum drawn to 756 mm Hg, at which point water boils at 16°F (−9°C).

FIGURE 8.3
Hydrocooling vegetables. (Photo courtesy of A. Kader, University of California, Davis.)

lb (0.5 kg) of ice will cool 4 lb (1.8 kg) of produce 40°F (22°C). This cooling system has been used by some growers for muskmelon and cabbage and recently has been used to cool broccoli (Figure 8.4).

Air cooling by rapid circulation of refrigerated air through an insulated room is gaining acceptance, especially for field-packaged crops. Although the cooling time is longer than that achieved by hydrocooling or vacuum cooling, it is a satisfactory alternative and is the primary system for crops that require cooling only to 50 to 55°F (10 to 13°C).

FIGURE 8.5
Washing 'White Rose' potatoes. (Photo courtesy of A. Kader, University of California, Davis.)

The lowering of the temperature by all methods, except vacuum cooling, occurs rapidly at first, while the difference between coolant and product temperatures is greatest, and gradually slows as the temperature differential is reduced. Cooling efficiency is maximized when the air or water coolant circulates freely across the surface area of the product. Tight packing, packing box liners, or excessive load depth will restrict circulation and lead to uneven cooling. Vacuum cooling, dependent on evaporative cooling, more closely approximates a linear temperature drop over time.

Grading   Grading generally is preceded by brushing or washing (Figure 8.5) to improve recognition of defects. Specimens damaged by insects or disease organisms or bruised during or before harvest normally have increased respiration rates and also emit ethyl-

FIGURE 8.4
Cutaway of box of broccoli showing extent of icing. (Photo courtesy of A. Kader, University of California, Davis.)

FIGURE 8.6
Separating potatoes by size in a grading line. (Photo courtesy of A. Kader, University of California, Davis.)

ene. If not culled from produce to be shipped or stored, damaged areas constitute a site for pathogen-induced breakdown and decay that will spread rapidly to adjacent material. For long-distance shipping, produce is graded before shipment (Figure 8.6) and again before retail sale. In addition to the culling of off-types or damaged specimens, grading involves manual or machine sizing to create uniformity within a package and to create size categories appropriate for a specific use. Sizing may be by weight or by diameter of the product. Grading also involves culling on the basis of shape, color, maturity, and other factors affecting quality of the consumer's impression of quality.

Most produce now is graded on the basis of federal or state standards, although grading is required only where mandated by marketing order or for certain exports. Otherwise, the use of grading standards is voluntary. These standards are useful for all produce, however, because they enable brokers and growers to compare prices for the same grade, contribute to orderly marketing, and improve effectiveness of advertising. An inspection service is provided by the USDA.

Storage Facilities for ensuring proper storage of vegetables give a grower flexibility in marketing while maintaining fresh market quality of the produce. Some commodities are stored to extend the marketing season (potato, onion, cabbage); other commodities actually may be improved in quality by storage (sweet potato, parsnip). The primary purposes, however, are preservation of quality and maintenance of price stability.

An effective storage must provide uniform temperature and humidity control and constant air circulation through the stored produce. Common storage provides cooling by exchanging interior for exterior air without artificial refrigeration, using a fan or flue system. Night temperatures after fall harvest normally are low enough to provide adequate cooling by air exchange for the crops most frequently stored—potato, onion, cabbage. The storage structure may be above ground, constructed of cinder block and insulated with rigid foam or similar material, or below ground, using the cool soil temperature to maintain interior uniformity.

Refrigerated storage is replacing common storage in most areas, as it provides precise control of temperature and often of humidity, regardless of the exterior environment.

FIGURE 8.7
Mixed load of endive, celery, broccoli, and strawberries for long-distance shipping. (Photo courtesy of A. Kader, University of California, Davis.)

Relative humidity in storage seldom is maintained above 95 percent, mainly because of the fear of increased pathogen activity. Storage of produce at high (near 100 percent) humidity, however, has been suggested for many fruits and vegetables, based on tests showing no increase in decay and noticeable improvement in produce weight, color, and crispness following storage. It is argued that the relative humidity within a box or pallet load probably is near 100 percent; yet, with adequate air circulation, decay is minimal. Such a system would require increased attention to sanitation and grading.

Transit   Long-distance shipment (Figure 8.7) is storage in transit, and the environmental requirements of the carrier do not differ from those of farm storage. Refrigerated/heated insulated rail cars or trucks have replaced carriers using ice as the refrigerant, although top-icing (or slurry icing) is used as a supplement for several crops (e.g., broccoli) to maintain quality. The air leakage of freight cars is variable, and CO_2 accumulates in those with little air exchange, causing physiological defects and losses in product marketability of some crops (Figure 8.8).

Packaging   Packaging refers (1) to bulk containers designed to move and to protect produce and to provide a measure by weight, count, or volume of produce delivered, and (2) to unit containerization, which preserves quality and cleanliness of the product in units of weight or volume convenient for the buyer (Figures 8.9 to 8.11).

Bulk shipping containers vary widely in material and size (Table 8.9). Burlap, plastic mesh, and some plastics provide little protection but excellent visibility and aeration. Wirebound crates, wood boxes, and molded plastic provide the greatest protection, and telescoping fiberboard cartons are intermediate. Crops to be top-iced usually are packed in wirebound crates or bushels or in mesh bags. Those commodities previously cooled, requiring no icing, often are packed in cardboard boxes or,

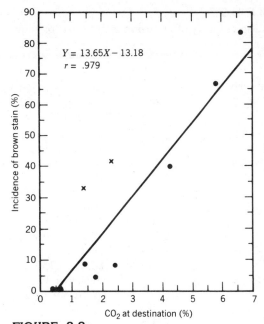

FIGURE 8.8
Relationship between the carbon dioxide concentration at destination in eight conventional cars of lettuce and the incidence of brown stem. r is significant at 1% level; x = brown stain in two cars modified to provide low CO_2 and high CO levels. (From J. K. Stewart, J. M. Harvey, M. J. Ceponis, and W. R. Wright, USDA Marketing Res. Rep. 937 (1972).)

if additional strength is needed, in nailed boxes. Size of container may be based on weight (as for potato and onion), count (number of melons or tomatoes or heads of lettuce), or volume (bushels of spinach or peas). Standardization of shipping containers would improve marketing efficiency, but there has been no inclination to change from current practice.

Consumer packages were first used to prepack potatoes, using 5- or 10-lb (2.2- or 4.5-kg) mesh bags. Polyethylene bags then became popular, offering excellent visibility. However, polyethylene does not allow transmission of water vapor through the film, and rapidly transpiring produce fogs the interior of the package and increases the risk of decay.

TABLE 8.9
Size and capacity of common containers for vegetables

Commodity	Container (contents in in.^{3})	Net weight lb	Net weight kg
Artichoke	7-in. special box (1588)	20–25	9.1–11.4
	Artichoke box (2212)	28–35	12.7–16.0
Asparagus	Pyramid crate (1844)	30–33	13.6–15.0
Bean, snap	Bushel (2150)	28–30	12.7–13.6
Beet, topped	Sack	50	22.7
	Bushel	50–56	22.7–25.4
Beet, bunch	½ crate (2212)	35	16.0
Broccoli	Pony crate (3650)	42–45	19.1–20.5
	14-bunch crate (1825)	20–22	9.1–10.0
	12-bunch crate (1720)	19–20	8.6–9.1
Brussels sprouts	Drum (1929)	25	11.4
	Bushel (2150)	23–25	10.4–11.4
Cabbage, green	Bushel (2150)	31–34	14.1–15.5
	1¾-bu W/B^a crate (3770)	55–60	25.0–27.3
	1⅕-bu W/B crater (2574)	38–41	17.3–18.6
	Carton (3763)	50–52	22.7–23.6
Cabbage, storage	Sack	50	22.7
Carrot, topped	Sack	50	22.7
	Bushel	50	22.7
Cauliflower	WGA crate (5331)	61–63	27.7–28.6
	Catskill W/B crate (3700)	42–44	19.1–20.0
	Long Island W/B crate (4510)	52–55	23.6–25.0
	Western pony (3360)	39–40	17.8–18.2
	Florida crate or carton 12s trimmed (2272)	24–28	10.9–12.7
Celery	16-in. W/B crate or box (3177)	55–60	25.0–27.3
	Carton (1361)	24–26	10.9–11.8
	Carton (1760)	29–31	13.2–14.0
	Sturdee crate (3003)	60	27.3
Chinese cabbage	16-in. crate (3177)	61–63	27.7–28.6
	Bushel (2150)	48–50	21.8–22.7
	1⅓-bu crate (2389)	53–55	24.0–25.0
Corn, green	Bushel (2150)	35–40	16.0–18.2
	Sack	40–50	18.2–22.7
	W/B corn crate (2166)	35–40	16.0–18.2
Cucumber	Bushel (2150)	48–50	21.8–22.7
	L.A. lug (1252)	28–30	12.7–13.6
	Carton (867)	19–20	8.6–9.1
	1⅓-bu W/B crate (2389)	53–55	24.0–25.0
Eggplant	Bushel (2150)	25	11.4
	L.A. lug (1252)	19–21	8.6–9.6
	1⅓-bu W/B crate (2389)	37–39	16.8–17.8
Escarole, endive, and chicory	Bushel (2150)	25	11.4
	1⅓-bu W/B crate (2389)	28	12.7
	WGA crate (5331)	60–62	27.3–28.3
	24-qt basket (1610)	18–19	8.2–8.6
	W/B crate (3064)	33–35	15.0–16.0

TABLE 8.9 (*Continued*)
Size and capacity of common containers for vegetables

Commodity	Container (contents in in.^{3})	Net weight lb	Net weight kg
Greens (all)	Bushel (2150)	18–23	8.2–10.4
	1⅓-bu W/B crate (2389)	20–26	9.1–11.8
	1⅖-bu crate (3082)	26–33	11.8–15.0
	1¾-bu crate (3770)	32–40	14.5–18.2
	WGA crate (5331)	45–57	20.4–25.9
Lettuce, Iceberg	Cartons (2867)	40–45	18.2–20.4
	W/B crate (3064)	43–48	19.5–21.8
Lettuce, Boston	W/B crate (2180)	18–20	8.1–9.1
Lettuce, greenhouse	24-qt basket (1610)	10	4.5
Lettuce, Cos	W/B crate (2180)	20	9.1
Lettuce, Bibb	12-qt basket	5	2.3
Mushroom	4-qt basket	3	1.4
	12-qt basket	9	4.1
Onion	Sack	50	22.7
	Bushel (2150)	50–57	22.7–25.9
Onion, green	16-qt basket (1075)	9	4.1
	1-doz bunches	3	1.4
	Crate, 10-doz bunches	30–32	13.6–14.5
	W/B crate (3064)	35	16.0
Parsley	Bushel (2150)	20–25	9.1–11.4
	16-qt basket (1075)	10–12	4.5–5.5
	8-qt basket (538)	5–6	2.3–2.8
	1⅓-bu W/B crate (2389)	22–28	10.0–12.7
Parsnip	Bushel (2150)	45–50	20.4–22.7
Pea, green	Bushel (2150)	26–30	11.8–13.6
Pepper, green	Bushel (2150)	25–30	11.4–13.6
	1½-bu crate (3225)	37–45	16.8–20.4
	Carton (2284)	27–32	12.3–14.5
	L.A. lug (1252)	15–17	6.8–7.7
	1⅓-bu W/B crate (2389)	28–33	12.7–15.0
Potato	Sack	100	45.4
	Sack	50	22.7
	Bushel (2150)	60	27.3
Radish, topped	16-qt basket	15	6.8
	Bulk bag	40–50	18.2–22.7
Radish, bunch	Bushel (2150)	35	16.0
Squash, summer	Bushel (2150)	44	20.0
	½-bu (1075)	22	10.0
	1⅓-bu W/B crate (2389)	49	22.3
	Lug (1252)	25	11.4
	⅝-bu carton (1195)	24	10.9
Squash, winter	Bushel (2150)	50	22.7
	Bulk	—	—
Sweet potato	Bushel (2150)	50	22.7
	Carton (1750)	41	18.6
	Crate (2150)	50	22.7
	W/B crate (2368)	50	22.7

TABLE 8.9 (Continued)
Size and capacity of common containers for vegetables

Commodity	Container (contents in in.³)	Net weight	
		lb	kg
Tomato	W/B or nailed crate (2672)	60–65	27.3–29.5
	Lug box	30–33	13.6–15.0
	W/B crate (1890)	42	19.1
	Bushel (2150)	50–60	22.7–27.3
	Carton (1822)	40	18.2
	8-qt basket or carton (538)	12	5.5
Turnip and rutabaga, topped	Sack	50	22.7
	Bushel (2150)	25–34	11.4–15.5
	½ bu (1075) (white)	25	11.4
Turnip, bunch	Bushel (2150)	25–34	11.4–15.5
	Crate (5000)	60–80	27.3–36.4
	1¾-bu W/B crate (3770)	45	20.4

Source: United Fresh Fruit & Vegetable Assoc., Washington, D.C.

[a]W/B, wirebound; WGA, Western Grower's Association; L.A., Los Angeles.

New plastic films were developed offering specific gas exchange properties (Table 8.10) superior in maintaining produce quality. In general, high respiring vegetables are preserved best when packaged (Figure 8.12) in films with high gas and water vapor transmission rates. Films not providing sufficient transmission may be suitable if perforated. Although these films enhance shelf life, they are not a substitute for refrigeration.

TABLE 8.10
Characteristics of several classes of plastic films

Packaging film	Characteristics
Oriented polystyrene	High water vapor and gas transmission
Oriented polyvinyl chloride	High water vapor transmission; low CO_2 transmission at low temperature
Polyethylene	High CO_2 permeability at high temperature, lower at low temperature; low water vapor transmission
Oriented polypropylene	Low gas transmission and low water vapor transmission
Nonoriented polypropylene	Low gas transmission and low water vapor transmission

FIGURE 8.9
Packaging potatoes in jute sacks. (Photo courtesy of A. Kader, University of California, Davis.)

FIGURE 8.10
Cantaloupes palletized with netting before cooling and transport. (Photo courtesy of A. Kader, University of California, Davis.)

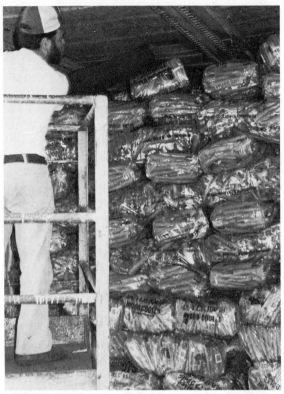

FIGURE 8.11
Carrots prepackaged in polyethylene bags to be top-iced prior to shipment. (Photo courtesy of A. Kader, University of California, Davis.)

Waxing In place of packaging, waxes are applied to several vegetables to minimize water loss and consequent shriveling, principally by sealing the stem scar. Cucumbers, rutabagas, peppers, tomatoes, cantaloupes, and sweet potatoes have been waxed, although cucumbers and rutabagas are the crops most frequently treated. Wax does not reduce decay: if commodities are not free of pathogens prior to waxing, the coating actually may increase decay. The wax layer applied to roots is rather thick; for cucumbers and other vegetable crop fruit, a thin film is applied to ensure exchange of carbon dioxide and oxygen.

FIGURE 8.12
Trimming and film-wrapping lettuce in the field. (Photo courtesy of A. Kader, University of California, Davis.)

MARKETING

Today's market practices constitute an integrated system designed to move produce efficiently, with little or no loss of quality, from the farm to the consumer. Marketing begins at harvest and includes preparation for sale as well as transit and distribution. It encompasses quality control, pricing, and competition within specific legal and organizational frameworks.

Following World War II, food marketing in the United States changed rapidly. Small grocery stores were replaced by chain stores that offered reduced prices, a variety of foods, convenience, and reliable supplies. Technological improvements in maintaining quality, including methods of packaging, pest control, and storage and transit, exposed each geographical area to increasing national and, later, international competition.

Major chains, such as A & P, First National, and those that followed, were organized to purchase produce for their member stores through a single agent (broker). Orders therefore were for large volumes, and from areas of the country that could ensure steady supply. Operators of small- to medium-size farms were not equipped individually to deal with this system. Therefore, in response to these changes in retailing, some truck farms increased in size while others either joined cooperatives to achieve economic stability or marketed through produce centers. Farms serving local consumers, although affected by chain store pricing, maintained profitability through improvements in direct marketing.

Cooperatives

The cooperative movement historically has gained strength in times of economic stress. The first vegetable marketing federation of any significance was the Michigan Potato Grower's Exchange (1918), formed to solve marketing problems following a year of production surpluses. Organization of this and subsequent marketing organizations was based on guidelines (Rochdale Principles) adopted previously by the National Grange.

1. The organization (association) would be a stock company.
2. Members (farms) would hold shares.
3. Membership would require ownership of one or more shares.
4. Each member would sign the rules of the association.
5. Each member would be required to purchase a minimum amount of goods and services from the association each year.
6. There would be a forced limit on the number of shares one member could own.
7. All sales would be in cash only.
8. Competitive pricing would be practiced.
9. Interest on capital would be payed.
10. Allocation of net profits would be made to members according to their share holdings.

Cooperatives may designate acreage to be planted by each member, grades to be sold, volume of harvest, cultivars planted, and other factors affecting market acceptance. They may act only as marketing cooperatives or also may provide member growers with supplies and production/harvesting services. The ultimate goal is to enhance economic returns to the members: control of quality and of volume shipped provides effective stabilization of harvest income, and grower costs of production can be reduced by cooperative purchases of supplies and/or equipment.

National Marketing System

The flow of most produce packed for interstate commerce is graphed in Figure 8.13. Individual farms may supply a marketing cooperative or be of sufficient size to act as a grower–shipper. Produce orders are negotiated between brokers operating on behalf of the grower–shipper and the buyer (wholesaler, chain store, terminal market). These orders specify volume, grade, and other conditions of

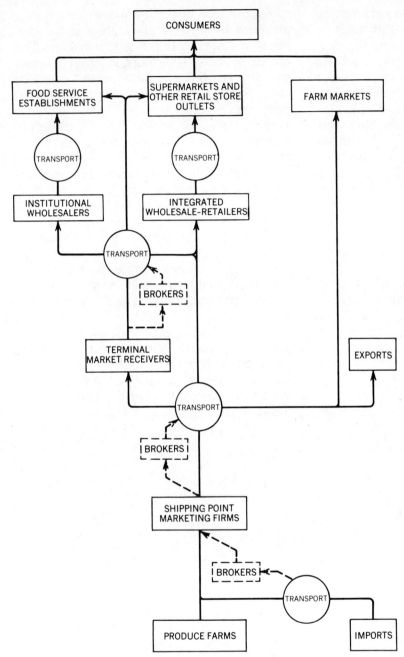

FIGURE 8.13
Schematic of the U.S. marketing system.
(*Source: Produce Marketing Almanac* (1983).)

sale. Produce received by wholesalers may be regraded and packaged prior to distribution to retail outlets. Each agent handling the produce adds to the final purchase cost. As shown in Figure 8.14, the share of consumer cost due to marketing has been increasing relative to that of farm value.

Within this structure, the produce remains the property of the grower or shipper until the buyer inspects and accepts the shipment. The grower thus assumes risk, even though others are responsible for specific handling operations, unless legal contracts provide a sharing of that risk.

Growers who market nationally must estimate the period of time when their produce can be marketed most profitably. In some instances, plantings can be adjusted to avoid severe competition at harvest from produce shipped from California, Mexico, Texas, or Florida.

Direct Marketing

Many vegetable growers prefer to act as their own retailer, often through a roadside stand, but also through a farmer's market or pick-your-own (PYO). There are advantages in direct marketing, but limitations as well. A grower selling directly to a consumer assumes total risk but has complete control of quality until final sale. Transit costs are nil, and other marketing costs are less than those involved in long-distance shipping. Direct marketing operations require multiple talents: a grower must not only be skilled in scheduling and managing plantings for continual harvest, but also must be adept in managing a retail business with its associated problems. The rewards per acre for a well-managed roadside market, however, should exceed those of wholesale suppliers. Poor organizational skills or inability to attract a loyal clientele will reduce profit margins considerably.

The growth of interest in direct marketing in many states was indicated by surveys in 1979 and 1982, showing a gain of approximately 41 percent over the 3-year period. Direct marketing is most profitable in or near highly urban areas and seems to be most popular in the eastern half of the United States (Table 8.11).

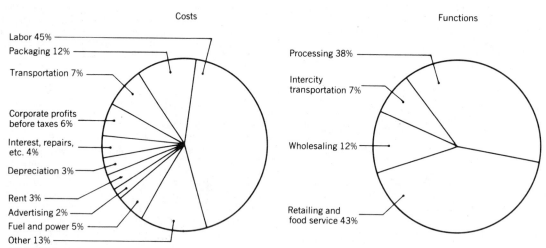

Costs

Labor 45%
Packaging 12%
Transportation 7%
Corporate profits before taxes 6%
Interest, repairs, etc. 4%
Depreciation 3%
Rent 3%
Advertising 2%
Fuel and power 5%
Other 13%

Functions

Processing 38%
Intercity transportation 7%
Wholesaling 12%
Retailing and food service 43%

FIGURE 8.14
Components of the farm-food marketing bill. (*Source: Handbook of Agricultural Charts*, USDA Agric. Handb. 609 (1982).)

TABLE 8.11
Extent of direct marketing within a sample of states (1982 survey)

State	Roadside market		Pick-your-own		Farmer's market (No.)
	Est. No.	Total sales ($)	Est. No.	Total sales ($)	
California	2000	22,000,000	225[a]	2,900,000	65
New York	1200	90,000,000	150	22,500,000	95
Pennsylvania	1157	68,550,500	120	11,550,000	125
Michigan	1023	—	—	—	—
Texas	910	246,016	450	178,000	5
New Jersey	783	65,000,000	46	10,000,000	14
North Carolina	750	27,700,000	103	3,950,000	52
Massachusetts	675	30,375,000	20	2,216,000	46
Oregon	650	5,000,000	25	—	12
Connecticut	640	—	10	—	12
Ohio	600	35,000,000	55	6,000,000	15
Missouri	450	3,800,000	100	2,120,000	50
New Hampshire	450	13,500,000	5	3,430,000	15
Illinois	400	25,000,000	50	—	50

Source: From Jane Lieberth, Spilling the Beans about Roadside Marketing. Copyright © 1982 by American Vegetable Grower. Reprinted by permission.

[a] California pick-your-own includes fruit with vegetables.

Processing

Crops designated for processing are contracted by legal agreement between the processor and grower. Contracts provide some stability in income, although the price generally will be less than that obtainable by fresh market sales. The points addressed in a contract are summarized in Table 8.12. Similar contracts are common within the seed industry and occasionally are used for certain fresh market sales, particularly those involving stored vegetables (potato, onion, cabbage) as a means of guaranteeing a price.

Legal Aspects of Marketing

Agricultural production fluctuates within local geographical areas and nationally, depending on acreage planted, technology applied, and environmental conditions. Both shortages and surpluses can occur from time to time, but the latter cause most of the economic stress for growers. Prior to 1930, produce marketing was uncontrolled. There was no recourse for growers or buyers victimized by fraud, and there was no mechanism for dealing with erratic market volume. During the 1920s and 1930s, protective legislation was enacted that today remains the primary framework for orderly marketing.

Perishable Agricultural Commodities Act

Of the various laws and regulations enacted to regulate marketing, the most significant for the produce industry is the Perishable Agricultural Commodities Act (PACA), passed in 1930. This act was designed to encourage fair trade and prevent fraud in interstate or foreign commerce and applies to commission merchants, dealers, or brokers. Growers marketing the produce they grow themselves are not included; however, produce marketed through a cooperative is included under the

TABLE 8.12
Recommendations for contracts for fresh or processed vegetables

1. Contracts should be an agreement to buy and sell, not bailment (where grower owns no interest or title to seed or crop grown).

2. Responsibilities for planting, culture, harvest, and delivery should be specified.

3. Buyer's obligation to weigh, sample, and keep records should be stated.

4. Services by buyer and charges for each should be stated.

5. List basis of grading, number of grades, and price schedule. Top grade should get incentive price.

6. Basis and right of rejection should be stated.

7. Waste should not be specified as a fixed percentage. It should be determined by actual amount.

8. Ensilage ownership rights should be stated (this may involve solid waste disposal).

9. Provisions for passed acreage should be included.

10. Packer's rights and charges and the method of repayment for advances of seed, fertilizer, and pesticides should be specified.

11. Disclaimer clauses must be specified to exempt buyer from liability for damage by elements, insects, diseases, seed or plant warranty, chemical injury, etc.

12. Conditions under which the buyer can assume ownership of the crop should be stated, as, for example, when it is not cared for properly.

13. Right of cancellation and by what specified date should be included.

14. A contingency clause to define responsibilities of each party should be included.

act. PACA prohibits rejection of produce by a buyer without reasonable cause (falling prices are not a reasonable cause; quality poorer than advertised would be); failure to deliver without reasonable cause; false and misleading statements with respect to market demand or other conditions prompting a sale that otherwise the grower would not have made; false statements with respect to the conditions or grade of the produce by either buyer or seller; incorrect accounting on consignments; failure to pay promptly for goods purchased or received on consignment; and altering federal inspection certificates. All dealers and handlers must be licensed, and violations of any of the provisions may lead to fines or revocation of license.

Marketing Agreements and Orders   Efforts to stabilize agricultural marketing within a very depressed economy culminated in the passage of the Agricultural Marketing Agreement Act of 1937, subsequently amended to allow establishment of minimum quality and maturity standards. This act authorized voluntary control (marketing agreements) and regulatory control (marketing orders). In either case, a program would be enacted only following a formal request and vote by affected growers.

A marketing agreement is a voluntary contract between a handler of farm products and the Secretary of Agriculture. A marketing order is a grower program, approved by two-thirds vote of the growers in the area affected (regional or national) and then issued by the Secretary of Agriculture to apply to all handlers in that area. The types of provisions in either program may include quality regulation, rate of flow regulation, acreage allocations, standardization of containers and packs, a self-taxing system (check-off) to finance research and/or promotion, use of federal inspection, and legal action against violators. Relatively few vegetables have been covered by agreements and orders, particularly since 1946. Potato marketing has been regulated most frequently.

In addition to these marketing acts, consumer protection legislation in recent years has ensured accuracy in product labeling, and both the Environmental Protection Agency and the Food and Drug Administration regulate the safety and purity of foods.

SELECTED REFERENCES

Arthey, V. D. (1975). *Quality of Horticultural Products.* Wiley, New York.

Bourne, M. C. (1980). Texture evaluation of horticultural crops. *HortScience* **15:**51–57.

Goddard, M. S., and S. E. Gebbhardt (1979). Nutritional importance of fresh fruits and vegetables. In *Proceedings, 1979 Illinois Vegetable Grower Schools.* University of Illinois Coop. Ext. Serv. Dept. Hort. Ser. 12.

R. E. Hardenburg, A. E. Watada, and C. Y. Wang (1986). *The Commercial Storage of Fruits, Vegetables, and Florist and Nursery Stocks,* USDA Agric. Handb. 66. U.S. Govt. Printing Office, Washington, D.C.

Ivie, G. W., D. L. Holt, and M. C. Ivey (1981). Natural toxicants in human foods: Psoralens in raw and cooked parsnip root. *Science* **213:**909–910.

Kramer, A. A., and B. A. Twigg (1966). *Fundamentals of Quality Control for the Food Industry.* AVI, Westport, Conn.

Krochta, J. M., and B. Feinberg (1975). Effects of harvesting and handling on fruits and vegetables. In *Nutritional Evaluation of Food Processing,* (R. S. Harris and E. Karmas, eds.). AVI, Westport, Conn.

Moline, H. E. (ed.) (1984). *Postharvest Pathology of Fruits and Vegetables: Postharvest Losses in Perishable Crops,* Northeast Regional Res. Publ. NE-87.

National Academy of Science (1973). *Toxicants Occurring Naturally in Foods.* Washington, D.C.

Olday, F. C. (1979). Influence of soil fertility on vegetable quality. In *Proceedings, 1979 Illinois Vegetable Grower's School,* University of Illinois Coop. Ext. Serv. Dept. Hort. Ser. 12.

Ryall, A. L., and W. J. Lipton (1972). *Handling, Transportation and Storage of Fruits and Vegetables,* Vol. I, *Vegetables and Melons.* AVI, Westport, Conn.

Stevens, M. A. (1970). Vegetable flavor. *HortScience* **5:**95–98.

USDA (1954). *Marketing,* Yearbook of Agriculture. U.S. Govt. Printing Office, Washington, D.C.

van den Berg, L., and C. P. Lentz (1978). High humidity storage of vegetables and fruits. *HortScience* **13:**565–569.

Vittum, M. T. (1963). Effect of fertilizers on the quality of vegetables. *Agronomy Journal* **55:**425–429.

Yamaguchi, M. (1983). *World Vegetables.* AVI, Westport, Conn.

STUDY QUESTIONS

1. In what ways do soil properties contribute toward the eventual quality of a product at harvest time?

2. How do refrigeration, high humidity, and aeration within a storage maintain high quality of most cool season crops? Which factor is the most important and why?

3. What factors would argue for the use of vacuum cooling rather than hydrocooling?

4. Why are there so few marketing orders in operation in the vegetable industry?

5. Can an independent commercial grower market produce in neighboring states without being licensed under PACA? Explain.

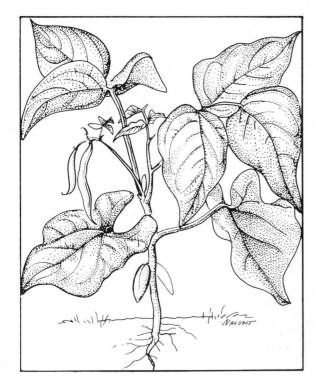

9
Classification of Vegetables

BOTANICAL CLASSIFICATION

Plant relationships that reflect common or divergent evolutionary pathways are invaluable to botanists and others studying plants. These relationships also can be useful knowledge for producers, since management systems may be governed by botanical similarities. For example, crop rotation as a method of pest management must involve botanically unrelated plants. Otherwise, disease and insect levels may increase, because related plants tend to attract the same pathogens and insect pests. The botanical relationship between a crop and weed also is significant, the latter plants often serving as an overwintering host of serious insects and diseases.

The botanical classification is a hierarchical one, based upon successive levels of morphological relationships. Families represent large natural groups, identified by one or several major features shared by all genera (Figures 9.1 and 9.2). In a similar way, genera and species are distinguished, each including successively reduced levels of heterogeneity. Species may be divided further into subspecies, reflecting differences not sufficient to constitute separate species, or certain plants may be identified by group (as in *Brassica oleracea*, Gemmifera Group) based on distinctive appearance and/or use. Botanical varieties are used to distinguish specific forms of a species that otherwise are identical; for example, corn is separated into several botanical varieties based on characteristics of the kernel.

The grouping of vegetables into genus

It is useful to organize vegetables and their cultivars into groups with certain similarities. The similarities may relate to use, appearance, morphological features, sensitivity to environmental factors, type of life cycle, or other criteria. Each classification creates a descriptive terminology for vegetables, reflecting adaptation, general cultural needs, or management systems. Some classifications are more useful than others, but each reduces the large number of different crop plants to logical associations.

FIGURE 9.1

Plant inflorescence types: (*a*) spike; (*b*) raceme;
(*c*) panicle; (*d*) cyme; (*e*) umbel; (*f*) composite.
(From *Crop Production*, by S. R. Chapman and
L. P. Carter. W. H. Freeman & Company. Copyright © 1976.)

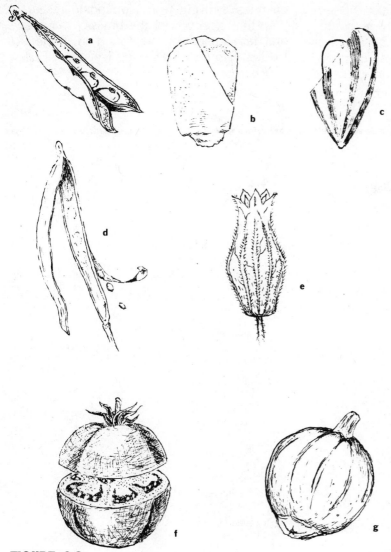

FIGURE 9.2
Several common fruit types: (*a*) legume or pod;
(*b*) caryopsis or grain; (*c*) achene; (*d*) silique; (*e*)
capsule; (*f*) berry; (*g*) pepo. (From *Crop Production*, by S. R. Chapman and L. P. Carter. W. H.
Freeman & Company. Copyright © 1976).

and species, the further identification of subclasses, and the placement of these groups into families are presented in Table 9.1. For some vegetable taxa, there is some inconsistency in the literature, reflecting nomenclature changes evolving from continuing research. With the exception of the brassicas and several family names, the listing in Table 9.1 follows *Hortus Third* (Bailey and Bailey, 1976).

TABLE 9.1
Botanical classification of vegetables

Monocotyledoneae
  Alliaceae (onion family)
    Allium ampeloprasum
      Porrum Group   Leek
      Ampeloprasum Group   Elephant garlic
    Allium cepa
      Cepa Group   Onion
      Aggregatum Group   Multiplier onion, Shallot
      Proliferum Group   Egyptian onion
    Allium fistulosum   Welsh onion, Spanish onion, Japanese bunching
    Allium sativum   Garlic
    Allium schoenoprasum   Chive
  Araceae (arum family)
    Colocasia esculenta   Taro (dasheen)
  Gramineae (grass family)
    Zea mays var. *praecox*   Popcorn
    Zea mays var. *rugosa*   Sweet corn
  Liliaceae (Lily family)
    Asparagus officinalis   Asparagus
Dicotyledoneae
  Apiaceae (Umbelliferae) (parsley family)
    Anethum graveolens   Dill
    Anthriscus cerefolium   Chervil
    Apium graveolens var. *dulce*   Celery
    Apium graveolens var. *rapaceum*   Celeriac
    Daucus carota var. *sativa*   Carrot
    Foeniculum vulgare var. *azoricum*   Florence fennel
    Pastinaca sativa   Parsnip
    Petroselinum crispum var. *neopolitanum*   Italian flat parsley
    Petroselinum crispum var. *tuberosum*   Turnip-rooted parsley
  Asteraceae (Compositae, composites)
    Cichorium endivia   Endive/escarole
    Cichorium intybus   Chicory
    Cynara cardunculus   Cardoon
    Cynara scolymus   Globe artichoke
    Helianthus tuberosus   Jerusalem artichoke
    Lactuca sativa   Lettuce
    Scolymus hispanicus   Spanish oyster plant
    Scorzonera hispanica   Black salsify
    Taraxicum officinale   Dandelion
    Tragopogon porrifolius   Salsify

TABLE 9.1 (*Continued*)
Botanical classification of vegetables

Brassicaceae (Cruciferae, mustard family)
    Amoracia rusticana — Horseradish
    Barbarea verna — Upland cress
    Brassica juncea — Mustard greens
    Brassica napus
        Napobrassica Group — Rutabaga
        Pabularia Group — Siberian kale
    Brassica oleracea
        Acephala Group — Scotch kale, collards
        Alboglabra Group — Chinese kale
        Botrytis Group — Cauliflower, heading broccoli
        Capitata Group — Cabbage
        Gemmifera Group — Brussels sprouts
        Gongylodes Group — Kohlrabi
        Italica Group — Sprouting broccoli
    Brassica campestris
        Chinensis Group — Pak choy (leafy)
        Pekinensis Group — Pe tsai (heading)
        Perviridus Group — Spinach mustard
        Rapifera Group — Turnip
        Ruvo Group — Broccoli raab
    Crambe maritima — Sea kale
    Lepidium sativum — Garden cress
    Raphanus sativus — Radish
    Rorippa nasturtium-aquaticum — Watercress
Chenopodiaceae (goosefoot family)
    Atriplex hortensis — Orach
    Beta vulgaris
        Crassa Group — Garden beet
        Cicla Group — Swiss chard
    Spinacia oleracea — Spinach
Convolvulaceae (morning glory family)
    Ipomoea batatas — Sweet potato
Cucurbitaceae (gourd family)
    Citrullus lanatus — Watermelon
    Cucumis anguria — Gherkin
    Cucumis melo
        Inodorous Group — Honeydew and casaba melons
        Reticulatus Group — Netted melons (Persian and muskmelon)
        Cantalupensis Group — True cantaloupe
    Cucumis sativus — Cucumber
    Cucurbita maxima — Winter squash
    Cucurbita mixta — Cushaw squash
    Cucurbita moschata — Winter crookneck squash
    Cucurbita pepo var. *pepo* — Pumpkin, acorn squash
    Cucurbita pepo var. *melopepo* — Summer squash
    Sechium edule — Chayote

TABLE 9.1 (*Continued*)
Botanical classification of vegetables

Leguminoseae (pea family)	
Cicer arietinum	Garbanzo bean
Glycine max	Soybean
Phaseolus acutifolius var. *latifolius*	Tepary bean
Phaseolus coccineus	Scarlet runner bean
Phaseolus limensis	Thick-seeded lima bean
Phaseolus limensis var. *limenanus*	Bush lima bean
Phaseolus lunatus	Butter (sieva) bean
Phaseolus vulgaris	Common bean
Pisum sativum	Garden pea
Pisum sativum var. *arvense*	Field pea
Pisum sativum var. *macrocarpon*	Edible-pod pea
Vicia faba	Broad bean
Vigna aconitifolia	Moth bean
Vigna angularis	Azuki bean
Vigna mungo	Urd bean
Vigna radiata	Mung bean
Vigna umbellata	Rice bean
Vigna unguiculata	
subsp. *unguiculata*	Cowpea
subsp. *sesquipedalis*	Yardlong bean (asparagus bean)
subsp. *cylindrica*	Catjang bean
Malvaceae (mallow family)	
Abelmoschus esculentus	Okra
Martyniaceae (martynia family)	
Proboscidea louisianica	Martynia
Polygonaceae (buckwheat family)	
Rheum rhabarbarum	Rhubarb
Rumex acetosa	Garden sorrel
Rumex patientia	Spinach dock
Solanaceae (nightshade family)	
Capsicum annuum	
Cerasiforme Group	Cherry pepper
Conoides Group	Cone pepper
Fasiculatum Group	Red cluster
Grossum Group	Bell pepper
Longum Group	Cayenne, chili pepper
Capsicum frutescens	Tabasco pepper
Lycopersicon esculentum	Tomato
Lycopersicon pimpinellifolium	Current tomato
Physalis pruinosa	Husk tomato
Solanum tuberosum	Potato
Solanum melongena	Eggplant
Tetragoniaceae (carpetweed family)	
Tetragonia tetragonoides	New Zealand spinach
Valerianaceae (valeriana family)	
Valerianella locusta	Corn salad
Basidiomyceteae	
Agaricales	
Agaricaceae	
Agaricus brunnescens	Mushroom

CLASSIFICATION BASED ON EDIBLE PART

Among the vegetables, some are classified according to the edible part of each plant (e.g., root crops, leafy crops). To a limited extent, this grouping, shown in Table 9.2, has some value in crop management and marketing. As examples, (1) there are similarities among crops from which the roots are harvested— their quality is affected by soil fertility and water management and by soil texture; (2) fruiting crops present different developmental and marketing problems than do crops in which flowering and fruit set do not or should not occur during the production cycle; and (3) leafy crops are very perishable and require rapid chilling after harvest to preserve quality. Classification by edible part thus provides a grower or handler with several broad plant groups which imply specific cultural or handling techniques.

TABLE 9.2
Classification of vegetables based on edible part

Underground portion		
Root	Beet, carrot, sweet potato, horseradish, parsnip turnip, radish salsify, rutabaga, scorzonera, celeriac, turnip-rooted parsley	
Tuber	Potato, Jerusalem artichoke	
Corm	Taro	
Bulb	Onion, leek, garlic	
Above-ground portion		
Stem	Asparagus, kohlrabi	
Petiole	Rhubarb, celery	
Leaf	Lettuce, spinach, cabbage, Chinese cabbage, onion, mustard greens, cress, brussels sprouts, kale, collards, garlic chive, New Zealand spinach, Swiss chard, endive, corn salad, chicory, dandelion, parsley, watercress	
Flower	Globe artichoke, broccoli, cauliflower, broccoli raab	
Fruit	Tomato, pepper, eggplant, snow pea, snap bean, okra, squash, melon, sweet corn, husk tomato, cucumber	
Seed	Dry bean, dry and green peas, lima bean, chickpea	

CLASSIFICATION BASED ON USE

Certain vegetables are categorized as "greens" and others as "salad crops" based on primary use. Greens and salad crops both include only leafy vegetables, but the kind of use—boiled or raw—separates the several crops further. Classification by use is not definitive, however, since many crops may be used in several ways. As an example, a number of vegetables (broccoli, onion, and others) used raw in salads are not included among those crops normally considered salad crops. Use provides an image of certain kinds of plants, but, as a classification, it is applicable only to a limited number of vegetables.

CLASSIFICATION BASED ON LIFE CYCLE

Most vegetable crops are grown as annuals. Many of these crop plants are true annuals, completing their life cycles within 1 year. In addition, however, there are biennials and perennials mostly grown as annuals. Few vegetables are cultivated as perennials in temperate climates, but these few present distinctly different cultural requirements.

The biennial species present an interesting group, since they tend to be sensitive to temperature regulation of flowering (vernalization). Occasionally, this regulation is modified by changes in daylength (Table 9.3).

CLASSIFICATION BASED ON TEMPERATURE

The vegetable crops originated and evolved in widely different parts of the world, extending from the tropics to Siberia. Throughout the migration of each species to new parts of the

TABLE 9.3
Classification of some vegetables based on life cycle[a]

Perennial

Artichoke	Asparagus	Chicory (T_W)
Chive	Dandelion	Eggplant
Garlic (T_C)	Horseradish	Husk tomato
Jerusalem arti-	Lima bean (large)	Onion, aggregate
choke	Potato	(T_C)
Pepper	Sweet potato (P_S)	Rhubarb (T_C)
Scarlet runner	Watercress	Taro
Tomato		

Biennial

Beet (T_C)	Broccoli	Brussels sprouts
Cabbage (T_C)	Carrot ($T_C P_L$)	(T_C)
Celery (T_C)	Celeriac (T_C)	Cauliflower
Chinese cabbage	Corn salad	Chervil (T_C)
($T_C P_L$)	Kale (T_C)	Fennel (T_C)
Collards (T_C)	Onion (T_C)	Kohlrabi (T_C)
Leek (T_C)	Parsnip (T_C)	Orach (T_C)
Parsley (T_C)	Swiss chard	Rutabaga (T_C)
Salsify (T_C)	(T_C)	Turnip
Winter radish		

Annual

Amaranth	Bean	Broccoli
Chickpea	Cucumber	Endive ($T_W P_L$)
Garden cress	Lettuce	Lima bean
Mung bean	($T_W P_L$)	(small)
New Zealand	Muskmelon	Mustard
spinach	Okra	Peas
Pumpkin	Radish ($T_C P_L$)	Soybean
Spinach ($T_C P_L$)	Squash	Sweet corn
Watermelon		

[a] $T_{C,W}$: Temperature (cool, warm) exposure induces reproductive growth. $P_{L,S}$: Specific photoperiod (long, short) induces reproductive growth. TP, Interaction of temperature and photoperiod induces flowering.

world, selection for local adaptation resulted in significant modifications in crop performance. These modifications, however, did not change the inherent adaptation of plants to certain temperature regimes. Tomato is a warm season crop, even though it is grown over a wide range of latitudes. All vegetables, therefore, can be separated broadly into two groups: warm season and cool season crops. Classification by temperature adaptation is perhaps the most useful crop grouping, since the distinct characteristics of a specific group have

management implications. The characteristics for cool season crops have been summarized by Knott (1955) as follows:

1. Hardy or frost tolerant.
2. Seeds germinate in cold soil.
3. Plants absorb water efficiently from cold soil.
4. Produce is stored at cold temperature.
5. Plants tend to be shallow rooted.
   a. Frequent water is required.
   b. Additional fertilizer, especially nitrogen, is required.

Cool season crops also normally have relatively small plant size, and a number of them are susceptible to premature seeding (bolting). The cool season crops include most root crops (except sweet potato) and crops normally used for salads and greens (leaf and stem tissue). Crops bearing edible fruit predominantly are warm season crops (Table 9.4). Although there are certain exceptions (as an example, sweet corn, a warm season crop, is handled as a cool season crop in storage), these statements are rather broadly applicable.

OTHER CLASSIFICATIONS

Crops may be classified according to sensitivity to various environmental factors (Tables 9.5 to 9.7), including pH, salinity, tolerance to specific nutrient levels, and affinity for certain levels of soil moisture. Such information is especially useful in management decisions and in diagnosing problems in plant growth.

The discussion of the principal vegetables in subsequent chapters follows a classification based largely on custom, evolving from the organization of Thompson (1949). Some plants are grouped botanically (cole crops, solanaceous crops, cucurbits, etc.), and others by edible part (root crops, alliums, tuber and tuberous rooted crops), by use (leafy salad crops, greens), or by life cycle (perennials).

TABLE 9.4
Classification of vegetables based on optimum temperature

Cool season crops

Hardy

Asparagus	Broad bean	Broccoli
Brussels sprouts	Cabbage	Chive
Collards	Dandelion	Garlic
Horseradish	Kale	Kohlrabi
Leek	Mustard	Onion
Parsley	Pea	Radish
Rhubarb	Rutabaga	Sorrel
Turnip	Watercress	

Half-hardy

Amaranth	Artichoke	Beet
Carrot	Cauliflower	Celery
Celeriac	Chervil	Chicory
Chinese cabbage	Corn salad	Endive
Fennel	Garden cress	Jerusalem artichoke
Lettuce	Orach	Parsnip
Potato	Salsify	Scorzonera
Sea kale	Swiss chard	

Warm season crops

Tender

New Zealand spinach	Snap bean	Soybean
Sweet corn	Tomato	Southern pea

Very tender

Cucumber	Eggplant	Lima bean
Muskmelon	Okra	Pepper
Pumpkin	Squash	Sweet potato
Taro	Watermelon	

Source: Based on Knott (1955).

Several unique crops, such as sweet corn, are discussed separately, others as miscellaneous minor crops.

CLASSIFICATION OF CULTIVARS AND CULTIVAR GROUPS

A large number of cultivars, old and new, are offered by seedhouses, particularly for vegetables grown in large volume. Most cultivars have been developed for a specific use or at least contain combinations of traits desired for a particular growing system. For most vegetable crops, the cultivars have been categorized on the basis of economically important traits, including shape and size, use (processing or

TABLE 9.5
Classification of some vegetables for susceptibility to chilling damage when stored at moderately low but nonfreezing temperatures

Most susceptible	Moderately susceptible	Least susceptible
Asparagus	Broccoli	Beet
Snap bean	Cabbage	Brussels sprouts
Cucumber	Carrot	Cabbage, stored, savoy
Eggplant	Cauliflower	Kale
Lettuce	Onion, dry	Parsnip
Okra	Parsley	Rutabaga
Pepper	Peas	Salsify
Potato	Radish	Turnip
Squash, summer	Spinach	
Sweet potato	Squash, winter	
Tomato		

Source: From R. E. Hardenburg, A. E. Watada and C. Y. Wang *The Commercial Storage of Fruits, Vegetables, and Florist and Nursery Stocks.* USDA Agric. Handb. 66 (1986). Used by permission.

TABLE 9.6
Classification of vegetables for rooting depth

Shallow (18–24 in.)	Moderately deep (36–48 in.)	Deep (>48 in.)
Broccoli	Bean, green	Artichoke
Brussels sprouts	Beet	Asparagus
Cabbage	Carrot	Bean, lima
Cauliflower	Chard	Parsnip
Celery	Cucumber	Pumpkin
Chinese cabbage	Eggplant	Sweet potato
Corn	Muskmelon	Tomato
Endive	Mustard	Watermelon
Garlic	Pea	
Leek	Pepper	
Lettuce	Rutabaga	
Onion	Squash, summer	
Parsley	Turnip	
Potato		
Radish		
Spinach		

Source: From O. A. Lorenz and D. N. Maynard, *Knott's Handbook for Vegetable Growers.* Copyright © 1980 by John Wiley & Sons, Inc., New York, p. 139. Reprinted by permission.

TABLE 9.7
Classification of vegetables by sensitivity to pH

Slightly tolerant (pH 6.8–6.0)	Moderately tolerant (pH 6.8–5.5)	Very tolerant (pH 6.8–5.0)
Asparagus	Bean	Chicory
Beet	Bean, lima	Dandelion
Broccoli	Brussels sprouts	Endive
Cabbage	Carrot	Fennel
Cauliflower	Collards	Potato
Celery	Corn	Rhubarb
Chard	Cucumber	Shallot
Cress	Eggplant	Sorrel
Chinese cabbage	Garlic	Sweet potato
Leek	Gherkin	Watermelon
Lettuce	Horseradish	
Muskmelon	Kale	
New Zealand spinach	Kohlrabi	
Okra	Mustard	
Onion	Parsley	
Orach	Pea	
Parsnip	Pepper	
Salsify	Pumpkin	
Soybean	Radish	
Spinach	Rutabaga	
Watercress	Squash	
	Tomato	
	Turnip	

Source: From *Knott's Handbook for Vegetable Growers*, O. A. Lorenz and D. N. Maynard. Copyright © 1982 by John Wiley & Sons, Inc., New York, p. 84. Reprinted by permission.

fresh, local or shipping), quality (specific gravity, sugar content, texture, color), pungency, or adaptation. It is more important to become familiar with cultivar types than with individual named phenotypes, since only a few introductions continue over many years to outperform those subsequently released by plant breeders. For the most part, the descriptions in the following chapters are of cultivar groups, identifying those features deemed important for a specific use. These criteria may change as new germ plasm is introduced: cultivar groups evolve subgroups that may, in time, replace the current models for cultivar classification. Such changes are most likely for groupings based on morphological traits and least likely for those based on quality.

SELECTED REFERENCES

Bailey, L. H., and E. Z. Bailey (1976). *Hortus Third.* Macmillan Co., New York.

Knott, J. E. (1955). *Vegetable Growing.* Lea & Febiger, Philadelphia.

Lorenz, O. A., and D. N. Maynard (1980). *Knott's Handbook for Vegetable Growers.* Wiley, New York.

MacGillivray, J. H. (1953). *Vegetable Production.* McGraw–Hill (Blakiston), New York.

Simmonds, N. W. (ed.) (1976). *Evolution of Crop Plants.* Longmans, London.

Thompson, H. C. (1949). *Vegetable Crops.* McGraw–Hill, New York.

STUDY QUESTIONS

1. Why is it important to have an understanding of the botanical relationships among vegetable crops?
2. In an evolutionary sense, why would one expect a high proportion of cool season crops be biennials or produce flower stalks in response to temperature change?

10
Perennial Crops

sensitive to low temperature and can be grown only in a few areas in the United States. Sweet potato is an exception, since it is grown commercially in several northern and northeast states. This crop, however, is grown as an annual and is discussed in Chapter 16.

ASPARAGUS

Historical Perspective and Current Status

Asparagus originated in areas bordering the Mediterranean Sea and was considered a delicacy by the ancient Greeks. The characteristics of modern cultivars apparently do not differ from those described by the elder Cato in 200 B.C. and 200 years later by Pliny. In addition to its popularity as a food in early times, asparagus was believed effective in preventing bee stings, heart trouble, dropsy, and toothache. In 1949, scientists found that asparagus contained rutin, a substance that strengthens capillary walls, thereby preventing hemorrhaging. The content of rutin in edible young spears, however, is not high, tending to increase with maturity of canes in late summer. Nutritionally, asparagus is a source of vitamins A and C (Table 10.1).

Commercial production of asparagus is important in many parts of Europe and Asia and in Australia and New Zealand. In the United States, production is centered in California, Washington, and Michigan (Table 10.2). Acreages in eastern and central states have declined in recent years, and production is now relatively low. Asparagus acreage in eastern Canada has increased.

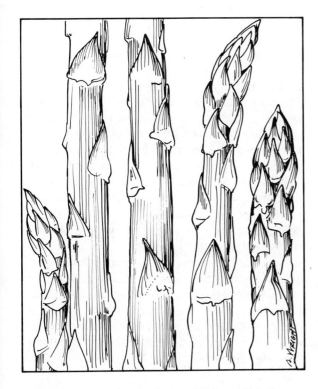

Asparagus　　Rhubarb　　Globe artichoke

The perennial crops represent taxonomically diverse families, but there are similarities in production. All are relatively long-lived, and most of those indigenous to temperate zones are adapted to cool seasons. In addition to asparagus, rhubarb, and globe artichoke, discussed in this chapter, several other vegetables (e.g., Jerusalem artichoke, horseradish, potato) botanically are perennials but are grown as annuals or otherwise have unique production requirements.

There are a number of perennial vegetables grown in the topics, most of which are

TABLE 10.1
Nutritional constituents of some perennial vegetables[a]

Crop	Water (%)	Energy (cal)	Protein (g)	Fat (g)	Carbohydrate (g)	Vitamins					Minerals				
						A[b] (IU)	C[c] (mg)	Thiamine (mg)	Riboflavin (mg)	Niacin (mg)	Ca (mg)	P (mg)	Fe (mg)	Na (mg)	K (mg)
Asparagus	91.7	26	2.5	0.2	5.0	900	33	0.18	0.20	1.5	22	62	1.0	2	278
Globe artichoke	85.0	20	2.7	0.2	2.3	160	11	0.06	0.08	0.8	53	78	1.5	43	430
Rhubarb	95.0	16	0.6	0.1	3.7	100	9	0.03	0.07	0.3	96	18	0.8	2	251

Source: National Food Review (1978), USDA.

[a] Data per 100 g sample.

[b] 1 IU = 0.3 μg vitamin A alcohol.

[c] Ascorbic acid.

Classification, Growth, and Development

Asparagus (*Asparagus officinalis*), a monocot, is a member of the family Liliaceae. It is very winter hardy and, in disease-free soils, tolerates such environmental stresses as heat, drought, and salinity. Although commercial production is not uncommon in subtropical areas of the world, asparagus is considered to be a cool season crop with 75 to 85°F (24 to 29°C) day and 55 to 66°F (13 to 19°C) night temperatures favoring productivity and longevity.

Asparagus is a dioecious herbaceous perennial, 4 to 6 ft (1.2 to 1.8 m) tall in established plantings. Both sexes occur in approximately equal numbers in plantings of traditional cultivars, and male and female plants differ in several secondary traits. The most significant differences relate to yield (Table 10.3). Females produce spears somewhat larger in diameter (generally preferred because of superior quality) than those from male plants, but males produce a greater number of spears for an overall per acre yield superiority. Male plants also show greater longevity than females.

Flowers of both sexes are rather inconspicuous, although almost every cane will bear many. Male blossoms are slender, bell-shaped, and a greenish white in color (Figure 10.1). Each flower has an aborted ovary and well-developed anthers bearing orange pollen. Female flowers are smaller than those on male plants and contain vestigial, functionless anthers and a well-developed ovary, style, and three-feathered stigma. Occasionally, a plant

TABLE 10.2
Production of asparagus for fresh market and processing, 1980 to 1981

State	Production			
	1980		1981	
	1000 cwt	MT	1000 cwt	MT
California	631	28,622	819	37,150
Washington	41	1,860	69	3,130
Michigan	70	3,175	53	2,404
U.S. total	845	38,329	1,040	47,174

Source: Agricultural Statistics (1984), USDA.

TABLE 10.3
Effect of sex on spear number and size and on total yield of asparagus

Measurement	Sex	Mean performance
Number of spears	Male	9.5
	Female	6.3
Mean spear weight (g)	Male	26.1
	Female	28.8
Yield (g)	Male	243.0
	Female	176.0

Source: From A. A. Franken (1970) *Euphytica* **19**:227–287. Reprinted by permission of the author.

FIGURE 10.1

Asparagus flowers: (*a*) female; (*b*) male; (*c*) perfect. Asparagus plants normally are male or female, but occasional perfect flowers appear on male plants. (Photo courtesy of Iowa State University.)

will produce a few to many perfect flowers which have been useful in cultivar breeding programs. Because of the predominance of dioecy, plants must cross-pollinate, and the offspring, as a consequence, show substantial genetic variation in vigor and, to some extent, in appearance.

The asparagus fruit (Figure 10.2) is a red berry containing from one to several small, nearly round black seeds with hard seed coats. The seeds are primarily endosperm tissue, with a small incompletely differentiated embryo. As the seed germinates, a radicle (primary root) appears first, followed by the primary stem. As subsequent shoots develop, the primary shoot senesces and dies.

The development of asparagus is unique in several other respects. Along the edible spears (stems) are triangular, papery bracts, the only true leaves (Figure 10.3). These bracts have no photosynthetic function; instead, photosynthesis is active in the modified stems (**cladophylls**) that constitute the vigorous fern growth.

FIGURE 10.2

Asparagus berries, each containing one to five seeds, and cladophylls comprising the fern. The berries abscise from the plant, and subsequent seedlings pose a weed problem each year. (Photo courtesy of University of New Hampshire.)

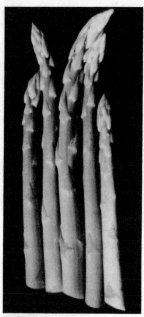

FIGURE 10.3

Asparagus spears. The small bracts at each node on the spear are the true leaves but are not photosynthetically active. Photosynthesis is centered in the cladophylls. (Photo courtesy of University of New Hampshire.)

The carbohydrates stored in crown and roots support spear growth in the spring. The crown begins to develop shortly after seed germination when short rhizomes form at the base of the stem. Each rhizome contains a bud for new spear growth, and as growth continues, the number and size of new rhizomes and spears increase. The majority of buds form the year prior to harvest, and their size is highly correlated with subsequent spear size. Eventually, the crown becomes rather massive. In an established asparagus bed, crowns 12 to 20 in. (30 to 50 cm) in diameter are not uncommon.

Concurrently, an extensive fleshy storage root system is produced, extending outward and downward; many roots extend to a depth of 3 ft (0.9 m), and depths to 10 or more ft (3 m) have been recorded. These roots arise from the original primary and secondary roots and also adventitiously from rhizome tissue. Fibrous roots, those active in nutrient and water uptake, develop each year from the fleshy storage roots.

Edible spears develop from crowns, when cells already formed in buds elongate. Elongation proceeds first at the lowest internode and then within successive internodes, resulting in rapid spear growth. As spears develop, cell walls in the pericycle and vascular bundles gradually become lignified, starting at the base of the spear.

As spears are harvested, the stored carbohydrates become depleted, and vigorous fern growth must be encouraged during the balance of the growing season to restore crown and root reserves. Inferior fern growth, resulting from an excessively long harvest, disease, or poor management, reduces the amount of carbohydrate translocated to the crown, and subsequent yields will decline.

Crop Establishment and Maintenance

It is expensive to establish a permanent asparagus bed, and no cash returns are generated until the third year of growth. It is therefore important to establish the plants properly to ensure a long-lived bed. Asparagus stands are established from crowns, transplants, or direct seeding. If crowns are used, 1-year-old stock is preferred. Although crowns can be purchased, the risk of soilborne disease is less if crowns are produced by the grower. To produce healthy crowns, seeds first should be soaked in a 20 percent solution of sodium hypochlorite for 1 to 2 h, rinsed in water, dried, then, after the danger of frost has passed, sown thickly at a depth of $1\frac{1}{2}$ in. (3.7 cm) in sterilized outdoor beds of light-textured soil. In soils 42 to 50°F (5 to 10°C), germination will take up to 4 to 6 weeks, decreasing to 10 days at 70°F (21°C).

Prior to seeding, a complete fertilizer (1–1–1 ratio in a low-fertility soil) is added to the seedbed, then tilled deeply. Excessive nitrogen fertilizer must be avoided, since it promotes shoot and top growth at the expense of root and crown development, and development is favored by a ratio of 75 percent nitrate-N:25 percent ammonium-N. Day temperatures of 70 to 75°F (21 to 24°C) and night temperatures of 65°F (18°C) are recommended for seedling growth. As this growth proceeds, seedlings appearing weak or off-type must be rogued, leaving only vigorous specimens to develop crowns. Final spacing should be approximately 2 in. (5 cm) between plants. Early in the following spring, crowns must be dug before shoot growth occurs and placed in 35 to 40°F (2 to 4°C) storage until transplanted to a permanent location. For direct seeding, the same seeding procedure would be followed, except the rate would be light, and the soil would be fertilized as described for a permanent bed.

In recent years, permanent asparagus beds have been established successfully with transplants. Approximately 3 to 4 months prior to field planting, treated asparagus seed is sown 1 in. (2.5 cm) deep in containers in a greenhouse, and individual plants then are established in peat pots or in Speedling-type trays filled with a soilless medium amended with controlled release fertilizer. Before field planting, poor seedlings are rogued, and the remaining plants form sufficient crown reserves following field planting to endure winter freezes and to provide energy for spring growth.

Soils for the permanent asparagus bed must be deep, preferably 8 ft (2.4 m) or more, well drained, and friable to accommodate the plant's extensive root system. Loams, sandy loams, sands, and mucks have been used with success, although mucklands no longer are planted with this crop. Clay soils, with poor permeability, should be avoided.

Asparagus is only slightly tolerant to pH extremes. Well in advance of planting, often in the preceding fall, soils should be limed to bring the pH to between 6.0 and 6.8 and plowed deeply, preferably turning in a green manure crop and/or a liberal application of animal manure. Nitrogen added at this time will enhance breakdown of organic matter. Before planting, the soil should be plowed again, fertilized as indicated by soil test, and harrowed. Rates and ratios of fertilizer differ widely, depending on locale, soil type, and previous cropping history. In the eastern production areas, for example, rates of 1000 to 2000 lb/acre (1120 to 2240 kg/ha) of 1–2–1 or 1–2–2 normally are recommended. North Carolina recommendations include a general application of superphosphate, with nitrogen and potash applied at planting time, plus a banded application of superphosphate directly beneath the newly planted crowns. California growers use a preplant incorporation of 1–2–2 in the furrow. Phosphorus is considered essential for early aggressive root establishment and is thus prominent in most recommendations. Further, once the plants have been established, it is difficult to incorporate large amounts of phosphorus into the root zone. Asparagus requires higher amounts of boron than many other vegetables, and borax or solubor should be applied to fields known to be deficient. Normally, a rate of actual boron of 2 to 3 lb/acre (2.2 to 3.4 kg/ha) is sufficient to maintain an adequate level for several years.

The permanent planting is established by placing crowns or plants in a furrow. The top of the crown should be approximately 6 in. (15 cm) below the surface of the soil; transplants are not placed quite as deep. Many growers use a furrow–ridge system in which a wide furrow with a small ridge at the bottom is formed (Figure 10.4). This configuration is believed to favor root development, as the crown or transplant is placed on the ridge with roots oriented down on either side. After the crowns or plants are set, sufficient soil is drawn in the furrow to cover the crown. Then, as new growth proceeds during the season, the remaining soil gradually fills the furrows to the soil line. Spacings of 12 to 36 in. (30 to 90 cm) between plants and 42 to 60 in. (1.1 to 1.5 m) between rows are recommended. Several of the new clonal and hybrid cultivars require

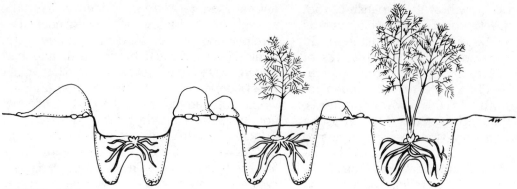

FIGURE 10.4
System for planting crowns. Crowns are placed on the ridge within the furrow, and soil is pulled over the crown gradually as the spears develop.

greater space than recommended for 'Mary Washington' and similar types. If wide rows or beds are desired, twin-row planting can be used, still keeping the crowns 12 to 36 in. (30 to 90 cm) apart.

Once asparagus is established, productivity and longevity depend on maintenance of fertility and on pest control. Each spring, prior to spear emergence, dead cane growth should be removed and burned. This practice largely was abandoned during the 1950s and 1960s, but experience suggests that it may help in reducing foliar disease problems. The field then should be disked only to a depth that will avoid damaging crowns. The disking reduces perennial weed populations and, in conjunction with an herbicide, can maintain a field relatively weed free for the entire season. Disking may damage the buds of old asparagus crowns, as they expand upward over time from their original 6-in. (15-cm) planting depth. Herbicides provide an excellent weed control alternative, and several have been registered for asparagus plantings. To control quack grass and other perennial weed species, glyphosate may be used prior to 7 days before harvest begins. This material is not selective and under no conditions should contact asparagus tissue. After harvest, *with all spears and stubs removed below the soil line,* glypho-

sate may be particularly effective, since active weed growth is more susceptible than early spring growth to the chemical. Volunteer asparagus seedlings also can be controlled at this time. Preemergence applications of diuron and simizine have been successful in suppressing some weed species but will not control quack grass.

Manure may be applied prior to spring growth, before the field is disked, or, more commonly, it is applied after the cutting season, before fern growth is allowed to resume. At either time, the organic matter should be well decomposed. Following each cutting season, fertilizer should be applied in bands on both sides of a row.

The most important nutrient to provide yearly is nitrogen. A normal harvest will remove 16 lb (7.3 kg) of nitrogen, 4 lb (1.8 kg) of phosphate, and 10 lb (4.5 kg) of potash, assuming a yield of 2 tons (1.8 MT). Expected uptake by a plant is only one consideration in fertilizing for optimum growth. Soil texture, inherent fertility, pH, organic matter content, and natural climatic patterns influence leaching and availability of nutrients.

Eastern maintenance fertilization programs for asparagus often include such ratios as 10–10–10, or 5–10–10 plus supplemental sidedressing of ammonium nitrate. Average rates of nutrients applied under California conditions are 142–52–22 lb (64–24–10 kg) of nitrogen–phosphate–potash. Fertilizer is applied before or after harvest, frequently both, applying half of the total rate before and after cutting. Soils inherently acidic should be limed prior to tillage to maintain pH at 6.0 to 6.8.

In tropical climates, asparagus is cut throughout the year, preserving crown vigor by allowing some spears to develop full fern growth, or twice yearly (the "mother plant" system). The total length of cutting season may be as much as 7 months. In such areas, fertilization practices and other management systems would be modified to ensure continuous vigorous fern growth.

Cultivars

In the early 1900s, J. B. Norton was commissioned to develop asparagus cultivars free from rust (*Puccinia asparagi*). Two notable selections were released as a result of this work: 'Mary Washington' and 'Martha Washington.' The 'Mary Washington' cultivar gained widespread popularity for its tolerance to rust and for its productivity. A number of strains since have been selected from 'Mary Washington,' including 'Paragon,' 'California 500,' 'Roberts Strain,' 'Viking,' and many others.

Because of inherent variability within most asparagus cultivars, these selections were substantially different from the original 'Mary Washington' in adaptation and yield.

In many current production areas, rust again has become a problem, largely within previously tolerant cultivars. The increasing infection may occur because these cultivars, propagated by open pollination, have lost tolerance, or it may reflect a change in the pathogen. Increases in asparagus diseases, especially *Fusarium* and rust, have renewed interest in varietal development. At the same time, tissue culture technology and development of homozygous male plants have made it possible to develop all-male hybrids. Although no truly disease-resistant cultivars, either standard or hybrid, have been developed, there are several that tolerate *Fusarium*-infested soils, and many also are tolerant to rust.

The current all-male cultivars are not uniform F_1 hybrids, since fully inbred parents are not used as parents. They are hybrid for the sex factor, somewhat more uniform than standard cultivars, and several show hybrid vigor (Table 10.4). Efforts to develop pure inbred parents by culturing anthers (haploid microspores) may lead to uniform vigor and other traits of value to the industry. Development of

TABLE 10.4
Characteristics of several U.S. asparagus cultivars

Cultivar	Rust tolerant	*Fusarium* tolerant	All male	Other features
Standard				
Mary Washington	×			
Martha Washington	×	×		
Viking 2k	×	×		
Emerald	×			No purple pigment
Hybrid				
Rutgers Beacon	×	×		
Centennial	×	×	×[a]	
UC 157	×			Smooth spear
Brock's Hybrid	×			High yield
Jersey Giant	×	×	×	High yield
Greenwich	×	×	×	High yield

[a] Small percentages of female plants may appear.

genetically uniform clones by tissue culture also has shown promise, yielding aggressive plants with superior productivity (Figure 10.5), but costs of such plants increase the initial investment in a permanent bed.

Disease and Insect Pests

Diseases The major diseases of asparagus are rust (*Puccinia asparagi*), *Fusarium* root rot (*Fusarium oxysporum* f.sp. *asparagi*), *Fusarium* crown rot (*Fusarium moniliforme*), *Phytophthora* spear rot (several *Phytophthora* species), *Botrytis* blight (*Botrytis cinerea*), *Stemphylium* leaf spot (*Stemphylium vesicarium*), and virus (Figure 10.6).

Asparagus Rust Asparagus rust is a typical rust fungus, with all spore stages occurring on asparagus. Some members of the Alliaceae,

such as onions and chives, also may become infected. The fungus overwinters as teliospores in old canes. If canes are left in place, the teliospores germinate in the spring and produce basidiospores which move by air currents to emerging spears and cause infection. However, rust is not usually noticeable during harvest. Later, small raised lesions (the aecia) appear on the lower part of the cane. These lesions gradually become orange and release aeciospores that reinfect canes and fern during moist periods. Approximately 14 days later, tan blisters appear throughout the foliage, these releasing urediospores. The urediospores continually reinfect asparagus tissue until fall. Late in the season, the telia develop, producing the overwintering teliospores. Infection is favored by warm weather with heavy dew, fog, or light rain. The orange pustule stage is the most obvious, and this and subsequent infection results in a significant

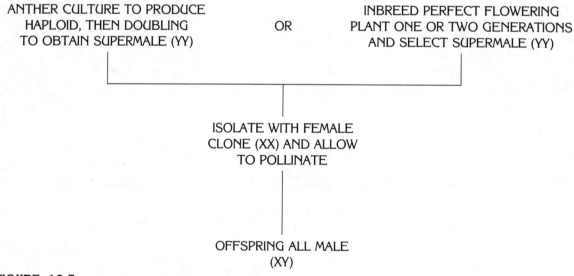

FIGURE 10.5
Methods for creating all-male cultivars. Obtaining inbred lines from haploid tissue has several theoretical advantages: it is a relatively rapid procedure and can lead to true F_1 hybrids. Practically, it has been difficult to obtain sufficient numbers of superior inbred lines by this method.

FIGURE 10.6
Diseases of perennial crops: (*a*) *Fusarium* symptoms on mature asparagus canes; (*b*) rust on asparagus fern; (*c*) crown rot of rhubarb. (Photos courtesy of University of New Hampshire.)

loss of photosynthetic area. Entire canes can be killed, and the long-range effect is a gradual reduction in carbohydrate translocated to crown and roots and a consequent decrease in yield. In several years of light to moderate infection, the economic loss will be noticeable. Control can be achieved through tolerant cultivars and proper management practices. Clean cutting and burning old canes each year will reduce inoculum levels and prevent infection for a considerable period of time. If fungicides are required, zineb is the only effective material, and applications should begin when fern growth resumes after harvest.

Fusarium   Crown and root rot is a serious disease complex affecting asparagus throughout the world. Infection causes a rapid decline in productivity. Of the two fungi responsible, the most serious usually is *Fusarium oxysporum* f.sp. *asparagi*, responsible for root decay. *Fusarium moniliforme* is responsible for crown rot and seems to follow prior infection by *F. oxysporum* on well-established or old crowns. The first symptoms of root and crown rots are shriveling of spears after emergence in the spring. Examination of the plant reveals root systems to be reddish brown and hollow, not the solid white of healthy roots. If crowns are cut, a reddish brown discoloration is evident. In the field, symptoms are sporadic at first, gradually increasing over several years to include most plants. Infected crowns persist over several seasons, but yields decline. Both fungi are relatively unaggressive species but may grow as saprophytes or parasites. They colonize old roots and crowns, gaining entry by wounds or directly through root tips. Asparagus harvest and insect damage by cutworms, grubs, wireworms, and asparagus miner also may provide access. Infection occurs particularly on plants growing under environmental stress. Control of root and crown rots is difficult. Although breeding efforts have produced tolerance in some cultivars, this tolerance is not reliable and may reflect primarily vigor of a particular cultivar. Cultural control is complicated by the presence of spores of *F. oxysporum* in virtually every seed lot and every crown. Infection can be delayed or minimized by (1) selecting tolerant (vigorous) cultivars, (2) growing plants from treated seed on sterilized medium, (3) using field locations not previously planted with asparagus, and (4) avoiding stress (drought, poor drainage, poor fertility, low pH, soil compaction, excessive competition, pest damage, excessive cutting season).

Phytophthora Spear Rot   Spear rot ("slime"), was reported on asparagus in California in 1938. It develops from infection of spears at or near the soil level by species of *Phytophthora*. Under moist conditions, the lesions become slimy, primarily because of infection by secondary organisms. In dry weather, lesions become light brown, and the spear will shrivel. The spears will be crooked if the lesion affects only one side of the spear. Mature storage roots may become brown and hollow. Samples in California have shown the organism to be active in a majority of production fields. During periods of high moisture, the disease may reduce yields substantially.

Botrytis Blight   Botrytis blight is characterized by browning of the lower foliage. This browning occurs most frequently during warm humid weather when the understory does not dry adequately during the day. The causal organism is *Botrytis cinerea*, a fungus responsible for decay in many small fruits and vegetables. The fungal spores, spread by wind and rain within the fern canopy, infect injured and dying fern, producing tan lesions. Subsequent sporulation will infect emerging spears, often causing complete blight. Botrytis blight is difficult to control when environmental conditions favor its development. Zineb spray can retard progression somewhat.

Stemphylium Leaf Spot   Leaf spot or purple spot (perfect stage, *Pleospora alii*) has increased in several asparagus production areas. Plant debris provides the overwintering site and the prime source of inoculum. Spores also may be carried on seed. At sporulation in the spring, the fungus penetrates spears at the stomates, resulting in small, oval, purple-margined surface lesions. These lesions reduce marketability but do not affect internal spear appearance. Infection of fern growth causes severe necrosis and loss of photosynthetic area, reducing the yield in subsequent years. The disease is favored by high humidity. Sprays and elimination of plant debris by burning and tillage provide economic levels of control.

Viruses   Virus diseases have been reported to reduce asparagus yields by as much as 20 percent. In tests in Washington, three viruses have been discovered, one of which also has been isolated from seed in other areas. In tests with two viruses, each alone caused a slight decline in vigor; plants infected with both viruses suffered severe decline and death after a year or two in the field. At least one virus seems to predispose plants to increased *Fusarium* infection. It is possible that tissue culture might be used to provide virus-free plants; otherwise, prevention of aphid infestations is the most effective control for these diseases.

Insects   Damage from insect infestation is confined to asparagus beetles, tarnished plant bug, alfalfa plant bugs, asparagus miner, aphids, and several soil larvae (Figure 10.7).

Asparagus Beetle   The asparagus beetle is the most common insect problem and may involve either of two species: the common asparagus beetle (*Crioceris asparagi*) and the twelve-spotted asparagus beetle (*C. duodecimpunctata*). Both are bright colored, easily recognizable pests that migrate from crop refuse when spears begin to emerge in the spring. Within a few days after feeding, the common asparagus beetles lay eggs at the tip of the spears, whereas oviposition by twelve-spotted beetle occurs as berries are setting, with eggs attached to the fern. Hatching occurs within 3 to 14 days, and small, dull green or gray larvae emerge and feed on young tissue. After 2 to 3 weeks, the mature larvae drop to the ground, forming pupal cells beneath the soil surface. In 1 to 2 weeks, adults again emerge to begin a new cycle. Damage can be extensive. Insect-feeding scars render spears

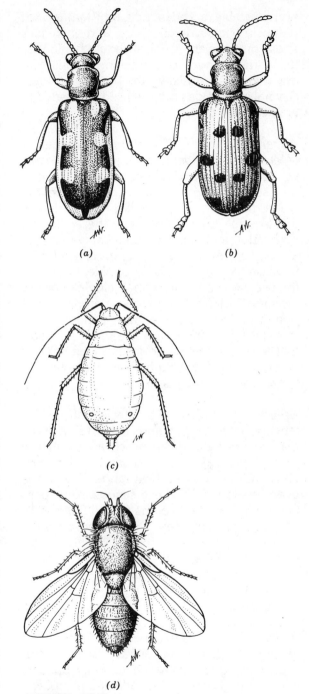

(a)   (b)

(c)

(d)

FIGURE 10.7
Insects of asparagus: (*a*) common and (*b*) spotted asparagus beetles; (*c*) asparagus aphid; (*d*) asparagus miner (adult fly).

unmarketable, and destruction of fern growth reduces photosynthetic area and increases stress on the plant. Control should begin with the elimination of overwintering debris and any wild asparagus near the production area. Several insecticides may be used as needed to eradicate the pest.

Asparagus Miner　The asparagus miner fly (*Melanogromyza simplex*) is small and black with clear wings and appears early in the cutting season. The flies lay eggs beneath the epidermis of asparagus stalks, and the larvae that emerge tunnel just under the epidermis both below and above ground. Larvae eventually pupate, and the entire life cycle takes 6 to 9 weeks to complete. It is possible, therefore, for two cycles to occur in a single growing season. The insect is not considered an economic problem, but the most serious effect may be in the increased *Fusarium* infection that seems to follow. No insecticides provide certain control, although several directed at beetle control may help.

Aphid　There are two species of importance: European asparagus aphid (*Brachycolus asparagi*) and green peach aphid (*Myzus persicae*). The latter is the primary vector of virus and has a broad host range. The European asparagus aphid is a serious insect pest in northwest sections of the United States but, although present, has caused recognizable losses only in certain areas in the east. The asparagus aphid, a very small, narrow, powdery green winged or wingless insect, injects toxins which, under high insect populations, cause growth malformations. Newly formed buds sprout immediately, giving a bushy growth of thin, stunted spears, in effect converting the plant to an annual. Because the overwintering crowns have few, if any, buds to start growth in the spring, affected plants are lost or, at best, severely weakened. Disyston is effective as a control for the asparagus aphid but is not suitable for green peach aphid. Either malathion or diazinon is required for con-

trolling the latter. Maintenance of a clean bed during the cutting season will prevent first generation females from reproducing.

Soil Insects　Cutworms cause damage to spears prior to emergence and cause spears to curl and stunt. It also is thought that the damage may encourage disease infection. Wireworms and grubs feed on the roots of grasses, and asparagus planted in a field previously in sod may suffer some damage. Any insecticide should be applied in advance of planting, not after the plants are established.

Harvest and Market Preparation

Asparagus harvest normally commences the third year following seeding, although there is evidence that light harvests during the second year are not harmful. The extent of early cutting depends on the vigor of growth and environmental factors that might stress the plants, predisposing them to disease or decreasing carbohydrate reserves. The length of any cutting season, regardless of age of the bed, can affect subsequent yield. In those areas with a long growing season, an extended harvest period still may leave ample time for plant recovery. The normal harvest period for a fully developed asparagus bed is 8 weeks.

Although mechanical harvesters have been tested with some success, most asparagus is harvested either by cutting spears below the ground with a special knife or by snapping above ground, both hand operations. Harvest aids that transport workers along the rows to facilitate cutting have improved efficiency. Cutting below ground maintains a very clean bed, eliminating fern growth attractive to beetles and other insects. It does create a wound below ground, however, that might provide a point of entry for pathogens, and the harvested spears must be trimmed prior to use to remove woody white tissue. Some growers therefore harvest spears by snapping at or above the soil level. This procedure leaves a stub that may send out axillary shoots that attract insect pests. These stubs also impede future harvest somewhat.

The spear length should be approximately 8 to 10 in. (20 to 25 cm) at cutting, with spear tips tightly closed. An emerging spear may reach that length in 2 days under warm temperatures and good soil moisture. In some areas, a slight ridging of the soil over the row is recommended midway through the cutting season. This procedure seems to enhance spear size, perhaps by reducing soil temperatures near the crown surface.

In Europe, blanched, or white, asparagus is preferred. Spears are blanched by ridging 10 to 12 in. (25 to 30 cm) of soil over the row at emergence. Long knives then are used to cut the spear when the tip first breaks through the ridge. Blanched spears have a mild flavor but lack the nutritional value of green asparagus.

Once spears have been harvested, they must be cooled immediately to 32 to 36°F (0 to 2°C). Cooling preserves the sugar content which, along with fiber content, is important in quality. Losses in sugar and increases in fiber take place most rapidly during the 24 h after harvesting, and although cooling will not prevent these quality changes, it does delay them markedly. In general, white asparagus deteriorates more rapidly than green. Cooling can be accomplished with hydrocooling, which also serves to clean the product, or by refrigeration.

Harvested spears are prepared for market by grading, sizing, and bunching. Grades are based on freshness, length and diameter of the stalk, color of spear, tightness of the spear tip, and extent of bruising. Spears of a large diameter are considered to be superior in quality. Studies have shown less fiber, on a percentage basis, in large-diameter spears than in thin ones.

Sizing is based on spear diameter (Table 10.5), and each bunch is trimmed to a standard length of 7 to 10 in. (18 to 25 cm), using bunching and cutting equipment. The spears are tied in bunches weighing 2 to 2.5 lb (0.9 to 1.1 kg) and packed vertically in pyramid crates

TABLE 10.5
Asparagus size grades

Category	Size	
	in.	mm
Very small	$<\frac{5}{16}$	<9
Small	$\frac{5}{16}-\frac{8}{16}$	9–12
Medium	$\frac{8}{16}-\frac{11}{16}$	12–17
Large	$\frac{11}{16}-\frac{14}{16}$	17–22
Very large	$>\frac{14}{16}$	>22

holding 12 bunches. The bottom of the crate is lined with moistened paper which helps to preserve quality of spears in transit.

Asparagus is a very perishable crop, and storage is not recommended. If market conditions demand, spears may be held at 32 to 36°F (0 to 2°C) and 95 percent relative humidity for no longer than 2 to 3 weeks. Even under these conditions, quality will deteriorate somewhat.

RHUBARB

Historical Perspective and Current Status

Rhubarb is native to cool regions of central Asia. Although records of its use in China date to 2700 B.C., the derivation of the name rhubarb (from the Greek *rha,* meaning Volga, and the Latin *barbarium,* meaning barbarian) infers its use by people living in Siberia. The early value of rhubarb was as a medicine. Acceptance as a food occurred much later, and it was not introduced into European diets until the seventeenth century, first in Great Britain. It reached the United States from Italy during the late 1700s and rapidly gained popularity in the New England area.

Rhubarb, or "pie plant," is prized for use in pies, tarts, and sauces. It also is used in making wine, and certain forms containing high levels of oxalic acid have been used in tanning hides. Nutritionally, rhubarb is 95 percent water and contributes fair amounts of vitamins A and C and moderate levels of calcium and potassium (Table 10.1). The raw product is not high in fats or carbohydrates, but its acidity (pH 3.1 to 3.2) normally must be offset by sugar in pies or sauces that adds considerably to the caloric value.

Only petioles (stalks) (Figure 10.8) of rhubarb are edible. Although medicinal use once involved roots and leaves, it is now understood that constituents within the leaf render it toxic to humans. This toxicity has been ascribed to high levels of oxalic acid, which do exist, but it is likely that other substances are primarily responsible. The petioles contain lower levels of oxalic acid than the leaves, and it is primarily in an insoluble form.

Rhubarb is produced in areas of the world where summer temperatures average less than 75°F (24°C), or in the winter less than 39°F (4°C). Within the United States and Canada, commercial production includes both field and forcing crops. Most of the outdoor crop is processed (frozen) for use in pies. Small amounts are sold as consumer packages of frozen product, and even less is sold fresh, mostly at roadside stands. The forced rhubarb crop is exclusively a fresh market commodity, and it commands an excellent price. Costs of forcing, however, have resulted in a decline of this industry in recent years.

Commercial production is concentrated in three states—Washington (275 acres), Oregon (200 acres), and Michigan (200 acres) (111, 81, and 81 ha, respectively). Of these states, Washington and Michigan also produce a forced crop [125 and 80 acres (51 and 32 ha), respectively]. Small commercial acreages for fresh market are located in many northern and central states.

FIGURE 10.8
Rhubarb leaf and petioles. (Photo courtesy of University of New Hampshire.)

Classification, Growth, and Development

Rhubarb, *Rheum rhabarbarum* (formerly *R. rhaponticum*), is a member of the Polygonaceae, or buckwheat family. Within this family are several garden plants and a number of weeds, including wild buckwheat, red sorrel, curly dock, knotweed, Pennsylvania smartweed, and others. It is the only species of *Rheum* cultivated for food.

Rhubarb is a cool season perennial, very winter hardy, and resistant to drought. The crop is produced from crowns consisting of fleshy rhizomes and buds from which are derived enlarged fibrous roots. Following a season of growth, the rhubarb crown becomes dormant, and temperatures below 39°F (4°C) are required to stimulate bud break and subsequent growth. The first shoots to appear in the spring are edible petioles and leaves, and these emerge sequentially as long as temperatures remain cool. As temperatures increase, this top growth is suppressed, occasionally appearing as almost complete dormancy in periods of extreme heat. With declining temperatures of late summer, foliage growth resumes.

Vernalization at 39°F (4°C) or below is necessary for flower production.

In most regions of production, flower stalks will be produced in early summer and can abort a harvest season if not removed. The flowers are bisexual, greenish white, borne on panicles on hollow, erect stalks. Seeds are characteristically winged achenes and highly heterozygous.

Crop Establishment and Maintenance

Rhubarb is established from crown divisions, eyes, buds, or sets. Each crown piece should contain at least one bud. Although seeds are sold, seed propagation generally is unsatisfactory because of extreme variation among resulting seedlings. Crowns may be divided in late fall and placed in storage until the following spring. Storage should be at 32°F (0°C) or slightly below, with protection against desiccation. Crowns must be planted as early in the season as soil conditions permit.

In preparing for planting, soils are plowed deeply following liberal applications of animal manure. Planting conditions are improved by applying manure, adjusting pH, and seeding a cover crop of winter rye or bromegrass the preceding fall. Prior to planting, the area should be treated with a complete fertilizer (1–1–1 or 1–2–1) and disked. Michigan recommends 1000 lb/acre (1120 kg/ha) of 20–20–20. In other areas, adjustments in the rates of phosphate and potash may be suggested by soil tests. Preplanting fertilization is important, since it is difficult to supplement phosphorus and, to some extent, potassium levels in the root zone in subsequent years.

The crowns are planted in rows 4 ft apart and normally 4 ft apart (1.2 × 1.2 m) in the row. Trenches approximately 6 in. (15 cm) deep are cut for each row with tractor-mounted shovels. The field then is marked at 4-ft (1.2-m) intervals across the trenches to establish a planting grid. Crown pieces are placed at the intersection of trench and mark, with the top of the bud 2 in. (5 cm) below the soil surface. Soil is pulled over the crown, firmed, and irrigated as needed. If a crop is to be harvested mechanically, spacing within the row normally is decreased to 18 in. (45 cm).

In subsequent years, fertility is maintained by a split applications of commercial fertilizer. The major application is made as the shoots are from 2 to 10 in. (5 to 25 cm) high and should be placed in a broad band over the row. In some production areas, nitrogen alone is applied (either urea or ammonium nitrate) at a rate determined by soil type and site. In such sites, a complete fertilizer (1–1–1 or 1–2–1) is substituted every second or third year. On soils not inherently fertile, a complete fertilizer application would be warranted every year. Following harvest, a light application of nitrogen [50 lb/acre (56 kg/ha)] is recommended to stimulate regrowth.

Crowns may have a longevity of many years, but because of disease and crowding, it is normal to reset a bed after 4 to 5 years of production. All crowns diseased or showing poor growth should be discarded at this time.

Before establishing a rhubarb bed, the field should be weed free. Treatment with glyphosate prior to seeding the cover crop in the fall preceding planting eliminates quack grass, a host of the potato stem borer and a serious pest in some production areas. There are few herbicides registered for use within an established rhubarb bed. Paraquat may be used prior to crop emergence to eliminate weed topgrowth. Subsequent weed control, however, is confined largely to cultivation. The grid planting system makes it possible to cross-cultivate, thereby reducing weed infestations considerably. It is important to eliminate weeds in and near a rhubarb field, particularly those weeds botanically related, since they provide overwintering hosts for disease and insect pests.

Cultivars

A range of cultivars is available for commercial and home production, categorized as green, pink, or red. Color may be confined to the epidermis or may extend through the entire petiole. A deep red petiole is popular among consumers but is often accompanied by poor growth and yield. The most productive cultivars include 'Canada Red,' 'Chipman,' and 'Valentine' (red); 'Strawberry,' 'MacDonald,' and 'Sutton' (pink); and, for forcing, 'Victoria,' a green type that produces a light pink under forcing conditions.

Disease and Insect Pests

Diseases   *Crown Rot*   Crown rot, or foot rot, is the most serious disease problem affecting rhubarb. It most often is caused by the fungus species *Phytophthora cactorum* and *P. parasitica,* particularly in the eastern United States. A *Pythium*-induced crown rot with similar symptoms has been identified in western production areas. Crown rot fungal activity produces lesions at the base of the petioles, causing a sudden collapse of the entire leaf. Infection penetrates the crown, and secondary organisms then cause rapid deterioration. The disease is common in heavy and poorly drained soils. Chemical control measures are not effective. Periodically, crowns should be dug and sorted, eliminating those showing infection.

Other Diseases   Under conditions of high humidity, rhubarb can be affected by *Botrytis cinerea,* causing gray mold, both in the field and in storage. *Rhizoctonia* crown rot, also favored by moist conditions and mild temperature, begins on petioles just below the soil line. Neither of these organisms becomes active unless the soils are poorly drained and air circulation through the field is impeded.

Insects   *Curculio*   The curculio (*Lixus concavus*) is the primary insect pest of rhubarb and occurs wherever rhubarb is grown. The insect is a rust or yellowish snout beetle that bores holes in petioles and crowns, but the larvae feed on curly dock. The most important control method, therefore, is the elimi-

nation of the alternate host, curly dock, from and around the rhubarb bed.

Harvest and Market Preparation

Harvest of rhubarb stalks should be delayed until the second year following planting to permit accumulation of root reserves. During the second year, a light harvest may be taken, and normal harvest may begin in the third year. Stalks may be clean cut, taking the entire plant at the soil line, or harvested sequentially by pulling the stalks at prime size over a period of 4 to 6 weeks. In areas of long growing seasons and vigorous plant growth without summer dormancy, a second harvest may be possible in late summer.

As stalks are cut, leaves are removed and left in the field. Stalks may be washed, although moisture does cause the petiole ends to "broom" or split. Stalks should be approximately 18 in. (45 cm) in length, trimmed, and bunched for fresh sale. The storage duration is 2 to 4 weeks at 32°F (0°C).

Yields depend on soils and crop management, but each crown should average 2.5 lb (1.1 kg).

Forcing Rhubarb

Commercial forcing of rhubarb has been a significant enterprise in Washington and Michigan and in several areas of Canada. In the late fall, crowns of established plants are dug and covered lightly with straw to prevent drying. After a period of 2 to 3 weeks of chilling or freezing temperatures [29 to 43°F (−2 to 6°C)], the roots are moved to a dark cellar, packed close together, and covered with 2 in. (5 cm) of soil. Moisture is provided, and a temperature of 60°F (16°C) is maintained by an electric or flue bottom heating system. From 4 to 6 weeks of heat are required before the first stalks are harvested. Because of the dark conditions, the stalks are quite long, bright pink, and very tender.

A structure 100 × 30 ft (31 × 9.2 m) should accommodate 1500 crowns, each producing approximately 2.5 lb (1.1 kg) of stalks. Following the harvest, crowns would be discarded.

GLOBE ARTICHOKE

Historical Perspective and Current Status

Globe artichoke was cultivated thousands of years ago in the central and western Mediterranean region of Europe, believed to be its place of origin. The earliest form cultivated was a leafy plant known as cardoon, used for edible roots or, more often, for petioles cooked as a green or used raw in salads. Artichoke became a popular luxury food in second century Rome, but was little mentioned during the Dark Ages. The first record of the modern type bearing edible flower heads came from Naples, Italy, in 1400 A.D.

In the United States, plantings were made by the French Settlers in Louisiana and by the Italians in California. It is now accepted as a delicacy in many parts of Europe, Africa, and North and South America. Nutritionally, artichoke is not a valuable crop (Table 10.1). Although low in protein, the levels of vitamins and minerals rank artichoke seventh among the top vegetables; but, in spite of its increasing popularity, per capita consumption is rather low.

The globe artichoke is used primarily as a luxury vegetable. It may be boiled, steamed, fried, stuffed, or marinated and is a popular hors d'oeuvre. A small percentage of the crop is processed, often in a marinade spiced with garlic. In addition to use as food, it has been reported to have medicinal benefits, including reduction of blood clotting and capillary resistance, neutralization of some toxic substances, and amelioration of gastrointestinal processes.

Of the more than 270,000 acres (109,350 ha) of artichokes worldwide, most (56 percent) is located in Italy, with smaller acreages in France (14 percent) and Spain (12 percent).

Ninety percent of U.S. production is located near Castroville, California, in Monterey County. This region produces from 65 to 100 million lb (29.5 to 45.4 million kg) for fresh market and a small amount for processing.

Classification, Growth, and Development

The globe artichoke, *Cynara scolymus,* is a tender herbaceous perennial restricted to cool, moist climates typical of the central California coast. The plants are coarse and thistle-like, reaching a spread of 4 to 6 ft (1.2 to 1.8 m) and a height of 4 to 5 ft (1.2 to 1.5 m). Root systems are deep, commonly penetrating to a depth of 6 ft (1.8 m). The leaves are large, deeply lobed, and somewhat pinnatifid, often whitish on the underside. Flower buds, the edible "chokes," terminate the main stems and branches and consist of a fleshy receptacle (the "heart") covered with broad, somewhat spiny bracts. The top growth dies back to the ground line each spring after the winter crop has been harvested, and new growth quickly renews the plant from the crown.

Crop Establishment and Maintenance

The artichoke plant cannot withstand a freeze, but it does tolerate a light frost. Temperatures below 30°F (−1°C) may kill the buds, and excessively warm days will impair quality as flower bud bracts open and become tough. Thus, the limits of adaptation are quite narrow.

Air temperature and moisture are the primary concerns in determining productivity of artichokes; soils are not a limiting factor. In California, artichoke fields are located on either heavy loams or light clay soils that are deep, well drained, and fertile. Prior to planting, the soil must be plowed deeply, working in substantial amounts of organic matter. Ammonium nitrate, at a rate of 60 to 100 lb/acre (67 to 112 kg/ha), normally is applied prior to final disking, and lime should be added if the pH falls below 6.5.

Artichokes are propagated vegetatively from old crowns, generally in the summer or fall in California. Basal stem/root pieces, called stumps, are planted on ridged rows 9 to 10 ft (2.7 to 3.0 m) apart at an in-row spacing of $4\frac{1}{2}$ to 8 ft (1.4 to 2.4 m). Two or three stumps are placed at each location or hill 6 to 8 in. (15 to 20 cm) deep, covered with soil, and provided with periodic irrigation to foster rapid establishment and early growth. Both vegetative and reproductive shoots will develop from buds on the stumps. There is a tendency, particularly if all stumps at a hill survive, for over-stimulation of vegetative shoots at the expense of reproductive growth. As each shoot develops, it forms its own root system, becoming an independent plant that subsequently spawns new plants by producing buds and shoots, and the crown and shoot area increase rather quickly. In Europe, these shoots with attached roots (called offshoots), obtained from old stumps, are used for propagation. In California, stumps have been preferred for some time, since early growth is thought to be encouraged by the additional carbohydrate reserves in root pieces, and the offshoots also require considerable care in handling (Ryder et al., 1983).

Once each year the foliage is cut 1 to 2 in. (2.5 to 5 cm) below ground level. This cutting enables growers to exercise some control over the time of flower bud development. For fall and spring harvests, fields would be cut between mid-April and June. For the summer harvest, the cutting would be late summer or fall. The cutting debris may be dried and burned or tilled into the soil as organic matter. During the harvest season, mature flower stalks that have completed bud production are removed. Growers believe this process prolongs productivity, but its value has been questioned.

A replanting of an existing bed using stumps reduces harvest substantially for 1 year. Thereafter, yields increase to a peak at 3 to 5 years, and bed life is from 5 to 10 years. Offshoots planted annually, early each summer, will improve yield in the first harvest season, in part reflecting the opportunity for increased planting density.

Each year, fertilizer is applied to the field before buds begin to form. The removal of nutrients during an average harvest approximates 75, 20, and 146 lb (34, 9.1, and 66.3 kg), respectively, of nitrogen, phosphorus, and potassium. In California, only ammonium nitrate is applied to artichokes, since soil reserves of phosphorus and potassium are ample.

Artichokes require rather frequent furrow irrigation. Although the preferred cool and moist environment restricts water loss by evaporation or transpiration, soils must contain ample, but not excess, water. Lack of moisture will cause development of loose flower buds of inferior quality. Weekly applications usually are required.

Weeds are controlled with diuron or simizin applied prior to weed emergence and by cultivation.

Cultivars

The most commonly grown cultivar is 'Green Globe.' This cultivar is typical of the Italian type with long and slightly pointed buds. French types, with a round or flattened end, are not used in the United States. The success of 'Green Globe' over a period of many years and the classification of artichoke as a minor crop have suppressed interest in genetic improvement. Only one new cultivar, 'Magnifico,' has been released in recent years, and it is grown on relatively few acres.

While seeds of 'Green Globe' are available from seedsmen, they tend to yield variable offspring. Many plants from seed are inferior, and, in spite of efforts to reduce the incidence of such traits as spiny bracts, the variability is unacceptable by commercial growers. For market and home gardeners, seed propagated artichokes might be appropriate. In some areas, there is interest in growing them from seed as annuals. In order to produce seedstalks, however, some vernalization of the plant is necessary.

Disease and Insect Pests

Diseases   The most important diseases are curly dwarf, caused by a virus, and botrytis, caused by *Botrytis cinerea.* Curly dwarf symptoms include leaf curl, a dwarfing of the plant, and reduced production of buds, many of them misshapen. The virus is transmitted easily during the replanting process if infected stumps are not culled carefully from propagative material. Insect transmission of the disease also may occur. Botrytis appears as a brownish fungus growth and decay on tissue damaged by insects or frost when the weather becomes warm and moist. In some areas, this disease is serious enough to discourage production. The best control for botrytis is low humidity and good air circulation, factors that do not enhance quality of the crop. Chemical control is not effective.

Insects

Artichoke Plume Moth   The plume moth (*Platyptilia carduidactyla*) is the most common pest of artichoke. It is a brown-buff moth with narrow wings. The yellowish larvae with black heads feed on leaves, stems, and developing buds (Figure 10.9). It is the latter feeding that can cause serious losses. Because there are several generations within a growing season, larvae can be found at any time. Control is enhanced by sanitation, by elimination of infested plant material, and by chemicals. Insecticides such as ethyl and methyl parathion once were used regularly, but the insect population became resistant to frequent applications, and other insect pests also seemed to increase. Some of the synthetic pyrethroids apparently are effective and have replaced parathion. There is interest and apparently some progress in biological control of this insect.

FIGURE 10.9
Artichoke plume moth larva feeding on flower bud. (Photo courtesy of N. DeVos, U.S. Department of Agriculture Research Station, Salinas, Calif.)

Aphids   Several species of aphids will feed on artichoke. One, the artichoke aphid (*Myzus braggii*), is yellowish green; the other, *Myzus fabae,* is a black species. Both cause serious damage by sucking sap from the underside of leaves, but both are controlled effectively with parathion or thiodan.

Harvest and Market Preparation
Weekly harvests (Figure 10.10) begin in the fall and extend to the spring. Each stem will bear several buds, but the terminal bud tends to be the largest and most desirable. At the appropriate developmental stage, full size but still with tightly appressed bracts, the buds

FIGURE 10.10
Moving harvested artichokes to a packing shed. (Photo courtesy of A. Kader, University of California, Davis.)

are cut, leaving 1½ to 2 in. (4 to 5 cm) attached stem, and placed in a hamper or picking container. The buds later are graded and packed according to bud size. Buds that are overmature, as indicated by lack of compactness, are very fibrous and must be culled.

During the winter, light frosts may cause development of bronze color at the bud tips or outer bracts. This color does not impair quality; in fact, the virtues of such "winter kissed" buds, considered to be quite flavorful, are advertised prominently during the marketing season.

Artichokes should be cooled to 40°F (4.4°C) after harvest and may be stored for a few weeks at 32°F (0°C) and 90 to 95 percent relative humidity. Cooling normally is achieved by hydrocooling, although both air and vacuum systems have been reported. The buds normally are packed immediately after grading for market in cardboard boxes containing from 48 to 125 chokes. These boxes are shipped under top-icing to market.

SELECTED REFERENCES

Ellison, J. H., D. F. Scheer, and J. J. Wagner (1960). Asparagus yield as related to plant vigor, earliness and sex. *Proceedings, American Society for Horticultural Science* **75**:411–415.

Franken, A. A. (1970). Sex characteristics and inheritance of sex in asparagus (*Asparagus officinalis* L). *Euphytica* **19**:227–287.

Robbins, W. W., and H. A. Jones (1925). Secondary sex characters in *Asparagus officinalis* L. *Hilgardia* **1**:183–202.

Ryder, E. J., N. E. DeVos, and M. A. Bari (1983). The globe artichoke (*Cynara scolymus* L). *HortScience* **18**:643–653.

Sims, W. L., V. E. Rubatzky, R. H. Scaroni, and W. H. Lange (1977). *Growing Globe Artichokes in California*, University of California Coop. Ext. Leaflet 2675.

Takatori, F. H., F. D. Souther, J. I. Stillman, and B. Benson (1977). *Asparagus Production in California*, University of California Sci. Bull. 1882.

Thompson, R. C., S. P. Doolittle, L. L. Danielson, and H. C. Mason (1958). *Asparagus Culture*, USDA Farmer's Bull. 1646.

Thornton, R., W. Ford, R. Dyck, W. Cone, R. Parker, and O. Maloy (1982). *Washington Asparagus Production Guide*, Washington State University Coop. Ext. EB-0997.

Wiebe, J. 1967. Physiodormancy requirements of forcing rhubarb. *Proceedings, American Society for Horticultural Science* **90**:283–289.

Wukasch, R. T. (1982). *Insect Pests of Asparagus*, Ontario, Canada Factsheet 82-061.

Wukasch, R. T. (1982). *Diseases of Asparagus*, Ontario, Canada Factsheet 82-045.

Zandstra, B. H., and D. E. Marshall (1982). A grower's guide to rhubarb production. *American Vegetable Grower*, December, pp. 8–10.

STUDY QUESTIONS

1. What are the most effective methods for minimizing field infection of asparagus by *Fusarium*?

2. Assess the relative merits of crown, transplant, and direct seeding establishment of asparagus.

3. Both rhubarb and asparagus produce copious seed. Why is rhubarb vegetatively propagated and asparagus seed propagated?

4. Contrast European and American systems for establishing artichokes. What are the advantages and disadvantages of each?

11
Potherbs or Greens

productive plants in terms of nutritional value per unit area, in part because they grow rapidly, allowing several crops or harvests in a season. Although some of the constituents are lost during cooking, they still contribute important amounts of vitamins A and C and of several minerals. Most leafy greens are rather easy to grow and, because of their short growing season, are useful for succession cropping with other vegetables.

In recent years, some of the potherbs, especially spinach, have been used raw in salads. Not only do they add flavor, but they excel in vitamins and minerals in comparison to the green salad vegetables, lettuce, celery, and endive.

SPINACH

Historical Perspective and Current Status

Spinach, the most important of the greens in the United States, is native to Iran and was cultivated by the Persians 2000 years ago. Records of its use are meager, but it is believed that cultivation of the crop developed during the period of Greek and Roman civilizations. It was introduced to China in 647 A.D. and apparently was transported across North Africa to Spain by the Moors in 1100. The crop was known in Germany in the thirteenth century only in the prickly seeded form. Smooth-seeded spinach, used exclusively in today's commercial production, was not described until 1552. The colonists introduced the crop to the New World, and it was listed in American seed catalogs in 1806.

Spinach	Collards	New Zealand spinach
Kale	Swiss chard	Amaranth
Beet greens	Mustard	Dandelion

Worldwide, there are many plants from which harvested leaves and stems are boiled for eating. The most important of these is spinach, but significant acreages also are planted with kale, collards, and mustard. Swiss chard, beet greens, New Zealand spinach, and dandelion are popular with home gardeners and are grown on a limited scale by market gardeners.

Greens are among our most nutritious vegetables on a fresh weight basis (Table 11.1) and some are among the world's most

TABLE 11.1
Nutritional constituents of greens[a]

Crop	Water (%)	Energy (cal)	Protein (g)	Fat (g)	Carbo-hydrate (g)	Vitamins					Minerals				
						A[b] (IU)	C[c] (mg)	Thia-mine (mg)	Ribo-flavin (mg)	Niacin (mg)	Ca (mg)	P (mg)	Fe (mg)	Na (mg)	K (mg)
Spinach	91	26	3.2	0.3	4.3	8,100	51	0.10	0.20	0.6	93	51	3.1	71	470
Kale	83	53	6.0	0.8	9.0	10,000	186	0.16	0.26	2.1	249	93	2.7	75	378
Collards	85	45	4.8	0.8	7.5	9,300	152	0.16	0.31	1.7	250	82	1.5	—	450
Mustard	90	31	3.0	0.5	5.6	7,000	97	0.11	0.22	0.8	183	50	3.0	32	377
Chard	91	25	2.4	0.3	4.6	6,500	32	0.06	0.17	0.5	88	39	3.2	147	550
Turnip greens	90	28	3,0	0.3	5.0	7,600	139	0.21	0.39	0.8	246	58	1.8	—	—
Beet greens	91	24	2.2	0.3	4.6	6,100	30	0.10	0.22	0.4	119	40	3.3	130	570

Source: National Food Review (1978), USDA.

[a] Data per 100 g sample.

[b] 1 IU = 0.3 μg vitamin A alcohol.

[c] Ascorbic acid.

Spinach is used both as a cooked green (potherb) and as a raw ingredient in salads. It contains a high level of vitamins and minerals, particularly vitamin A, calcium, phosphorus, iron, and potassium, and has moderate levels of protein (Table 11.1). All constituents of spinach are not of nutritional benefit, however. Oxalic acid in spinach and related greens (beet greens and Swiss chard) reacts with calcium to form calcium oxalates. Excessive oxalic acid may interfere with calcium absorption in human metabolism, a condition particularly serious for infants. Levels of oxalic acid are substantial in all spinach cultivars, although apparently less in savoy than in smooth leaf types.

An additional problem relates to the accumulation of nitrogen in the nitrate form, especially in spinach fertilized heavily with ammonium nitrate and grown under high temperatures and low light intensity. Nitrates convert to nitrites in digestion, and nitrites will oxidize hemoglobin to form methemoglobin. This substance can lead to methemoglobinemia, a disorder of humans and ruminants. Nitrates also can form carcinogenic nitrosamines. These toxic constituents in spinach do not present a risk when the crop is grown with proper fertilization and is consumed as part of a balanced diet.

Spinach is produced on approximately 35,000 acres (14,175 ha) in the United States, but significant acreage is confined to only a few states. The major fresh market commercial areas are California, mostly in Ventura County, and in the Winter Garden region of the lower Rio Grande Valley of Texas. California is responsible for almost one-half of the total fresh market spinach [total of 1,426,000 cwt (64,683 MT) in 1981], and Texas for one-third of the total, although the Texas acreage exceeds that of California (Table 11.2). Maryland, New Jersey, and Colorado also have significant acreages. For the processing industry, California is responsible for one-third of the total of over 152,000 tons (137,894 MT), with Arkansas and Oklahoma accounting for most of the remaining acreage.

Spinach is adapted to spring, fall, and winter production in different regions of the country. In some areas, a single planting may be harvested twice. As a result, average yields per acre differ widely, with California the most productive (Table 11.2).

Classification, Growth, and Development
Spinach, *Spinacia oleracea,* is a member of the Chenopodiaceae, a family including table beet, Swiss chard, sugar beet, and the

TABLE 11.2
Average yields of spinach in the leading production states, 1981

State	Average yield	
	lb/acre	kg/ha
California	14,845	16,626
Colorado	7,818	9,381
Maryland–Virginia	7,714	9,257
Texas	5,547	6,656
New Jersey	4,024	4,829

Source: Agricultural Statistics (1982), USDA.

common pigweed. In the vegetative growth phase, spinach is characterized by a compact rosette of ovate to triangular succulent leaves that may be crinkled (savoy) or smooth. The transition to reproductive growth (Figure 11.1) is accompanied by rapid stem elongation (bolting) and development of lateral stems from the axils of narrow, pointed leaves of the central stem. Flower clusters (spikes of panicles) terminate both main and lateral stems and tend to open first at midplant, progressing toward the tip and the base.

Sex expression in spinach is variable. It is predominantly dioecious, but there are also monoecious plants with varying degrees of maleness. Rosa (1925) has characterized the sex types as follows:

1. **Extreme males** Precocious, nonleafy or bracted, early male flowers, small plants.
2. **Vegetative males** Leafy but small plant, not as early, male flowers.
3. **Monoecious** Both flower types, later flowering, leafy.
4. **Female** Large plant, late flowering, female flowers.

Male plants normally die following flowering, female plants following ripening of seed. All spinach types become reproductive in response to specific daylength and temperature conditions. As daylength approaches 15 h, seedstalk initiation takes place, accelerated by exposure to 40 to 50°F (4.4 to 10°C) temperature. Once initiation of flower buds has occurred, however, actual elongation and flowering is favored by 57 to 68°F (14 to 20°C) or higher. Generally, the higher the postinduction temperature, the more rapid the bolting. Conditions favoring bolting occur frequently in northern states, but spinach can be fall planted and overwintered to produce an early spring crop, before warm summer temperatures accelerate seedstalk development. Because the monoecious and female forms are not as sensitive to low temperature as are extreme males and vegetative males, they predominate in commercial cultivars, particularly those listed as "long-standing."

Spinach has a deep tap root, with branch roots in the top 6 to 10 in. (15 to 25 cm) of soil. Branching is profuse, with roots extending outward up to 1 ft (30 cm) before growing downward. Although the taproot and many branch roots can extend to several feet, most of the feeder system is located in the top 2 to 4 in. (5 to 10 cm) of soil.

Crop Establishment and Maintenance

Spinach can be grown successfully in a variety of soils, ranging from light sandy loams

FIGURE 11.1
Spinach plant showing rosette form of vegetative growth. (Photo courtesy of O. S. Wells, University of New Hampshire.)

of Virginia and New Jersey (preferred for early production) to the silts and clay loams of Texas. High-quality crops also are grown on muck soils. Because the crop is very sensitive to soil acidity, the pH must not fall below 6.0. Spinach grown on acidic soils often will show symptoms of aluminum toxicity. Alkaline soils are acceptable but the fertility of such soils must be managed carefully to ensure top yields and quality. All soils should be well drained, either naturally or by forming raised beds.

All spinach is direct seeded, with little or no subsequent thinning, either broadcast or in rows on wide beds. In California, fresh market producers seed in rows or narrow bands near the shoulders of beds spaced 40 in. (1 m) on center. This system facilitates cutting the whole plant for bunching. Processors generally seed four rows on each half of a raised bed using a multiple shoe planter. Spinach for polyethylene bag packs (bulk harvested leaves) also is planted using this system. Row spacing varies between 5 and 16 in. (12.5 to 41 cm), depending on cultural system, requiring from 4 to 15 lb of seed/acre (4.5 to 16.8 kg/ha). In-row spacings range from 3 to 5 plants/ft (10 to 16/m) for fresh production to 8 to 10 plants/ft (26–32/m) for processing.

Emergence is related to temperature. In 40°F (4.4°C) soils, germination and emergence may require 3 weeks, but emergence should occur in 5 days at 75°F (24°C). If soil temperatures exceed 85°F (29°C), seeds may become dormant, and emergence and consequent stand will be erratic.

Normal soil preparation procedures are used for spinach. Limestone, if needed, should be applied prior to disking. In most areas, fertilizer is applied in bands, 3 in. (7.5 cm) directly below the seed. This practice is particularly helpful in soils high in pH, since the nutrients remain available to the plants over an extended time. A crop of 10 tons (9 MT), considered a good processing yield, removes approximately 100, 12, and 100 lb (45, 5.5, and 45 kg), of nitrogen, phosphorus, and potas-

sium, respectively, most of this removal occurring in the 21 days before harvest. Texas and California growers apply primarily nitrogen, since the soil levels of phosphorus and potassium are high. Eastern production areas must use a complete fertilizer. New York recommends 75 to 100 lb/acre (84 to 112 kg/ha) each of nitrogen and phosphate and up to 200 lb/acre (224 kg/ha) of potash, followed by a sidedress of 30 to 60 lb/acre (34 to 68 kg/ha) of nitrogen 3 to 4 weeks after seeding. Spinach is moderately sensitive to low boron; however, from 1 to 2 lb of actual boron/acre (1.1 to 2.2 kg/ha) may be required periodically for deficient soils.

Because of the shallow feeder root system, soil moisture must be maintained, and cultivation must be shallow to avoid damage to these roots. Irrigation is essential in most production areas. It is not uncommon to seed in a dry soil, then apply furrow or overhead irrigation immediately to start germination. Four to five additional applications of $1\frac{1}{2}$ to 2 in. (3.8 to 5 cm) of water would be needed to mature a crop under low-rainfall conditions.

Spinach does not compete well with weeds, also related to the shallow feeder roots. In addition, leafy weeds are difficult to remove from the harvested crop. Many growers rely on preemergence herbicides for weed control. Properly applied, they are very effective against a wide spectrum of grasses and broadleaf species, and the need for potentially damaging cultivation can be eliminated. Treflan and CDEC are two materials commonly used.

Cultivars
Until the mid- to late 1950s, most spinach cultivars grown in the United States were from Europe, even though these cultivars were poor in yield and susceptible to downy mildew. As a result, the U.S. spinach industry was not a robust one. Efforts beginning in 1947 to transfer mildew resistance into adapted types also revealed the merits of F_1 hybrids. The flowering habit characterizing spinach has made it possible to develop inbred lines and, eventually,

hybrids that have proven to be superior in vigor, uniformity, bolting resistance, and disease resistance and are adapted to mechanization. Although several old standard cultivars ('Bloomsdale Long Standing,' 'Viroflay') still are available in the seed trade, there is no justification for using them. Superior hybrids are available in savoy (crumpled leaf) form, preferred for fresh market, semisavoy for both fresh and processing markets, and flat or smooth leaf forms, generally used in processing. Hybrid cultivars that are somewhat upright in growth have been developed for mechanical harvesting. Well over 75 percent of the acreage devoted to spinach production is planted with hybrid cultivars, a dramatic change in a 10-year period. Savoy types are preferred for the fresh market because of superior form and dark green color. The smooth leaf type is generally superior in yield and is more easily cleaned; hence, cultivars of this type are preferred by processors.

Disease and Insect Pests

Diseases   A crop such as spinach, in which the leaves are consumed both raw and cooked, must be maintained free of defects. Even slight damage reduces marketability. Three diseases are found in most production areas, and three others are prevalent primarily in western states.

Downy Mildew   Downy mildew, or blue mold, is caused by the fungus *Peronospora spinaciae*. This organism has worldwide distribution and is particularly severe in the southern states and in coastal regions. Just prior to sporulation, the leaves develop light yellow areas. The fungus sporulates profusely in high relative humidity. The spores overwinter on plants in mild climates, but a secondary inoculation may result from windblown conidia. The best control measure is the use of resistant cultivars, but carbamates can be used effectively where chemical control is needed.

Fusarium Wilt   Wilt, caused by the fungus *Fusarium oxysporum* f.sp. *spinaciae,* has typical vascular discoloration, particularly in the crown area. The plants become pale green, and leaf margins roll inward as the plant begins to wither and eventually die. The fungus is both seedborne and soilborne and is common and persistent where soil temperatures are warm. If a problem exists, fields should be rotated and seeds treated. It is imperative that seeds be free from the organism. Most *Fusarium* diseases of vegetables can be controlled with resistance. Development of resistance in spinach, however, has been slow.

Mosaic   Mosaic is one of the most serious spinach field diseases. The causal organism is the cucumber mosaic virus, transmitted by the green peach aphid. The symptoms first appear on young leaves as a mottle, becoming yellow before the leaf dies. Virus eventually moves into the older leaves which also succumb. The concentration of virus in the plant seems to increase in long days and under high light intensity. Several cultivars resistant to virus have been released, and careful insect control is essential where susceptible cultivars are grown.

Curly Top   The virus disease, curly top, is common on beet and several other crops, including spinach. In western spinach fields, it can result in significant losses. Symptoms first appear as vein clearing on young leaves, followed by curling of leaf margins. Young plants usually will die; old plants may survive, but will be stunted. No resistance is yet available in cultivars, but it is the best long-range hope for control. Control of the beet leafhopper vector will reduce disease, although this insect is difficult to eradicate from production areas.

Beet Yellows   Yellows has been severe on spinach in California. It, too, is spread by the green peach aphid but also can be transmitted mechanically. Plants become yellow and, if infected early in growth, stunted. The disease can account for crop losses up to 6 tons/acre (13.4 MT/ha). Insect control is the only deterrent.

White Rust   Infection by *Albugo occidentalis* causes white pustules on the underside of leaves. These pustules may be scattered or grouped, and surrounding tissue turns brown and dies. The problem is found in the southwestern United States, and the recommended control is maneb applied at 7- to 10-day intervals. No resistant cultivars have been developed.

Storage Rot   Rot caused by the bacterium *Erwinia carotovora* is the most serious postharvest disease. This disease develops rapidly in bulk storage, especially if there is heat accumulation in the container. The result is a foul-smelling, wet rot that quickly destroys the entire lot. Careful grading and washing and cleanliness and proper cooling will minimize the incidence of this disease.

Insects

Green Peach Aphid   The green peach aphid (*Myzus persicae*) (Figure 11.2) is perhaps the most serious insect because it transmits several diseases that can destroy a large percentage of a crop. Infestations can be controlled, however, with thorough applications of malathion or diazinon.

Other Insect Pests   Other spinach insects include the cabbage looper, the cucumber beetle, and the leaf miner (Figure 11.2), all of which are leaf feeders, causing considerable damage. The descriptions and specific control measures are discussed for primary crop hosts.

Harvest and Market Preparation

Spinach matures in 37 to 70 days, most within 40 to 50 days. There is no single stage of maturity preferred for fresh market sales;

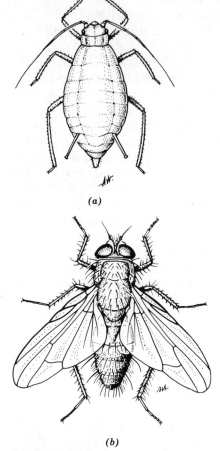

(a)

(b)

FIGURE 11.2
Insect pests of spinach: (*a*) green peach aphid; (*b*) spinach leaf miner (adult fly).

instead, a grower must judge yield and market price. As long as bolting can be avoided, each day of growth will add to the total yield. If, at the same time, market price is falling because of increased supply, the grower may lose income by delaying harvest. An acre of fresh market spinach at marketable size will yield 4 to 8 tons/acre (9 to 18 MT/ha); processing yields, because of the close spacing, are somewhat higher.

All processing spinach and, increasingly, acreages for fresh market are harvested mechanically. For fresh market, the plant is cut just below the crown and above the root. For processing spinach and the "poly bag" fresh pack, leaves are cut just above the growing point to allow regrowth. The harvested product then is washed, trimmed as needed, graded, and packed. Most fresh market spinach is shipped in wire-bound crates or in bushel baskets for regrading, washing, and repacking in polyethylene bags at the destination; some also is shipped in crates containing 2-dozen bunches. Spinach is sold as bunches, loose, or, most commonly, in consumer-size polyethylene bags.

Because spinach has a high surface area and high respiration rate, it must be cooled rapidly to prevent weight loss and decay. Satisfactory cooling can be achieved in 10 min by a vacuum system, usually applied after bulk packing and washing. Hydrocooling takes longer than vacuum cooling but is more feasible for small market operations. Following hydrocooling, excess water should be removed by centrifuging. The product is shipped under snow ice to preserve freshness.

Quality in spinach is related to bright green color, tenderness, and flavor; cleanliness (freedom from grit) and appearance are major concerns of the consumer. Discolored or wilted leaves or those with mechanical damage not only will detract from appearance but also will reduce shelf life.

KALE AND COLLARDS

Historical Perspective and Current Status

Kale and collards are the oldest forms of cabbage, native to the eastern Mediterranean region of Europe. The use of kale as a food is believed to date to 2000 B.C. or earlier. Theophrastus described a savoyed form of kale in 350 B.C. Through the travels of traders and nomads, many parts of the world were introduced to these greens.

In the United States, the collard is popular in southern diets, while the use of kale is much more widespread. Both crops excel in food value, with kale superior to most vegetables in protein, vitamin, and mineral content. On a fresh weight basis, kale is among our most nutritious vegetables (Table 11.1). One serving meets all adult daily requirements of vitamins A and C and 13 percent of the calcium requirement.

Collards and kale are adapted to a wide area, but commercial production is concentrated along the eastern shore. Collards, which tend to be heat tolerant, are grown from Washington, D.C., southward. Kale, which is cold tolerant, is concentrated from the Norfolk area of Virginia northward to Long Island, New York, and can be grown over a long season. Small plantings of kale can be found in many parts of the country.

Classification, Growth, and Development

Collards and the most commonly grown kales belong to the same taxonomic group, *Brassica oleracea,* Group Acephala in the family Brassicaceae, yet each plant is quite distinctive. Collards are characterized by broad, flat, or slightly furrowed leaves, whereas most popular kale cultivars have very curled, almost fringed leaves (Figure 11.3). Neither crop forms a head. Both are biennial, producing a rosette form of vegetative growth on a short stem. A vigorous collard plant may reach a height of 3 to 4 ft (0.9 to 1.2 m). Kale is somewhat smaller. After exposure to cold temperature, approximately 40°F (4.4°C) or less, for a period of 8 to 10 weeks, reproductive growth is initiated, and subsequent exposure to mild temperatures results in rapid elongation of the stem and eventual flowering. Siberian kale, a dwarf perennial with glaucus blue to purplish leaves, is classified as *Brassica napus,* Group Pabularia.

Crop Establishment and Maintenance

Both kale and collards can be transplanted but generally are direct seeded. Heavy, but friable, loams with a pH of 6.5 to 6.8 will support

FIGURE 11.3
Production fields of kale (*a*) and collard (*b*). (Photos courtesy of E. A. Borchers, Virginia Truck and Ornamentals Research Station, Virginia Beach.)

planting, a split application of one-half at seeding and one-half early in the spring is recommended.

Seeds normally are drilled ¼ in. (0.6 cm) deep as soon as the soil can be prepared. For spring plantings, in which young plants are to be harvested, the normal spacing is 6 in. (15 cm) in the row and up to 36 in. (0.9 m) between rows. Row spacing of plants from which only leaves are to be harvested as they reach marketable size should be increased. If fall seeding, in-row spacing is increased to 15 to 18 in. (38 to 45 cm). For small market gardens, a grower may seed at a close spacing, using thinnings from the row as an early crop.

Control of weeds and of insect and disease pests is identical to the systems described for cabbage. Because these crops are short season, often followed by different crops in the same year, weed control through cultivation is preferred.

Cultivars
Little varietal choice is available for either kale or collards. Two general types of kale are grown for market: curly leaf, the most widely grown, and smooth leaf. Of the curly leaf forms, Scotch kale is rather light green, with very ruffled, finely divided leaves, and may be dwarf or tall, with the dwarf form preferred. Siberian kale is a deep blue-green with leaves less curled than Scotch types.

The most common collard cultivar is 'Vates,' although several others are available. Efforts are underway to improve bolting resistance, cold and heat tolerance, and upright growth habit. The latter has been shown to reduce soil rots on leaves and it facilitates mechanical harvest.

Harvest and Market Preparation
Collards can be harvested by cutting young plants, similar to mustard greens. Large plants can be cut off, or the lower leaves can be removed during the season.

Kale may be thinned, with small plants bunched and marketed. Usually, however, en-

substantial yields and therefore are preferred. Soil preparation is the same as for cabbage (Chapter 12). Kale and collards are considered as relatively heavy feeders, since little of the plant is returned to the soil. Based on a yield of 10,000 lb (4540 kg) of greens, both crops require 40, 12, and 40 lb (18, 5.5, and 18 kg), of nitrogen, phosphorus, and potassium, respectively. Nitrogen is especially important in promoting leafy top growth, and a complete fertilizer (similar rate and ratio as recommended for cabbage) at planting, normally banded under the seed, would be followed by a side-dressing of 15 to 30 lb/acre (17 to 33 kg/ha) of nitrogen 1 month after seeding. For fall tire plants are harvested about 40 days after seeding, or, on widely spaced plantings or in tall cultivars, the lower leaves are harvested.

Leaves of both crops should be young and tender. Flavor of kale is especially sweet after a frost, and many prefer to harvest at that time. After harvest, the leaves or small plants should be washed, graded, and bunched or packed. Containers used for marketing include bushel baskets, crates, and cartons (12 or 24 bunches) or wirebound crates. Shipment is under ice to preserve freshness. When necessary, kale and collards can be stored for 10 to 14 days at 32°F (0°C) at 90 to 95 percent relative humidity. It is preferable to market these crops promptly.

SWISS CHARD AND BEET GREENS

Swiss chard is considered to be an ancient vegetable, and early civilizations utilized the roots as a medicine. The first records of cultivation place the origin of Swiss chard in the Mediterranean area, perhaps Italy. From this leafy plant was selected the swollen root form, and the table beet has become a far more important vegetable than its predecessor. Although large acreages are not common, both crops are grown widely to supply local markets.

Swiss chard and beet greens are very similar in cooked appearance, flavor, and food value. A normal helping of Swiss chard provides 87 percent of the average adult daily requirements of vitamin A and 25 percent of the required vitamin C. As in the related spinach, both Swiss chard and beet greens contain high levels of minerals (Table 11.1).

Both Swiss chard and beet greens belong to the family Chenopodiaceae and both are *Beta vulgaris*. Swiss chard is classified as Group Cicla. This form is leafier than the beet, with prominent enlarged midribs. It succeeds quite well regardless of soil type, daylength, or temperature and is therefore popular among home gardeners. Both Swiss chard and beet greens develop identically in early vegetative growth, forming a loose rosette of leaves attached to an insignificant stem. In the beet, however, the root soon enlarges and becomes a storage organ for carbohydrates, and during this period, leaf growth within the rosette is reduced. Swiss chard leaves sustain vigorous growth throughout the season. Both plants are biennials, flowering after a chilling period of 6 to 12 weeks at 50°F (10°C) or less.

The production techniques, including disease and insect control, are the same as those outlined for the garden beet (Chapter 14). Soil moisture, proper fertility, and effective control of leaf diseases and insects are important in maintaining leaf quality. Leaves are of best quality just when fully expanded or slightly earlier. Beet greens may be harvested as thinnings from a young crop or as individual leaves from young plants.

The tops of mature beets tend to be rather tough. Swiss chard, in contrast, remains succulent throughout the season as long as the leaves are harvested at proper size. Both greens are prepared for market by washing thoroughly, grading, and bunching. Storage is not recommended, but both commodities can be kept for short periods (see spinach).

MUSTARD GREENS

There are several species of mustard used for spice, flavoring, oilseed, fodder, and greens or salads. All are Brassicaceae, native to central Asia and the Himalayas, and several have become commercially important in many parts of the world.

There are four major species:

Sinapsis alba	White Mustard
Brassica juncea	Brown Mustard
Brassica nigra	Black Mustard
Brassica carinata	Ethiopian mustard

Of these species, *S. alba*, *B. juncea*, and *B. nigra* are the most pungent, and *B. juncea* and *B. nigra* are important oilseed crops.

The species predominantly used as a green is *B. juncea.* Its leaves are strong flavored and pungent, although the inner leaves are relatively mild and quite suitable for raw salad use. As in the other greens, nutrient levels are substantial (Table 11.1). Mustard greens are grown in many parts of the world. In the United States, most plantings are small, concentrated in Texas, California, Florida, Georgia, Louisiana, Mississippi, Tennessee, Arkansas, and Alabama.

Mustard is an annual cool season plant with its early growth in a basal rosette. Leaf form can vary among cultivars. Some have large leaves; others have leaves that are broad toward the apex. Within both forms are cultivars with curled and smooth leaf margins.

The crop normally is seeded in rows 12 to 18 in. (30 to 45 cm) apart or in twin rows on beds spaced 40 in. (1 m) on center. Sandy loam soils with a pH between 5.5 and 6.8 are preferred. Soils are prepared as for the other greens, and a 15–10–15 fertilizer would be used on low-fertility soils, applying 150 lb/acre (168 kg/ha) of nitrogen as a band at seeding. Three to four weeks after emergence, the young plants are thinned to reduce crowding, and the thinnings may be marketed. In approximately 7 weeks, the crop should be ready for harvest. Plants are cut by hand, washed, and packed. To preserve quality, mustard must be cooled immediately to 32°F (0°C) at 90 to 95 percent relative humidity. It is packed for transit in the same way as spinach.

NEW ZEALAND SPINACH

Unlike many of the greens, New Zealand spinach is a warm season crop with very wide adaptation. It is native to east Asia (Japan, Australia, New Zealand) and was introduced to England in 1771. Food value is excellent, with an average serving providing 72 percent of the adult daily requirements of vitamin A and 27 percent of vitamin C. Its flavor is comparable to that of spinach, but without the astringency.

Production statistics are combined with those of spinach, but acreage is limited and found primarily in market gardens in areas not suited for spinach growth.

New Zealand spinach (*Tetragonia tetragonoides*) is a member of the carpetweed family, Tetragoniaceae. Growth is procumbent and rapidly spreading (Figure 11.4), and plants eventually may reach 3 to 4 ft (0.9 to 1.2 m) in diameter. The stems are much branched, bearing thick, dark green, somewhat triangular succulent leaves. The young tops are harvested for boiling, and each harvest encourages new branching. A crop thus may be harvested repeatedly through the season.

In its early growth, New Zealand spinach is entirely vegetative. As it begins to develop, however, it soon produces flowers from the leaf axils. These flowers are quite inconspicuous but are considered by growers to be undesirable in the market product. Experiments by Kays and Austin (1975) were successful in limiting flower production and increasing leafiness in plants by application of growth regulators.

There are no active breeding efforts to improve New Zealand spinach. Seed is sold as a type of spinach, not under a specific cultivar

FIGURE 11.4
New Zealand spinach. As the shoot tips are harvested, branching is encouraged, maintaining yield throughout the season. (Photo courtesy of University of New Hampshire.)

name. Seeding should be delayed until the danger of frost has passed, preferably after the soil has warmed to 50°F (10°C). Soils should be prepared as for spinach, adjusted to a pH of 6.0 to 6.8, and fertilized as for spinach, using a band or broadcast application. Nitrogen is the most important need and in some areas may be the only nutrient applied. Substantial potash is required for soils testing low in potassium.

The crop is direct seeded, often using seed presoaked for 2 to 3 h. Seeds should be placed 1 in. (2.5 cm) deep at an in-row spacing of 4 to 6 in. (10 to 15 cm). Later, the plants would be thinned to 12 to 18 in. (30 to 45 cm) apart. Cultivation for weed control is feasible for a young planting, but as the plants develop their spreading habit, cultivation becomes difficult and generally ineffective. Insect and disease pests are minimal in New Zealand spinach.

When the plants have developed sufficient size, perhaps 10 to 12 in. (25 to 30 cm) in diameter, branch tops may be harvested. The tops are cut approximately 3 in. (7.5 cm) in length, washed, graded, and packed and should be sold as soon as possible after harvest.

MISCELLANEOUS GREENS

To a limited extent for commercial growers, more frequently for home gardeners, other greens are grown to satisfy a small, but steady, demand. The most common is the dandelion (*Taraxicum officinale*), a perennial weed plant in the Asteraceae. Like many of the greens, its leaves are arranged in a whorl, and the forms used for food have been selected for their leafiness and freedom from bitterness.

Dandelion production in the United States is centered primarily in five states: Texas, Florida, New Jersey, New York, and Illinois. The leaves are an excellent source of vitamin A (14,000 IU/100 g sample), vitamin C (35 mg/100 g sample), calcium, and several other

minerals. It will succeed in a wide range of soil types within a pH range of 5.0 to 6.8. Plants normally are established at a spacing of 12 to 24 in. (30 to 61 cm) between rows and 10 to 12 in. (25 to 30 cm) within rows. In the second year of growth, plants are tied to promote blanching, then cut while young to reduce bitterness. All plants are harvested by cutting below the whorl to keep the plant intact. The leaves then are washed, graded, and cooled as other greens.

Other plants used as greens have included lamb's-quarter, wild cress, upland cress, orach, dock, and several of the Brassicaceae (turnip, cabbage, rutabaga, kohlrabi, brussels sprouts). Some of these species normally are weeds; others are recognized by a use or feature more prominent than those associated with potherbs.

Amaranth (*Amaranthus* species) is grown both as a leafy green vegetable and for grain in subtropical and tropical climates. The grain amaranth, *Amaranthus caudatus,* is an important crop in some areas of Africa, whereas the vegetable amaranth, *A. tricolor,* is grown predominantly in Southeast Asia. Vegetable amaranth has been offered in the United States under the cultivar name 'Tampala,' and it has been as acceptable as spinach in cooked quality, but not as a raw product. Amaranth is not as high in vitamin A as spinach, but other constituents are comparable. The crop is either direct seeded or transplanted in fertile soils and reaches a height of 8 ft (2.4 m) at maturity. The young leaves or plants are harvested 3 to 4 weeks after transplanting, either as whole plants or as individual leaves.

SELECTED REFERENCES

Campbell, T. A., and J. A. Abbott (1982). Field evaluation of vegetable amaranth (*Amaranthus* spp.). *HortScience* 17:407–409.

Kays, S. J., and M. E. Austin (1975). Use of growth regulators for increased quality of New Zealand spinach. *HortScience* 10:416–417.

Longbrake, T., S. Cotner, J. Larsen, and R. Roberts (1973). *Keys to Profitable Spinach Production,* Texas A&M University System Ext. Serv. Fact Sheet L-1076.

Rosa, J. T. (1925). Sex expression in spinach. *Hilgardia* **1**:259–274.

Ryder, E. J. (1979). *Leafy Salad Vegetables.* AVI, Westport, Conn.

Sackett, C. (1975). *Spinach. Fruit and Vegetable Facts and Pointers.* United Fresh Fruit & Vegetable Assoc., Washington, D.C.

Webb, R. E., and C. E. Thomas (1976). Development of F$_1$ spinach hybrids. *HortScience* **11**:546.

STUDY QUESTIONS

1. What are the conditions that cause spinach to bolt? Are there cultural systems that can be used to reduce the impact of bolting?

2. How and why do fertilization practices affect the quality of spinach?

3. Of the greens, which are most suited to warm climates? Do high temperatures alter quality of all greens? How can quality be ensured in crops subjected to high temperatures?

12
Cole Crops

Cabbage	Cauliflower	Chinese cabbage
Broccoli	Kohlrabi	Brussels sprouts
	Broccoli raab	

The family Brassicaceae includes a number of vegetables of worldwide importance, and most of these are native to Europe, the Middle East, or Asia. Two species, *Brassica oleracea* and *B. campestris,* appear to be the source of most edible crops in this family. The several forms of *B. oleracea* grown commercially in the United States and Europe are naturally distributed along the cliffs bordering the Mediterranean Sea and along the Atlantic coast of Europe. Evidence suggests that *B. campestris* originated from the Mediterranean through the Near East, Iran, Afghanistan, and Pakistan. Al-

though there are exceptions, *B. oleracea* forms tend to be biennial, with woody basal stems, and *B. campestris* forms are annual or biennial with fleshy basal stems.

The group of vegetables classified as cole crops includes cabbage, brussels sprouts, broccoli, cauliflower, kohlrabi, Chinese cabbage, and broccoli raab. Of these plants, all but Chinese cabbage and broccoli raab are botanical forms of *B. oleracea*. Although there is some disagreement as to the taxonomy of Brassicaceae, Chinese cabbage and broccoli raab currently are placed in the species *B. campestris*.[1,2]

The cole crops are classified as a unit because of their close taxonomic relationships and the similar cultural requirements. All are hardy, cool season crops and all are susceptible to the same insect and disease pests. There are differences among them in sensitivity to environmental stress and, of course, in use. As a group, the cole crops constitute a very significant commercial enterprise.

Nutritionally, the cole crops are quite variable in vitamin and mineral content (Table 12.1). A common characteristic of all cole crops is the presence of **glucosinate** which, in the presence of myrosinase, hydrolyzes to form several goitrogens, including isothiocyanates (mustard oils), thiocyanate, nitriles, and goitrin. These substances contribute to the characteristic odor and flavor and, in substantial amounts, can diminish the supply of iodine to the thyroid or interfere with thyroxine production, causing glandular enlargement (goiter). General levels of the most active substances in cultivated plants are low in

[1] L. H. Bailey and E. Z. Bailey (1976) *Hortus Third.* Macmillan Co., New York.

[2] N. W. Simmonds (ed.) (1976) *Evolution of Crop Plants.* Longmans, London.

TABLE 12.1
Nutritional constituents of the cole crops[a]

| Crop | Water (%) | Energy (cal) | Protein (g) | Fat (g) | Carbo-hydrate (g) | Vitamins | | | | | Minerals | | | | |
						A[b] (IU)	C[c] (mg)	Thia-mine (mg)	Ribo-flavin (mg)	Niacin (mg)	Ca (mg)	P (mg)	Fe (mg)	Na (mg)	K (mg)
Cabbage	92	24	1.3	0.2	5.4	130	47	0.05	0.05	0.3	49	29	0.4	20	233
Cauliflower	91	27	2.7	0.2	5.2	60	78	0.11	0.10	0.7	25	56	1.1	13	295
Broccoli	89	32	3.6	0.3	5.9	2500	113	0.10	0.23	0.9	103	78	1.1	15	382
Brussels sprouts	85	45	4.9	0.4	8.3	550	102	0.10	0.16	0.9	36	80	1.5	14	390
Chinese cabbage[d]	95	14	1.2	0.1	3.0	150	25	0.05	0.04	0.6	43	40	0.6	23	253
Kohlrabi	90	29	2.0	0.1	6.6	20	66	0.06	0.04	0.3	41	51	0.5	8	372

Source: National Food Review (1978), USDA.

[a] Data per 100 g sample.

[b] 1 IU = 0.03 μg vitamin A alcohol.

[c] Ascorbic acid.

[d] Heading form.

comparison to wild forms. There is some evidence that the goitrogens also may contribute allelopathic effects: the breakdown products, particularly in sandy soils, have been reported to inhibit root growth of subsequent crops.

Of the cole crops, cabbage is the most important, followed by broccoli, cauliflower, and brussels sprouts. Kohlrabi, Chinese cabbage, and many Asian vegetables in the Brassicaceae do not have as wide appeal but are gaining in sales as the popularity of ethnic foods increases.

CABBAGE

Historical Perspective and Current Status

Wild cabbage, a leafy winter annual native to Europe, grows along the coast of the North Sea, English Channel, and northern Mediterranean. This form is the most likely ancestor of kale, and kale is believed to have been the first domestic form of cabbage. The Greeks cultivated it as early as 600 B.C., and it was described by Theophrastus in 350 B.C. Early use of cabbage was medicinal, a treatment for gout, stomach problems, deafness, headache, and hangovers. A soft-headed form was reported by Pliny in ancient Rome, and apparently the Saxons and Romans both cultivated this type and may have introduced it to the British Isles. The hard-headed types were not mentioned until the ninth century. Wherever introduced, cabbage quickly was adopted and cultivated.

Cabbage is not particularly high in vitamins and minerals, but, because of the volume consumed, it contributes substantially to the average adult daily requirements. The several forms of heading cabbage have somewhat different nutritional constituents (Table 12.2), with the savoy type slightly superior to others.

TABLE 12.2
Differences in nutritional constituents among cabbage cultivar types[a]

Cabbage type	Water (%)	Vitamin A[b] (IU)	Vitamin C[c] (mg)	Ca (mg)
Green	92.4	130	47	49
Red	90.2	40	61	42
Savoy	92.0	200	55	67

Source: National Food Review (1978), USDA.

[a] Data per 100 g sample.

[b] 1 IU = 0.03 μg vitamin A alcohol.

[c] Ascorbic acid.

Cabbage is an important fresh market and processing crop in most areas of the world, particularly in Germany, France, Italy, Belgium, Japan and other countries of Asia, and Australia. In the United States, it is grown for fresh market in most states, but there are several areas that are major suppliers for a particular season.

Winter   Texas, Florida, California
Spring   Florida, California
Summer   New York, Wisconsin, California
Fall   New York, Texas, California

Other production areas include Georgia, North Carolina, Michigan, and New Jersey. Fresh market production in the United States totals approximately 19,800,000 cwt (898,128 MT). Most of the processing crop, 230,800 tons (209,381 MT) annually, is confined to New York and Wisconsin. In total volume, cabbage is our fourth most important vegetable, after potato, lettuce, and tomato. It ranks eighth in contribution to sales.

Classification, Growth, and Development

Cabbage (*Brassica oleracea,* Capitata Group) is a cool season biennial forming a terminal head. It is highly tolerant of freezing temperatures, less so to excessive heat. The optimum for growth is 59 to 65°F (15 to 18°C).

In early developmental stages, the cabbage plant shows no tendency to head, but as leaves become large and growth accelerates, new leaves arising from the very short stem curve and cup inward, overlapping to cover the growing point. New leaves developing inside become crumpled and densely packed as the head develops. The hardness of the head relates to smoothness and number of leaves (the smoother the leaf, the more dense the head), to angle of leaf branching from the stem, and, to some extent, to head shape. Head size develops slowly at first, accelerating

during mid- to late growth until the head constitutes over one-half of the plant's total weight at harvest.

The root system is fibrous and finely branched. In early growth, roots are confined to the top 12 in. (30 cm) of soil but later may extend to a depth of 5 ft (1.5 m) and up to 3½ ft (1.1 m) in lateral spread. Many feeder roots are located within 2 in. (5 cm) of the soil surface. Because of the extensive root system and the demand for nutrients, and perhaps related to allelopathic activity, cabbage is considered to be hard on the soil, and other vegetables subsequently grown on this land may show reduced growth.

Although the ancestral cabbage was an annual, the heading form is strongly biennial. In areas of mild winters, it will flower and complete its life cycle in the second year. Temperature triggers the transition to reproductive growth. In field production, bolting losses are more common in early cabbage than in a late crop. Seedlings are not as sensitive to induction as large plants, and some cultivars are more susceptible than others. A cabbage plant exposed to 50 to 55°F (10 to 13°C) (threshold temperature) for several months will initiate a seedstalk. The percentage bolting will increase with increasing exposure to temperatures at or below 40°F (4.4°C). If the plant then is exposed to a 70°F (21°C) environment, the stem will emerge rapidly, splitting the center of the head. The reproductive growth is characterized by rapid elongation of the central stem, eventually bearing long racemes of whitish yellow flowers, each with four petals arranged as a cross. These flowers are highly self-incompatible, and seeds produced by natural pollination can be genetically quite variable.

Crop Establishment and Maintenance

Cabbage is produced on a variety of soils, both mineral and organic. The preferred pH is 6.5 or slightly higher, although plants will tolerate soils within a range of 5.5 to 6.8. Sands and sandy loams are most suitable for early cabbage, and main crop or late production should be confined to heavy mineral soils or muckland.

The crop may be established by direct seeding or, more commonly, by transplanting. If direct seeded, the seeds are planted at a spacing of 1 to 2 in. (2.5 to 5 cm). Later, the seedlings are thinned to the desired density. For most areas, the seeding rate would be 1 lb/acre (1.1 kg/ha). Emergence in soils of high impedance can be reduced or delayed substantially, affecting uniformity of the crop. Soils planted with cole crops previously may be infested with the clubroot and *Fusarium* wilt organisms, and fumigation before seeding would be recommended.

Transplants have been produced traditionally in the south, where the seeds are sown thickly in rows 12 to 16 in. (30 to 40 cm) apart in outdoor seedbeds, dug at the appropriate seedling stage, and shipped to northern growers. There is a risk of both soilborne and seedborne disease transmission in such plants, and many cabbage growers prefer to produce their own transplants. As far north as New York, outdoor seedbeds can be used to grow plants for the late fall storage crop or for processing crops. These plants normally are well hardened when dug, and transplanting may be delayed until field conditions are suitable. Greenhouse grown plants must be planted promptly; otherwise, plants become soft and leggy.

A good transplant can be grown in a forcing structure at 60 to 65°F (16 to 18°C) in 4 to 8 weeks. After the seedlings emerge, they are transferred to a soilless medium in peat pots or Speedling trays. Plug flats are seeded directly, using automated equipment that meters one to two seeds per cell. Final crop yield does not seem to be affected by the size of the transplant container. Adequate moisture and fertilizer, the latter often supplied by a controlled-release formulation, are necessary for early growth, but excess nitrogen will favor top growth at the expense of a well-developed root system.

Fields should be well tilled after fertilizer and lime application. Because cabbage generally is a full-season crop, growers may need to split their fertilizer application, applying some as a broadcast prior to final disking, the remainder as a sidedressing as plants near the heading stage. Nitrogen and potassium are especially critical elements. A yield of 30,000 heads of cabbage will take up approximately 100 lb (45 kg) of each, while only 25 lb (11 kg) of phosphorus would be removed. Application rates and ratios, however, vary widely among different soils. In Florida, actual amounts of nitrogen, phosphate, and potash applied range from 120, 160, and 160 lb/acre (134, 179, and 179 kg/ha) for irrigated mineral soils to 0, 120, and 180 lb (0, 134, and 202 kg) for muckland. From one to three applications may be made. Potash is not required on many California soils, and ratios of 1−1−0 are common. In eastern states, 1−2−2 ratios are recommended for low-fertility soils, applying 75 to 125 lb/acre (84 to 140 kg/ha) actual nitrogen. Regardless of area and soil, excessive nitrogen must be avoided, since it can lead to rapid head expansion and consequent splitting. Calcium and boron levels also must be maintained to avoid physiological problems (tipburn and hollowheart, respectively). Cabbage is sensitive to low soil boron and occasionally may require corrective applications of 3 to 4 lb/acre (3.6 to 5 kg/ha) actual boron. In some areas, sulfur is added. The distinctive odor of cabbage harvested from soils high in sulfur tends to be stronger than that of cabbage grown in soils moderately low in this element.

Plants are established either on raised beds or on flat terrain and may be field planted as soon as the soil can be prepared. Temperatures as low as 20°F (−7°C) for brief periods will not cause serious damage to transplants that have been hardened sufficiently.

Spacing between and within rows is related to soils and to the earliness of the crop. Cultivars intended for early production may be 10 to 18 in. (25 to 40 cm) apart in the row; late cultivars (normally large plants) may require 18 to 28 in. (40 to 71 cm). In California, cabbage is spaced 12 to 14 in. (30 to 35 cm) apart in twin rows on 40 in. (1 m) raised beds, with the twin rows approximately 14 in. (35 cm) apart. In areas not using raised beds, single rows are spaced from 36 to 42 in. (0.9 to 1.1 m) apart. Some cultivars have a tendency to develop axillary heads (small heads resembling brussels sprouts), particularly under wide spacing and high fertilization. This tendency may be especially noticeable in direct seeded cabbage but is rare when transplanted. Increases in planting density, a reduction in fertilization rate, and/or change in cultivar may eliminate this problem.

At least 1 to $1\frac{1}{2}$ in. (2.5 to 3.8 cm) of water per week is required for uninterrupted growth. Erratic soil moisture not only restricts the uptake of nutrients, thereby depressing yield, but also may cause physiological problems, most notably tipburn and cracking. Regular water supply, preferably at 60 to 100 percent of field capacity, is important as heading begins. Absorption by roots in the top 12 in. (30 cm) of soil is very active at this time, and good management dictates both shallow cultivation and regular irrigation to avoid root damage and plant stress.

Weeds must be controlled by cultivation or applications of herbicide or use of black plastic mulch. Plastic may be used for transplanted crops, applying the material before field planting. Most commercial acreage, however, is cultivated and/or treated with herbicides. As plants develop, the risks of root damage outweigh the benefits of cultivation, and for that reason herbicides are preferred. A range of materials is available for preplant, preemergence, and postemergence or posttransplant applications. Current materials include trifluralin, DCPA, and nitrofen. Any material must be used in accordance with label directions.

Several cole crops are susceptible to certain nonpathogenic stresses. In cabbage, the most serious disorders are **cracking**, **tipburn**, and **black speck**. Cracking of a developing head is caused by an erratic water supply, often high soil moisture following a dry period. The rapid growth of young leaves inside the head, brought on by the sudden increase in soil moisture, especially those soils high in nitrogen, cannot be accommodated by the relatively inelastic outer leaves. The problem can be avoided in most instances through careful fertilizer and water management. Tipburn is a disorder that also can relate to management practices, but it may appear regardless of cultural techniques. The cause of tipburn is complex, but it involves calcium and water relations of the plant. Originally, it was related to inadequate calcium, either because of erratic water supply and/or inadequate soil calcium. However, even with sufficient leaf calcium, sudden increases in growth rate seem to encourage tipburn. Mobility of calcium within the plant thus is also involved. Root pressure flow has been found to be a more critical factor than transpirational pull. When root pressure is high, as in humid weather, little, if any, tipburn will develop. Some cultivars appear to be somewhat resistant to tipburn and should be used by growers encountering this problem.

In some areas, a physiological disorder called black speck may appear. Its cause is not understood, but it may appear in storage or transit, occasionally in the field. The disorder is characterized by few to many small sunken brown to black specks on leaves throughout the head and occurs more on some cultivars than others.

Cultivars

Cabbage cultivars have been classified by several criteria to distinguish appearance (Figure 12.1), production season, or market use. Table 12.3 lists and describes the major groups classified by both type and market designation. The latter grouping is considered the most useful. **Danish** is the primary storage cabbage, with excellent quality and solid heads. It also is known as ballhead or 'Hollander' cabbage and is the most important group of cabbage grown. Some is used for

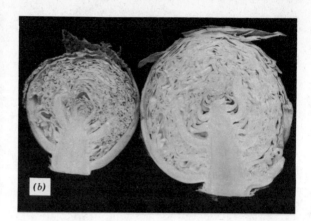

FIGURE 12.1
Cabbage cultivar types: (*a*) young developing Danish and pointed or Wakefield heads; (*b*) savoy (left) and domestic (right). (Photos courtesy of University of New Hampshire.)

pickling (sauerkraut), but most is marketed directly from storage. **Domestic** cultivars are not as compact as Danish and do not have the storage longevity of Danish types. **Red** cabbage is especially attractive for salad use, and the red color can be introduced into any of the several head types. Therefore, some red cabbage cultivars may be marketed directly, whereas others are stored. **Savoy** and **pointed** cabbages are suited only for the fresh market and, in the case of pointed cultivars, as an early spring cabbage crop. The savoy type makes an attractive salad cabbage. Some savoy cultivars have large, but soft, heads. Heads of pointed cabbage heads usually are small.

TABLE 12.3
Principle groups of cabbage cultivars

Type classification	Characteristics and use	Market classification
Wakefield	Early, small pointed heads; white; for fresh market use, not recommended for storage	Pointed
Copenhagen Market	Early to midseason, round; heavy bloom, short stem; for fresh and processing use; moderately long storage life	Domestic
Flat Dutch	Broad, flat solid head, light green; early to late; mostly for processing, some fresh; good storage capacity	Domestic
Savoy	Round, loose heads of blistered dark green leaves; little bloom; fresh market; poor in storage	Savoy
Danish Ballhead	Round, very firm heads with heavy bloom; few outer leaves; leaves light green; superior fresh market and storage type	Danish
Alpha	Very small, pointed heads; for fresh market; not stored	Pointed
Volga	Subglobe shape, steel blue leaves; outer leaves curl outward; late; no commercial use	n/a
Red	Red color on leaf surface; size and shape and density resemble Danish or domestic types; fresh market and storage	Red

Hybrid cultivars have become popular because of their improved uniformity and reliability and improved seedling vigor. Hybrid development is complicated by genetic self-incompatibility, but successful F_1 combinations have been achieved.

Disease and Insect Pests

Diseases   Of the pathogenic diseases, five are especially troublesome in all production areas. These are *Fusarium* yellows, clubroot, black rot, black leg, and downy mildew (Figure 12.2). Other diseases, occurring less frequently or only in specific geographical areas, include black blight (ring spot), *Alternaria,* soft rot, and mosaics.

Fusarium Yellows   Yellows is a vascular disease caused by the fungus *Fusarium oxy-* *sporum* f.sp. *conglutinans.* The first symptoms appear as yellow foliage, often predominantly on one side of the plant. The leaves become distorted and gradually turn brown and drop prematurely. The vascular area shows discoloration characteristic of the *Fusarium* wilt diseases. Once a soil is infested, the only effective control is the use of resistant cultivars. Much of the early work with *Fusarium* diseases was focused on genetic studies of resistance in cabbage, and highly stable single-gene resistance has been introduced into many standard and hybrid cultivars.

Clubroot   Clubroot, also a soilborne disease, is one of the most serious crucifer problems worldwide. Caused by the fungus *Plasmodiophora brassicae,* it becomes active as soil temperatures rise and is more aggressive in acidic soils than in those slightly alkaline.

FIGURE 12.2
Cole crop diseases: (*a*) clubroot; (*b*) blackleg; (*c*) black rot. (From *Diseases of Plants*, slide set 31, Diseases of Cauliflower, Broccoli, and Kale. Reproduced by permission of the American Phytopathological Society.)

The resting spores germinate, producing swimming zoospores that enter young plants by way of root hairs or wounds. Within the plant, they multiply and move to other parts of the root system. The disease stimulates cell division and enlargement, giving rise to knots or clublike growths that restrict uptake of water and nutrients. Above ground, the disease may not be apparent until some wilting appears during periods of strong sun. Gradually, recovery from this wilting does not occur, and the plant stunts or dies. Clubroot is transmitted by infected transplants, by equipment to which infested soil adheres, by windblown dust, or by manure from animals fed infected plants. Control is difficult, for there are no resistant cultivars, and the organism is very persistent in the soil. The recommended measures include (1) maintenance of soil pH at 7.3 or above; (2) rotation with nonsusceptible crops; (3) use of clean transplants; and (4) sanitation, disinfecting all tillage equipment. Of these measures, pH control is perhaps the most effective, rotation the least effective.

Black Rot   A worldwide disease of crucifers, black rot affects every major cabbage-producing area in Europe, Japan, and the United States. It was first identified in 1891 on rutabagas in Iowa, and the disease soon infested cabbage fields from Wisconsin to New York. It is caused by the bacterium *Xanthomonas campestris.* The bacteria usually infect the plant through pores at the leaf margins; thus, the early symptoms appear as yellow segments at leaf edges, gradually spreading inward to the center of the leaf. The edges of the necrotic areas are bounded by veins on the midrib, giving the diseased areas a characteristic "V" appearance. Gradually, the necrotic areas brown and dry up. Large veins and the midrib frequently become black, and this discoloration may extend to the root. In areas of cool temperatures, the disease may go unnoticed even up to harvest. Under warm, humid conditions, the plant may show distinct symptoms 10 to 14 days after infection. The organism is seedborne and can be spread by infected transplants and by equipment, animals, spattering rain or irrigation, and people. Most cole crop seed is produced in environments that minimize disease, and seeds normally are hot water treated to ensure reliability. However, it is essential, as well, to practice sanitation and to rotate crops in a minimum 3-year cycle to reduce inoculum levels.

Black Leg   Black leg, caused by the fungus *Phoma lingam,* affects all cole crops and cruciferous root crops. It is more active than black rot at low temperature and therefore is observed frequently in cool production areas. The fungus has an imperfect stage and a perfect stage. The disease is most commonly identified with the imperfect stage (*P. lingam*) in which it survives in seed and infected plant debris. As infected seeds germinate, the fungus quickly infects seedlings, causing lesions bearing pycnidia, dark-colored fruiting bodies. In high moisture, the pycnidia absorb water and expand in a gelatinous coil. Rainfall or irrigation then can spread the disease readily to healthy plants. The fungus also produces a perithecium which gives rise to the perfect stage ascospores. These spores are expelled into the air and have the potential for wide dispersal. Most infection occurs by transplant time. The first symptoms usually appear 2 to 3 weeks before field planting when plants become covered with circular black spots. These enlarge and develop gray centers, and eventually the entire plant is affected. Diseased plants usually become stunted. The control is similar to that described for black rot. In addition, weed control, especially of cruciferous weeds, and isolation from other cole crops can minimize disease. Transplants should not be clipped or topped to harden off and should not be dipped in water prior to transplanting.

Downy Mildew   Downy mildew is a fungus disease characterized by the existence of many races. It predominates in humid coastal regions, causing damage both to young seedlings and to produce in transit. The causal organism, *Peronospora parasitica,* is favored by temperatures of 50 to 60°F (10 to 16°C). It overwinters on perennial plants or on infected plant debris. The fungus penetrates the leaf, growing between leaf cells and sending fruiting branches above the leaf surface. These branches bear asexual spores (conidia) which are carried on wind currents to healthy plants. As infection proceeds, purple irregular spots develop on the leaves, stems, and seed pods. The spots enlarge, becoming yellow-brown on the upper surface while the mildew grows on the undersurface. Control is provided by appropriate fungicides and by fall plowing and a 3-year rotation to eliminate infected plant refuse. Since irrigation can spread the disease, overwatering should be avoided.

Alternaria   Several species of *Alternaria* infect cabbage, causing brown or black lesions on leaves, stems, and seed pods. The fungal spores (conidia) are spread rapidly during moist weather and are carried by air currents to adjacent fields. Seeds can be infected during maturation. Therefore, reliable, clean seed must be used for planting. Carbamate fungicides also help in controlling this disease.

Black Blight   Black blight, or ring spot, is caused by the fungus *Mycosphaerella brassicola.* It appears as brown lesions with concentric black specks containing fruiting bodies. As the spores germinate, they penetrate the leaf causing dark spots that eventually coalesce to give the black blight appearance. The disease is most prevalent along the Pacific coast, particularly in wet weather. Control is achieved through carbamate fungicides and by strict sanitation.

Other Diseases   There are other diseases of cabbage that occasionally cause losses. These include rhizoctonia, bacterial soft rot (particularly in storage), black leaf spot, and mosaic viruses. Some of the control measures applied routinely for the major diseases will minimize development and spread of these sporadic problems.

Insects   Of the insects affecting the Brassicaceae, aphids, the larvae of several moths or butterflies, and the cabbage root fly are some of the most serious pests (Figure 12.3). Three larvae are encountered frequently: cabbage maggot, imported cabbage worm, and cabbage looper. Infestation of a crop can occur very early in the season, and if feeding is unchecked, heavy losses will result.

Cabbage Maggot   The adult fly of the cabbage maggot (*Hylemyia brassicae*) deposits eggs just below the soil surface on young plant stems. As the maggots emerge, they feed on the surface of roots, tunneling through them, thus depriving the plants of a water and nutrient absorbing system. Characteristic brown streaks develop in the fleshy roots or basal part of the stem. A standard, effective control is diazinon drench at the time of transplanting (with transplant water) or in the furrow during direct seeding.

Cabbage Worms   Imported cabbage worm and cabbage looper both appear as small green larvae, the former of the butterfly *Pieris rapae,* the latter of *Trichoplusia ni.* The looper is identified easily by its looping mode of locomotion. Both insects are widespread, and their cycles may overlap. Early in the season, small butterflies can be seen hovering around cruciferous foliage, occasionally landing briefly to deposit eggs. It is difficult to control egg laying, but the larvae can be controlled by the biological insecticide *Bacillus thuriengensis* or by such materials as methomyl.

Aphids   Aphids (*Brevicoryne brassicae*) can infest plants in large numbers, but the pest is controlled effectively with pesticides (diazinon, malathion, parathion). This insect is not considered to be as serious as the looper or imported cabbage worm, although damage can be extensive if not controlled.

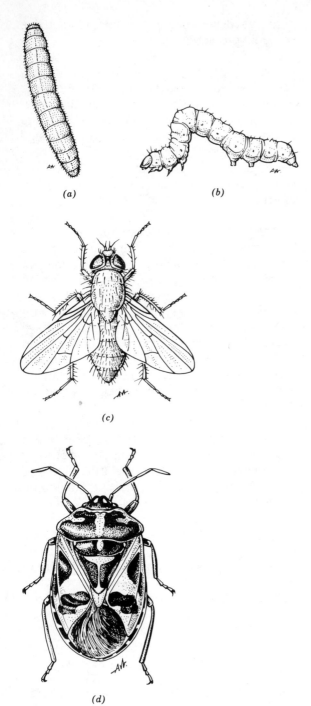

(a)

(b)

(c)

(d)

FIGURE 12.3
Cole crop insects: (a) imported cabbage worm; (b) cabbage looper; (c) cabbage maggot fly; (d) harlequin bug.

Harvest and Market Preparation

Cabbage maturity depends on cultivar, location, and season. In New York, early cultivars should be ready for harvest in 85 to 90 days from seeding, whereas late cultivars mature in 95 to 125 days. In Florida, a transplanted crop matures in 65 to 100 days, but 95 to 125 days is required for those direct seeded.

Heads should be harvested when firm and well sized. A grower has some flexibility to delay harvest and allow head size to increase or to harvest early, sacrificing some yield but perhaps gaining a price advantage. Delays may increase the risk of split heads and field rots; and if the harvest is too early, the heads may be soft. Late rains also may soften the heads. Because marketability and storability are directly related to head density, a grower cannot risk delaying harvest if marketable head size and density have been attained.

Cabbage for fresh market is cut selectively by hand and trimmed to the desired number of leaves. It can be field packed or moved in bulk to a packing shed for grading and packaging. Heads intended for storage and/or processing generally are harvested all at once, either mechanically or by hand, then loaded into pallet boxes for transit.

The optimum storage temperature for cabbage is 32°F (0°C), but 32 to 36°F (0 to 2°C) is an acceptable range. If temperatures drop below 32°F (0°C), some damage will occur as cell fluids freeze. A relative humidity of 90 to 95 percent should be maintained. Cabbage placed in storage should be mature and disease free and should not have been exposed to prolonged frost or cold. Heads bruised prior to storage present an increased disease risk in storage. The two most serious disease problems in storage are soft rot (*Erwinia carotovora*) and gray mold (*Botrytis cinerea*), both of which can spread rapidly from head to head. Suspected infection and poor heads must be eliminated through careful grading and strict sanitation.

Cabbage for market usually is packed in wirebound crates, fiberboard cartons, nailed crates, or mesh bags. After grading and packing, the heads are cooled to 32 to 36°F (0 to 2°C) at 95 percent relative humidity. Crushed ice is used to maintain quality in transit.

Cabbage yields vary widely. Red and savoy forms produce $4\frac{1}{2}$ to 18 tons/acre (10 to 40 MT/ha), whereas green types range from 26 to as high as 44 tons/acre (58 to 99 MT/ha).

BROCCOLI

Historical Perspective and Current Status

The broccoli most familiar to producers and consumers is the sprouting form, now known as Calabrese or Italian green. It is believed to be the first of the cole crops to evolve from wild cabbage and was cultivated by the ancient Romans. However, sprouting broccoli production was not introduced to other geographical areas until early in the sixteenth century, when it became known in England as "Italian asparagus" or "sprout cauliflower." In the United States, it was grown in the early 1800s, but its popularity is a recent phenomenon.

Sprouting broccoli is the most nutritious of the cole crops, especially in vitamin content, calcium, and iron (Table 12.1). It is marketed fresh and frozen, the latter alone or as an ingredient in prepared vegetable mixes. It also is used raw in salads.

The total fresh market production in the United States averages 5,600,000 cwt (254,016 MT). Broccoli for processing approaches 137,000 tons (124,286 MT). The total area harvested in 1983 was 91,800 acres (37,179 ha). Although many areas of the country have commercial acreages, at least 75 percent of the total production for fresh and processing use is located in California. It is very popular with home gardeners, because it is easy to grow and can be harvested over a period of several weeks.

Classification, Growth, and Development

Sprouting broccoli is identified as Group Italica in *Brassica oleracea*. Although broccoli includes both a heading form and a sprouting form, it is the latter that is included in Group Italica. The heading form is a strong biennial; sprouting broccoli generally is considered an annual.

Branched clusters of flower buds are formed at the apex of 2- to $2\frac{1}{2}$-ft (0.6- to 0.8-m) plants. These buds are dark green and are tightly packed. When these terminal sprouts are harvested, axillary shoots develop. Although an annual plant, broccoli is sensitive to chilling. Broccoli exposed to 40°F (4°C) will initiate flower primordia earlier than those grown at 70°F (21°C). Temperature sensitivity increases as the plants develop beyond seedling size.

Crop Establishment and Maintenance

Among the cole crops, broccoli is relatively tolerant of environmental stress. It is adapted to a range of soil types and can tolerate heat to a greater degree than cauliflower and brussels sprouts. As in the other cole crops, deficiencies of boron and other trace elements will reduce productivity and damage market quality. Boron stress will cause browning of some buds in a sprout, and deficiencies of manganese or magnesium will cause vein clearing and yellowing of the old leaves. Whiptail also can occur in molybdenum-deficient or acidic acids. Soil pH and fertility for broccoli culture should be maintained as carefully as for cabbage or cauliflower. Broccoli stressed by extreme temperature and/or inadequate soil nitrogen may develop button heads in which the terminal shoot fails to develop.

Substantial increases in growth rate take place during the transition to reproductive growth and as the terminal head nears maturity. Fertilization practices must ensure adequate nutrients, especially nitrogen, at those times. However, excessive nitrogen must be avoided as it can increase the incidence of hollow stem.

Broccoli generally is established by transplants in the eastern and northern United States, although direct seeding is used in California and Oregon. Field soil preparation and techniques for plant production, weed control, and irrigation are identical to those outlined for cabbage. Plant spacing must be adjusted for the vigor of the cultivar and may range from 10 to 30 in. (25 to 76 cm) in rows $2\frac{1}{2}$ to 3 ft (0.8 to 0.9 m) apart. A twin-row system similar to that described for cabbage is common in western areas.

Cultivars

There are no major subgroups of cultivars described for broccoli. For many years, the most popular cultivar was 'Italian Green Sprouting' or 'Calabrese.' In recent years, the number of new introductions has increased markedly, many of them F_1 hybrids. Some of the popular cultivars include 'Green Duke,' 'Green Comet,' 'Waltham 29,' 'Bravo,' and 'Premium Crop.' Of these, all but 'Waltham 29' are F_1 hybrids, now preferred for their improved yield and uniformity.

Disease and Insect Pests

Insect and disease pests are the same as those described for cabbage, and the same control measures are appropriate. Of the insect pests affecting cole crops, aphids are particularly troublesome on broccoli. They lodge within the dense bud clusters, sucking plant sap from the young leaves, and eradication of an established infestation is difficult. As sprouts are forming, careful and timely applications of insecticides (malathion, diazinon, thiodan) are critical in preventing aphid feeding.

Harvest and Market Preparation

The apical cluster of buds constitutes the most significant portion of total yield. These compact sprouts, cut with 6 to 8 in. (15 to 20 cm) of stem before the individual buds begin to open, may reach 6 to 7 in. (15 to 18 cm) in diameter, each weighing up to 1 lb (0.4 kg). Subsequent to harvest of the main head, axillary buds will form. These sprouts develop to 1 to 4 in. (2.5 to 10 cm) in diameter, but generally are sufficient for marketing. Several harvests of these secondary shoots may be feasible, particularly if environmental conditions remain optimum for continuous, rapid growth, and these harvests may constitute 50 percent of the total yield.

The rate of head development is accelerated by heat. The flower buds in the head develop very rapidly, and the period of time when sprouts are at peak quality therefore is very brief under high temperature. Heads should not be marketed if some of the buds have opened to expose yellow petals, or if the normal deep green color of the stem or buds has faded. Overmature sprouts will be woody in texture. If harvesting under high temperatures, it is wise to cut just before full maturity to help maintain high quality, and subsequent harvests should be scheduled at 2- to 5-day intervals. If temperatures are cool, the interval between harvests may be 4 to 7 days.

The cut shoots are very perishable and must be cooled as soon as possible to 32°F (0°C). The broccoli is graded, cleaned, trimmed to a uniform length of 6 to 8 in. (15 to 20 cm), tied in bunches of $1\frac{1}{4}$ to $1\frac{1}{2}$ lb (0.5 to 0.7 kg), and packed in waxed cartons (14 to 18 bunches/carton). They should be stored no longer than 1 to 2 days without top-icing; in transit, crushed ice is blown in and around the boxes to prolong fresh quality. Some prewrapping is done, but it is most frequently packaged at point of delivery.

CAULIFLOWER

Historical Perspective and Current Production

Cauliflower was described by Boswell as the "aristocrat of the cole crops."[3] It, like the other coles, is a native of Europe, apparently

[3] V. Boswell (1949) Our vegetable travelers. *National Geographic* (August).

evolving from sprouting broccoli. The oldest record of cauliflower used as a crop dates to the sixth century B.C.

Nutritionally, cauliflower is not rich in vitamins and minerals, nor is per capita consumption high. Therefore, it is not an important contributor to adult dietary needs. Its flavor, however, is very delicate and its popularity has grown in recent years.

The total U.S. production includes approximately 3,792,000 cwt (172,005 MT) cut for fresh market and 85,510 tons (77,575 MT) for processing, of which 74 percent is produced in California's coastal valleys. New York and Oregon each produce only 10 percent of California's output. The other production areas include Texas, Washington, and the mountain region of Colorado. The processing industry primarily is confined to freezing, with a small volume for pickles.

Classification, Growth, and Development

Cauliflower is classified in the Group Botrytis of *Brassica oleracea*. The early to midseason forms are weak biennials and may produce flowers in the first year. The late form, actually a heading broccoli, is strongly biennial.

The edible part of the plant is termed a curd (Figure 12.4) and, in summer types, consists of a mass of immature or undifferentiated shoot apices. In winter types, the curd consists of immature differentiated flower buds, thus resembling broccoli. The summer type does not require cold temperature to form the curd, whereas the winter type does. Winter types therefore must be grown where there is sufficient exposure to vernalizing temperatures [<50°F (10°C)]. A well-formed curd of summer cauliflower should be creamy white and dense, with a smooth texture. Winter types characteristically have a purple curd. Although 60 to 65°F (15 to 18°C) is preferred to attain high quality, if exposed to prolonged cold weather, the curd will loosen as internodes elongate.

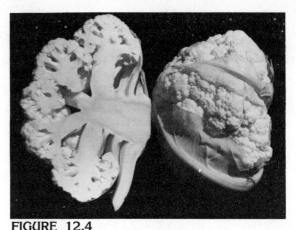

FIGURE 12.4
Cauliflower curd. Leaves of most cultivars must be tied over the developing curd to ensure purity of white color. (Photo courtesy of University of New Hampshire.)

Crop Establishment and Maintenance

Although the systems for growing cauliflower are similar to those outlined for cabbage, the plant is very sensitive to variation in environment. The growing season should be cool, 60 to 65°F (16 to 18°C), but not less than 40°F (4.5°C), relative humidity should be moderately high, and well-drained soils that are fertile and moist with good organic matter are preferred. Soil pH must not fall below 5.5, as plant growth will decline, in part because of unavailability of some micronutrients.

The specifications for soil preparation are similar to those described for cabbage. Organic matter should be incorporated into a soil, as either animal manure or a green manure, sufficiently in advance of planting to allow some decomposition. Following a broadcast application of lime and fertilizer appropriate for the soil, two subsequent nitrogen sidedressings often are applied—one 2 to 3 weeks after transplanting, a second as curds are developing. There is a significant response to nitrogen, particularly at moderate planting densities. Under high population density, stimulation of growth by nitrogen fertilization is counterproductive (Table 12.4). Transplanted cauliflower may be given a high analysis

TABLE 12.4
Effects of nitrogen fertility and plant populations on 'Snow Crown' cauliflower and 'Southern Comet' broccoli yields

Crop and plants/ha	Nitrogen rate (kg/ha)			Significance[a]
	56	112	224	
Cauliflower				
Marketable curd weight (g/curd)				
72,000	0	200	154	NS
36,000	0	372	327	L[c]Q[c]
24,000	0	504	590	L[c]Q[c]
Significance	NS	L[b]	L[b]	
Marketable yield (MT/ha)				
72,000	0	13.1	10.8	NS
36,000	0	12.8	11.7	L[c]Q[c]
24,000	0	11.9	11.7	L[c]Q[c]
Significance	NS	NS	NS	
Cull yield (MT/ha)				
72,000	11.8	8.6	10.3	NS
36,000	8.9	5.4	5.4	NS
24,000	6.5	2.1	1.0	L[c]Q[c]
Significance	L[b]	L[b]	L[b]	
Broccoli				
Marketable head weight (g/head)				
72,000	31.8	99.9	140.7	L[c]
36,000	45.4	131.7	217.9	L[c]
24,000	55.5	145.3	281.5	L[c]
Significance	NS	L[b]	L[c]	
Marketable yield (MT/ha)				
72,000	2.1	7.1	10.1	L[c]
36,000	1.2	4.8	7.6	L[c]
24,000	1.3	3.4	6.7	L[c]
Significance	NS	L[c]	L[c]	
Cull yield (MT/ha)				
72,000	4.2	1.8	0.6	L[b]
36,000	3.0	1.1	0	L[b]
24,000	2.5	1.2	0.4	L[c]
Significance	NS	NS	NS	

Source: From R. J. Dufault and L. Waters, Jr. (1985) *Hort-Science* 20:127–128. Reprinted by permission of the authors.

[a] NS, not significant; L, linear; Q, quadratic.

[b] Significant, $P = .05$.

[c] Significant, $P = .01$.

starter fertilizer which could eliminate the need for the first sidedressing in some soils. If large amounts of phosphorus are required, banding may be recommended in place of broadcast.

In addition to basic fertilizer needs, cauliflower is very sensitive to several trace element deficiencies. **Whiptail,** a disorder caused by inadequate molybdenum, is characterized by straplike, crinkled leaves and a deformed growing point. Occasionally, only a leaf midrib develops, and sprouting may be stimulated at the base of the plant. Whiptail is especially common in soils below pH 5.5 and has been more serious on some cultivars than others. The problem can be corrected by proper liming and/or by applying 1 lb/acre (1.1 kg/ha) of a foliar spray of ammonium molybdate. **Browning,** common along the Atlantic seaboard, relates to a deficiency of boron or to its unavailability in limed soils. This condition first develops within the stem and the center of the curd at the base of curd branches. It is not visible, however, until the curds show external discoloration. Such curds are unmarketable, and the stem may be hollow, although hollow stem also can arise from excessive nitrogen fertilization. Leaves of boron-deficient plants also are sometimes blistered and yellow. Fields of cauliflower showing browning will need 1 to 3 lb/acre (1.1 to 3.4 kg/ha) of actual boron, more if the soil is organic.

Ricing, or "ricey," is a disorder appearing when nitrogen and temperature favor very rapid head formation. The curd appearance becomes velvety as small white flower buds develop, reducing marketability. Proper management of soil moisture and fertility as heads develop can minimize the occurrence of this disorder.

Inadequate nitrogen, or other growth stresses, may cause **buttoning,** the development of small curds on young plants. These buttons never develop to mature heads. **Blindness** describes those instances in which no curd is formed. The plants become dark green with thick leaves. This disorder may have several causes—poor fertility, insect damage, disease, genetic irregularities, or cold temperature.

Cauliflower is a transplanted crop in the east and north, direct seeded in the west. In areas with a long growing season, transplants are grown in outdoor beds. These beds are fumigated after thorough tilling and are seeded thinly in rows 12 to 14 in. (30 to 35 cm) apart. Greenhouse-grown plants are direct seeded in plug trays or seed flats. Small seedlings from the latter are transferred to Speedling flats or to peat pots for eventual field planting. Assuming good survival, 3 to 4 oz of seed should be sufficient for an acre of cauliflower (210 to 280 g/ha).

Field spacing should be 30 to 42 in. (0.7 to 1.1 m) between rows, with in-row spacing adjusted according to the normal plant size of the cultivar grown. Commercial spacings range from 15 to 36 in. (30 to 91 cm). Some growers plant in twin rows on raised beds, with water provided at the edges and, in some systems, through a trench along the center of the bed. Eastern growers use overhead irrigation. At least 1 to $1\frac{1}{2}$ in. (2.5 to 3.8 cm) of water is necessary each week for optimum growth and quality. Larger amounts would be needed in arid climates and for sandy soils.

Cultivation should be shallow to avoid root injury, particularly as plants begin to head. Weed control is achieved predominantly through herbicides, not cultivation. The materials used for cauliflower are the same as those recommended for cabbage.

As the curd begins to develop, the creamy white color and smooth texture can be attained only if the curd is screened from direct sunlight. Cauliflower left uncovered will discolor due to activation of peroxidase by sunlight, and the curd also will loosen in the sun's heat. Discoloration is prevented by blanching. For most standard cultivars, blanching involves tying the large outer leaves over the top of the head and securing them with raffia or a rubber band. In the self-blanching cultivars, the curds are shielded from the sun by inwardly folding, overlapping leaves. Curds develop within the tied leaves in 5 to 15 days,

depending on air temperature. Different-color ties are recommended when the blanching operation involves different dates. Those tied on the same date should be ready for harvest at the same time.

Cultivars

Cultivars normally are grouped by head size and density. The major types are super snowball, snowball, and winter cauliflower (heading broccoli). The super snowball type is an early dwarf with medium-size leaves. Their somewhat flattened curds soften if not harvested promptly. Snowball types are characteristically larger than the super snowball and somewhat later, although maturity varies considerably among cultivars. The leaves are long and erect, and the curd is large, rounded, and very dense. In general, the later the maturity, the larger the curd.

The winter cauliflower includes a range of cultivars maturing during the winter months. These types normally are grown where winters are mild, and the plants become large in size.

Within these three classes of cultivars are new F_1 hybrids and/or self-blanching introductions. The F_1 hybrids offer improved uniformity, and the self-blanching cauliflower, used to a considerable extent on western processing acreage, eliminates the labor required for tying. Self-blanching cultivars do not have the purity of curd whiteness demanded in the market, however, and there is a higher cullage rate than with standard cultivars.

Disease and Insect Pests

Cauliflower is susceptible to the same insect and disease pests as cabbage, and the same control measures are recommended.

Harvest and Market Preparation

Cauliflower heads should be cut when they have reached approximately 6 in. (15 cm) in diameter, or 2 to 2.5 lb (0.9 to 1.1 kg), and the leaves are still vigorous and green. The curds must be compact with no evidence of discoloration or loosening. The plants are cut below the head, and the spreading leaves are removed. Those leaves enfolding the head are trimmed to 2 in. (5 cm) above the curd as protection. In the west, mobile conveyer units, the largest covering 30 rows at a time with a crew of 40, may be used, with trimming and packing done in the field. Otherwise, pallets of harvested heads are moved to a packing shed for trimming, grading, precooling, and packing. Much of the western pack is wrapped in perforated film, then placed in cardboard cartons. Eastern growers traditionally have used wirebound crates without wrapping individual heads. All heads should be cooled soon after harvest to 32 to 34°F (0 to 1°C), and long-distance shipments are top-iced or slurry-iced to preserve freshness.

BRUSSELS SPROUTS

Historical Perspective and Current Status

Brussels sprouts, a relatively new crop, had its origin in Belgium in the 1500s. Within 200 years, however, this form of cabbage was an item of international commerce. The French settlers grew the crop extensively in the delta region of Louisiana, but the major production area now is on the west coast.

Total fresh market production is about 743,000 cwt (33,702 MT), produced on 6100 acres (2470 ha), most of it in California. New York (Catskill region), once a major production area, now produces only a small volume. Production for freezing has increased, but brussels sprouts is not considered a major processing crop.

Nutritionally, brussels sprouts are not high in vitamin A, but the crop is a fair source of calcium, ranking fourth among all vegetables, and of vitamin C.

Classification, Growth, Development, and Culture

Brussels sprouts is classified as *Brassica oleracea,* Group Gemmifera. The plant lacks an apical head, but instead produces small axillary heads or sprouts along an elongated stem (Figure 12.5). In eastern states, the axillary heads may not develop uniformly in size and maturity because of the sequential development of leaves and buds along the stem. At any given time, the most mature sprouts are those at the base of the plant. In the coastal areas of the far west, environmental conditions minimize this problem. Removal of the vegetative apical bud approximately 3 weeks before harvest will stimulate uniform sizing and maturity of the axillary sprouts.

FIGURE 12.5
Brussel sprouts. Each axillary bud develops a small head, with basal sprouts maturing slightly earlier than those near the plant apex. (Photo courtesy of Harris—Moran Seed Company, Rochester, N.Y.)

The plant requires a long growing season and does not produce good-quality sprouts in excessively warm weather. It is a strong biennial: large plants (at least 15 leaves) will flower given sufficient exposure to temperatures below 50°F (10°C).

The crop is grown under the same cultural systems as the other cole crops.

Cultivars

All brussels sprouts cultivars can be classified as tall or dwarf. The dwarf forms produce sprouts that are more densely packed than those on tall cultivars. The most popular cultivars are 'Jade Cross' (F_1), 'Jade Cross E,' and 'Long Island Improved.' Of the two 'Jade Cross' hybrids, 'Jade Cross E' is the taller, providing more space for even development of axillary sprouts. 'Long Island Improved' is an old open-pollinated cultivar popular among eastern growers, but its use has declined in recent years in favor of the hybrids.

Harvest and Market Preparation

For many years, the brussels sprouts crop was harvested repetitively, cutting only the mature sprouts. With the introduction of hybrid cultivars and mechanical disbudders, a once-over harvest now is the accepted practice.

High-quality sprouts should be bright green and firm. Those that are off-color, wilted, or puffy are characterized by poor flavor and woody texture. The marketable size is approximately 1 to 1½ in. (2.5 to 3.8 cm). The sprouts are highly perishable, virtually a nonstorable commodity. They should be cooled as quickly as possible to 32°F (0°C) after harvest. After grading and washing, fresh sprouts are packed in quart-size containers, frequently overwrapped to preserve freshness. If wrapped, ventilation is essential, since brussels sprouts have a high respiration rate. Failure to provide good gas exchange will result in poor flavor.

KOHLRABI

Historical Perspective and Current Status

Kohlrabi evolved in northern Europe and was described first in the 1500s. By the end of that century, it was known in Germany and in countries bordering the Mediterranean. Its use in America is relatively recent.

Kohlrabi is similar to red cabbage in its vitamin and mineral content. Compared to turnip and rutabaga, which it resembles in flavor, it is superior to turnip, but rutabaga provides much more vitamin A.

Production statistics are not reported for kohlrabi, but it is not an uncommon vegetable at roadside stands and local markets. It can be grown on a wide range of soil types, and only 55 to 65 days is required to mature a crop from seed.

Classification, Growth, Culture, and Marketing

Kohlrabi is classified as *Brassica oleracea*, Group Gongylodes. It is a biennial, requiring a period of cold temperature [less than 50°F (10°C)] to induce flowering. As the plant develops, the white, green, or purple stem enlarges just above the soil surface. Since early growth tends to be a loose rosette of leaves, the subsequent enlarged stem has petioles attached at regular intervals on the globe surface, giving kohlrabi a rather unique appearance.

Kohlrabi is direct seeded ½ in. (1.2 cm) deep and thinned to a spacing of 2 to 3 in. (5 to 7.5 cm) between plants. All of the cultural systems described for cabbage apply, except that the short season may reduce fertilizer needs somewhat. For market gardens, a planting interval of 2 to 3 weeks will ensure a steady supply.

The best quality is obtained with rapid, unchecked growth provided by mild temperature and a constant supply of nutrients and moisture. Under those conditions, the optimum 3-in. (7.5 cm) globe size will be obtained quickly, and flavor and texture will be delicate and smooth. Erratic growth, caused by inadequate water or other stresses, results in woody texture and strong flavors.

Cultivars include 'Purple Vienna,' 'White Vienna,' and the hybrid 'Grand Duke.' The plants are pulled at optimum stem size, and the basal root and unswollen stem are removed. After cleaning and grading, they normally are bunched with tops on for local sale. With tops attached, quality cannot be maintained for more than a few days, even at the recommended 32°F (0°C) temperature in 90 to 95 percent relative humidity. If topped, storage can be extended to 4 weeks.

CHINESE CABBAGE

Historical Perspectives and Current Status

Chinese cabbage probably originated in Asia, and it was domesticated in several areas independently in approximately 2000 B.C. Original forms of *Brassica campestris* most likely were used for oilseed; the leafy forms were selected later. Chinese cabbage was known in many areas of Asia in the fifth century, but it was not introduced into the United States until the late nineteenth century.

Numerous types of Chinese cabbage are extremely important in the diets of Asians. It is used raw in salads, cooked as a green, and fermented to preserve it for later use. The Korean dish *kimchi* includes either fermented Chinese cabbage or fermented radish, along with hot peppers, garlic, and other ingredients. In the United States, Chinese cabbage is grown throughout the year in California, Florida, and Hawaii and in the spring and fall in New Jersey. The predominant form grown is Group Pekinensis, which is used as a salad or cooked vegetable. It is not appreciably different from green cabbage in food value.

Classification, Growth, and Development

There are two important types of Chinese cabbage cultivated in the United States. Both are members of *B. campestris.* The heading form, termed **pe tsai,** is Group Pekinensis. Its heads are cylindrical to round with light green leaves. The nonheading form, **pak choy,** is classified as Group Chinensis and has shiny, dark green leaves with white petioles (Figure 12.6). The latter sometimes is termed Chinese mustard because of its growth habit.

Both crops are annuals, with flowering inhibited by temperatures of 78 to 90°F (26 to 32°C) under natural or short days. Flowering is accelerated by exposure to 40 to 50°F (4.4 to 10°C) and long days, and the longer the cold exposure, the more uniform the flowering. Spring planting in northern areas therefore is difficult, since the plants are exposed to periods of cold temperature and increasing daylength. Late summer plantings in northern states, seeding the crop in late July or early August, will provide reliable yields of high-quality heads.

FIGURE 12.6
Pak choy cabbage (mustard cabbage) is characterized by white petioles and midribs and a lack of heading. (Photo courtesy of Harris–Moran Seed Company, Rochester, N.Y.)

Crop Establishment and Maintenance

The production techniques differ somewhat for the two forms of Chinese cabbage. The pe tsai (heading) form requires a fairly long season (60 to 100 days), compared with the short 40- to 50-day season for pak choy. Fertile, well-drained soils with good water-holding capacity and a pH range of 5.5 to 7.0 are suitable for both types of Chinese cabbage. Moderate amounts of nitrogen and potassium promote optimum growth; excess nitrogen has been shown to reduce the number of marketable heads.

In California, the cultural system and spacing for pe tsai are the same as those used for lettuce and cabbage, using a bed system to accommodate surface irrigation. Hawaiian growers plant on beds 24 to 48 in. (0.6 to 1.2 m) wide and 6 to 8 in. (15 to 20 cm) high to improve drainage. Most of the crop is transplanted, and size of transplant does not alter final head yield (Table 12.5). Final spacing between plants should be 12 to 16 in. (30 to 40 cm): the greater the spacing, the higher the percentage of grade 1 heads (Table 12.6). The pak choy type may be broadcast or seeded in rows 12 to 15 in. (30 to 38 cm) apart and thinned after several weeks to 8 to 12 in. (20 to 30 cm) between plants. Cultivation for both types should be shallow to avoid root damage.

Cultivars

The standard cultivars of the pe tsai type have been 'Michihli,' a tall heading cabbage with dark green leaves, and 'Won Bok,' in which the heads are shorter and thicker than 'Michihli.' Both are being replaced by F_1 hybrids, including 'Jade Pagoda' (Michihli type) and 'Springtime,' 'Summertime,' and 'Wintertime' (three Won Bok types differing in maturity date and head size). Few cultivars are avail-

TABLE 12.5
Transplant size of 'Nagoaka 55' Chinese cabbage and final head weight as influenced by diameter of transplant container

Transplant size (age weeks)	Container diameter (cm)				
	2.5	3.75	5.0	7.5	Mean
Transplant fresh weight (g/plant)					
3	3.5	5.7	6.5	13.2	7.2 d[a]
4	4.8	9.3	14.3	23.1	12.9 c
5	6.0	9.8	17.4	35.1	17.1 b
6	7.4	11.5	23.4	48.8	22.8 a
Mean	5.4 d	9.1 c	15.4 b	30.1 a	
Head weight (kg)					
3	2.4	2.6	2.3	2.4	2.4 a
4	2.3	2.6	2.6	2.7	2.5 a
5	2.4	2.3	2.5	2.7	2.5 a
6	2.0	2.2	2.3	2.4	2.2 b
Mean	2.3 b	2.4 ab	2.4 ab	2.5 a	

Source: From B. A. Kratky, J. K. Wang, and K. Kubojiri (1982) *Journal of the American Society for Horticultural Science* **107**:345–347. Reprinted by permission of the authors.

[a] Mean separation within transplant age or within container diameter by Duncan's multiple range test, $P = .05$.

TABLE 12.6
The effect of plant spacing of 'Nagoaka 55' Chinese cabbage on head weight, percentage salable heads, and percentage grade 1 heads

Plant spacing (cm)	Head weight (kg)	Salable heads (%)	Grade 1 heads (%)
1975–1976			
28	0.85 a[a]	88 a	48 a
33	1.04 b	95 b	78 b
38	1.28 c	96 b	94 c
43	1.50 d	98 b	95 c
1976–1977			
28	0.70 a	48 a	14 a
33	0.88 b	70 b	28 ab
38	1.06 c	84 bc	44 bc
43	1.25 d	95 c	65 cd
48	1.44 e	99 c	86 d

Source: From B. A. Kratky, J. K. Wang, and K. Kubojiri (1982) *Journal of the American Society for Horticultural Science* **107**:345–347. Reprinted by permission of the authors.

[a] Mean separation within columns and years by Duncan's multiple range test, $P = .05$.

able for mustard cabbage. 'Lei Choy' was developed in California as a slow bolting strain of pak choy and was described as having celery-like stalks. With the increasing popularity of Chinese cabbage, new cultivars representing different types can be expected.

Disease and Insect Pests

Diseases   Chinese cabbage is susceptible to some of the same problems as the other cole crops, black leg and black rot being two notable examples. In the southeast United States, turnip anthracnose (*Colletotrichum higginsianum*) also can become serious. This disease is seedborne in radish and overwinters in plant refuse. It appears on Chinese cabbage leaves as water-soaked spots that dry and fall out, giving a "shot hole" effect. Turnip anthracnose is controlled with carbamate fungicides.

Insects   Of the insect pests, the diamondback moth (*Plutella maculipennis*), turnip aphid (*Hyadaphis pseudobrassicae*), and striped flea beetle (*Phyllotetra vittata*) constitute the major concerns. The larva of the diamondback moth is more serious in Asia than in the United States. All of the insects are controllable with insecticides.

Harvest and Market Preparation

The pe tsai crop is harvested by hand at the base, keeping the entire head intact. Outer leaves are trimmed, and the heads are packed in cartons or wooden crates and vacuum cooled. The pak choy type can be harvested as single leaves or as a complete plant.

BROCCOLI RAAB

Broccoli raab is not a member of *B. oleracea,* but was derived as a selection from turnip. It is classified as *Brassica campestris* Group Ruvo. Before the flowers have opened, the leaves with the seedstalks are cut and used for greens. It is not a major commercial crop, but small acreages can be found in market gardens, especially those supplying urban ethnic groups.

There are two forms of broccoli raab—rapine, or spring raab, and rappone, or fall raab. Other than season of maturity, there is no difference in appearance and flavor. Both go to seed very rapidly.

In some areas of moderate climate, raab is planted in the fall and overwintered to produce an early spring crop. Rappone seems superior to rapine for these fall plantings. In most areas, however, both are spring planted for early or late summer harvest. The harvest system is similar to that used for mustard greens, and the leaves and seedstalks are washed carefully and bunched. Raab is very perishable and must be marketed immediately.

SELECTED REFERENCES

Davis, H. R., F. M. R. Isenberg, and R. B. Furry (1978). *The Control of Natural-Air Cabbage Storage Environment,* New York State College of Agriculture and Life Sciences, Cornell University, Coop. Ext. Inf. Bull. 137.

Nieuwhof, M. (1969). *Cole Crops.* World Crops Books, Leonard Hill, London.

Palzkill, D. A., T. W. Tibbitts, and B. Esther Struckmeyer (1980). High relative humidity promotes tipburn on young cabbage plants. *HortScience* **15:**659–660.

Ryder, E. J. (1979). *Leafy Salad Vegetables.* AVI, Westport, Conn.

Shear, C. B. (1975). Calcium related disorders of fruits and vegetables. *HortScience* **10:**361–365.

Webb, R. E. (1981). *Growing Cauliflower and Broccoli,* USDA Farmer's Bull. 2239.

Yamaguchi, M. (1983). *World Vegetables.* AVI, Westport, Conn.

STUDY QUESTIONS

1. Why does tipburn occur first on the inner leaves of a cabbage head?

2. Given a field of wilted, yellowed cabbage or other cole crop, indicate the possible causes and how you would determine which causes were not involved.

3. Indicate how quality of the harvested product is affected by fertilization and irrigation practices in cabbage, cauliflower, and kohlrabi.

4. Explain the reasons for storage longevity or lack of it in the different cole crops.

13
Leafy Salad Crops

crops throughout the year. Concurrently, the popularity of salads has increased. Lettuce, in particular, has become one of the most important vegetables in per capita consumption and in farm value.

Leafy salad crops have been defined as those from which leaves and associated parts are harvested for use as raw vegetables. This definition could include cabbage or spinach, for example; yet, according to predominant use, cabbage and spinach would be considered a cole crop and a potherb, respectively.

There are four major leafy salad crops—lettuce, celery, endive, and chicory—and a number of lesser vegetables used primarily for salads, flavoring, or garnishing foods. The latter include, among others, watercress, garden cress, parsley, chervil, dill, fennel, and corn salad. The major leafy salad crops are important worldwide. The others may be significant commercial commodities only in certain geographical areas. This chapter attempts to focus on the major salad crops with some reference to those of lesser commercial volume.

Lettuce	Chicory	Parsley
Celery	Cress	Chervil
Endive	Watercress	Sea kale
	Corn salad	

Most leafy salad crops are considered "high-risk" vegetables. Their succulent leaves and petioles are susceptible to damage by pests, by poor production and handling systems, and by environmental extremes over which a grower may have little or no control. Technological improvements have decreased the risk substantially, and improvements in transportation, precooling technology, and packaging have enabled consumers to enjoy leafy salad

LETTUCE

Historical Perspective and Current Status

Both cultivated and wild lettuce species are native to Europe and Asia. Although lettuce has been cultivated for over 2000 years, it is likely that early use was confined to seed oil. Cultivation as a vegetable crop was described by Hippocrates (343 B.C.), Aristotle (356 B.C.), and others, indicating that it quickly became popular once introduced. Subsequently, it was

TABLE 13.1
Nutritional constituents of the major leafy salad crops[a]

| | | | | | | | | Vitamins | | | | Minerals | | | | |
Crop	Water (%)	Energy (cal)	Protein (g)	Fat (g)	Carbo-hydrate (g)	A[b] (IU)	C[c] (mg)	Thia-mine (mg)	Ribo-flavin (mg)	Niacin (mg)	Ca (mg)	P (mg)	Fe (mg)	Na (mg)	K (mg)
Celery	94	17	0.9	0.1	3.9	240	9	0.03	0.03	0.3	39	28	0.3	126	341
Witloof chicory	95	15	1.0	0.1	3.2	Trace	—	—	—	—	18	21	0.5	7	182
Endive (curly)	93	20	1.7	0.1	4.1	3300	10	0.07	0.14	0.5	81	54	1.7	14	294
Lettuce[d]	96	13	0.9	0.1	2.9	330	6	0.06	0.06	0.3	20	22	0.5	9	175
Parsley	85	44	3.6	0.6	8.5	8500	172	0.12	0.26	1.2	203	63	6.2	45	727

Source: National Food Review (1978), USDA.

[a] Data per 100 g sample.

[b] 1 IU = 0.3 μg vitamin A alcohol.

[c] Ascorbic acid.

[d] Crisphead type.

carried to western Europe and was growing in the New World by 1494.

Lettuce is not especially rich in vitamins and minerals (Table 13.1), ranking 26th among the major vegetables. However, because of the volume consumed, it has become the 4th most significant crop nutritionally. Leaf lettuce and romaine contain substantially higher levels of vitamins A and C than the other types, and somewhat higher mineral content (Table 13.2).

Lettuce production, worldwide, is important, both as a field crop and as a forcing crop. In the United States, lettuce is grown commercially on 225,000 acres (91,125 ha). All areas of the country include some production, but the major commercial acreage is centered in just two states—California and Arizona. Of the total U.S. production of crisphead ("Iceberg") lettuce, 70 percent is grown in California and 16 percent in Arizona, all under irrigation. California produces lettuce throughout the year— in its coastal valleys during the summer, in interior desert valleys in the winter, and in the San Joaquin Valley in the spring and fall. The largest single production area is centered in Salinas, accounting for 25 to 30 percent of all U.S. head lettuce. Arizona's center of production is the Yuma Valley, shipping 9 percent of the U.S. crop.

Crisphead lettuce is produced commercially in other areas, most notably Florida, Michigan, Wisconsin, New Jersey, and Colorado. In some of these and other areas, however, market garden production also is important, supplying local consumers with butterhead, romaine, and leaf lettuce in addition to the crisphead. On a minor scale, greenhouse production and, more recently, hydroponic systems within controlled environments also provide a quality product during the winter months. The total U.S. lettuce production in 1983 was 63,276,000 cwt (2,870,199 MT), predominantly the crisphead type, with a U.S. dollar value of $755,496,000.

TABLE 13.2
Differences in calcium and vitamins A and C among four lettuce types[a]

Lettuce type	Vitamin C (mg)	Vitamin A (IU)[b]	Calcium (mg)
Crisphead	6	330	20
Butterhead	8	970	35
Cos (romaine)	18	1900	68
Leaf	18	1900	68

Source: National Food Review (1978), USDA.

[a] Data per 100 g sample.

[b] 1 IU = 0.3 μg vitamin A alcohol.

Classification, Growth, and Development

Lettuce is a member of the family Asteraceae (Compositae), closely related to chicory, endive, sunflower, aster, artichoke, chrysanthemum, and several troublesome weeds. Cultivated lettuce (*Lactuca sativa*) differs from the wild species (*L. serriola*) in its broad lower leaves and its tendency to form a head.

Lettuce is a typical cool season annual crop. Germination is favored by uniformly cool temperatures [optimum 65 to 70°F (18 to 21°C)]. Temperatures of 79°F (26°C) and above inhibit germination severely in some cultivars, less so in others. It is known that red light promotes germination, far-red inhibits it, and that there is an interaction between light and temperature. Imbibed seed becomes thermodormant at temperatures above 80°F (27°C), and this thermodormancy is intensified by lack of light. In some production areas, particularly desert locations, thermodormancy due to high soil temperature results in poor stands. Periodic sprinkle irrigation to lower temperature dissipates germination inhibitors adjacent to the seed, thereby increasing rate of emergence. This irrigation also reduces salt concentration in the seed zone.

The lettuce root system is characterized by a rapidly developing taproot. The first lateral roots grow horizontally, very close to the soil surface, spreading to 6 to 18 in. (15 to 45 cm) before turning downward. These and other laterals eventually may reach a substantial depth when plants are mature. During the period of vegetative growth and harvest, however, the major part of the feeder root system is close to the surface.

Vegetative growth occurs as a rosette of leaves on a short stem. In some forms of lettuce, heading is similar to that of cabbage, with tightness of head relating to internode length and leaf characteristics. The first leaves on a lettuce plant are narrow. As light intensity and daylength increase, leaf width increases. Heading begins as these wide leaves cup and curl inward, overlapping to contain the new growth. Nonheading forms are characterized by a tight rosette of leaves in which the leaf curvature is predominantly away from the center.

Lettuce types differ in sensitivity to daylength. Some are day-neutral plants; others are long-day plants. Flowering in both is accelerated by high temperature, and many lettuce cultivars bolt quickly in northern growing areas in June and July, when prolonged periods of high temperature are common.

Crop Establishment and Maintenance

Lettuce is grown as transplants or is direct seeded. Because seeds are small and chaffy, it is difficult to establish a uniform stand by direct seeding without thinning or blocking. Several planting systems, including block or clump planting, seed tapes, and/or precision seeding, have been utilized to enhance uniformity (Chapter 5).

In some northern areas, lettuce sown in forcing structures in late winter or early spring will mature in the field sufficiently early to escape high summer temperatures, thereby avoiding excessive losses from tipburn, brown rib, and bolting. In all areas, transplanting can achieve a uniform stand. Although transplanting does require additional management, the systems for handling plants have improved. Speedling or plug tray plant growing systems, which are highly automated, provide at minimal cost the volume needed for large acreages.

Regardless of the planting system, the grower tries to achieve a final density of approximately 26,000 to 30,000 plants/acre (64,246 to 74,130/ha).

The soils available for lettuce production determine the market objective. Early crops require a sandy loam, and mucks and clay loams are suited for midseason or late crops. Plantings on northern organic soils normally are delayed in comparison to those on sands, thereby exposing the plants to summer heat; but the water-holding capacity of muckland favors steady growth and reduced tipburn, regardless of temperature.

The pH of acidic soils should be adjusted to 6.5 to 7.0. In some western areas, where alkalinity is a problem, soils are acidified with potassium sulfate. Most eastern soils are acidic and require regular lime applications.

Based on a yield of 420 cwt (19 MT), lettuce removes 95 lb of nitrogen, 28 lb of phosphorus, and 208 lb of potassium (43, 13, and 94 kg, respectively). New York growers apply 50 to 100 lb/acre (56 to 112 kg/ha) of nitrogen and phosphate and 50 to 150 lb/acre (56 to 168 kg/ha) of potash. On mineral soils, half of the rate is broadcast after plowing, and half is banded 2 in. (5 cm) below and to the side of the seed. On muck soils, all is broadcast after plowing. In California, the average rate of application to mineral soils is 160 lb/acre (178 kg/ha) of nitrogen, 90 lb/acre of phosphate (100 kg/ha), and 50 lb/acre (56 kg/ha) of potash.

Approximately 80 percent of the growth of lettuce occurs during the 3 to 4 weeks prior to harvest. It is at this time that nutrition is critical, and some additional fertilizer applications may be warranted. In mineral soils, it may be necessary to add nitrogen periodically during the season to ensure steady growth. California growers apply nitrogen after thinning and again several weeks before harvest. For muck soils, additional nitrogen may be needed only during cool, wet periods in the spring and fall when decomposition of organic matter is slow. During the summer, sufficient nitrogen is released by decomposition to meet crop requirements.

Mineral soils should be plowed after applications of lime and manure and then disked, fertilized, and disked again to develop a fine seedbed. Forming raised beds, particularly where surface irrigation is used or where drainage of excess rainfall may be needed, is a common practice in the west and southwest.

Lettuce is spaced 12 to 15 in. (30 to 38 cm) apart in single or twin rows. In western areas (Figure 13.1), the crop is grown as twin rows on raised beds that are 40 in. (1 m) on

FIGURE 13.1
Lettuce production in California. Twin rows are precision seeded on each bed and given overhead irrigation to cool the soil and maintain moisture for germination. (Photo courtesy of University of New Hampshire.)

center. Spacing is achieved by precision seeding with pelleted seed or by blocking. In some market gardens, up to five rows may be close-planted on a bed, with between- and in-row spacing 12 to 15 in. (30 to 38 cm).

Irrigation is essential for high-quality lettuce. Even in areas with adequate total rainfall, irrigation is necessary to avoid physiological problems in mature heads. Surface irrigation is preferred, since it reduces disease spread.

Weed control is achieved predominantly through cultivation and herbicides, although black plastic may be used successfully. Cultivation must be shallow to avoid damage to the feeder roots, since root damage reduces growth and quality of the head. Herbicides have not been completely reliable, but most commercial growers utilize them. Materials applicable to lettuce include benefin or bensulide preplant incorporated. Glyphosate may be applied to weedy fields before soil preparation. Some growers may fumigate to eliminate both weeds and soilborne diseases.

Lettuce yield and marketability are reduced by several physiological disorders relating to environment. The disorders most frequently encountered include tipburn, brown rib, and puffiness.

Tipburn symptoms on lettuce are not unlike those of cabbage, but the problem can be very severe. The disorder occurs most frequently when average temperatures exceed 70°F (21°C), especially when temperature changes occur abruptly and when soil moisture is low. It is considered as a translocation problem, although nutrition is involved.

Because tipburn occurs at or near harvest, it may be difficult to detect. It affects the margins of midsize to large leaves, causing rapid necrosis and sites for secondary infection. Research by Tibbitts and coworkers (see Collier and Tibbitts, 1982) has linked tipburn to the rupture of laticifers, the latex-bearing duct system, releasing latex into the surrounding tissue. Whether this rupture is a cause or an effect of tipburn is not entirely clear.

Calcium nutrition does seem to be a factor in severity of tipburn. In tests, lettuce foliage treated with calcium nitrate or calcium chloride did not develop symptoms, and the incidence was reduced by applications of calcium to the soil. Other factors also are involved, however. Temperature and fluctuations in moisture, both soil and air, that affect plant turgor are related to the incidence of tipburn. Lettuce subjected to a cloudy, humid period followed by dry, sunny weather will show a marked increase in tipburn. Nutrients other than calcium, including nitrogen, magnesium, and boron, can play a role. A growth rate that is rapid by reason of temperature, soil moisture, and nutrients may encourage tipburn, particularly if preceded by reduced growth rate. Under these high growth rates, calcium may be available, but its slow mobility actually may create a deficiency within the plant.

Control of tipburn seldom is complete. Some cultivars show resistance; none is immune. All lettuce should be grown with careful management of soil moisture and fertility.

Brown rib may occur under those environments that favor tipburn. It is not clear that the disorder is related to tipburn, but the corrective action is the same. Brown rib appears as brown streaking on the midrib.

Puffiness develops when high temperature, rain, and high humidity occur near harvest. The affected heads are soft and loose and may, in some instances, resemble early stages of bolting. In some cultivars resistant to bolting, however, puffiness also has been observed. The best preventative is selection of cultivars adapted to specific growing conditions.

Spiraling of lettuce heads apparently results from the fusing of leaf margins of one or more outer wrapper leaves which forces the head to expand upward. There are cultural differences in the frequency of this disorder.

Cultivars

All lettuce cultivars are standard open-pollinated types. Resistance to disease is a major consideration in selecting a cultivar. Other important traits include head color, size, earliness, uniformity and shape, bolting tendencies, and susceptibility to tipburn. There are six recognized types of lettuce, several of which are shown in Figure 13.2.

Crisphead has large, heavy leaves that are brittle in texture, tightly folded, and tough enough to withstand long-distance shipping without serious damage. Outer leaves are green whereas the interior of the head is white or yellowish. Within this type, several general groups have been described (Ryder, 1986).

Imperial	Light to medium green leaves with serrated or wavy edges; texture relatively soft
Great Lakes	Cultivars somewhat variable in appearance and widely adapted; leaves brittle with ruffled margins and bright green to yellow with prominent ribs
Empire	Leaves light green, deeply serrated, and crisp; heads often slightly elongate or conical; flat ribs
Vanguard	Dull green, yellow interior, softer texture than 'Great Lakes'; leaves with scalloped margins; green butt and flat ribs

FIGURE 13.2
Lettuce types: (*a*) crisphead; (*b*) leaf; (*c*) butterhead. (Photos (*a*) and (*b*) courtesy of E. J. Ryder, U.S. Agricultural Research Station, Salinas, Calif. Photo (*c*) courtesy of Harris–Moran Seed Company, Rochester, N.Y.)

Ryder also notes that cultivars have been developed combining desirable traits of several of these groups, gradually evolving new types.

Butterhead is characterized by its loose heads of crumpled leaves that have a very soft, buttery texture. Veins and midribs are less prominent than in crisphead types. Both greenhouse and outdoor cultivars are available, the latter somewhat less tender than those produced in the greenhouse. In the United States, there are two groups:

Boston type   Relatively large heads; light green leaves

Bibb type   Small heads with dark green leaves

European cultivars are separated into summer and greenhouse types, the latter small with soft heads. Butterhead lettuce is of high quality, but is not suited for shipping. It is, however, popular for local sales.

Romaine (Cos) develops elongate heads of long leaves with heavy midribs. Outer leaves are coarse in appearance and dark green. Inner leaves are fine textured and light green. Romaine is a very high quality lettuce generally suited to local sales, although some is shipped from Florida and the west coast.

Leaf lettuce is popular in market and home gardens and constitutes approximately 5 percent of the total lettuce crop. The leaves, variable in shape, margin, and color, form a compact rosette. Most leaf-type cultivars withstand greater environmental variation than heading forms, and quality, particularly nutritional quality, is superior.

Stem lettuce is sold as 'celtuce' in the United States. Although not a significant crop in this country, it is popular in the Orient. The stems of this type enlarge during growth and are peeled and used as a cooked vegetable.

Latin lettuce leaves are somewhat elongate, but more leathery than romaine. This type is grown primarily in the Mediterranean area and in South America. Only one cultivar has been sold in the United States ('Fordhook').

Disease and Insect Pests

Diseases *Sclerotinia Drop* *Sclerotinia* drop, caused by the soilborne fungi *Sclerotinia sclerotiorum* and *S. minor,* is most active under cool moist conditions and can become especially severe in fields repeatedly planted to lettuce. *Scelotinia* drop is a common problem in those areas where heavy rains may occur during the growing season. The fungi first attack lower leaves in contact with the soil, and a cottony growth develops. Black sclerotia appear in this growth, providing a means for the organism to persist in unfavorable environments. Under favorable conditions, the entire plant may become infected, and up to 40 percent loss can occur in a given field. Some cultivars may show a measure of resistance, but the disease is difficult to control, since it can infect a number of different crops. Rotation is suggested to avoid severe infestation of a field (Figure 13.3).

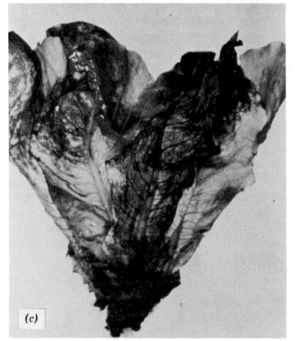

FIGURE 13.3
Lettuce diseases: (*a*) *Sclerotinia* drop; (*b*) big vein; (*c*) gray mold. (Photos (*a*) and (*b*) courtesy of E. J. Ryder, U.S. Agricultural Research Station, Salinas, Calif. Photo (*c*) from *Diseases of Plants*, slide set 35, Diseases of Onion, Lettuce and Celery. Reproduced by permission of The American Phytopathological Society.)

Botrytis Rot *Botrytis* rot, or gray mold, is caused by the soilborne fungus *Botrytis cinerea.* Brown lesions develop near the soil line, on either the stem or the leaf bases (Figure 13.3). Infection then progresses along the stem and through successive leaves. The disease therefore may develop first only on one side of a plant. This organism has a broad host range and grows on decaying vegetable matter in the soil. The control measures described for control of *Sclerotinia* drop apply to *Botrytis.*

Downy Mildew   Downy mildew is a common field disease in California and occurs in greenhouses elsewhere. The causal organism is the fungus *Bremia lactucae*. The first symptoms of infection are scattered light green to yellow areas on upper surfaces of outer leaves. Shortly thereafter, white hyphae and conidia can be seen on the lower leaf surface, under the yellow spots. These areas gradually become brown, and secondary organisms then may cause serious losses. The fungus attacks only *Lactuca*, overwintering on debris and wild lettuce. The intensity of the disease may be reduced by weed control in and around the field, by elimination of crop debris, and by avoiding excessive irrigation or areas of high humidity. Carbamate sprays can control the spread of the disease, although their effectiveness is improved in dry locations. Some cultivar resistance is available, but new races of the fungus soon render the resistance ineffective.

Powdery Mildew   Mildew caused by *Erysiphe cichoracearum* produces powdery lesions on both surfaces of the leaf. It is confined to California and Arizona, and some of the butterhead cultivars are resistant. The fungus can be controlled with applications of sulfur.

Bottom Rot   The bottom rot fungus (*Pellicularia filamentosa*) penetrates the base of old leaves under moist conditions, producing brown lesions. It then invades successively younger leaves, eventually consuming the entire head. In New York muckland, up to 30 percent of the crop is lost to this disease, and, under very wet conditions, the losses may exceed 75 percent. Infected plants are susceptible to secondary infection by soft rot (*Erwinia carotovora*). The only control is through the use of fungicides, but to be effective, these materials must reach the leaf base. No cultivars have shown resistance to bottom rot.

Corky Root   Corky root (Figure 13.3) is a recent disease caused by a bacterium of the *Corynebacterium* group. The bacterium or its toxin affects roots of many cultivars, causing a reduction of root growth and severe stunting of the top. Management techniques that improve soil drainage and aeration seem to reduce the incidence of corky rot, and some cultivars (e.g., 'Greenlake' and 'Montello') seem to be more tolerant than others.

Mosaic   Several viruses (lettuce mosaic virus, broad bean wilt virus, cucumber mosaic virus, and others) are present in all lettuce production areas (Table 13.3). The major vector is the green peach aphid (*Myzus persicae*) and there is some transmission in seeds (3 to 8 percent). Plants infected when young seldom will reach marketable size. They become mottled, showing vein clearing, recurved leaves, and ruffled margins. As the disease develops, necrotic areas appear, and the plant tends to yellow and become stunted.

Zink *et al.* (1956) showed that the level of field infection is proportional to the percentage of seed transmission. An indexing procedure for all seed lots has reduced field infection from a level once approaching 85 percent to the current 11 percent or less. As a result, losses have been reduced by 95 to 100 percent. Prior to the use of indexed seed, yields averaged 352 cartons/acre (869/ha). An additional 126 cartons/acre (311/ha) was obtained using indexed seed.

Other preventative measures also are important. Old plants and plant debris must be destroyed by plowing and disking, and wild lettuce and other virus hosts should be eliminated in production fields. Alternate hosts include lamb's-quarter, pigweed, chickweed, endive, chicory, garden pea, and spinach. Genetic resistance has been introduced into some cultivars. Plants with this resistance may still contain mosaic virus, but there is little effect on plant performance.

Aster Yellows   Caused by a mycoplasma, this disease has a very broad host range. Affected leaves of lettuce become curled, yellow or white, occasionally with small brown spots

TABLE 13.3

Percentages of lettuce (*Lactuca sativa*) plants infected with broad bean wilt virus (BBWV), cucumber mosaic virus (CMV), or lettuce mosaic virus (LMV) in surveys of Oswego County, New York, 1975 to 1980

Year	Number of samples	BBWV	CMV	LMV	BBWV + CMV	BBWV + LMV	CMV + LMV
1975	102	26	47	6	15	0	6
1976	175	23	37	0	35	1	4
1977	164	15	45	0	32	0	8
1978	180	10	38	2	49	1	0
1979	200	21	56	4	14	0	5
1980	250	19	27	0	54	0	0

Source: From R. Provvidenti, R. W. Robinson, and J. W. Shail (1984) *HortScience* **19**:569. Reprinted by permission of the authors.

of dried latex along the margin. Leaf veins may appear somewhat translucent. Plants infected after head formation develop twisted, deformed heart leaves, and the heads are soft.

Aster yellows is transmitted by leafhoppers, and insect control is an important means of reducing infection. Weed control in and near the field and isolation from other host crops can help to reduce the inoculum.

Big Vein   This disease results in vein clearing (Figure 13.3) that gives the leaf vein a prominent appearance. It is introduced into the plant by *Olpidium brassicae,* a soilborne fungus, and is most common in summer production areas in the west. The causal virus resides in the root system, but leaf growth is most noticeably affected. The plants do not die, but quality, head firmness, and appearance suffer. Control is difficult. Soil treatments have helped to reduce the incidence somewhat.

Insects   Several insects damage lettuce (Figure 13.4), either directly by feeding or indirectly through disease transmission. The most common pests are green peach aphid, cabbage looper, corn earworm, beet armyworm, and leafhopper.

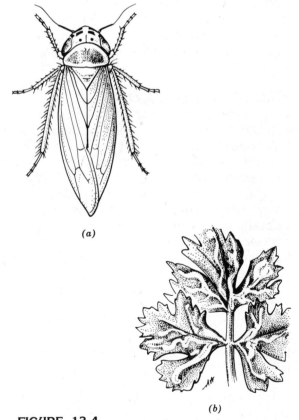

(a)

(b)

FIGURE 13.4
Insects of leafy salad crops: (*a*) six-spotted leafhopper; (*b*) damage caused by celery leaf miner.

Green Peach Aphid The green peach aphid (*Myzus persicae*) is the sole transmitter of lettuce mosaic. In itself, it reduces vigor and yield by sucking plant sap. The insect has a wide host range and so is a frequent problem. Control can be achieved with several organic phosphates (see asparagus, Chapter 10).

Six-Spotted Leafhopper The six-spotted leafhopper (*Macrosteles divisus*) is the vector of aster yellows. It is capable of migrating rather substantial distances, and control therefore is difficult. Inoculation of some strains is suppressed at high temperature. Organic phosphate insecticides will eradicate leafhoppers on contact, but will not discourage migration into the field.

Larval Insects The cabbage looper, corn earworm, and armyworm are larval feeders that can be controlled with carefully timed sprays (see cabbage and sweet corn, Chapters 12 and 20).

Greenhouse Production

Lettuce is well adapted to greenhouse conditions. However, the costs of production and competition throughout the year from areas shipping field-grown lettuce have retarded the development of a large forcing industry in the United States. In Europe, forced lettuce is an important commodity, and some excellent cultivars have been developed specifically for that industry. In many greenhouses, it is intercropped with tomatoes or cucumbers to enhance utilization of space.

The major production requirements for forced lettuce are a sterile, light, friable growing medium, careful control of plant pests, uniform watering, often with a nutrient solution, and temperature control. The crop is established by transplanting at close spacing in beds [7 × 7 in. (18 × 18 cm) or 9 × 9 in. (23 × 23 cm)]. A trickle irrigation system maintains uniform soil moisture without wetting the foliage, thereby minimizing disease. Temperatures of 45 to 50°F (7 to 10°C) night and 60 to 65°F (16 to 18°C) day are considered optimum.

The cultivars most successful under greenhouse conditions include several European 'Boston'-type butterheads and standard outdoor leaf cultivars of the 'Grand Rapids' group.

The most troublesome insects in greenhouses are whitefly and aphid, and mildew is the most serious disease. These disease and insect problems must be monitored daily. The closed environment of a greenhouse favors rapid life cycles of many greenhouse pests, and appropriate control sprays must be applied promptly.

Harvest and Market Preparation

Harvest of crisphead lettuce (Figure 13.5) is begun when the plants reach a suitable size and the market price is reasonable. The heads are hand cut just below the lowest leaf. Loose and damaged leaves are removed, and, when necessary, the product is washed prior to packaging. A fully mature head is one that yields slightly to pressure. Overmaturity is characterized by yellowing of the leaves, cracked ribs, and bitter flavor, and such heads are susceptible to postharvest disorders. Se-

FIGURE 13.5
Harvesting lettuce. The heads are cut by hand, trimmed, film-wrapped, and crated on mobile conveyers. (Photo courtesy of A. Kader, University of California, Davis.)

lective mechanical harvesting has not been as effective as hand harvest, in large measure because head firmness is difficult to sense mechanically. Lettuce heads do not reach maturity uniformly, and experienced labor is able to select those ready for harvest by applying hand pressure.

There are three general types of market preparation.

1. **Naked pack** The heads are cut, trimmed, and packed in two tiers of 12 heads each in cardboard boxes.
2. **Source wrapped** After being cut and trimmed, heads are wrapped, heat-sealed, and placed 24/carton.
3. **Bulk pack** Cut heads are loaded into bulk containers and transported to a central packing shed where they are cored, shredded, washed, and cooled.

In western lettuce areas, several harvests are taken from an individual field. In eastern truck farms, a single harvest may be feasible. For local market, the planting dates are staggered to extend the season. The crop reaches maturity in 55 to 120 days, depending on cultivar and growing temperature. Spring and summer crops in the north are early; midwinter plantings in the south take almost twice the time to mature.

Cooling is predominantly by the vacuum method (refer to Chapter 8), particularly for head lettuce shipped to distant markets. Hydrocooling is common for market gardens. The shelf life of crisphead lettuce is approximately doubled by reducing air storage temperature from 50°F (10°C) to 41°F (5°C) or from 41 to 35°F (5 to 2°C).

Most lettuce is shipped by refrigerated truck. There are several transit and storage problems in head lettuce, some due to pathogen infection (bacterial soft rot and gray mold), others related to physiological changes.

Russet spotting, a disorder in which olive brown spots appear on the lower midribs of outer leaves, is related to excessive ethylene in the storage room. It may be especially severe on overmature lettuce or lettuce produced in desert areas.

Rib blight and **brown rib** also occur on outer head leaves, causing yellowing or tan discoloration. The cause is not known, but the disorder seems to occur most often at high temperature.

Brown stain is characterized by numerous small lesions with sunken centers and brown margins. It usually appears on the leaf surface near the lower part of the midrib. The lesions may coalesce as the disorder becomes severe. It apparently is caused by excess carbon dioxide in storage and is thought to be related to metabolism of phenolics. Some cultivars seem to be more susceptible than others.

Pink rib describes a diffuse discolored area at the midrib base, often on overmature heads. Bacterial infection often accompanies this disorder.

Most of these storage and transit problems can be minimized by carefully managing the environment, ventilating to avoid gas buildup, avoiding overmature heads, and maintaining

TABLE 13.4
Effects of packaging on weight loss of iceberg lettuce that was vacuum cooled after harvest, shipped from the west coast to the eastern market, and stored for 2 weeks at 3°C

Packaging treatment[a]	Weight loss (%)		
	After cooling	After shipping	After shipping + 2 weeks at 3°C
A	2.1 a[b]	2.3 a	5.1 a
B	0.7 a	0.7 a	1.7 b
C	1.5 a	1.8 a	4.8 a
D	1.4 a	1.6 a	3.3 ab

Source: From C. Y. Wang, R. T. Hinsch, and W. G. Kindya (1984) *HortScience* **19**:584. Reprinted by permission of the authors.

[a] A = naked; B = 0.013-mm polyethylene with 10-mm slits; C = 0.032-mm polyethylene with 285 perforations/cm²; D = 0.025-mm polyethylene with 112 perforations/cm².

[b] Mean separation in columns by Duncan's multiple range test, $P = .05$.

uniform temperature and humidity. Use of perforated polyethyene wraps for lettuce in storage or transit reduces weight loss and storage defects, thereby increasing salability (Table 13.4).

CELERY

Historical Perspective and Current Status

Celery was most likely derived from wild plants occurring in marshy sites bordering the Mediterranean in southern Europe and North Africa, and similar habitats in southwest Asia. Homer's *Odyssey* mentions selinon (celery) grown in 850 B.C., but this early form, later called smallage, was a leafy plant, quite pungent and bitter, used exclusively for medicines. Celery was first reported to be used as a food in the late 1500s, but primarily as a flavoring in cooking. Fresh use began in the eighteenth century, and its use as a vegetable in the United States was recorded as 1806.

Celery now is one of the leading salad crops consumed in the United States. It is produced throughout the year in California and seasonally in other areas. California accounts for two-thirds of the domestic production, mostly in the coastal valleys south of San Francisco. Winter production is centered in southern California, near San Diego. Florida accounts for 20 percent of the U.S. celery crop, most produced on organic soils in south Florida. New York and Michigan grow late summer and fall crops, also largely on organic soils, and other states with suitable soil and temperature have small acreages.

The major value of celery is its flavor and crispness: nutritionally, it is not especially valuable, although it is somewhat better than head lettuce (Table 13.1).

Classification, Growth, and Development

Celery (*Apium graveolens* var. *dulce*) is a member of the family Apiaceae (Umbellife-

rae). The most common form bears approximately 5 to 12 thick petioles in a tight bunch or head. A less popular form is the turnip-rooted celery, known as celeriac (var. *rapaceum*).

Celery is a biennial cool season crop, but it is sensitive to prolonged cold temperature. Growth through the first year normally produces a tight cluster of petioles and leaves attached to a very compressed stem. Early vegetative growth is quite spreading, but leaves arising from the apex of a short stem form compact, elongated heads (Figure 13.6). Celery requires a relatively long uniform sea-

FIGURE 13.6
Celery head. Note insignificant stem; most of the head is composed of thickened petioles. (Photo courtesy of University of New Hampshire.)

son to reach marketable size, and suitable growing sites therefore are concentrated in areas with mild winter climates or those modified by oceans or lakes.

Although a true biennial, celery may bolt after exposure to temperatures below 50°F (10°C) for as little as 10 to 14 days. The longer the exposure, or the lower the temperature, the higher the percentage of flowering. Transplants hardened by withholding water or by top-pruning are less sensitive than those not hardened. Hardening by cold exposure is not recommended, although seedlings are not as susceptible to bolting as plants with substantial top growth. Once seedstalks have been initiated, temperatures of 52 to 70°F (13 to 21°C) will accelerate seedstalk development. The individual flowers, borne on umbels, are small and inconspicuous.

The root system of celery is spreading and fibrous with many feeder roots close to the soil surface. Because of this surface root system, the plant will suffer from close cultivation and from competition from weeds. It also is very sensitive to fluctuating water supply.

Crop Establishment and Maintenance

A celery crop is established from transplants grown in outdoor seedbeds or in suitable plant containers (Speedlings, plug trays, peat pots, or cubes) in forcing structures. Seed germination can be poor in all vegetables within Apiaceae. The reasons for poor germinability may be several: damage to the embryos from lygus bug feeding, failure of embryos to develop properly, or light/temperature dormancy. The latter, however, is the major factor. Seeds are most sensitive to heat when germinated in the dark. When exposed to far-red light, germination has not been affected appreciably by high temperature. Seeds treated with gibberellin (GA_4 + GA_7) in laboratory tests germinated as though exposed to far-red light.

Seed priming is suggested as a practical way to reduce thermodormancy. Seeds are imbibed in water for a period of 1 to 4 days at 68°F (20°C), followed by 6 to 10 days at 34°F (1°C). Soaking for 20 h in 1 percent K_3PO_4 + 100 ppm benzyl adenine followed by a wash and careful drying also has been successful. Preconditioning seeds with osmotic materials, such as polyethylene glycol, or KNO_3 plus tripotassium orthophosphate, then careful drying also has improved germination. Such osmotic treatments permit some water imbibition and germination but inhibit radicle emergence until after seeding. Cultivars vary in thermodormancy, but all have been improved in germinability with a priming or preconditioning treatment.

To grow transplants in outdoor beds, seeds (preferably coated) are sown with a vacuum seeder in a very fine seedbed of friable, well-drained loam or its equivalent. Uncoated seeds may be scattered and lightly scratched in and covered with wet burlap to ensure adequate and constant moisture for germination. Flat-grown plants are seeded mechanically in conical-celled containers. Shading is recommended for all seedings to reduce soil temperature during germination. For field- or flat-grown transplants, the tops are pruned periodically to harden the plants. After 6 to 8 weeks of growth in outdoor beds, or 5 to 6 weeks in forcing structures, the plants should be ready for transplanting. In one test, container-grown transplants were superior in establishment and early growth, compared to those transferred bare root, but the latter did not produce a taproot, and, as a consequence, increased water levels were required to ensure adequate moisture for the predominantly shallow, fibrous roots.

Occasionally, direct field seeding has been practiced in California. For this system, pelleted seed is sown at 2-in. (5-cm) spacing, later thinned to 6 to 8 in. (15 to 20 cm). Gel seeding, using seed allowed to imbibe water for 8 days prior to mixing in gel, has resulted in substantial stand improvements in tests in England.

Celery is well adapted to muck soils but

also is grown on well-drained mineral soils. All soils should be limed to a pH range of 6 to 7, and mineral soils should have organic matter added prior to plowing. Fertilizer normally is applied in split treatments, with one-half applied before planting, the remainder 4 weeks before harvest. The latter treatment has encouraged rapid head development in comparison to those not sidedressed. The preplant application may be broadcast or banded. A crop of celery will remove approximately 195 lb of nitrogen, 50 lb of phosphorus, and 435 lb of potassium (88, 23, and 197 kg, respectively), based on a yield of 50 tons (45 MT). Nitrogen is especially important in the month preceding harvest, and heavier applications of phosphate and potash may be required in muck soils than in mineral. Nutrient sprays are used occasionally to supply magnesium, boron, or calcium, since serious disorders develop if these elements are deficient.

Chlorosis occurs in cultivars that are inefficient with respect to magnesium uptake and metabolism. This inefficiency is due to a single recessive gene, but high calcium in the plant will intensify the disorder. The problem can be eliminated by using efficient cultivars (those that tolerate reduced soil magnesium) and by applying high-magnesium lime to neutralize soil pH. Foliar applications of $MgSO_4$ also will alleviate the problem.

Cracked stem (Figure 13.7), sometimes called brown check, is related to a deficiency of boron. The reaction of cultivars also is controlled genetically, with inefficient cultivars recessive to efficient. The disorder begins as a crack of the inner petiole surface below the first leaflet and extends partway down the stalk. The exposed tissue becomes brown and susceptible to disease infection. Excessive nitrogen and potassium will intensify the susceptibility to low boron. The problem is common along the eastern coastal plain where soils are naturally low in boron. Occasional soil treatment with borax or foliar sprays may be used to correct the deficiency.

Blackheart (Figure 13.7), caused by a deficiency of calcium, causes the youngest leaves

FIGURE 13.7
Blackheart of celery (*a*), caused by insufficient calcium or poor mobility of available calcium; cracked stem (*b*) develops from boron deficiency. (Photo (*a*) from *Diseases of Plants*, slide set 35, Diseases of Onion, Lettuce and Celery. Reproduced by permission of the American Phytopathological Society. Photo (*b*) courtesy of D. N. Maynard, Florida Agricultural Research and Education Center, Bradenton.)

to become brown and water soaked, and, after drying, the leaves turn black. The disorder is very similar to tipburn in lettuce and cabbage, appearing when weather conditions favor a sudden growth surge and available calcium cannot meet plant requirements. Improved fertilization and regular irrigation reduce the incidence, and calcium sprays (calcium nitrate or calcium chloride) on foliage also have been effective.

Celery may be planted in single rows or in twin rows, either on flat terrain or on ridges. In many areas, growers use two rows, 14 in. (35 cm) apart, on beds set 40 in. (1 m) on center. If in single rows, the spacing normally is 24 in. (61 cm). Plants within rows are 6 to 8 in. (15 to 20 cm) apart.

Irrigation is essential for the continued steady growth required for crispness and flavor. In the major production areas, the frequency of irrigation is at 7- to 14-day intervals in cool seasons, up to two per week in warm seasons. In mineral soils, furrow irrigation is preferred, although sprinklers may be necessary when the plants are very young. A crop of celery will mature in 80 to 120 days from transplanting, about 30 days longer if direct seeded.

Cultivars

There are two major groups of celery cultivars: golden, termed self-blanching, and green. Yellow celery has a mild flavor but is considered inferior to the green type in quality. Within the golden group, there are two subtypes: golden self-blanching and golden plume. For best flavor, both should be blanched by piling soil around the heads or by shielding the stalks from the light with boards or paper. Green celery, termed pascal, may be divided into three types:

Summer pascal	Less susceptible to bolting than Utah, but smaller with a less compact head
Utah	The preferred standard, with large dense heads, susceptible to bolting
Slow bolting	Less susceptible to bolting than Utah but with fewer petioles and less attractive

Celeriac, a botanical variety (*rapaceum*) within *Apium graveolens,* is very much like celery in its growth habit. It develops a thick root and stem, up to 3 to 4 in. (7.5 to 10 cm) in diameter, which is peeled, cut, and boiled or used in flavoring soups and meats. Unlike celery, the petioles remain narrow and slender and seldom are used. The plant excels in all

areas in which celery is produced successfully. Little, however, is grown in the United States, and only a few cultivars are available: 'Large Smooth Prague,' 'Giant Prague,' and 'Alabaster.'

Disease and Insect Pests

Diseases   Celery often is produced in areas of rather high humidity. This humidity favors rapid development of several foliage diseases (Figure 13.8). In addition, several soilborne fungi and viruses can reduce productivity.

Late Blight   Late blight may be caused by two species of *Septoria.* These are *S. apii-graveolentis,* which produces small chlorotic spots, and *S. apii,* which is responsible for large spots. As the lesions caused by *S. apii-graveolentis* develop, the centers become necrotic, and the spots coalesce. Pycnidia may arise from the lesions. This form is most common in the north, where it develops during late season rains. *S. apii* is favored by somewhat higher temperatures and is prevalent in subtropical areas. This disease begins as a yellow fleck, progressing to collapse of the tissue as the spots become quite large. The *Septoria* species are seedborne, and infected seed will

FIGURE 13.8
Early blight of celery. (From *Diseases of Plants,* slide set 35, Diseases of Onion, Lettuce and Celery. Reproduced by permission of the American Phytopathological Society.)

provide substantial seedbed inoculation. The fungus also can overwinter on plant refuse. The recommended control measures include rotation to eliminate plant refuse, seed treatment [118 to 120°F (48 to 49°C) for 30 min], and protective sprays.

Early Blight   Early blight is a destructive disease worldwide. The causal organism, *Cercospora apii,* is favored by high humidity and by temperatures higher than those optimum for late blight. In the northern states, the disease precedes late blight, whereas it generally follows late blight in Florida. The fungus causes round yellow spots that enlarge up to 0.4 in. (1 cm) in diameter. These spots darken to a gray color as the conidia develop. On the petioles, the lesions become elongate. The control technique is similar to that recommended for late blight, since it, too, is seedborne and can overwinter on plant debris. Seeds stored for 3 years appear to be free of infection, but germination also can be somewhat reduced.

Fusarium Yellows   Yellows, caused by *Fusarium oxysporum* f.sp. *apii,* is a soilborne vascular disease that appears in mid- to late summer. Favored by moderate soil temperature, the organism infects the young roots, reducing movement of water and nutrients. Resistant cultivars have been the most reliable control, although there is evidence that new races of the fungus may infect otherwise resistant cultivars.

Pink Rot   Pink rot results from infection by a common soil inhabitant, *Sclerotinia sclerotiorum,* and occasionally by two lesser species of *Sclerotinia.* In celery seedbeds, it may cause damping-off. In maturing plants, a pink basal rot develops, covered by white mycelia that eventually develop black sclerotia. Because it may develop after cutting as well as during growth, it can become a serious transit disease. Control is difficult. There are no resistant cultivars, and chemical control is ineffective. In areas where it is possible, flooding for 6 weeks

before or after a cropping season seems to reduce the problem.

Basal Stalk Rot   During moist weather, the soilborne organism *Rhizoctonia solani* may cause sunken, brick-red lesions on the stalks at the base of the plant. Control is difficult once the disease appears. Improving soil drainage by growing the plants on ridges may help in preventing stalk rot.

Virus Diseases   In California, the western celery mosaic occurs in all production areas. The symptoms are vein clearing, mottling, and, eventually, necrotic spots and twisting of the leaves. Plants infected early in growth become stunted. The vector is any of several species of aphid, and insect control is the only method of reducing the disease. Eliminating celery growth for 3 to 10 weeks helps to break the insect life cycle, and periodic sprays must be used to prevent aphid buildup. Southern celery mosaic occurs in Florida. Symptoms of the disease are mottled leaves with yellow veins, and petioles that curl down and outward show discolored vascular tissue. The alternate host for the virus is the ornamental *Commelina nudiflora* (wandering Jew), and aphids are the primary vector.

Insects   Aphids   Eleven species of aphid can transmit western celery mosaic. Among them, the green peach aphid, melon aphid, and cotton aphid are the most common types. Fields should be inspected frequently for infestations, and repeated spraying with one of several insecticides may be necessary. Most aphid species migrate to other plant hosts during their life cycle; therefore, adjacent fields or border areas also should be inspected and sprayed if insects are found.

Larvae   There are four species of leaf-feeding worms occurring on celery:

Fall armyworm	*Spodoptera frugiperda*
Black cutworm	*Agrostis ipsilon*

| Cabbage looper | *Trichoplusia ni* |
| Celery leaf tier | *Oeobia rubigales* |

The fall armyworm, cutworm, and cabbage looper are general feeders described elsewhere (Chapters 12 and 20). The celery leaf tier is a $\frac{3}{4}$-in. (2-cm) green worm that feeds on leaves and stalks and will roll the leaves with webs. All of these pests may be controlled with *Bacillus thuriengensis,* acephate, permethrin, endosulfan, and other materials.

Other Insects   Other insects occasionally damaging celery include spider mites, tarnished plant bug, and leaf miner (Figure 13.4). Organic phosphates, such as parathion, are used to control plant bug and mites. Leaf miners, which have become frequent pests in Florida and other regions, are controlled most effectively by systemic pesticides.

Harvest and Market Preparation
Commercial celery is cut by hand or by machine (Figure 13.9). Prior to cutting, the plants may be topped to $14\frac{1}{2}$ to 16 in. (37 to 40 cm), or may be trimmed to length after cutting. The self-propelled mobile packing unit is the most popular harvest system. Conveyers take the harvested heads past graders, who

FIGURE 13.9
Mechanized harvesting of celery. (Photo courtesy of University of New Hampshire.

then pack by size in crates. In other areas, the harvested crop may be moved to packing sheds for grading and packing. In California, celery is packed 2, $2\frac{1}{2}$, or 3 dozen heads per crate, each crate holding approximately 60 lb (27 kg). Florida celery is packed in seven size grades, from $1\frac{1}{2}$ to 8 dozen per crate.

Prepackaging in film is restricted primarily to celery hearts and may be done at the shipping or receiving points. Celery hearts are derived from plants with stalks smaller than regular celery and are trimmed to 8, 10, or 12 in. (20, 25, or 30 cm) in length.

California celery generally is vacuum cooled. In Florida, most celery packed in wirebound crates is hydrocooled, whereas the prepackaged product is vacuum cooled. Properly cooled, celery can be stored for 2 to 4 weeks at 32°F (0°C) and 95 percent relative humidity, a shorter period at 40°F (4.4°C). In contrast to celery, celeriac can be stored for 3 to 4 months under those conditions.

Quality in celery is related to flavor, which should be sweet (without bitterness), and to crispness of the stalk. For the grower, such factors as head size and compactness, presence of side shoots, petiole size (width, thickness, height), and pithiness also are important. The level of quality very often is directly related to management practices that ensure steady growth and little or no environmental stress.

ENDIVE AND CHICORY

Historical Perspective and Current Status
Endive and chicory are especially popular salad vegetables in Europe and have increased in popularity in the United States. The two plants are closely related, and the names occasionally are interchanged (Belgian endive, a synonym for witloof chicory). Endive is believed to be native to Egypt or India. During the thirteenth century, it reached northern Europe, and it was described in Germany, England, and France during the 1500s. The first written

record of this crop in the United States was in 1806.

Endive now is an important market garden crop. The largest acreage is located in Florida [5600 acres (2268 ha) in 1982], this area growing approximately 40 percent of the U.S. total shipped. Most of the Florida production is located on organic soils south of Belle Glade, reaching the market in late winter or early spring. North central Florida produces some endive for the April–May market. The total U.S. crop for shipping, however, is less than half of that of Italy.

The food value differs among cultivar types. Curly endive contains 3300 IU of vitamin A and 10 mg of vitamin C. Plain-leaf endive (escarole) is much higher, with 14,000 IU of vitamin A and 100 mg of vitamin C. All endive contains moderate amounts of minerals (Table 13.1).

Chicory probably originated in the Mediterranean area. The first mention of chicory culture was in 1616 in Germany. Cultivation began in England in 1686 and in France in 1826. Chicory culture in the United States did not begin until later.

In Europe, chicory is grown for greens and salad use, but particularly for the forced heads (chicons) of the witloof or Belgian endive type. The centers of production of witloof chicory in Europe are France, Holland, and Belgium. In the United States, chicory most often is used as a leafy salad vegetable. Although the forced chicons are considered a gourmet food, the food value is quite poor in comparison to the green leaves of salad chicory (Table 13.1).

Classification, Growth, and Development

Both endive (*Cichorium endivia*) and chicory (*C. intybus*) are members of the Asteraceae. Endive is a cool season annual, but chicory is classed as a perennial, although the leafy forms are grown as annuals. Early growth of both crops appears as a rosette of leaves. Later, as temperature and daylength increase,

both can become reproductive. However, endive plants exposed to cold temperatures for a substantial duration become reproductive much sooner than those exposed only to moderate temperature. This transition to flowering is accelerated further by long days.

Crop Establishment, Maintenance, and Pest Control

The general cultural needs of endive and chicory grown for salad use are not appreciably different from those of leaf lettuce. Adequate nitrogen, phosphorus, and potassium and regular irrigation are needed to produce a high-quality product. Soil and pH requirements are the same as for lettuce.

Most endive and leaf chicory plantings are in single rows in the north and in twin rows of 40-in. (1-m) beds in California. Spacing between rows is 12 to 15 in. (30 to 38 cm), with in-row spacing at 8 to 12 in. (20 to 30 cm). Some growers use a square [14 × 14-in. (35 × 35-cm)] planting arrangement. Pest control procedures are similar to those described for lettuce.

Cultivars

The types of cultivars of endive and chicory are based upon appearance and use. The following outline identifies each cultivar type and its primary features.

I. Endive
  A. Leaves broad, coarse, and crumpled; plant medium large, deep-hearted; inner leaves well blanched — Escarole
  B. Narrow leaf; leaf margins curled and deeply cut; plant broad in diameter, with creamy inner leaves — Curly endive
II. Chicory
  A. Leaves used as a salad or green, or for forcing

1. Green type: leaves dark green, narrow and notched; petiole and leaf harvested for potherb — Radichetta

2. Forcing type: resembles Cos lettuce, but smaller; leaves, narrow on broad stalks, may be used for salads; roots are enlarged as in IIB; forcing produces tight, blanched heads — Witloof, belgian endive

B. Roots enlarged, may be ground for use as coffee adulterant or substitute; leaves resemble dandelion and can be used in salad — Magdebourg chicory

Harvest and Market Preparation

Endive and green chicory are harvested when heads are compact, bright colored, and well sized. The plants are cut at the base, and damaged or diseased leaves are trimmed. The trimmed heads are washed and cooled, by either vacuum cooling or hydrocooling, to 32°F (0°C). The heads then are packed in wire-bound crates or in cartons or baskets for transit.

Forced Chicory

Witloof chicory (Figure 13.10) is very exacting in its cultural and environmental requirements. In the preforcing stage, it should be grown on deep, fertile, and light-textured loams. The objective in this stage of growth is to produce clean, well-shaped roots. Fertilizer applications should be low in nitrogen to avoid excessive leaf growth, but high in phosphorus and potassium to stimulate root development.

Seed is sown directly in the field in late April or May at a heavy rate in rows 12 in. (30 cm) apart. Both single-row and twin-row systems are used. Several weeks after emergence, the plants are thinned to 6 in. (15 cm) between plants. The final stand is critical, since it controls ultimate root size.

FIGURE 13.10
Blanched heads (chicons) of witloof chicory. (Photo courtesy of University of New Hampshire.)

Natural rainfall is ample in most production areas, and weed, disease, and insect control are as described for lettuce. At harvest, the roots should be 6 to 8 in. (15 to 20 cm). The plants are lifted with a plow or knife and piled or windrowed to dry. Roots appropriate for storage [$1\frac{1}{4}$ to 2 in. (3 to 5 cm) in diameter; $3\frac{1}{2}$ to 4 oz (99 to 127 g)] are selected and trimmed to uniform length. The roots then are stored at 36°F (2°C) and removed for forcing as needed.

For forcing, a bed is prepared with fine-textured, well-tilled soil or artificial medium.

Roots are placed upright in a pit 8 in. (20 cm) deep, tightly packed, then covered with fine light soil. The top of the root should be 8 in. below the soil surface. The area then is covered with straw and curved metal sheeting, and heat [40 to 70°F (4 to 21°C)] is applied. At 65°F (18°C), approximately 18 days is required to produce a mature crop of chicons. If the temperature is excessive, the chicons may be poor in quality.

Some growers do not cover the roots with soil but instead provide sprinkle irrigation to maintain the desired humidity. A crop grown this way will be easier to harvest and clean than one grown in the standard system.

Quality is judged by weight, size and tightness of chicons and mildness of flavor. Ideally, the length should be five times the thickness, between $4\frac{1}{2}$ and 8 in. (11 and 20 cm) long and 1 in. (2.5 cm) thick, and heads must be pure white with only two outer leaves showing. Quality is affected by vernalization prior to forcing and by the sugar changes during storage. Excessive vernalization results in long cores in chicons, which suggests a transition to reproductive growth. Comparative studies also have shown core size to be correlated with reducing sugars.

The chicons are very perishable and must be handled carefully. They are normally packed without washing in paper-lined boxes and cooled rapidly to near 32°F (0°C) and must be kept at temperatures at or just above freezing in 95 percent relative humidity to preserve fresh quality.

MINOR LEAFY SALAD CROPS

Parsley (Apiaceae, *Petroselinum crispum*)

Parsley may be a salad ingredient in small amounts, but it is most widely used as a flavoring or garnish. The plant is strongly biennial, is native to the Mediterranean region, and was popular in early times among Greeks and Romans. Its value was reputed to be in aiding digestion and in suppressing odors of onion and wine. Both curled- and flat-leaf types were described by Theophrastus in 322 B.C.

Parsley is produced commercially in Texas, California, New Jersey, Florida, and New York and in small amounts in other states. It is an excellent source of vitamins A and C and of calcium and protein, and also contains moderate amounts of potassium, iron, sodium, and phosphorus (Table 13.1). The characteristics and culture of parsley are similar to those of celery, and it is susceptible to the same pests.

Parsley cultivars can be grouped into five types: plain leaf, celery leaf or neapolitan, curled, fern leaf, and turnip rooted. The most common and attractive is the curled leaf, which contains the subtypes double curled ('Moss Curled'), evergreen, and triple curled. These subtypes are distinguished by the degree of leaf curling, coarseness of the leaf, and plant growth habit. The plain-leaf parsley ('Dark Green Italian') has deeply cut leaves, but no curling or fringing. Of the turnip-rooted forms, the only cultivar is 'Hamburg.'

Parsley can be harvested over a long period of time, cutting outer leaves only. The leaves are either packed loose or bunched. Leaves should be washed thoroughly before packing in crates or baskets. Much of the crop goes to institutional trade (restaurants and hotels) and to dehydration. Under storage at 32°F (0°C) and high humidity, parsley can be held for up to 2 months.

Chervil (Apiaceae, *Anthriscus cerefolium*)

Chervil is an annual cool season crop more common in Europe and Asia than in the United States. It is probably native to Europe and was cultivated in England in 1597. The cultivars utilized are curled-leaf forms, with a growth habit and cultural systems similar to those of parsley. Chervil will bolt easily, especially if the seed is vernalized and plants are grown in long days. The leaves are harvested approximately 6 to 10 weeks after planting for flavoring and garnishing use.

Watercress (Brassicaceae, *Rorippa nasturtium-aquaticum*)

The most common watercress used is the diploid species. It is a cool season crop that requires flowing water of uniform temperature, preferably derived from deep limestone wells. The plant morphology is described as three segments. The above-water portion consists of an aerial stem with leaves, but no roots. The stem segment under water has foliage and adventitious roots that develop from leaf axils. These adventitious roots do not anchor the plant but are suspended in the water. The anchoring root system is the basal plant segment. Both anchoring and suspended root systems absorb nutrients, although the latter component is the more important.

Beds normally are placed side by side, oriented north to south; each bed has a water input and output with sufficient slope to ensure water movement. A layer of silt overlaid with $\frac{1}{2}$ in. (1.3 cm) gravel is applied to the bed floor to provide anchoring.

Seed is sown on dry beds at a rate of 1 lb/700 yd^2 (7.7 kg/ha). Beds then are moistened slightly, but water level is not increased until plants begin to grow. A thick stand is preferred, since it produces a uniform product and helps to reduce weed growth. Fertilization is not required, for the constantly renewed water provides sufficient nutrients for growth. During the summer months, watercress growth is predominantly above water, whereas it remains underwater during the winter. The summer crop is harvested at water line or slightly above, and the stubble then is rolled to keep it under water where it will regenerate new growth. Winter crops are taken by pulling up one-third of the plants. The remaining plants will develop and fill in the harvested areas.

The primary diseases are crook root, a waterborne fungus (*Spongospora subterranea* f. *nasturtii*), and turnip mosaic. Crook root infection is indicated by swollen and curved roots that become stunted and eventually break from the anchor roots. Control is achieved by providing zinc in the water. Turnip mosaic is spread by the green peach aphid, cabbage aphid, and/or bean aphid. The disease can be minimized by seed propagation and careful insect control.

The most important insect pest in many production areas is the mustard beetle, which will damage leaves. Control is achieved through insecticides and by using concrete beds that prevent insect overwintering. In Hawaii, the diamondback moth larvae, cyclamen mite, and cabbage aphid constitute the major pests. Each can be controlled with applications of insecticides.

Cress (Garden Cress, Brassicaceae, *Lepidium sativum*)

Cress, also known as pepper grass, is a cool season annual plant. It is native to Iran and eventually spread to India, Egypt, and Greece. It has been cultivated in England since 1548 and was first mentioned in the United States in 1806.

There are four types of cress in cultivation: common, curled, broadleaf, and golden. The plant grows rapidly, often going to seed as daylength and temperatures increase. A rich soil and cool growing temperature provide the most suitable environment. The leaves are harvested for garnishing, adding pungency to salads and resembling radish in flavor. **Upland cress** (*Barbarea verna*) is a member of the same family as cress and is used in the same way. It tends to be somewhat more bitter than garden cress.

Sea Kale (Brassicaceae, *Crambe maritima*)

Sea kale is a fleshy perennial grown for the blanched shoots. It is a large-leafed plant reaching a height of 12 to 24 in. (30 to 61 cm). Sea kale is grown from seeds or shoots and dug in October to be forced, similar to witloof chicory. The blanched shoots are harvested after 4 to 5 weeks at 50°F (10°C) and may be eaten raw or cooked. Very little is grown or consumed in the United States.

Corn Salad (Valerianaceae, *Valerianella locusta*)

Corn salad, also known as lamb's lettuce or fetticus, is popular in Europe and was once more widely grown in the United States than is true today. This cool season annual plant produces a rosette of spatulate leaves that reach a height of 4 to 12 in. (10 to 30 cm). The mild-flavored leaves are cut and used in salads.

Dill (Apiaceae, *Anethum graveolens*)

Dill is an annual herb used as a flavoring for pickles and other foods and is similar to other Apiaceae in most cultural requirements. The plant reaches 3 to 4 ft (0.9 to 1.2 m) and therefore requires wider spacing than related crops. Seeds, flowers, and foliage may be used for flavoring.

Fennel (Apiaceae, *Foeniculum vulgare* var. *dulce*)

Fennel, also called Florence fennel or finnochio, is a short-lived perennial cultivated as an annual. The bases of the leaves become enlarged and bulblike and, when blanched, have the texture of celery. The cultural requirements are similar to those described for celery. It is used primarily for flavoring.

SELECTED REFERENCES

Collier, G. F., and T. W. Tibbitts (1982). Tipburn of lettuce. *Horticultural Reviews* **4:**49–65.

Raleigh, G., and P. L. Minotti (1972). *Lettuce Production in Muck Soils,* New York State College of Agriculture and Life Sciences, Cornell University, Coop. Ext. Bull. 1194.

Ryder, E. J. (1979). *Leafy Salad Vegetables.* AVI, Westport, Conn.

Ryder, E. J. (1986). Lettuce breeding. In *Breeding Vegetable Crops* (M. J. Bassett, ed.), pp. 433–474. AVI, Westport, Conn.

Ryder, E. J., and T. W. Whitaker (1980). The lettuce industry in California: A quarter century of change, 1954–1979. *Horticultural Reviews* **2:**164–207.

Seelig, R. A. (1964). *Endive–Escarole–Chicory. Fruit and Vegetable Facts and Pointers.* United Fresh Fruit & Vegetable Assoc., Washington, D.C.

Seelig, R. A. (1980). *Celeriac. Fruit and Vegetable Facts and Pointers.* United Fresh Fruit & Vegetable Assoc., Washington, D.C.

Seelig, R. A. (1981). *Parsley. Fruit and Vegetable Facts and Pointers.* United Fresh Fruit & Vegetable Assoc., Washington, D.C.

Seelig, R. A. (1982). *Belgian Endive (Witloof). Fruit and Vegetable Facts and Pointers.* United Fresh Fruit & Vegetable Assoc., Washington, D.C.

Sharples, G. C. (1981). Lettuce seed coatings for enhanced seedling emergence. *HortScience* **16:**661–662.

Whitaker, T. W., E. J. Ryder, V. E. Rubatsky, and P. V. Vail (1974). *Lettuce Production in the United States,* USDA Agric. Handb. 221.

Wittwer, S. H., and S. Honma (1979). *Greenhouse Tomatoes, Lettuce and Cucumbers.* Amer. Vegetable Grower, Willoughby, Ohio.

Zink, F. W., R. G. Grogan, and J. E. Welch (1956). The effect of the percentage of seed transmission upon subsequent spread of lettuce mosaic virus. *Phytopathology* **46:**662–664.

STUDY QUESTIONS

1. The leafy salad vegetables are considered high-risk crops to produce. What are the various problems that create that risk?

2. Contrast the differences in management of lettuce and celery grown on organic and mineral soils.

3. Seed germination is erratic in some of the leafy salad crops. What is the physiological basis for this problem and why do such practices as seed priming improve germination?

4. For what diseases of the leafy salad crops has resistance been attained in new cultivars?

14

Root Crops

Development of all root crops occurs most successfully in regions where cool night temperatures slow the respiration rate, enhancing retention of stored carbohydrates.

CARROT

Historical Perspective and Current Status

Carrots are believed to be of middle Asian origin, probably evolving from a wild plant resembling Queen Anne's lace. Seeds found by archaeologists in the Swiss Lake dwellings indicate that cultivation of carrots may have begun 2000 to 3000 years ago. It is believed that early use of the plant was medicinal, prescribed for curing stomach problems and treating wounds, ulcers, and liver and kidney ailments. The first roots consumed, probably purple in color and large and rough, were grown in Afghanistan around 600 A.D. Eventually, yellow types were isolated and grown in Syria in the ninth or tenth century. The crop was introduced to China in the late thirteenth century and to Europe 100 years later.

The first orange types appeared in The Netherlands during the seventeenth century and, because of improved color and flavor, quickly became popular in Europe. Settlers brought the crop to North America, where it was adopted by Indians and colonists.

Of all of the root crops, carrot is the most important. It is grown worldwide and is the major single source of vitamin A in U.S. diets, contributing approximately 14 percent of the adult daily requirements (Table 14.1). In addition to vitamins and minerals, raw carrots also provide important dietary fiber.

Carrots for fresh market are grown on an estimated 49,000 acres (19,845 ha) with a total production of 14,478,000 cwt (656,722

Carrot	Radish	Rutabaga	Parsnip
Beet	Turnip	Horseradish	Salsify

The root crops are cool season predominantly biennial vegetables. The seeds of most root crops are relatively small, and emergence of some is affected by soil crusting, requiring irrigation or anticrustant treatments to ensure a good stand. Horseradish, a perennial root crop, is propagated asexually from high-yielding clones.

Root crops range from excellent to relatively poor in food value (Table 14.1). The long storage life of these vegetables contributed to their utility and to their nutritional importance among early civilizations, and several root crops still contribute significant dietary value.

TABLE 14.1
Nutritional constituents of the major root crops[a]

Crop	Water (%)	Energy (cal)	Protein (g)	Fat (g)	Carbo-hydrate (g)	A[b] (IU)	C[c] (mg)	Thia-mine (mg)	Ribo-flavin (mg)	Niacin (mg)	Ca (mg)	P (mg)	Fe (mg)	Na (mg)	K (mg)
Beet	87	43	1.6	0.1	9.9	20	10	0.03	0.05	0.4	16	33	0.7	60	335
Carrot	88	42	1.1	0.2	9.7	11,000	8	0.06	0.05	0.6	37	36	0.7	47	341
Parsnip	79	76	1.7	0.5	17.5	30	16	0.08	0.09	0.2	50	77	0.7	12	541
Radish	95	17	1.0	0.1	3.6	10	26	0.03	0.03	0.3	30	31	1.0	18	322
Rutabaga	87	46	1.1	0.1	11.0	580	43	0.07	0.07	1.1	66	39	0.4	5	239
Turnip	92	30	1.0	0.2	6.6	Trace	36	0.04	0.07	0.6	39	30	0.5	49	268
Salsify	78	13	2.9	0.6	18.0	10	11	0.04	0.04	0.3	47	66	1.5	—	380

Source: National Food Review (1978), USDA.

[a] Data per 100 g sample.

[b] 1 IU = 0.3 μg vitamin A alcohol.

[c] Ascorbic acid.

MT). For processing, 463,000 tons (420,033 MT) is produced on approximately 30,000 acres (12,150 ha). The total estimated value of the crop in 1983 was $206,802,000. California is the leading producer of both fresh and processing carrots, followed by Texas and Michigan for fresh market and Washington for processing. The substantial fresh market production in home gardens or small market gardens is not included in national production statistics, but this local production constitutes a significant economic contribution.

Classification, Growth, and Development

The carrot (*Daucus carota* var. *sativus*) belongs to the family Apiaceae (Umbelliferae). Although both annual and biennial forms exist in Europe, the cultivars grown in the United States are biennial. As the plant emerges, it is characterized by two straplike seed leaves followed by the typical tight rosette of finely cut leaves. The predominant root is a tap with fine, fibrous side branches. The taproot enlarges gradually and at first contains two areas each of primary xylem and phloem. Cambium fragments between xylem and phloem eventually join, becoming circular, and produce secondary xylem toward the center of the root, secondary phloem toward the outside. At maturity, the two tissues are distinguishable as a

light-colored core (xylem) and the deep orange cortex (phloem) (Figure 14.1). The ma-

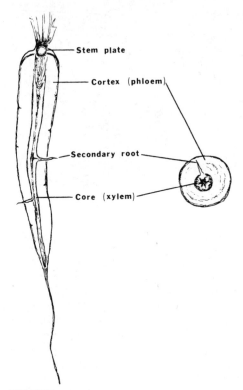

FIGURE 14.1
Longitudinal and cross sections of carrot roots during vegetative growth. The stem becomes prominent as reproductive growth begins.

ture root "peel" is periderm, of which suberin and waxlike substances are constituents. The periderm functions in a manner similar to the cuticle of a leaf.

As in the other Apiaceae, carrots contain essential oils that impart a characteristic aroma. The seed endosperm is especially rich in oils, and these oils have been found to contain potent germination inhibitors.

Carrots require a period of up to 6 to 8 weeks at temperatures below 50°F (10°C) for flower induction, and induction is accelerated under long days (Figure 14.2). Seedlings with a slender taproot are less susceptible to cold than those with roots larger than $\frac{1}{4}$ in. (6 mm) in diameter. The red (anthocyanin) carrot cultivars are reported to respond primarily to long day, regardless of temperature. Once initiated, the branched seedstalk elongates rapidly as temperature increases, each branch bearing compound umbels.

Optimum root growth occurs at 60 to 70°F (16 to 21°C). Although top growth is not affected visibly by temperatures well outside this range, root color, texture, flavor, and shape may be adversely affected.

FIGURE 14.2
Young carrot plants. (*a*) The stem has elongated, indicating a change to reproductive growth. (*b*) Vegetative plant. (Photos courtesy of University of New Hampshire.)

Crop Establishment and Maintenance

Soils must be free of stones or debris and should be porous and friable in texture and well drained to ensure symmetrical, straight roots. Deep, uniform, light-textured loams and organic soils are preferred for the long, tapered cultivars, whereas short, blunt types succeed as well in heavy loam soils.

The optimum pH range is 5.5 to 7.0. Lime, along with organic matter, should be applied during the fall prior to planting. Organic matter applied immediately before seeding will promote plant growth, but will hinder development of straight, unbranched roots.

To prepare a field for planting, plow deeply and disk to eliminate soil clods. Organic soils may be rotovated following deep plowing. In heavy loams, the first plow and disk operation may be followed by a flood irrigation and a second plow and disk. Soils tending to compact must be chiseled thoroughly. A compacted soil layer impedes root growth and can cause severely misshapen roots (Figure 14.3).

Fertilizers may be broadcast prior to final disking or banded at seeding. On soils receiving furrow irrigation, fertilizer normally is broadcast before raised beds are formed. If a band is used, it is placed 1 to 3 in. (2.5 to 7.5

cm) beside the seed, between the seed row and the irrigation furrow, at a depth of 1 to 4 in. (2.4 to 10 cm). In some areas, trace elements, especially boron and, to a lesser extent, manganese, constitute part of the soil management program. In these instances, growers must monitor soil fertility and availability carefully through soil and tissue tests. Addition of 2 to 3 lb of boron/acre (2.2 to 3.4 kg/ha) will be required every 2 to 3 years for deficient soils.

Based upon a yield of 500 cwt (22 MT) the roots remove 80 lb of nitrogen, 20 lb of phosphorus, and 200 lb of potassium (36, 9, and 91 kg, respectively.) Top growth removes 65, 5, and 145 lb (20, 2, and 66 kg), respectively, of nitrogen, phosphorus, and potassium. Because of the wide range of soil types used for carrot production, fertilizer rates and ratios are diverse, generally low in nitrogen on muck soils with moderate levels on mineral soils. Typical application rates in New York are 50 to 75 lb/acre (56 to 84 kg/ha) of nitrogen, 50 to 100 lb/acre (56 to 112 kg/ha) of phosphate, and 50 to 200 lb/acre (56 to 224 kg/ha) of potash. Half is broadcast before plowing, the remainder as a band at planting. A sidedressing of 30 lb/acre (34 kg/ha) of nitrogen is recommended 4 to 6 weeks after seeding. In California, the averages are 120, 95, and 40 lb/acre (134, 106, and 45 kg/ha), respectively, of nitrogen, phosphate, and potash. The rates recommended in Florida are generally higher, averaging 170, 145, and 205 lb/acre (190, 162, and 230 kg/ha), respectively, of nitrogen, phosphate, and potash. Abundant nitrogen may stimulate excessive top growth at the expense of root development, and maturing roots may split.

Stand establishment is a common problem in carrot production, affected by soil crusting, soil temperature (thermodormancy), and soil moisture. After seeding, the soil to a depth of $\frac{1}{2}$ in. (1.3 cm) must not dry; otherwise, imbibition of water will be slowed, germination inhibitors leaching from seeds will not dissipate, and the stand will be uneven. In addition, adequate soil moisture will reduce crusting and, to

FIGURE 14.3
Carrots grown in heavy soil with stones and clods. (Photo courtesy of University of New Hampshire.)

some extent, soil temperature. Several pre-conditioning or priming seed treatments have enhanced stand establishment (see Chapter 13).

Most commercial growers scatter seed $\frac{1}{2}$ to $\frac{3}{4}$ in. (13 to 19 mm) deep in a band 3 to 4 in. (7.5 to 10 cm) wide to eliminate the need to thin. Single rows may be 12 to 24 in. (30 to 61 cm) apart; twin rows would be sown near the edges of raised 20-in. (50-cm) beds spaced 40 in. (1 m) on center. Seeding rates vary from 25 to 60/ft (82 to 197/m), the heavy rates in band seeding or where soil conditions are somewhat unfavorable and the low rates for large processing carrots or where carrots are seeded in narrow rows.

Uniform soil moisture is a critical factor in producing crisp, sweet carrots. Periods of soil water stress will slow root growth, producing thickened cells and woody texture. Flavor also becomes bitter, or at least not as sweet as in carrots grown under ideal conditions. At least $1\frac{1}{2}$ in. (3.8 cm) of water is needed weekly, more in arid areas and light soils.

The development of deep orange color, reflecting increased carotenoids and nutritional value, is affected by cultivar, age of the root, and temperature within which growth occurred. Young roots are light in color, becoming darker orange with age. Maximum color develops at 60 to 70°F (16 to 21°C), with roots becoming lighter colored at temperatures above or below the optimum range. Root shape also is affected by temperature: temperatures over 70°F (21°C) reduce root length, whereas those 60°F (16°C) or slightly below favor long roots.

Competition stress imposed by weeds also will interfere with normal carrot growth. Commercially, weeds are controlled primarily with herbicides. Stoddard solvent, a very light petroleum solvent, may be applied at 50 to 100 gallons/acre (468 to 935 liters/ha) after two true leaves have appeared but not after the roots have developed to a pencil size. Roots of large plants sprayed with Stoddard solvent tend to retain a distinctive petroleum taste.

Other materials available for weed control include trifluralin (preplant incorporated) and linuron [postemergence, when carrots are 3 in. (7.5 cm) high].

Cultivars

The cultivars of carrots are grouped largely according to shape and use (Figure 14.4). Of the cultivars available, those most recently introduced have substantially higher levels of vitamin A than old cultivars and should be used where adapted.

Chantenay Type 'Chantenay'-type carrots are 2 to $2\frac{1}{2}$ in. (5 to 6 cm) in diameter at the shoulder, with a short, conical shape. The length is $4\frac{1}{2}$ to $5\frac{1}{2}$ in. (11 to 14 cm), with a medium to large neck. Color is medium to light orange, with a light to red core. Texture is coarse in the raw product. The 'Chantenay' type is the most widely grown carrot and is popular in processing because of the relatively small amount of waste.

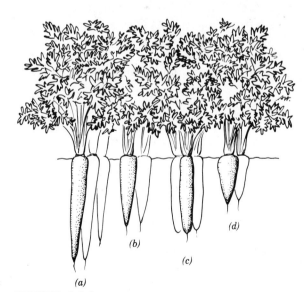

FIGURE 14.4
Carrot cultivar types: (*a*) Imperator; (*b*) Danvers; (*c*) Nantes; (*d*) Chantenay.

Imperator Type 'Imperator' has become the definition of the principal commercial fresh market cultivar. Carrots of this type are grown on up to 95 percent of the major fresh market acreages of Texas and California. The length is 7 to 8 in. (18 to 20 cm) with a shoulder diameter of $1\frac{1}{2}$ in. (3.8 cm) maximum and a long tapered tip. The root has a medium neck size, deep orange phloem, and light orange xylem. 'Imperator' carrots become woody at full maturity, but young roots have excellent quality. Subtypes such as 'Gold Pak' have become standards for color and texture.

Danvers Type Used for both fresh market and processing, the 'Danvers' (or half-long) type has a conical shape, top diameter of 2 to $2\frac{1}{2}$ in. (5 to 6 cm), and length of 6 to 7 in. (15 to 17 cm) at maturity. Flesh color is deep orange with a light center. Quality is excellent in young roots, but becomes fibrous with age.

Nantes Type 'Nantes' or derivatives of it are the principal home and market garden cultivars. The shape is nearly cylindrical with a blunt end, 6 to 7 in. (15 to 17 cm) long and $1\frac{1}{2}$ in. (3.8 cm) in diameter. The phloem is bright orange and constitutes a large percentage of root volume. The core often is indistinct, top growth is typically small, and quality is excellent.

Miniature Type Several cultivars are grown largely for baby carrots, sometimes termed Amsterdam types. These roots are $\frac{1}{2}$ to $\frac{3}{4}$ in. (13 to 20 mm) in diameter and only $2\frac{1}{2}$ to 3 in. (6 to 7.5 cm) long. Many are used for mixed vegetable packs in the freezing industry. In addition, a stump-rooted or beet-shaped carrot is available, averaging $1\frac{1}{2}$ in. (3.8 cm) in diameter and length. The round carrots may be successful in heavy soils, but are of best quality when harvested before full maturity. They must be hilled to prevent green shoulders.

Disease and Insect Pests

Diseases Carrots are largely free from major disease problems. Under certain conditions, losses may occur due to damping-off, blights, root rots, and yellows (Figure 14.5).

Carrot Blight Two fungi are responsible for carrot blight: *Cercospora carotae* and *Alternaria dauci*. These two organisms often co-exist, and the symptoms of the two diseases are easily confused. *Alternaria* is likely to attack old tissue and therefore may appear later in the season than *Cercospora*. Both organisms overwinter on crop refuse and on seed and cause yellowing and small dark brown elongate to round lesions on the foliage. Plants may be sprayed with chlorothalonil or carbamates, and a 2- to 3-year rotation is recommended to reduce inoculum levels. Both diseases may be troublesome in areas repeatedly planted to carrots.

Yellows The causal organism for carrot yellows is the same mycoplasma responsible for aster yellows. The disease organism has a wide host range and is transmitted by a number of leafhopper species. The first symptom is yellowing, occasionally vein clearing in the center of the plant. Many petioles are produced from the crown, giving a broom effect, and the roots may be malformed with prolific fibrous roots branching from the primary taproot. The leafhopper is the sole vector; hence, the most effective control consists of insect control and elimination of nearby plants serving as inoculum sources. Weed control in and near the fields is essential.

Root Rots Root rots seldom cause serious losses in the field. Foliage symptoms of black rot, caused by *Alternaria radicina,* are indistinguishable from those of alternaria blight. Lesions on the roots, however, may develop after roots are in storage, appearing as irregular to

FIGURE 14.5
Carrot yellows (*a*), showing a normal and an infected plant, and (*b*) *Cercospora* leaf spot on carrot foliage. (From *Diseases of Plants*, slide set 37, Diseases of Carrots, Eggplant, Peas, Pepper, Sweet Corn. Reproduced by permission of the American Phytopathological Society.)

circular dark green to black depressed areas showing some surface sporulation. Bacterial soft rot, caused by *Erwinia carotovora*, is characterized by a slimy, odiferous rot that develops rapidly under certain storage conditions. All storage rot diseases are minimized by careful handling and grading, and sanitation in storage facilities is imperative. The storage temperature should be maintained just above 32°F (0°C).

Insects The major insect pests of carrot (Figure 14.6) include the rust fly maggot, weevil, carrot caterpillar, and leafhoppers. In addition to the insect problems, nematodes can be serious, particularly in warm soils.

Rust Fly Maggot The carrot rust fly (*Psila rosae*) is a dark, shiny insect with a yellow head. It lays its eggs in the soil at the base of plants. The emerging yellowish maggots then

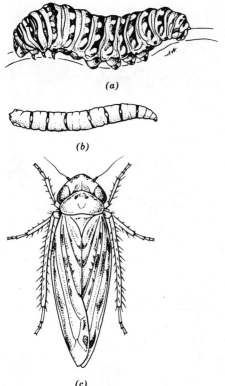

FIGURE 14.6
Insects of root crops: (*a*) carrot caterpillar (also known as parsley worm); (*b*) carrot rust fly larva; (*c*) beet leafhopper.

Carrot Caterpillar The carrot caterpillar (*Papilio polyxenes*), also termed celery worm, is a large larva, to 2 in. (5 cm) in length. It is green, with a black cross band on each segment and six yellow spots on the front margin. There are from two to three generations per year, and the damage can be substantial. Control by carbaryl is effective.

Leafhopper The most serious leafhopper problem is *Macrosteles divisus,* the species primarily responsible for transmitting aster yellows. The tiny insect produces a burned appearance on the foliage. To prevent disease, an insecticide must be applied when the first true leaves appear, with regular applications through the remainder of the growing season. Carbaryl and methoxyclor are recommended.

Harvest and Market Preparation

Carrots are harvested when they attain a size appropriate for the cultivar. For processing, the shoulder diameter ranges from $1\frac{1}{2}$ to 4 in. (3.8 to 10 cm) at harvest. Fresh market carrots are $\frac{3}{4}$ to $1\frac{1}{4}$ in. (2 to 3 cm). The roots are harvested by machine, either by mowing tops and lifting roots with a modified sweet potato harvester or, more frequently, by using carrot harvesters that loosen, lift, top, and load carrots into pallet boxes or trucks.

The roots then must be washed, graded, and sized. Most are packed in polyethylene bags [approximately 1 lb (0.5 kg)] for consumer sales. There is a small demand for carrots with tops attached. These roots normally are picked by hand after lifting, then washed and bunched, with approximately 12 carrots to a bunch. Topped carrots can be stored for up to 4 to 5 months at 32°F (0°C) and 90 to 95 percent relative humidity without noticeable loss of quality. Bunched carrots lose moisture rapidly and must be sold shortly after harvest.

Quality constituents of carrots include sugar content, color (both internal and external), freedom from fibrousness and strong flavor, and shape. The sugar level tends to increase in cold storage.

tunnel into the roots, causing mechanical damage and providing an avenue for root rot diseases. Parathion or diazinon, applied as a drench in the furrow at seeding, provides excellent control.

Weevil The carrot weevil (*Listronotus oregonensis*) is widely dispersed and common in such weeds as dock, plantain, and Queen Anne's lace. In carrot fields, the insect population builds slowly, and the white, legless larvae feed on the roots, causing zig-zag surface grooves or tunnels. Affected carrots may wilt and eventually die, or at least become unmarketable. Under good management, with appropriate insecticides for the major pests, the weevil should not become a major problem.

BEET

Historical Perspective and Current Status

Beets have been grown as a potherb throughout recorded history. The roots of wild beets, however, were used by ancient civilizations only for medicine. These wild forms did not resemble the modern enlarged beet. It was not until the sixteenth century that the fleshy, edible form ('Roman beet') was described, and it did not become popular until 1800. At that time, several cultivars had been described, including 'Flat Egyptian' (grown in the Near and Middle East) and 'Long Red' (popular in England).

Beets are believed to be native to the Mediterranean area of Europe and North Africa, and a secondary area of development was located in the Near East. Many members of the beet family are found in areas with elevated salt levels.

The total area of beets harvested in the United States is approximately 13,893 acres (5627 ha). The fresh market production of beets is estimated at 405,000 cwt (18,371 MT), with a processing volume of 203,900 tons (184,978 MT). It is ranked 21st among 22 principal vegetables. The major winter suppliers are Texas and California, and summer beets are produced in a number of locations in the northeastern United States and Canada. Processing acreage is located principally in New York, Michigan, and Wisconsin.

Nutritionally, beets are not a major source of vitamins and minerals, containing only low to moderate levels (Table 14.1). The crop is considered primarily a source of carbohydrate. Like spinach, it can accumulate elevated levels of nitrate in leaf tissue, and oxalic acid also is a constituent.

Classification, Growth, and Development

The beet (*Beta vulgaris*) is a member of the goosefoot family, Chenopodiaceae. Other common plants within this family include chard, spinach, and the common pigweed. Beet and Swiss chard are the same species, the latter assigned to Group Cicla. Both are biennials, with the vegetative phase characterized by a rosette of fleshy large ribbed leaves. The leaf petioles are attached to an insignificant stem forming the root crown.

Anatomically, the beet "root," actually enlarged hypocotyl, consists of alternating layers of secondary xylem (broad, dark bands) and phloem (narrow, light areas, Figure 14.7). In young hypocotyls, two cambium arcs are laid down between primary xylem and phloem. The arcs give rise to secondary xylem and phloem, and as each ring is formed, new cambium forms at the outer edge of the phloem. Each new cambium produces xylem toward the center and phloem toward the outside of the developing beet. All cultivars show this

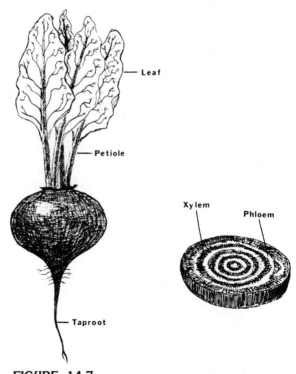

FIGURE 14.7
Morphology and development of table beet. Enlargement is due to secondary xylem and phloem.

zoning, although selection for color uniformity has improved appearance.

The actual root system is characterized by a prominent taproot that develops rapidly and can reach a depth of 10 ft (3 m). Just at the base of the swollen edible structure, lateral branches develop, some of which extend a substantial distance, both laterally and vertically. Because of the extensive root system, beets require a deep, moist soil. The root system is not especially dense, however, and close spacing does not increase competition to a level that detracts from root quality.

Optimum root development occurs at temperatures between 60 and 65°F (16 and 18°C). Excessive heat, especially at night, reduces storage of photosynthate, and growth, textural quality, and flavor are impaired. As biennials, beets are sensitive to prolonged exposure to temperatures below 50°F (10°C). After 30 to 60 days of exposure, the stem elongates, reaching a height of 4 ft (1.2 m), and flower buds appear. Flower induction is accelerated if exposed to a long day. Under most growing conditions, bolting is rare.

The beet seed, for most cultivars, is a seed ball or fruit consisting of several true seeds within dried, corky flower parts. As a result, precision seeding is of limited effectiveness. Distribution of the seed ball is facilitated by mechanical crushing and sizing to achieve more uniform seed. The monogerm trait, in which each seed ball contains only a single seed, has been introduced into table beets, but few cultivars have been released.

Beets contain two pigments. The deep red color is betacyanin, a pigment similar to anthocyanin. Betaxanthin, a yellow pigment, also is present. In red beets, the intensity of color reflects the amount of betaxanthin: low amounts result in deep red color. In yellow beets, betaxanthin predominates as an expression of a single recessive gene.

Crop Establishment and Maintenance

Table beets can be grown in a range of deep friable soils, including muck, sand, and sandy or silty loams. Organic matter should be applied to mineral soils prior to tillage, and pH should be adjusted to 6.0 to 6.8 regardless of soil type. Beets will tolerate alkaline soils but are sensitive to excessive acidity. Soils are plowed thoroughly to encourage root development, and the seed zone should be smooth and free of clods. Raised beds are used to improve drainage or to accommodate irrigation.

A crop of 20,000 lb (9.1 MT) will use 66 lb of nitrogen, 8 lb of phosphorus, and 80 lb of potassium (30, 3.6, and 36 kg, respectively). In eastern areas, a 1–2–2 ratio with 75 to 100 lb/acre (84 to 112 kg/ha) of nitrogen is common. In Florida, nitrogen is increased, whereas the California recommendations are similar to those for carrots. Fertilizer may be banded or broadcast at seeding, with the formulation depending on soil type. For muck soils, muriate of potash (0–0–60) should be broadcast to correct potash deficiencies, followed by a 1–2–2 or 1–1–1 ratio complete fertilizer. A side-dressing of ammonium nitrate also may be required, especially in sandy soils.

Boron deficiency causes development of black spots, especially within the phloem of the root, and young leaves tend to be malformed with an increase in red pigmentation. Periodic applications of 3 to 4 lb/acre (3.4 to 4.5 kg/ha) of boron are needed to alleviate these symptoms, and beets are quite tolerant of high boron levels. However, excessive application should be avoided, since other vegetables subsequently grown on treated areas may show toxicity symptoms.

Commercial plantings are in rows spaced 18 to 24 in. (45 to 61 cm) apart. Two rows per raised bed are preferred. Beets are hardy and may be seeded several weeks before the last spring frost, and throughout the summer to within 6 weeks of the last killing frost in the fall. Seeds germinate at a wide range of soil temperature [45 to 85°F (7 to 29°C)]. The seeds are placed 1 in. (2.5 cm) deep in sandy soils, $\frac{1}{2}$ in. (13 mm) in fine-textured soils at a rate of 10 to 12 lb/acre, or five to six plants/ft of row (11.2 to 13 MT/ha, 16 to 20 plants/m). Multi-

germ cultivars may require thinning to approximately 3 in. (7.5 cm) between plants. This
thinning must be done before the seedlings
exceed 2 in. (5 cm) in height. Commercial
growers do not thin, but space-plant the seed
balls sufficiently to avoid excessive competition. If small whole canning or pickling beets
are desired, the seed often is sown in bands
using a scattering deflector on the seeder.

Early in the season, weeds can be controlled by cultivation. As the beet develops,
however, excessive cultivation can destroy
some of the shallow feeder roots. Herbicides
therefore are preferred, and cycloate and pyra-
zon provide effective control of grasses and
broadleaf weeds, respectively. Both materials
must be applied before emergence of the
weeds or the beets.

Cultivars

Beet cultivars are described by shape,
color, and, to some extent, use. Table 14.2
outlines the descriptive categories. The most
popular type is the globe red. Currently, only a
few others are offered by seedsmen.

Disease and Insect Pests

Diseases   Diseases (Figure 14.5) are relatively minor problems in beet production. The
most common disease problems are leaf spot
and curly top virus, and occasional infestations
of downy mildew and root rot may develop.

Leaf Spot   The causal organism of leaf
spot, *Cercospora beticola,* attacks all forms of
Beta vulgaris and has been reported on other
plants as well. On leaves, many small circular
spots develop to $\frac{1}{16}$ in. (2 mm) in diameter.
Elongate lesions develop on the petioles. In
severe infections, the leaves drop prematurely,
but any infection reduces the photosynthetic
area, and, as a consequence, root growth and
quality suffer. The margins of lesions may
show more intense pigmentation than unaffected leaf areas. The center turns brown, then
gray as sporulation occurs. The preferred con-

TABLE 14.2
Cultivar groups of garden beets

Description	Cultivar
Table beets	
Flat shape	
Red	Flat Egyptian
Yellow	Burpee Golden
Top shape	
Light red	Light Red Crosby
Red	Crosby Egyptian
Dark red	Early Wonder
Yellow	Yellow Turnip
Globe shape	
Red	Detroit Dark Red, Pacemaker
Yellow	Yellow Tankard
Elongate	
Red	Long Dark Blood, Cylindra
Yellow	Long Yellow
Black	Long Black
Bunching beets,[a] flat shape, red	Green Top Bunching
Canning beets,[b] globe shape, red	Ruby Queen

[a] Grown for green tops as well as roots.

[b] Small globe with small crown.

trol is basic copper sulfate applied at 7- to 10-
day intervals or as needed. Although overwintering can occur on seed or in wild hosts,
infected plant debris is the major site. A rotation of 3 years therefore is effective in reducing
inoculum level.

Downy Mildew   Downy mildew, a disease
favored by cool, moist weather, is confined primarily to the Pacific coast. The organism,
Peronospora schachtii, affects all above-
ground parts of the plant. Affected tissue becomes light green on the upper surface, and
white conidia form on the underside. In severe
infestations, the crown may become infected,
and new growth then is infected as it emerges
from that crown. The spores do not survive
long periods of freezing temperatures. Survival
between seasons most often occurs as mycelia in roots or seed balls. The control procedures are the same as for cercospora spot,
and resistance has been reported in some cultivars.

Curly Top   The curly top disease is confined to production areas near and west of the Continental Divide. The causal virus has many host plants and therefore is difficult to eradicate. Among the vegetables, spinach, tomato, bean, muskmelon, squash, and pumpkin are susceptible, in addition to beet and chard. Infection at the seedling stage is lethal. Infection of established plants results in curling and blistering of young leaves and, eventually, in vein clearing and vein swelling. The virus is transmitted by the beet leafhopper from any of several wild hosts. Control must be directed toward eradication of alternate hosts and control of leafhoppers in and around the field.

Root Rots   Several root rot diseases occur sporadically, occasionally becoming severe. Most are included within the descriptive name black root. *Pythium* is the most serious organism, affecting young roots just after emergence. Plants may show healthy tops, but as the soil dries, the plants stunt, yellow, and die. The result is a very poor stand. Black root also can be caused by several other fungi (largely *Fusarium* and *Phytophthora*). Control is difficult. Seed treatment may help, but well-drained soils and favorable conditions at planting time are the most important factors in reducing incidence of disease.

Insects   The most common insect pests are leaf miners (see Figure 11.2 and Chapter 11), aphids, and leafhoppers, but extensive damage also may be caused by flea beetles and webworms.

Beet Leafhopper   The beet leafhopper (*Circulifer tenellus*) is the vector of curly top virus. This insect will feed on over 100 different plants, mostly in the western United States. It is a wedge-shaped insect (Figure 14.6), pale green to yellow, that lays eggs in the veins. The adults fly readily and can travel great distances. Organic phosphates are recommended for control.

Spinach Leaf Miner   The spinach leaf miner (*Pegomya hyoscyami*) causes blisterlike blotches on the leaves, and the larvae eat out the tissue between the upper and lower leaf surfaces. Leaves with miner "tracks" are unfit for marketing, and the root quality and size can be reduced. Leaf miners, generally distributed throughout the United States, overwinter as puparia, emerging early in the spring as two-winged flies. The first evidence of feeding is a narrow winding trail on the leaf, eventually expanding to show a blotchy appearance. Diazinon and malathion are effective insecticides, but timing and coverage are very important.

Harvest and Market Preparation
Fresh market beets normally are harvested when $1\frac{3}{4}$ to 2 in. (4.4 to 5 cm) in diameter. If bunched, they are washed thoroughly, graded by size, trimmed of damaged or dead leaves, then packed in crates. Topped beets are graded, cleaned, and packaged in perforated polyethylene bags. Topped beets may be stored at 32°F (0°C) and 90 to 95 percent relative humidity for up to 5 months. Bunched beets should be marketed immediately.

The diameter of beets for processing ranges from 1 to 4 in. (2.5 to 10 cm). Small sizes are used for whole pack, larger sizes for sliced or diced products. The payment to a grower differs according to size as well as condition, with small sizes commanding a higher price.

High-quality beets are produced with uniform soil fertility, cool temperature, and moisture throughout the growing season. Periods of growth stress result in fibrous roots, occasionally cracking, and poor flavor and internal color. High temperatures increase the light–dark zoning of most cultivars. Fall crops may be left in the field during light frosts but should be harvested before a hard freeze. Sweetness may be enhanced somewhat in roots exposed to moderately low temperatures.

RADISH

Historical Perspective and Current Status

China is considered the center of origin of radish, and wild forms exist there today. A number of types were introduced to middle Asia in prehistoric times. Radishes were a common food in Egypt before the building of the pyramids, and they were highly regarded by ancient Greeks. Roman writings described several forms and colors of radish at the beginning of the Christian era.

Columbus is credited with introducing radishes to the New World. However, the short season radishes, popular in today's U.S. market, were not grown until after the sixteenth century in Europe. Only the large, late winter types were cultivated.

Although radish is consumed worldwide, it contributes little to nutrition. Of the root crops, it is similar to beet and turnip in vitamins and minerals, but inferior to carrot, parsnip, and rutabaga. In the United States, it is used as a raw vegetable; elsewhere it also is fermented or cooked.

The major commercial acreages in the United States are along the coastal region of central California and in Florida, but radishes are produced in all states to some extent. California ships 410,000 cwt (18,598 MT) annually throughout the year. Florida has a winter crop on 24,700 acres (10,000 ha) of approximately 1,070,000 cwt (48,535 MT), and the northeast production supplies the early summer and fall markets. Summer radishes are supplied by Michigan, Minnesota, Ohio, and smaller acreages in several northeastern states. The total U.S. production is estimated at 1,220,000 cwt (55,339 MT).

Classification, Growth, and Development

The radish (*Raphanus sativus*) is a member of the Brassicaceae and includes both annual and biennial types. The spring and summer radishes (short season types) are annuals; the winter cultivars require up to 2 months to reach edible stage and may be annual or biennial.

Early growth of radish is characterized by a rosette of leaves attached to a small stem plate. In the spring types, top growth is not extensive, and edible roots develop within several weeks. Late cultivars of the summer type show substantially increased top growth, and root size can be very large. The warm temperature also increases top growth of some of these cultivars substantially, and growers may wish to select short-top forms at that time.

Chilling temperatures followed by a long photoperiod will initiate bolting of the biennial types. As temperatures moderate, seedstalk development is accelerated. The annual radishes flower rapidly after reaching edible size, if exposed to warm temperature. Once bolting has been initiated, the roots no longer are suitable for market.

Crop Establishment and Maintenance

The foraging root system is shallow and sparse and not an aggressive competitor. Suitable soils include muck, sand, sandy or silty loams, and clay loams, free from excessive debris and/or stones and well drained. The optimum pH ranges from 6 to 6.5, and most sites are fertilized with 1000 lb/acre (1120 kg/ha) of a complete 1–1–1 or 1–2–2 fertilizer, broadcast sufficiently in advance of seeding to support early rapid growth. Muck soils may require little or no nitrogen, but potassium should be applied prior to planting.

Radishes are very hardy and may be seeded well before the danger of frost has passed. A field committed to radishes may produce three or more successive crops, depending on length of growing season. The seeding rate is 15 to 20 lb/acre (17 to 22 kg/ha) depending on cultivar type and row spacing. Summer radishes frequently are grown in multiple rows 8 in. (20 cm) apart with only $\frac{1}{2}$ to 1 in. (1.3 to 2.5 cm) between plants. Large-rooted cultivars require greater space within

FIGURE 14.8
Radish roots showing surface cracks resulting from erratic water supply. (Photo courtesy of University of New Hampshire.)

and between rows. As seedlings emerge throughout the growing season, uniform soil moisture must be maintained. If subjected to periodic growth stress, caused by excessive heat or drought, the roots become tough and fibrous, occasionally with surface cracks (Figure 14.8), and the pungency increases.

Weed control normally is achieved through cultivation, since the season is short, with harvest often preceding the time of vigorous weed growth. The root system is shallow, however, and can be damaged by careless tillage. The best practice is to cultivate early with rotary hoes or similar equipment, removing the weeds when they are small.

Cultivars

Radish cultivars may be grouped according to season (summer, winter, all season),

and within each season by shape and color (Table 14.3).

Several cultivars of the summer type are available with resistance or tolerance to clubroot ('Saxafire,' 'Novitas') and with the globe shape desired by consumers. Winter radishes are most successfully grown as a fall crop, when short days and cool temperatures minimize bolting. Crops seeded in the spring usually show increased losses from premature seeding.

Disease and Insect Pests

Disease Of the diseases affecting radish production, several are described for the cole crops (Chapter 12), including black rot, *Pythium,* clubroot, *Fusarium* yellows, and *Rhizoctonia.* In general, the control methods are similar: avoid infested soils, use clean seed, rotate with noncruciferous crops, and, where applicable, use resistant cultivars. In addition to the diseases common in cole crops, radish also

TABLE 14.3
Classification of radish cultivars

I. Summer (approximately 25 days to maturity)	
A. Round	
1. Red	Scarlet Knight (F),[a] Cherry Beauty
2. Red/white	Sparkler
3. White or yellow	Golden Globe, Snowbelle
B. Half-long	
1. Red/white	French Breakfast
2. White	Icicle
II. All season (approximately 45 days to maturity, daikon type: all elongate, white)	Summer Cross Hybrid
III. Winter radish (approximately 60 days to maturity)	
A. Round	
1. Black	Round Black Spanish
B. Elongate	
1. Red	Chinese Rose
2. White	Chinese White

[a] F = *Fusarium* resistant.

may become infected with black root, downy mildew, and scab.

Black Root   The organism causing black root (*Aphanomyces raphani*) attacks a large number of the Brassicaceae. Black root first appears as blackening of the root, as uneven root development and as stunting of top growth. Secondary infection by soil rot organisms frequently follows. The disease is favored by high soil temperature and moisture and may be found in association with *Rhizoctonia solani.* Chemical controls are ineffective, and resistance is available only in a few red cultivars.

Downy Mildew   Downy mildew, caused by *Peronospora parasitica,* is important in cool humid weather. The upper surface of affected leaves shows irregular yellow spots, and the white mold growth bearing conidiospores appears on the underside. Occasionally, plants infected late in the season do not develop symptoms until the roots are packaged in polyethylene bags. Carbamate fungicides applied soon after seedling emergence, continuing at regular intervals until harvest, will minimize disease loss.

Scab   The organism causing scab on potato (*Streptomyces scabies*) also can infect radish roots and is favored by moderately high pH. Where the disease occurs, heavy irrigation in the first 2 weeks of growth and reduction of pH through application of sulfur will reduce infection. No cultivars are resistant to scab, although white cultivars appear to have some tolerance.

Insects   The major insect pests are root maggots, aphids, and flea beetles, although other insects found on cole crops may be found on radish occasionally.

Root Maggot   The root or cabbage maggot (*Hylemya brassicae*) is one of the most frequently encountered insect problems and generally is found where plants appear

stunted, or yellow, or where edible roots do not develop. The white larvae feed on the small and enlarging roots. Up to 80 percent of a crop may be lost where no precautions are taken. The insect overwinters as a pupa, emerging as a fly in the early spring. The flies deposit eggs on plants at the soil line, and, within a week, emerging maggots begin to feed. The recommended control is diazinon applied in the furrow at seeding. Several applications may be necessary in some areas.

Cabbage Aphids   Aphids (*Brevicoryne brassicae*) feed on a range of brassicas, causing leaves to curl, wilt, and die. Surviving plants often are stunted. In northern areas, the insect overwinters as eggs attached to cabbage leaves and stems; in southern areas, the life cycle continues throughout the winter. Sprays of malathion or diazinon will eradicate infestations, and elimination of debris will reduce the overwintering population.

Flea Beetles   The flea beetle predominant in radish (*Phyllotreta striolata*) can be recognized by a yellow stripe on each wing cover. This insect is similar in habits to other flea beetle species, except that the larvae mine leaves, causing a loss of photosynthetic area. Carbaryl and malathion applied at the proper time are recommended as controls.

Harvest and Market Preparation

Radishes are harvested as soon as they attain a size appropriate for the cultivar type. Commercially, radishes are lifted and topped mechanically. The roots then are washed, cooled, graded, and packaged in consumer- or institutional-size polyethylene bags. The bags are packed in cardboard containers and maintained at approximately 32°F (0°C) and 95 percent relative humidity until sold. Some radishes, principally for market garden sales, are bunched with tops on. These roots must be sold immediately; otherwise, excessive water loss will impair quality.

TURNIP AND RUTABAGA

Growth, Development, and Culture

Turnip and rutabaga are similar in flavor, but most common cultivars are distinguishable by foliage appearance and by root size, shape, and color. There are also differences in nutritional value (Table 14.1). Turnip (*Brassica campestris,* Group Rapaceum) generally has hairy leaves (although there are some new cultivars with smooth leaves) and a small white-fleshed root. Rutabaga (*B. napus,* Group Napobrassica) has smooth, large leaves with a large, yellow-fleshed root. The yellow color is due to carotenoids which enhance the vitamin A content, although levels are substantially lower than those of carrot. Turnip is a short season crop; rutabaga requires a full season. Rutabaga is less tolerant than turnip to high temperature, and production is concentrated in the northern states and in Canada. Turnip production is widely distributed, but commercially significant acreages are located in the southeast.

Rutabaga (38 chromosomes) probably evolved as a cross of cabbage (18 chromosomes) with turnip (20 chromosomes), followed by spontaneous chromosome doubling. The resulting amphidiploid (38 chromosomes, double the F_1 number) has been produced experimentally and resembles rutabaga. This evolution occurred in relatively recent history, probably in Sweden.

The biennial growth habit is similar to that of other root crops. Both turnip and rutabaga bolt if exposed to temperatures below 50°F (10°C) for 3 to 6 weeks. Vegetative growth appears as a rosette of leaves on a very small stem. Rutabaga has more robust growth than turnip and, as a consequence, requires more space, fertilizer, and moisture.

Soils should be prepared as for beets, adjusting to a pH of 6.0 to 6.8. For both turnip and rutabaga, a complete (1–1–1 or 1–2–2) fertilizer is broadcast prior to final harrowing, and a sidedressing may be applied 4 to 6 weeks after seeding, as roots begin to enlarge.

Steady, rapid growth results in sweet, tender flesh; erratic growth promotes strong flavors and a woody, fibrous texture. Weeds are controlled by cultivation and, for turnip, by pre-emergent application of DCPA.

Disease and Insect Pests

Disease Turnip and rutabaga are susceptible to the same diseases; however, the pesticides legally available for control are limited for rutabaga. Anthracnose, clubroot, downy mildew, and turnip mosaic virus affect both crops.

Anthracnose Anthracnose, caused by *Colletotrichum higginsianum,* is recognized by water-soaked spots on foliage and sunken dry spots on roots, particularly those of turnip. Root lesions are subject to secondary infection by soil rot organisms. Turnips may be sprayed with a carbamate (maneb, zineb); no pesticide is registered for controlling anthracnose on rutabaga.

Other Diseases Clubroot and downy mildew are caused by the same organisms that attack the cole crops, and the control measures described for cabbage (Chapter 12) are applicable to turnip and rutabaga. Turnip mosaic is a virus disease transmitted by the flea beetle. As with all mosaics in Brassicaceae, it is perpetuated in perennial plants. It causes a mosaic yellowing, stunting, and eventual death of the plant. Control of perimeter vegetation and of infestation by flea beetle will minimize the disease.

Insects Insect pests include aphids, leaf miners, cabbage maggot, and flea beetle. Control measures are as described for other brassicas.

Harvest and Market Preparation

At an appropriate size, the roots are lifted, topped, washed, and cooled. Of the two crops, turnip is not suited for long storage and normally is marketed without delay. Most turnips are packaged in vented polyethylene bags to preserve freshness. Rutabagas are waxed to retard weight loss in storage and transit, and turnips removed from storage for off-season marketing in the north also may be waxed. Rutabaga quality is poor if harvested before cold weather; consequently, seeding should be delayed until late spring, and roots should remain in the field as late in the fall as feasible.

PARSNIP

Historical Perspective and Current Status

Parsnip is native to the Mediterranean region, and wild forms were utilized by the Romans for medicine and for food. By the mid-sixteenth century, parsnip cultivation was widespread, and the crop became a staple for poor people in Europe. By the early 1600s, parsnips were grown extensively in the American colonies. Parsnip never has become an important commercial crop, but it is grown in many home gardens and market gardens.

The vitamin content of parsnip is not outstanding, although it provides reasonable levels of minerals. Roots have been reported to contain three phytotoxic mutagenic and photocarcinogenic furocoumarins that persist through normal cooking. Approximately 30 percent of these substances is in the peel, and levels rise dramatically if the roots are diseased. The furocoumarins are highly mutagenic in the presence of ultraviolet light—they will melanize skin and have been used to treat depigmented skin and also psoriasis. Workers handling parsnips frequently have reported dermatitis. However, no toxicological or dietary data have been published that would suggest that parsnip handling or consumption presents a health hazard.

Classification, Growth, Development, and Culture

Parsnip (*Pastinaca sativa*) is a member of the Apiaceae, similar to carrot in anatomy and development. It requires a long season and, because its quality is greatly enhanced by cold temperature and the roots can be overwintered in the field in northern climates, it is known as a winter vegetable. The best root quality is attained after several mild freezes or after a period of at least 2 weeks of cold storage.

Soil type, preparation, fertilization, and general culture are as described for carrot. Because of the long season, more than one side-dressing of nitrogen may be required, especially if rainfall has been above normal or the soil is sandy. Excessive nitrogen should be avoided, however, since it stimulates top growth and may increase susceptibility to foliage diseases.

In addition to the disease and insect problems described for carrot, parsnips also are susceptible to blight (*Pseudomonas marginalis*) and to a canker/leaf spot (*Itersonilla perplexans*) that is relatively serious as a storage disease. Parsnip blight is characterized by browning throughout the interior of the root, and canker appears as a reddish brown area developing where the rootlets had been attached to the taproot, generally near the crown. The canker organism also causes bright silvery to brown leaf spots. Both diseases are minimized by a 2- to 3-year rotation, and canker also may be reduced by adjusting soil pH to 6.5 to 7.0.

Two cultivars are listed in seed catalogs. These are 'All America,' described as having white texture, broad shoulders, and a relatively hollow crown, and 'Harris Model,' a more cylindrical, white-fleshed cultivar, with narrow shoulders and somewhat smaller than 'All America.'

Harvest and Market Preparation

Parsnips and carrots are harvested and marketed similarly. A period of cold storage allows starch to be converted to sugars. Home gardeners frequently store parsnips in sand outdoors, removing them throughout the winter, or they overwinter them in place in the garden. Parsnips for shipping are topped, graded, and packaged in vented polyethylene bags.

MISCELLANEOUS ROOT CROPS

Salsify Salsify (*Tragopogon porrifolius*) is a member of the Asteraceae and a biennial plant native to Europe, North Africa, and Asia. It is grown for its long, frequently branched taproot which is used in flavoring. The long, grasslike leaves form a rosette, and plants can reach 4 ft (1.2 m) in height. Salsify is very hardy, and the roots may be harvested well into the winter with no damage. Soil preparation and plant spacing are similar to turnip, but fertilization, irrigation, and weed control practices must be adjusted to extend through a long growing season. Little salsify is grown commercially in the United States.

Horseradish Horseradish (*Armoracia rusticana*) is a hardy perennial plant of the family Brassicaceae grown for its pungent root. Most of the commercial acreage is in the deep fertile soils of the midwest, particularly in the St. Louis, Missouri, area and in Wisconsin, which favor development of long straight roots. As the plant develops, it produces a rosette of leaves that resemble dock. A flower stalk of 2 to 3 ft (0.6 to 0.9 m) may be produced, but the seeds are not suitable for propagation. Instead, the side roots are used to provide the cuttings and rhizomes whereas the large primary root is harvested for processing or marketing. The propagative material is bundled and cut straight across the top, obliquely at the base, and stored until early spring. The cuttings then are planted, square end up, 3 to 4 in. (7.5 to 10 cm) deep in rows 3 to 4 ft (0.9 to 1.2 m) apart. The most rapid growth occurs late in the season, and the crop usually is left in the field beyond a killing frost before harvesting.

SELECTED REFERENCES

Cotner, S., T. Longbrake, and J. Larsen (1972). *Keys to Profitable Fresh Carrot Production,* Texas A & M University System Coop. Ext. Serv. Fact Sheet L-889.

Ivie, G., D. L. Wayne, L. Holt, and C. C. Ivey (1981). Natural toxicants in human foods: Psoralens in raw and cooked parsnip root (*Pastinaca sativa*). *Science* 213:909–910.

Kostewicz, S. R., and J. Montelaro (1979). *Radish Production,* USDA Leaflet 566.

Robinson, R. W. (1954). Seed germination problems in the Umbelliferae. *Botanical Review* **20:**531–550.

Rowe, R. C., H. S. Humaydan, M. L. Lacy, F. L. Pfleger, D. J. Pieczarka, C. T. Stephens, and P. H. Williams (no date). *Diseases of Radishes in the U.S.A.* North Central Reg. Ext. Pub. 126.

Ryall, A. L., and W. J. Lipton (1972). *Handling, Transportation and Storage of Fruits and Vegetables,* Vol. 1. AVI, Westport, Conn.

Simmonds, N. W. 1976. *Evolution of Crop Plants.* Longmans, London.

Weaver, J. E., and W. E. Bruner (1927). *Root Development of Vegetable Crops.* McGraw–Hill, New York.

Whitaker, T. W., A. F. Sherf, W. H. Lange, C. W. Nicklow, and J. D. Radewald (1970). *Carrot Production in the United States,* USDA–ARS Agric. Handb. No. 375.

Yamaguchi, M. (1983). *World Vegetables.* AVI, Westport, Conn.

STUDY QUESTIONS

1. Distinguish among the root crops in the anatomical development of the edible part.

2. For what reasons are organic soils well suited for production of root crops?

3. Distinguish between the handling methods for bunched root crops and for crops that are topped for marketing.

4. Describe the ways in which temperature affects development of carrots and parsnips and quality before and after harvest.

15
Alliums

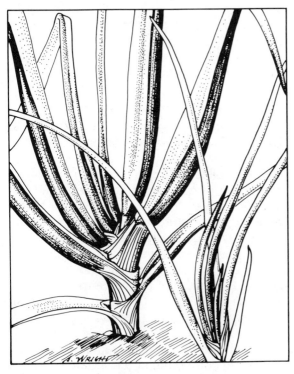

Onion	Leek	Shallot	Bunching and
Garlic	Chive	Green onion	multiplier onions

The commercial alliums include several related species in which the fleshy basal leaves constitute the edible part. All now are assigned to the family Alliaceae. Previously, the bulb crops had been classified as members of either the amaryllis family (Amaryllidaceae) or the lily family (Liliaceae).

All of the edible species of the genus *Allium* are believed native to Asia, including Iran, Pakistan, and Russia. The nutritional contributions of the different crops in this genus range from low in bulbing onion to moderate in chives and green onions (Table 15.1). Although the nutritional contribution of bulbing onions seems low on a fresh weight basis, the total contribution from an acre or hectare of land is substantial. The alliums do not contain starch, but carbohydrates include sucrose, glucose, fructose, and fructosan. Vitamin content is confined primarily to the tops. All bulb crops are valued for their pungency and variety of uses.

ONION

Historical Perspective and Current Status

Of all alliums, the bulbing onion (*Allium cepa*) is the most significant, both historically and in current production acreage. Archaeological discoveries date onion culture to at least 2800 B.C. Onions are mentioned in the Old Testament, and cultivars were described by Theophrastus and Pliny. Onions were brought to the New World by early explorers and were accepted quickly by Indians and settlers. As with all alliums, onions have been prized for their medicinal value. Gerarde, in *The Greate Herbal* (1596), stated

> *The juice of onions snuffed up into the nose purgeth the head . . . stamped with salt, rue and honey . . . they are good against the biting of a mad dog . . . annointed upon a bald head in the sun bringeth the haire again very speedily.*

General U. S. Grant reportedly would not move his army without onions, a testament to his belief that onions were invaluable in preventing dysentary and other ailments.

Currently, 38,762,000 cwt (1,758,244 MT) of onion bulbs is grown for fresh and processing use on 123,040 acres (49,831 ha). It is estimated that 3000 acres (1215 ha) of green onions is grown annually, largely as a market

TABLE 15.1
Nutritional constituents of the bulb crops[a]

Crop	Water (%)	Energy (cal)	Protein (g)	Fat (g)	Carbo-hydrate (g)	A[b] (IU)	C[c] (mg)	Thia-mine (mg)	Ribo-flavin (mg)	Niacin (mg)	Ca (mg)	P (mg)	Fe (mg)	Na (mg)	K (mg)
Chives	91	28	1.8	0.6	5.8	5800	56	0.10	0.18	0.7	69	51	1.7	—	250
Garlic	61	137	6.2	0.5	30.8	Trace	15	0.20	0.11	0.7	29	195	1.5	—	529
Leek	85	52	2.2	0.3	11.2	40	17	0.11	0.06	0.5	52	50	1.1	5	347
Bulb onion	89	38	1.5	0.1	8.7	40	10	0.03	0.04	0.2	27	39	0.5	10	157
Green onion	89	36	1.5	0.2	8.2	2000	32	0.05	0.05	0.4	51	36	1.0	5	231

Source: National Food Review (1978), USDA.

[a] Data expressed per 100 g sample.

[b] 1 IU = 0.3 μg vitamin A alcohol.

[c] Ascorbic acid.

garden commodity. Of the states with substantial acreages of dry bulb onions, California and Texas lead in production of sweet types, whereas the greatest percentage of the acreage in New York, Oregon, and other northern states is planted primarily with pungent storage cultivars (Table 15.2). Acreage increased steadily during the period 1974 to 1982 but decreased slightly in 1983. The average yield in 1974, 301 cwt/acre (34 MT/ha), rose to 356 cwt/acre (39 MT/ha) in 1985.

TABLE 15.2
Production of onions for fresh and processing use, 1982 to 1983

State	Production			
	1000 cwt		1000 MT	
	1982	1983	1982	1983
California	13,200	11,345	599	516
Texas	5,036	5,443	228	247
Oregon	4,821	5,292	219	240
New York	4,550	2,793	206	127
U.S. total	41,861	38,762	1,899	1,758

Source: Agricultural Statistics (1985), USDA.

Classification, Growth, and Development

Onion (*Allium cepa*) is a cool season biennial, tolerant of frost. Optimum temperatures for plant development are between 55 and 75°F (13 and 24°C), although the range for seedling growth is narrow, 68 to 77°F (20 to 25°C). High temperatures favor bulbing and curing.

The edible part of the onion is the bulb, composed of concentric, fleshy enlarged leaf bases or scales (Figure 15.1). The outer leaf bases lose moisture and become scaly; the inner leaves generally thicken as bulbs develop. The green leaves above the bulb are hollow and arise sequentially from a growing point on the stem at the centermost point at the base of the bulb. The stem is very small and insignificant during vegetative growth. After vernalization at temperatures below 50°F (10°C), the stem elongates rapidly, eventually producing compound umbels. Garner and Allard (1920) had first reported bolting to be related to length of day. However, long days do not induce reproductive growth but tend to accelerate development of the seedstalk once it has been initiated. This role of temperature in bolting was supported by the data of Thompson and Smith (1938) and subsequent research. Heath (1943) described the interaction of daylength and temperature in controlling rate of bolting in onions grown from sets, a response found to relate to the specific storage temperature level, length of temperature exposure, size of bulb exposed, and cultivar (Table 15.3). Set-size bulbs are affected much less than large bulbs, and a storage temperature of 40°F (4°C) generally will induce a

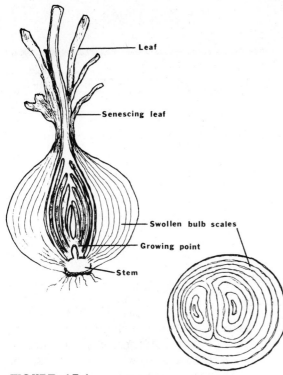

FIGURE 15.1
Longitudinal and crosssections of an onion bulb. Outer leaves senesce to form papery scales. Inner leaves thicken to form bulb.

higher frequency of bolting than 32°F (0°C) or 50°F (10°C).

Research by Magruder and Allard (1937) and Thompson and Smith (1938) showed bulb formation to be induced by light exposures at or exceeding a specific daylength (threshold). There are substantial differences among onion cultivars with respect to the threshold daylength required for bulbing (Table 15.4), and other environmental factors interact with daylength to modify the response. In all cultivars, bulbing accelerates with increasing temperature. Nutritional deficiencies (particularly nitrogen) near the threshold daylength also may induce early bulbing. Conditions favoring rapid bulbing occasionally may abort seedstalk development, even though induction of reproductive growth has occurred.

Temperature extremes not only affect rate of bulbing, but also can affect bulb shape. Thick and elongate necks are common in plants exposed to 21°F (−6°C) or below.

The onion root system is fibrous, spreading just beneath the soil surface to a distance of 12 to 18 in. (30 to 46 cm). There are few laterals, and total root growth is considered sparse and not especially aggressive. Therefore, in monoculture, onions tolerate crowding, particularly in loose, friable soils such as peat and muck. Competition from aggressive root systems (as from weed growth) severely limits onion growth.

The ultimate yield of onion is determined by the number of leaves that form prior to bulbing. Since bulbing in each cultivar is triggered by a specific daylength, early planting is the most effective method of improving bulb size and is a primary factor contributing toward yield per acre. Spacing also controls size. If, however, early planting coincides with cold air temperatures or cold, wet soils, the stand and, ultimately, the maturity of the crop will be erratic. Some cultivars of the Bermuda type also may bolt if substantial growth precedes a cold exposure.

TABLE 15.3
Average percentage of seedstalks developed from large, medium, and small sets of three varieties of onion after storage at various temperatures

Variety and size	Temperature (°C)				
	−1	0	4.4	10	15–21
Ebenezer					
Large	2.2	9.1	41.0	35.0	1.6
Medium	0.2	0.3	2.1	1.3	0.1
Small	0.0	0.0	0.0	0.1	0.1
Yellow Globe					
Large	33.4	42.7	80.0	68.6	18.5
Medium	4.7	4.9	21.6	14.1	2.7
Small	0.1	0.0	1.3	0.9	0.1
Red Wethersfield					
Large	40.4	68.1	84.4	78.9	24.6
Medium	8.3	9.7	26.6	20.6	6.3
Small	0.1	0.4	1.1	0.3	0.3

Source: Thompson and Smith (1938).

TABLE 15.4
Degree of bulb formation in several onion varieties
exposed to different daylengths[a]

Cultivar	Daily exposure to sunlight (h)						
	10	12	13	13.5	14	14.25	12.4–14.9
Yellow Bermuda	0–S[b]	G	G	G	G	G	G
Early Grano	0	G	G	G	G	G	G
Early Yellow Globe	0	0–F	S–G	G	G	G	G
Ebenezer	0	0	F–M	F–M	F–G	G	G
Australian Brown	0	0	S–F	F–M	F–G	G	G
Red Wethersfield	0	0	0–S	F–M	F–G	F–G	G
Southport Yellow Globe	0	0	0–S	S–F	F–G	F–G	F–G
Sweet Spanish	0	0	0–S	S–F	S–F	S–F	F–M

Source: From Magruder and Allard (1937).

[a] Recorded July 7.

[b] 0 = none; S = slight; F = fair; G = good.

Crop Establishment and Maintenance

Onions are grown successfully on any fertile well-drained, noncrusting soil, but the crop is especially adapted to muck. The optimum pH range, regardless of soil type, is 6.0 to 6.8, although alkaline soils also are suitable. Onions do not thrive in soils of pH below 6.0 because of trace element deficiencies or, occasionally, aluminum or manganese toxicity.

Three systems of planting may be employed: direct seeding, sets, and transplants. Direct seeding is preferred for most acreage and gives excellent results where the season is sufficiently long to provide early prebulbing growth. Sets and transplants are used in some areas to ensure large bulb size and uniform maturity, and either system may be used for early green onion production. Sets are small dry bulbs, approximately $\frac{1}{2}$ in. (12 mm) in diameter, produced the previous season by seeding thickly or growing under conditions favoring rapid bulbing. Transplants normally are grown in the south or in forcing structures and have three to five well-formed leaves at transplant time. Transplant leaves are pruned during growth prior to field setting, facilitating handling and increasing plant hardiness.

Prior to planting, soils should be plowed and disked sufficiently to eliminate debris and soil clods. In most commercial areas, beds 36 to 40 in. (0.9 to 1.0 m) wide are formed, and two to six rows are seeded or planted on the bed. If two rows, they may be two-line (twin) rows with plants staggered to achieve proper spacing and high population density.

Seeds are sown $\frac{1}{4}$ to $\frac{3}{4}$ in. (6 to 18 mm) deep in heavy mineral soils, deeper in light mineral soils and mucks. Excessively thick seedings of bulb onions may delay maturity; however, necks tend to be thinner than in sparse seedings, and bulbs are somewhat more globular in shape.

Using coated seed and precision seeding, the seeding rate can be adjusted easily for projected bulb size. For normal storage onions, as produced in the northern states and Canada, seeds are spaced 2 in. (5 cm) apart. Where small boiling, pickling, or pearl onions are desired, spacing would be reduced to 1 in. (2.5 cm) in the row. Large bulb size is promoted by spacings of 3 in. (7.5 cm) or more.

Fertilizer is applied either as a broadcast or, more commonly, as a band 2 to 4 in. (5 to 10 cm) directly below the seed, set, or transplant. Onion plants utilize substantial amounts of nutrients. Based on a yield of 40,000 lb (18 MT) of bulbs, the plants remove an average 145, 25, and 155 lb (66, 11, and 70 kg), respectively, of nitrogen, phosphorus, and potassium. Soils differ widely in fertilizer needs, depending on production history, soil type, and analysis. Muck soils frequently require an analysis high in potash but may need nitrogen only where soils are cold early in the season. Onions grown on new muck soils in New York receive 100 lb/acre (112 kg/ha) of nitrogen and phosphate and 150 lb/acre (168 kg/ha) of potash before planting. Mucklands previously used in production may need slightly greater amounts of nitrogen, but only 40 to 100 lb/acre (45 to 112 kg/ha) of phosphate and 40 to 150 lb/acre (45 to 168 kg/ha) of potash. Mineral soils in New York average 80 to 100 lb/acre (90 to 112 kg/ha) of nitrogen and 50 to 150 lb/acre (56 to 168 kg/ha) of phosphate and potash. Applications to mineral soils in California average 146, 83, and 34 lb/acre, respectively, of nitrogen, phosphate, and potash (163, 93, and 38 kg/ha).

On many mineral soils, one or two sidedressings of nitrogen are applied during a season. These sidedressings may be applied through the irrigation system. In Texas, fall-seeded onions receive phosphate only before seeding and receive nitrogen when active growth starts in the spring and twice thereafter. Insufficient nitrogen will induce early maturity and reduce bulb size; high nitrogen may increase bulb size and cause large necks and soft bulbs with poor storage quality.

If heavy fertilization rates are indicated by soil tests, the material should be incorporated thoroughly throughout the plow layer or, if banded, placed 6 in. (15 cm) to the side of the row.

Because muck soils are used extensively for onion production, minor element deficiencies, particularly zinc and copper, may be encountered. Suggested corrective rates are 10 lb/acre (11 kg/ha) of zinc or 15 to 25 lb/acre (17 to 28 kg/ha) of copper, applied every 2 to 3 years. Relatively high levels of sulfur are uti-

lized by onions, but corrective applications vary widely, according to soils, leaching losses, and presence of sulfur contaminants in the atmosphere. If applied, sulfur will acidify the soil, and liming rates should be adjusted accordingly.

Similar fertilizer ratios are recommended for green onions; however, because of the short growing period, the rates would be reduced.

As seeds germinate, the cotyledon first appears as an "elbow" or "knee" (Figure 15.2), as the shoot tip remains in the seed testa while the seedling elongates. It then pulls the seed above ground, and the first true leaf breaks through the capillary sheath.

Onions require even moisture throughout the growing season. Those fields that suffer a growth check may produce excessive numbers of doubles or splits, reducing the number of U.S. Grade 1 bulbs. Furrow irrigation is used on mineral soils; muck and light sandy soils are irrigated with overhead systems or by subsurface seep irrigation where the soil profile allows. Onions at the bulbing stage utilize substantial amounts of water, although excessive moisture must be avoided during the growing season.

Onions are not good competitors with weeds. Cultivation, if used, must be shallow to avoid root damage, and growers usually favor chemical control. Preemergent broadcast applications of DCPA and postemergence applications of 3 to 5 percent sulfuric acid solution or one of several organic compounds have been used with some success.

Cultivars

At one time, all onion cultivars were open-pollinated, and many of these cultivars are still offered by seed companies. The discovery of male sterility in onion led to a rapid transition to F_1 hybrids, made feasible because of the simplicity and low cost of seed production. Male sterility is a genic–cytoplasmic trait. Sterility is caused by a cytoplasmic factor, and male fertility can be restored in plants carrying

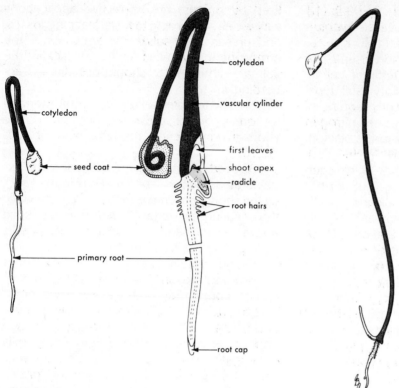

FIGURE 15.2
Germination of onion seed. (From *Botany. An Introduction to Plant Biology*, T. E. Weier, C. R. Stocking, M. G. Barbour, and T. L. Rost. Copyright © 1982 by John Wiley & Sons, Inc., New York, p. 324. Reprinted by permission.)

the sterility factor by introducing a single dominant gene. Any line carrying the sterile trait must cross-pollinate, and seeds harvested from male sterile plants isolated with a normal pollen-bearing parent will be hybrid. The advantage of hybrids is primarily in uniformity of bulb characteristics, not in yield or bulb size.

The cultivars of bulbing onions are grouped into short-, intermediate-, and long-day types (Table 15.5). Short-day onions (12- to 13-h threshold), also termed European types, are very mild, soft fleshed, and unsuitable for storage. These types are grown south of 35° latitude. Long-day onions (over 14½-h

threshold), termed American types, are very pungent and hard and store well. Long-day onions, if grown in the southern states, would not be exposed to the threshold daylength required for bulbing, and only green onions would be produced. In contrast, short-day types grown in the north bulb very quickly in the rapidly lengthening days and become little more than sets in size.

Short-day onions include the Bermudas and Grano-Granex types; long-day cultivars include yellow, white, and red globes. The intermediate-day cultivars (13½- to 14-h threshold), also relatively soft fleshed and used primarily for the fresh trade, are grown in areas of mild

TABLE 15.5
Classification of bulbing onion cultivars

Daylength	Scale color	Pungency	Representative cultivar
Short	Brown	Sweet	Awahia[a]
	Red	Sweet	Red Granex
	Red	Pungent	Red Creole[a]
	White	Sweet	White Granex, Crystal Wax[a]
	Yellow	Sweet	Grano,[a] Granex
	Yellow	Pungent	Yellow Creole[a]
Intermediate	Brown	Moderate	Cochise Brown[a]
	Red	Moderate	Stockton Early Red[a]
	White	Moderate	Fresno White[a]
	Yellow	Moderate	Rialto
Long	Brown	Pungent	Australian Brown[a]
	Red	Pungent	Carmen, Southport Red Globe[a]
	White	Pungent	White Lisbon,[a] Ivory
	Yellow	Sweet	Fiesta, Sweet Spanish[a]
	Yellow	Pungent	Autumn Spice, Downing Yellow Globe[a]

[a] Open-pollinated cultivar.

temperatures (lying between 32 and 38° latitude).

A specialty market has developed around three mild salad onions—'Vidalia,' 'Walla Walla,' and 'Maui'—so named for their production areas in Georgia, Washington, and Hawaii, respectively. These cultivars are soft-fleshed, nonstorage onions particularly adapted to these environments. There are also new cultivars being developed that combine adaptation to northern daylength with the mildness of the Granex type.

True pearl onions are classified as *Allium ampeloprasum* (Ampeloprasum Group) because they form just one storage leaf. In practice, short-day onion cultivars, such as 'Grano,' 'Crystal Wax,' and others, grown in northern latitudes, will develop pearl-size bulbs and be marketed as such. Most are used in pickling or in frozen mixtures of peas, broccoli, and other vegetables.

Green onions, scallions, and multiplier and bunching onions all are used in the immature stage. Green onions generally are bulbing-type, white cultivars harvested at the miniature bulb stage. Scallions are white cultivars of *A. cepa* that do not form bulbs. Multiplier onions

are cultivars of *A. cepa* of Group Aggregatum with white flesh and yellow or brown scales. These are distinguishable from the shallot by the latter's red scales and supposedly delicate flavor. The shallot can be used both in the immature stage and as a dry bulb. A cross of shallot, susceptible to pink root, with resistant *A. fistulosum* gave rise to the tetraploid cultivar 'Beltsville Bunching.' *A. fistulosum* includes the bunching onions, termed Japanese bunching or Welsh onion.

Disease and Insect Pests

Diseases Both field and storage diseases reduce profitability. Field diseases (Figure 15.3) include smut, downy mildew, pink root, smudge, leaf blight, and several basal rots. Storage diseases include some of the common field rots, botrytis neck rot, and bacterial soft rot.

Smut Smut is caused by the fungus *Urocystis cepulae* and attacks only immature tissue. Infected plants show gray streaks on the leaves, sheaths, and bulbs. Young plants affected by smut become twisted and ultimately

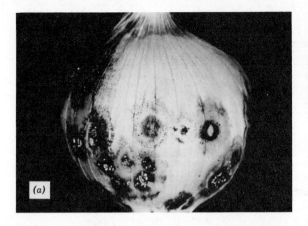

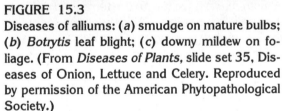

FIGURE 15.3
Diseases of alliums: (*a*) smudge on mature bulbs; (*b*) *Botrytis* leaf blight; (*c*) downy mildew on foliage. (From *Diseases of Plants,* slide set 35, Diseases of Onion, Lettuce and Celery. Reproduced by permission of the American Phytopathological Society.)

may die. The fungus overwinters in soil and, under the cool damp conditions of early spring, will infect immature tissue emerging from the seeds. Immature tissue enclosed within established tissue (transplants or sets) can become infected only from adjacent infected scales. Because smut is a disease confined to juvenile tissue, seed treatment provides an excellent control.

Downy Mildew Downy mildew (*Peronospora destructor*) also attacks young plants, appearing as white specks, usually confined to

the oldest leaves of young plants. A white mold develops rapidly in cool damp weather and progresses down the sheath, and plants eventually fall over and dry up. The fungus overwinters in bulbs and sets and on plant debris, and spores are carried long distances by air currents. For control, young plants can be treated with mancozeb at weekly intervals until bulbing begins.

Pink Root This disease, caused by *Pyrenochaeta terrestris,* is recognized by the color of newly infected roots. These roots gradually die, and new roots continue to become infected, resulting in poor bulb size. The fungus is soilborne and persistent. The most effective control is plant tolerance.

Smudge   *Colletotrichum circinans* is a fungus causing black or dark green areas on the bulb or neck, often appearing as concentric rings. The disease overwinters in sets and soil, and spores are spread by wind, splashing water, and tools and clothing. Infection is favored by 75 to 85°F (24 to 29°C) temperatures. Smudge occurs primarily on white cultivars and reduces their market value. Affected bulbs may shrink and sprout in storage. Proper selection of cultivars and careful curing will minimize the losses.

Botrytis Leaf Blight   Leaf blight, commonly termed blast, is caused by several *Botrytis* species. The disease first appears as white specks on leaves, expanding to cause a dieback from the leaf tips. Tops may be killed completely within a week, and entire fields may be affected. Frequently, blight follows previous damage from insects, disease, mechanical damage, or air pollution. Control is achieved through mancozeb sprays at approximately 7-day intervals.

Bulb Rots   Several rots may occur either in the field or in storage. Basal rot, caused by *Fusarium* species, results in a breakdown of inner scales. Outwardly, the bulb may appear normal. It eventually becomes soft, however, and will develop a water rot under moist conditions or a dry shriveled bulb in a dry environment. The disease is most severe in warm areas with poor soil drainage. *Botrytis* neck rot is an extension of the leaf blight disease and can become serious in storage. Field control of blight and careful grading will minimize storage losses.

Insects   Maggots and Thrips   Onion maggots and thrips (Figure 15.4) are responsible for most of the economic losses attributable to insects. The maggot (*Hylema antiqua*)

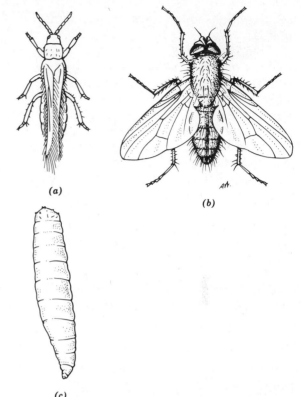

(a)

(b)

(c)

FIGURE 15.4
Insects of alliums: (*a*) onion thrips; (*b*) adult onion maggot fly; (*c*) onion maggot larva.

is primarily a pest of northern onions in which it may destroy up to 80 to 90 percent of the bulbs. In dry seasons, the maggot is of little importance, but in wet springs, particularly after a series of such springs, it will be destructive. The small larvae bore through the underground stem into the bulbs, causing plants to turn yellow. Small bulbs are completely mined out; large bulbs are damaged and often rot in storage. The insects overwinter as larvae or pupae in soil, old cull piles, or plant debris. The pupae transform to adults and emerge as small gray flies in late spring. The female adults lay eggs at the base of plants or in cracks in the soil. Eggs hatch in 2 to 7 days, and the maggots crawl down the leaves behind the sheath. Control can be achieved by applying diazinon as a drench during or after seeding. Subsequent treatments may be required.

Thrips (*Thrips tabaci*) are minute insects that cut or "rasp" the epidermis of leaves or stems and suck the plant sap, resulting in white blotches on leaves. Severe infestations result in leaf blasting and collapse. Bulbs become distorted and undersized. Infestations are more severe in dry seasons than in moist, and entire fields may be destroyed. The insect has many host plants, and adults and nymphs overwinter on plants or plant debris or in weeds bordering the field. Most of the insects are female, which can reproduce without a male. Eggs are thrust into the leaves and will hatch in 5 to 10 days.

Control of thrips is attained through diazinon spray at 7- to 10-day intervals. Up to six applications may be necessary, and good coverage is essential.

Soil Insects Insects such as cutworm and wireworm feed on young plants or roots and are discussed for other crops (Chapters 6 and 16). Although not considered a major onion pest, nematode infestations may be serious in some growing areas, particularly those with warm soils, causing light-colored foliage and leaftip burn, retarded growth, and poor yield (Table 15.6).

Harvest and Market Preparation

Onions are ready for harvest when the leaves collapse. For storage, onion tops should have broken over before harvest and the necks should be collapsed and dry. Storage bulb maturity can be accelerated by witholding irrigation water or by undercutting the root system. Bulbs for storage may be harvested when 50 percent or more of the tops have broken over, but the bulbs must cure and dry thoroughly before being placed in storage. Bulbs intended for immediate use can be undercut when 15 to 25 percent of the tops are down.

To harvest, a knife or lifter is drawn under a bed or row, cutting roots and loosening the

TABLE 15.6
Onion average bulb weight as affected by *Meloidogyne incognita*

Preplant nematode density (eggs + juveniles/500 cm³ soil)	Bulb weight (g)
0	58.4[a]
256–20,960	14.3

Source: From J. N. Corgan, D. L. Lindsey, and R. Delgado, (1985) *HortScience* **20**:135–136. Reprinted by permission of the authors.

[a] Significant difference by *t* test, $P = .01$.

soil, after which the bulbs may be dug or allowed to cure further before digging. Under dry conditions, bulbs may be left to cure in the field, either in place or in windrows. To avoid damage from direct sunlight, however, onions normally are placed in field containers and moved to a dry, shady location for subsequent curing.

The purpose of a curing period is to allow natural dormancy to develop and to dry the onion sufficiently to protect against disease organisms that may be present. A properly cured onion will have a dry shrunken neck and dry outer scales. The respiration rate also should be lower than that of an uncured bulb. Bulbs harvested fully mature are cured by exposure to temperatures up to 95°F (35°C) in low (less than 50 percent) relative humidity. Air movement must be provided at the rate of 1 ft³/min per cubic foot of onions (60 m³/h per cubic meter of onions). Immature onions require twice the rate of air exchange.

Following curing, the temperature of stored onions is lowered gradually to 32°F (0°C), or slightly higher, with the relative humidity at 60 to 70 percent. Air exchange in the storage facility is important to prevent any condensation on the bulbs. Also, when the bulbs are removed from storage, they should be conditioned for several days at 68°F (20°C) and 50 percent relative humidity.

In most operations, the tops and roots are removed during harvest. When this is not possible, they should be removed after curing, before storage or sale.

Freshly harvested onions are in a physiological rest period and will not sprout for a variable period of time (depending on cultivar). Storage will prolong this dormancy. Sprouting will increase in storage temperatures above 40°F (4.4°C), decreasing again as temperatures exceed 77°F (25°C). To reduce the frequency of sprouting after the rest period, onions may be field treated with maleic hydrazide (MH-30) at 2 to 3 lb/acre (2.2 to 3.4 kg/ha) when the tops are still green but beginning to senesce.

Cultivars intended for long-term storage should be firm with a thin dry neck, free from greening, root growth, sunburn, or freeze damage, and well covered with dry scales. Bulbs with fleshy, soft necks are susceptible to persistent rot, especially if storage humidity exceeds 70 percent.

Flavor in onion is associated with pungency (propyl disulfides and other disulfides) and with sugars (glucose, fructose, sucrose). Both sugar content and pungency are related to percentage dry matter, but cultivar differences, particularly between short- and long-day types, account for the major differences in flavor. Pungency and dry matter content are important quality attributes in onions for processing.

Onions normally are shipped in 50-lb (22.7-kg) mesh bags. The bulbs are graded by size, with jumbo and pearl sizes frequently used by processors. Those intended for international trade are packed in 25-kg bags.

Green onions (Figure 15.5) are pulled before bulbing, when the basal diameter exceeds $\frac{1}{4}$ in. (6 mm), and the roots are trimmed near the base. They should be washed free of soil, and discolored stalks should be discarded. The green onions are bunched and packed in ice to preserve crisp texture and quality. Vacuum cooling is possible but requires prewrapping in ventilated polyethylene bags to retard wilting. Storage life is limited to approximately 1 week at 32°F (0°C) and 90 to 95 percent

FIGURE 15.5
Green onions at marketable size. (Photo courtesy of University of New Hampshire.)

relative humidity. Shallots are harvested by hand when the bases are at least $\frac{1}{4}$ in. (6 mm) in diameter. The outer leaf is stripped off and the roots are trimmed before washing and bunching. If grown for dry bulbs, they are handled in a similar manner as onion bulbs.

LEEK

The leek (*Allium ampeloprasum,* Porrum Group) is a winter hardy biennial native to the eastern Mediterranean, although wild forms extend into western and southern Russia. They had been known in Egypt from biblical times and were imported by the Romans. Emperor Nero reputedly developed an enormous appetite for leeks, believing that they improved his singing voice. In the sixth century, the leek was considered a source of strength by the Welsh who wore them to distinguish themselves in battle.

Leeks are popular throughout Europe and China, and production has increased in the United States. The Chinese leek is *A. tuberosum,* a smaller plant than *A. ampeloprasum.* Cultural requirements for leeks are similar to those of onion, but the plant requires a longer growing season, and one somewhat cooler and wetter than required for onion. It may be transplanted, as onions, or established by direct seeding as early in the season as possible. Plants or seeds should be placed in a furrow 5 to 6 in. (13 to 15 cm) deep, gradually filling the furrow as the plant develops. This planting system, in effect, blanches the basal 5 to 6 in. (13 to 15 cm) of the stalk, producing an attractive white, thickened neck.

The leaves of leek are flat, similar to those of garlic, but the plant does not develop a hard bulb. Instead, the base thickens uniformly (cylindrically) to approximately 1 in. (2.5 cm) in diameter (Figure 15.6).

Leeks are bunched for market, similar to green onions. To preserve freshness, the newly harvested leeks are hydrocooled, or cooled by icing or by vacuum cooling, the latter only after wrapping with vented polyethylene to prevent desiccation. Transit and storage temperatures should be 32°F (0°C) with 95 percent relative humidity. Under these conditions, storage can be extended for 2 to 3 months.

GARLIC

Historical Perspective and Current Status

Garlic is native to middle Asia, with a secondary center in southern Europe, where it has been used for more than 2000 years. Reportedly, the Romans disliked the strong flavor, but they fed it to their laborers and soldiers for strength and courage. Egyptians had used

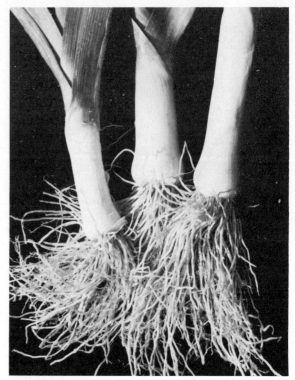

FIGURE 15.6
Leek differs from onion by its lack of bulb, thickened neck, and flat leaves. (Photo courtesy of University of New Hampshire.)

garlic for both cooking and embalming, and in early times, it had been reported to mend broken bones and to cure many ills, including tuberculosis, bronchitis, dropsy, and jaundice. Nutritionally, its value is marginal, since the volume of consumption is low. On a fresh weight basis, however, it is more nutritious than onion.

The use of garlic has expanded rapidly in recent years. Most of the U.S. garlic crop is grown in California, with lesser acreages in Oregon and Nevada and several southern states. Approximately 1,258,000 cwt (57,063 MT) is produced nationally on 15,380 acres (6229 ha), about half the amount consumed in the United States. It is used fresh and is dehydrated to produce garlic powder.

Classification, Growth, and Development

Garlic, *Allium sativum,* is a cool season hardy perennial, but it attains its best growth under conditions warmer and drier than those favoring onion growth. The edible part is termed a clove, a small bulb that, together with other cloves, forms an aggregate or cluster covered with the outermost leaf, a white or purplish papery sheath (Figure 15.7). The cloves also constitute the propagative part of the plant, since garlic rarely flowers and the flowers are sterile. Each clove is an axis with two mature leaves and a vegetative growing point. The outer leaf is a dry sheath, and the base of the inner leaf is thickened, constituting the bulk of the clove. The vegetative bud contains one or two leaf initials. Following maturation, the cloves are in a state of physiological rest that dissipates gradually. After rest, when environmental conditions favor growth, adventitious roots arise from the base of the axis, and the leaf initials become active, producing characteristic flat leaves.

As temperature and daylength increase, bulbing begins and is favored by uniform warm temperature. Bulbing is hastened in long photoperiods if cloves used for propagation have been exposed to 32 to 50°F (0 to 10°C) for the 1 to 2 months before planting, or, in warm climates, if they were planted in the fall. Bulbing may be suppressed completely if cloves have been exposed to temperatures of 77°F (25°C) or higher.

Crop Establishment and Maintenance

The soils for commercial production of garlic are rich, well-drained, mostly light sandy types, although any soil suitable for onion production may be used. Organic matter should be added each year. In California, most of the crop is planted from late August to early winter, since the plants withstand severe frost. The soil is prepared as for onion, forming 40-in. (1-m), or occasionally wider, beds. Two rows per 40-in. (1-m) bed are planted by hand or machine, using cloves stored at 40°F (4.4°C) for several months. The cloves are placed base down, 3 to 4 in. (7.5 to 10 cm) apart, 1 to 2 in. (2.5 to 5 cm) deep, similar to onion sets. Large cloves produce greater yields than small cloves, and yields are enhanced when planted early in the season, since size of the cluster is related to number of leaves initiated prior to bulbing.

Because of the restricted root system, soils must not dry excessively, and cultivation can sever feeder roots. For this reason, chemical weed control is preferred, using preemergence herbicides such as DCPA or other compounds effective on onions.

Cultivars

There are few cultivars of garlic, and some of them are not very distinct. In California, two cultivars—'Late' and 'Early'—are grown. Of the two, 'Late' is lower in yield but higher in keeping quality. Elephant garlic actually is a form of leek (*Allium ampeloprasum,* Ampeloprasum Group), but it forms cloves resembling garlic. The appearance and flavor predominantly resemble leek.

Disease and Insect Pests

Garlic is susceptible to some of the same diseases as onion but relatively few insect

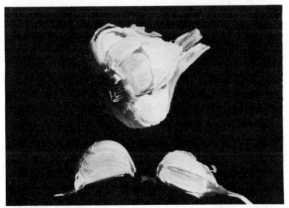

FIGURE 15.7
Garlic is an aggregate of bulblets (cloves) covered by a papery sheath. (Photo courtesy of University of New Hampshire.)

problems. Control methods for field pests would be the same as for onion. There are two problems incurred in marketing: blue mold rot, caused by *Penicillium* species, and waxy breakdown. The latter appears to be a physiological disorder resulting in gradual development of sunken yellow areas in the flesh, eventually affecting the entire clove. Blue mold rot occurs if garlic is harvested prematurely or if the cloves are stored under poor ventilation. Symptoms are light yellow lesions on cloves, eventually changing to blue-green as spores are produced. Proper harvest date and storage conditions should minimize the occurrence of this disease.

Harvest and Market Preparation

The crop is harvested when the leaves collapse. After the roots are undercut with a knife or lifter, the plants are pulled by hand and windrowed, using the tops to screen bulbs from the direct sun. After being dried for a week or longer, the plants are trimmed of roots and tops, cured further if necessary, graded, and packed in 50- to 100-lb (22.7- to 45-kg) mesh bags. Those harvested for dehydration are loaded in bulk in the field.

CHIVE

Chive (*Allium schoenoprasum*) is a perennial plant; its tender leaves are used as an herb or flavoring. The leaf growth resembles that of onion, but usable bulbs are not formed. Instead, the plant multiplies, producing many new plants (tillers) in a crowded tuft from the root zone. Flowers are produced each year, but seed set may be sparse in the crowded mass of plants.

Chive is thought to be native to the eastern Mediterranean and has been popular for hundreds of years among European gardeners. Commercial production in the United States is confined primarily to the northern states, and much of it is used in processing. Fresh market production is dominated by home gardeners and roadside markets. The crop may be established by seed or by small divisions (bulbs) of the plant mass. Cultural methods are similar to those described for onion.

SELECTED REFERENCES

Davis, H. R., R. B. Furry, and F. M. R. Isenberg (1979). *Storage Recommendations for Northern Grown Onions,* New York State College of Agriculture and Life Sciences, Cornell University, Coop. Ext. Serv. Inf. Bull. 148.

Garner, W. W., and H. A. Allard (1920). Effect of the relative length of day on growth and reproduction in plants. *Journal of Agricultural Research* **18:**533–606.

Heath, O. V. S. (1943). Studies of the physiology of the onion plant. I. An investigation of factors concerned in the flowering ("bolting") of onions grown from sets and its prevention. II. Effects of length of day and temperature on onions grown from sets. *Annals of Applied Biology* **30:**308–319.

Jones, H. A., and L. K. Mann (1963). *Onions and Their Allies.* Interscience, New York.

Longbrake, T. J., Larsen, S., Cotner, and R. Roberts (1974). *Keys to Profitable Onion Production in Texas,* Texas A & M University System Coop. Ext. Serv. Bull. MP-971.

Magruder, R., and H. A. Allard (1937). Bulb formation in some American and European varieties of onions as affected by length of day. *Journal of Agricultural Research* **54:**719–752.

Seelig, R. A. (1974). *Garlic. Fruit and Vegetable Facts and Pointers.* United Fresh Fruit & Vegetable Assoc., Washington, D.C.

Seelig, R. A. (1981). *Shallots. Fruit and Vegetable Facts and Pointers.* United Fresh Fruit & Vegetable Assoc., Washington, D.C.

Thompson, H. C., and O. Smith (1938). *Seedstalk and Bulb Development in the Onion (Allium cepa L),* Cornell University Bull. 708.

Yamaguchi, M. (1983). *World Vegetables.* AVI, Westport, Conn.

STUDY QUESTIONS

1. Describe the development of Granex onions seeded in Minnesota and of the cultivar 'Canada Maple' grown in south Texas.

2. Develop a chart showing the interaction of daylength and temperature on bulbing and bolting.

3. What environmental factors contribute to the development of high quality in processing onions?

16

Tuber and Tuberous Rooted Crops

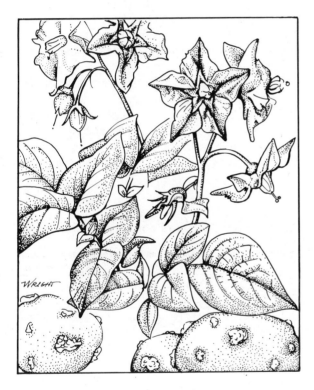

Potato   Sweet potato   Jerusalem Artichoke

POTATO

Historical Perspective and Current Status

The white ("Irish") potato, native to high altitudes of the Andes, has served as an important cultivated food since early civilization. Although little is known of the evolution of potato, pre-Incan pottery depicted tubers very similar to those grown today. The Incas, in addition to using fresh potatoes, developed a dried product, *chũno,* which could be stored for long periods. They also fermented the tubers to produce an alcoholic beverage.

During the 1500s, European explorers found potatoes growing from Chile to Colombia and eventually introduced them to Spain. The Spaniards called them "truffles," believing them to be a type of underground fungus. Within several years, the crop had been introduced to other European countries. These first forms of potato to reach Europe likely came from the northern Andes, a short-day area far different from northern Europe. The crop was not an immediate success, partly because of unadapted cultivars, partly because potato, a relative of nightshade, was believed to be poisonous or, at the least, responsible for such diseases as leprosy and scrofula.

It was not until long-day forms originating in Chile were introduced to Europe that commercial production began. These large-tubered types were so productive in northern Europe that government and church leaders promoted their cultivation as a staple food.

Of all countries growing potatoes, Ireland provided an ideal environment, and abundant production for two centuries supported substantial population gains in that country. Most of the Irish people were poor tenant farmers obliged to sell most of their profitable grain crops to pay the high rents charged by English landlords. Only the cheapest crop, potato, remained for Irish diets, and this vegetable was consumed at every meal. In 1845, however, late blight (*Phytophthora infestans*), which had appeared sporadically for several years, devastated the entire crop in Ireland, a disaster repeated in 1846. Although production of high-value crops continued, most of this produce was exported, even while the Irish people suffered from hunger, disease, and an unusually severe winter. Although environmental conditions improved potato yields in 1847, a complete failure again occurred 2 years later. Because of late blight and the dependency of the Irish people on this staple crop, 1.5 million people died during the period 1846 to 1851, and 1 million immigrated, mostly to the northeast United States (Heiser, 1969).

The potato had been introduced to the United States from Europe in the early 1700s and reportedly was grown first in Londonderry, New Hampshire, by a group of Presbyterian Irish in 1718. Late blight later became an occasional problem in North America, but not on the scale experienced in Europe. Nevertheless, the disease did have an effect, not only on productivity per acre, but also on the social fabric of the farming communities economically dependent on each potato crop. The quest for solutions to late blight led to notable contributions by two clergymen. The Rev. M. J. Berkeley had proposed a fungus as the cause of blight. Although his ideas were widely rejected, an accidental discovery, unrelated to Berkeley's proposal, led to the use of copper as a protective treatment and eventually corroborated his views. At about the same time, Rev. C. E. Goodrich, believing that blight was related to repeated vegetative propagation, grew true seed and selected from the resulting population a plant named 'Purple Garnet Chile.' This introduction was not a valuable cultivar; however, its eventual use as a parent led to a series of cultivars still considered to be some of the best ever grown: 'Russet Burbank' ('Idaho Russet'), 'Green Mountain,' 'Early Ohio,' 'Early Rose,' and many others.

Potato is now the leading vegetable grown. World production is concentrated in northern Europe, centered on latitude 50°N. This area produces far more than does the United States or other regions of the northern or southern hemispheres. Although there is some production in latitudes approaching the equator, it is successful only with adapted cultivars grown at high elevations.

United States potato production (Table 16.1), 333,911,000 cwt (15,146,203 MT), is confined largely to northern border states. Currently, Idaho leads in production, followed by Washington, Maine, and Oregon. Production is increasing in the western states, decreasing in the east. Of the total U.S. production, 294,679,000 cwt (13,366,639 MT) is fall harvested. Southern states grow potatoes in the winter months for fresh sales ("new" potatoes).

From 75 to 85 percent of the total dry weight produced by the potato plant is located in the tubers. On a fresh weight basis, starch content is 10 to 25 percent, depending on cultivar. In addition to its starch content, the potato serves as a source of moderate levels of

TABLE 16.1

Area, and production of potatoes in 1983 within the 10 leading states

State	Area harvested		Total production	
	1000 acres	1000 ha	1000 cwt	1000 MT
Idaho	312.0	126.4	83,694	3,796
Washington	104.0	41.7	54,080	2,453
Maine	94.0	38.1	22,090	1,002
Oregon	48.5	19.6	20,710	939
North Dakota	128.0	51.8	20,480	929
California[a]	56.2	22.8	19,949	905
Wisconsin	62.0	25.1	18,910	858
Colorado[b]	53.3	21.6	15,820	718
Michigan[b]	52.8	21.4	12,796	580
Minnesota[b]	67.6	27.4	11,639	528
U.S. total	1,242.5	503.2	333,911	15,146

Source: Agricultural Statistics (1985), USDA.

[a] Includes winter, spring, summer, and fall seasons.

[b] Includes summer and fall seasons; all other states reflect fall production.

TABLE 16.2
Nutritional value of white potato and sweet potato[a]

Constituent	White potato	Sweet potato
Water (%)	80	71
Energy (cal)	76	114
Protein (g)	2.1	1.7
Fat (g)	0.1	0.4
Carbohydrate (g)	17.1	26.3
Vitamin A (IU)	Trace	8800
Vitamin C (mg)	20	21
Thiamine (mg)	0.1	0.1
Riboflavin (mg)	0.04	0.06
Niacin (mg)	1.50	0.60
Calcium (mg)	7	32
Phosphorus (mg)	53	47
Iron (mg)	0.6	0.7
Sodium (mg)	3	10
Potassium (mg)	407	243

Source: National Food Review, December 1978. Economics, Statistics and Cooperatives Service (USDA).

[a] Data expressed per 100 g sample.

protein and minerals (Table 16.2). The biological value of the protein is high, however, and potato therefore ranks among the leading crops in usable protein. The food value tends to increase as tubers develop, and the constituents remain fairly constant after harvest if held at optimum storage temperature. Sugar increases or decreases, relative to starch, depending on the storage temperature.

Of the total U.S. crop harvested, 90 percent is sold, and 10 percent is used for seed or represents shrinkage. Fresh table stock utilizes 33 percent, whereas 50 percent is diverted to processing, primarily as chips and shoestrings (13 percent), frozen products (28 percent), canned products (1 percent), and dehydrated products (8 percent). Although potato skins once were discarded, they, too, are utilized, primarily as containers for various flavored stuffings in the fast-food industry. The per capita consumption of potatoes has grown from 107 lb (49 kg) in 1965 to 115.6 lb (52 kg) in 1982.

Classification, Growth, and Development

The potato (*Solanum tuberosum*) is a member of the Solanaceae. In its native habitat it is perennial, a cool season crop, but susceptible to frost or freezing temperatures. Maximum yields require an average growing season temperature between 50 and 59°F (10 and 15°C).

As the plant develops from seed pieces, stolons (botanically rhizomes) form first at the basal nodes of the stem, progressing acropetally. The stolons are lateral shoots with spirally arranged scale leaves. The terminus is characterized by a slight hook and an apical bud (Figure 16.1). The tuber develops just in back of this hooked tip. Behind the apical bud are a number of nodes, each of which has axillary and lateral buds, which become the eyes of the potato. These buds remain dormant during tuber growth. As on the stolon or leafy stem, the tuber buds (eyes) are arranged in a spiral fashion on the tuber and become numerous and densely packed at the apical end. On the proximal side of the eye is a curved scar, termed the eyebrow, which is the abscission scar of the detached leaf scale. Lenticels appear as small dots on the skin of the tuber. These structures enlarge and become prominent under wet soil conditions.

The tuber "skin" is the periderm. The fleshy parenchyma of the tuber is bounded by cortex, enclosing zones of phloem, xylem, and pith. Each of these tissues is parallel to the periderm, and each is connected to each eye and to the stolon (Figure 16.2).

Sprouting is inhibited for up to 90 days, depending on cultivar, after the tuber is separated from the mother plant. This rest period dissipates gradually, and only dormancy imposed by low temperature or a growth regulator, for example, maleic hydrazide or chloro-IPC (chlorpropham), will prevent sprouting. The rest period can be shortened with applications of gibberellic acid or ethylene, but subsequent growth may be abnormal. These mate-

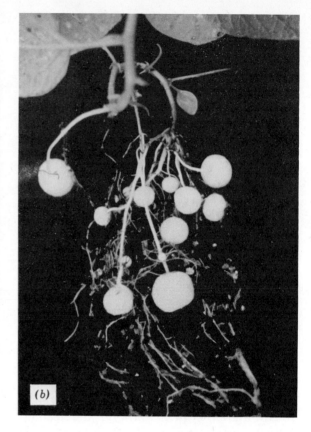

FIGURE 16.1
(a) Young stolons developing from lower nodes, showing apical hook. (b) Enlarged tubers at the tip of each stolon. (Photos courtesy of University of New Hampshire.)

rials appear to alter the auxin levels in and near the eye.

Sprouts arise first at the apical bud of the apical eyes. If seed pieces are cut from the tuber, those eyes located at the apical end of each seed piece are the first to sprout. This apical dominance is modified by temperature. At 59°F (15°C), apical dominance is complete; at 50°F (10°C), several buds develop simultaneously; but on tubers stored for several months at 34 to 41°F (1 to 5°C), then placed in high temperature, all eyes will sprout. As a seed piece sprouts, roots are formed from the bud axil in the eye and adventitiously on the new stem tissue. Stolons arise from the sprout below and occasionally above the soil line as

the plant develops, but tuberization does not occur until 30 to 60 days after planting, usually initiated by shortened daylength and/or cool night temperatures [optimum below 54°F

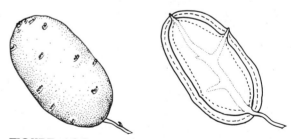

FIGURE 16.2
Diagram of potato tuber anatomy. Each eye is "attached" to each tissue layer within the tuber.

290 / **Tuber and Tuberous Rooted Crops**

FIGURE 16.3
By girdling a potato stem, thus preventing movement of carbohydrates to the stolons, aerial tubers will arise at the leaf axils. These tubers clearly show the features common to stem tissue. (Photo courtesy of University of New Hampshire.)

(12°C)]. The long bright days during the growing season favor photosynthesis and development of topgrowth. As tuberization begins, photosynthate, as sucrose, is translocating rapidly to the stolon tips (Figure 16.3) where it is converted rapidly to starch. It is believed that a critical concentration of starch in the stolon tip may trigger the tuberization process. Excessively warm night temperatures at this time increase respiration and reduce the photosynthate available for translocation, thereby delaying tuberization and reducing yield. Cool nights are essential for maximum production of top-grade potatoes.

The potato is a tetraploid, and some sterility exists among cultivars. Fully fertile cultivars flower and set fruit (seed balls) that resemble wild-type tomatoes. In only one instance has true seed been used to propagate potato commercially ('Explorer'), but seed propagation may have promise in tropical developing countries where maintenance of disease-free vegetative seed is a problem. Seed propagation normally results in extensive variation and in all instances will result in tubers of less than commercial size.

Crop Establishment and Maintenance

Potato management practices are directed primarily at improving yield, less toward attaining specific quality objectives. A productive potato crop depends on well-drained fertile soils, careful management, and the use of clean, certified seed. Reliable seed is critical, since potato is susceptible to a number of virus and bacterial diseases transmitted by asexual propagation. Serious outbreaks of ring rot and viruses stimulated development of certification programs, requiring growers intending to sell potatoes for seed to submit their crop to repeated inspection. Seed production fields in which disease is found are rejected if the number of infected plants exceeds a preestablished level. A crop grown from surplus table potatoes harvested the previous season normally will not attain normal yield and quality levels because of increased disease transmitted through the tubers.

To prepare seed for planting, certified stock is cut into pieces weighing approximately 2 oz (57 g). This size provides the necessary carbohydrate for aggressive early growth, yet it is not wasteful of seed tubers. Seed pieces substantially lighter than 2 oz produce weak plants that are poor competitors. Once cut, the exposed surface of the seed piece suberizes, forming a corky protective layer important in reducing decay. Temperatures of 50 to 55°F (10 to 13°C) with a high relative humidity (95 percent) generally favor suberization, and these conditions may occur naturally in soil in favorable years. Normally, however, growers precut and store seed pieces for 7 to 10 days to allow suberization to occur. As the seed is planted, a chemical treatment may be added to reduce wireworm and soil pathogen activity. Seed pieces exposed to high or to freezing temperatures will not suberize properly, and subsequent decay results in an erratic stand.

Potatoes may be seeded in a range of well-drained mineral or organic soils. Preferred soils are loams or sandy loams or mucks.

Most acreage is planted on raised ridges, either to accommodate surface irrigation or to provide improved drainage and aeration. The ridges also facilitate harvest.

Soils are plowed or chiseled to a depth of 8 to 12 in. (20 to 30 cm) and disked. If available, manure may be applied before plowing; however, most organic matter is supplied by a green manure crop. Fertilizer is broadcast and plowed down or applied as a band during seeding. If banded, the fertilizer must be placed 2 in. (5 cm) on either side and below placement of the seed.

Although nutrient use may differ among cultivars and differing environments a yield of 600 cwt (27 MT) of tubers will remove 395 lb (179 kg) of nitrogen, 46 lb (21 kg) of phosphorus, 563 lb (256 kg) of potassium, 44 lb (20 kg) of calcium, and 39 lb (18 kg) of magnesium. The levels of certain nutrients affect not only yield, but also maturity and quality. Phosphorus is especially important in early growth. Nitrogen will increase the number of stolons and foliage growth and increase tuber size. Excessive nitrogen may delay maturity, and the specific gravity (starchiness) of the tubers has been shown to be related to both nitrogen and potassium levels. Generally, liberal fertilization will reduce specific gravity, and tubers produced in organic soils also tend to show reduced starchiness. Each soil must be managed according to soil test and previous management history. In the acidic mineral soils of the northeast, a common formulation is 8–16–16–1.2 (N–P_2O_5–K_2O–$MgSO_4$), with the rate of nitrogen application 120 to 175 lb/acre (134 to 196 kg/ha). In Florida, the ratio is approximately 1–1–1, with a nitrogen application rate of at least 200 lb/acre (224 kg/ha). In North Dakota soils naturally high in potassium, a ratio of 18–46–0 is recommended, and California soils require a 2–1–1.

Potatoes are seeded mechanically, with the spacing depending on the characteristics of the cultivar and of the soil. Increasing population density will decrease tuber size; if a cultivar tends to produce large tubers (e.g., 'Ken- nebec') or if the soil supports vigorous growth, a close spacing may be warranted. Dry soils and/or small-tubered cultivars may dictate a wide spacing. As an average, 7 to 12 in. (17 to 30 cm) within rows and 34 to 36 in. (18 to 30 cm) between rows would accommodate most cultivars. Seed pieces normally are planted 3 to 4 in. (7.5 to 10 cm) deep.

Potatoes should be cultivated frequently to control developing weeds. However, this practice can damage the young stolons which lie close to the soil surface. Normally, rows are hilled periodically to cover emerging weeds, to stimulate new stolon development, and to prevent exposure of tubers to light. Herbicides are used frequently to avoid cultivation damage. Materials differ with respect to effectiveness on specific soil types and cultivars, and state recommendations must be followed carefully.

Most potato acreage is irrigated, even in areas with ample average rainfall. The distribution of moisture is important in producing tubers free from defects. Alternating periods of drought and moisture can result in "second growth" or knobby tubers, hollowheart, translucent end, and growth cracks.

Moisture level is only one of several environmental stresses that can cause physiological abnormalities (Figure 16.4), most of which impair market quality and/or yield.

Hollowheart   Hollowheart is not recognizable until the tuber is cut, showing a cavity in the center of the tuber. This disorder starts with the death of a small area of pith cells in the center of the tuber. As the tuber grows, the adjacent flesh cracks, and the hollow area expands. Hollowheart may predispose tubers to increased damage in shipping but otherwise does not affect eating quality. Its occurrence can be minimized by maintaining uniform soil water conditions, avoiding overfertilizing with nitrogen, and using cultivars less prone to this defect.

Ring and Net Necrosis   This disorder may develop in tubers that have been exposed to

FIGURE 16.4
Second growth in potato, caused by erratic soil moisture. (Photo courtesy of O. S. Wells, University of New Hampshire.)

freezing temperatures during or after harvest. Conductive tissue is affected, either the vascular ring (ring necrosis) or fine vascular elements (net necrosis), and the discoloration or blackening renders affected tubers unmarketable. Tubers normally show more damage at the proximal than at the apical end. Protection against freeze throughout harvest and storage will avoid the disorder.

Translucent End   Translucent end appears occasionally in some production areas and seems also to relate to environmental stress, particularly drought and heat. Affected tubers, often irregular in shape, develop a glassy appearance at the proximal end and may decay in storage. Analyses have shown high sugars and low solids in the translucent area, quite similar to the flesh in second growth. 'Russet Burbank' seems especially susceptible. Prevention includes maintaining 50 percent available soil moisture throughout the season and avoiding excessive applications of nitrogen.

Blackheart   Warm temperature and excessive soil moisture may give rise to a blackening of tissue in the center of the tuber. The disorder affects appearance to the consumer but does not imply pathogenic decay.

Blackspot   Also an internal browning, blackspot develops in vascular tissue 1 to 3 days after mechanical bruising occurs. Phenolics are suspected as the cause of discoloration. Some cultivars appear to have some resistance, and growing and storage conditions also seem to modify the symptoms.

Greening   Several environmental and management factors have been correlated with elevated levels of glycoalkaloids, including mechanical injury, premature harvest, excessive fertilization, and exposure to light (greening). Of these factors, greening is the most serious, since the green portions contain toxic alkaloid levels. The problem will not occur if plants are hilled to prevent exposure of tubers to sunlight and tubers are stored in darkness.

Cultivars

Potato cultivars are grouped into three major types which dictate their use (Table 16.3). These types are based on starch con-

TABLE 16.3
Classification of potato cultivars by use

Bakers	Boilers	Processing
Abnaki	Chippewa	Cascade
Alamo	Early Gem	Jewel
Chieftain	Green Mountain	Kennebec
Haig	Haig	Le Chipper
Jewel	Katahdin	Monona
Kennebec	Kennebec	Norchip
Norgold Russet	Norgold Russet	Nooksock
Raritan	Norland	Raritan
Russet Burbank	Red La Soda	Russet Burbank
Russet Sebago	Red Pontiac	Sebago
Russet Rural	Superior	Superior
	White Rose	

Source: Smith (1977).

tent, or specific gravity. Cultivars with a high specific gravity (above 1.08) are considered to be excellent for baking and for processing. Low-specific-gravity tubers, termed boilers, are preferred for that purpose because the starch does not slough off in the boiling water. Such cultivars also are not suitable for processing, because the yield of the finished product is low, compared to that of high-specific-gravity cultivars. Although the correlation between specific gravity and quality is useful, it is not complete. Other biochemical factors are involved in textural quality but are more difficult to measure. Among the processing cultivars, all high in specific gravity, some are best suited for specific products: chips, French fries, canned whole potatoes, or flaked (dehydrated) products.

In addition to specific gravity, several traits describing tuber appearance also are used to categorize cultivars:

Skin color	Commonly white or red, occasionally purple
Tuber shape	Oval, elongate, blocky, flat, or round
Skin	Clear or russetted
Eyes	Deep or shallow
Flesh	White (common) or yellow

Resistance or tolerance to specific defects or diseases is as important as productivity of a cultivar. Several of the important cultivars and their characteristics are listed in Table 16.4.

TABLE 16.4
Characteristics of leading potato cultivars

Cultivar	Rank	Market[a]	Maturity[b]	Vine growth[c]	Tuber color and shape	Resistance
Russet Burbank	1	F,P	M–L	LS	White, long oval, large russet	Scab
Norchip	5	P	E–M	MU	White, round to oblong, smooth	Scab, scurf
Kennebec	2	P	E–M	LU	White, oblong to elliptical smooth	Scab, late blight, net necrosis, slight mosaic
Red La Soda	9	F	E–M	MU	Red, round, medium deep eye	Moderate mosaic
Katahdin	3	F	M–L	LS	White, round, to elliptical	Mosaic, Leaf roll
Norgold Russet	6	F	E	MC	White, oblong russet	Scab, net necrosis, hollowheart, cracking
Superior	4	F,P	E–M	MS	White, oval, smooth	Scab
Red Pontiac	7	F	L	LS	Red, round to oblong, medium deep eye	—
Norland	8	F	E	MS	Red, oblong, smooth	Scab
Centennial Russet	10	F	E–M	MU	Oblong, russet, shallow eyes	Early and late blights, *Verticillium* wilt

Source: Smith (1977); University of California Coop. Ext. Leaflet 2684.

[a] F = fresh use; P = processing.

[b] E = early; M = medium; L = late.

[c] LS = large spreading; MS = medium spreading; MU = medium upright; LU = large upright; MC = medium compact.

Although many new potato cultivars have been introduced, the most important still include several that have been popular for many years: 'Russet Burbank' ('Netted Gem'), 'Katahdin,' 'Kennebec,' and 'Pontiac.' Of these, 'Russet Burbank,' known widely as 'Idaho Russet,' is the oldest.

Disease and Insect Pests

Diseases   Potato is susceptible to many field and storage diseases. Resistance has not been developed for many diseases, in part because of the polyploid nature of the plant. The seriousness of some of the diseases is exacerbated by vegetative propagation, and growers must use seed stock free of disease and maintain plant health by adhering to strict cultural and chemical control systems. The most important problems are rhizoctonia, scab, ring rot, early and late blights, wilt, viruses, and several storage rots (Figure 16.5).

Rhizoctonia   The causal organism, *Rhizoctonia solani,* is a permanent saprophytic inhabitant of most soils, and it can infect a range of crops. On potato, symptoms may appear at any stage of growth when soil moisture is high. Young sprouts may be infected before emergence, developing brown cankers that girdle the sprouts below the growing point. New shoots then arise from the eye, but these shoots become increasingly spindly as infection continues. Mature stems are infected at the soil line, a condition known as stem canker. Affected plants wilt and fall over. Stolons also can be infected, restricting translocation of photosynthate to affected tubers and leading to oversized tubers on stolons not infected. The tubers themselves may show "black scurf," small black sclerotia adhering to the

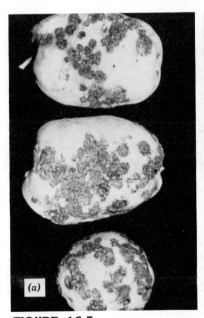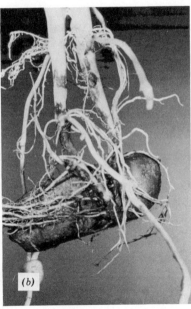

FIGURE 16.5
Potato diseases: (*a*) scab; (*b*) *Rhizoctonia*; (*c*) spindle tuber (virus). (From *Diseases of Plants,* slide set 34, Potato Diseases. Reproduced by permission of the American Phytopathological Society.)

tuber surface. The sclerotia are the overwintering structures of the fungus. Control of this disease is difficult. Selecting well-drained production sites and adapted cultivars may be the most effective measures.

Scab   Potato scab, typified by rough, sometimes raised, irregular scales on the tuber surface, is caused by *Streptomyces scabies.* Although the affected area can be trimmed in preparing tubers for cooking, scab does affect marketability. The scab organism is a common soil inhabitant and can attack other root crops. It is particularly aggressive in dry soils and those with a pH exceeding 5.4. The fungus is unaffected by livestock digestion and is disseminated by manure. Control measures include careful management of the pH, use of organic matter other than barnyard manure, and use of resistant cultivars. Rotation of potato with beets is discouraged.

Ring Rot   The most serious bacterial disease of potato is ring rot, caused by *Corynebacterium sepedonicum.* The first symptom, a gradual wilting of the foliage, usually appears as plants reach maturity. Vascular rings of the tuber discolor and exude a creamy ooze. Secondary rots frequently result in total loss. The bacteria are spread rapidly by seed piece cutting and cultivation equipment. The most effective control measures are use of certified seed and disinfestation of all equipment.

Early Blight   As potato plants begin to mature, early blight, caused by *Alternaria solani,* may develop on the leaves, appearing as dark circular spots with a "target board" appearance. These spots enlarge, and lesions eventually may appear on stems and tubers. The fungus overwinters in potato tubers or plant refuse and will become active in warm, humid weather. Inoculum levels can be reduced by disposing of infected debris, although some infection normally occurs on plants late in the season, regardless of sanitation procedures. This late infection has little or no effect on

yield, but early infection must be prevented. Preventative sprays with maneb or mancozeb are effective.

Late Blight   Although a devastating disease, late blight (causal organism, *Phytophthora infestans*) is active only in very specific environments. Average temperatures less than 77°F (25°C) combined with a very high humidity or constant mist or rain for a period of 5 days will favor pathogen activity. Under these conditions, water-soaked lesions appear inside the plant canopy, and white mycelium may appear on the underside of a leaf. Spores are spread rapidly by wind and rain and may infect tubers as they are dug.

Because of the well-documented environmental conditions in which the disease becomes active, it has been possible to develop effective IPM procedures for protecting a crop. Several states have installed computerized forecast services that provide immediate recommendations to growers. Based on warnings of likely infection, a protective spray (maneb, mancozeb, or copper fungicides) may be applied only as needed.

Wilts   Two groups of organisms may cause vascular wilt in potato: *Verticillium* and *Fusarium.* Early symptoms are yellowing of the foliage and curling or drooping of lower leaves, followed by vascular browning. Tubers infected with *Fusarium* wilt also may show pink discoloration. The fungi are soilborne and persist for many years. Although rotation may help, the diseases cannot be eliminated completely from an infested site.

Virus Diseases   There are a number of virus diseases, many of which are controlled largely through seed certification. These diseases include virus X, virus Y, virus A, leaf roll, spindle tuber, yellow dwarf, purple top, and others. Viruses X, Y, and A, termed latent viruses, may be carried in plants appearing to be healthy. Some yield reduction in infected plants has been demonstrated, however, and strains of the viruses can interact with cultivars and with each other to cause more extensive damage. Of the other virus diseases, leaf roll usually appears after midseason and gradually affects the entire plant top. Tubers of infected plants may develop net necrosis. Spindle tuber infection is recognized by the elongate tubers bearing many eyes. The plants also become erect with dark green foliage. Yellow dwarf infection is characterized by a rosette-type top growth with dark green foliage. Leaf margins roll upward and the leaf apex curves downward. Symptoms are most striking in warm weather. Purple top is common in the north when tuberization begins. Young terminal leaves roll upward, and may, in pigmented cultivars, become distinctly purple. In other cultivars, leaflets become yellow. Tubers also may become discolored and flabby. The virus organisms not considered as latent viruses usually affect crop yield severely. In all virus diseases, infected tubers are the source of inoculum, and aphids frequently are the vectors. Disease-free (certified) seed and effective insect control are the recommended preventative measures. Control of alternate weed hosts in and around the field also will reduce infection levels.

Storage Diseases   Late blight, *Verticillium* and *Fusarium* wilts, and ring rot can become serious storage diseases. In addition, soft rot (*Erwinia carotovora*), pink rot (water rot; caused by *Phytophthora erythroseptica*), and leak (*Pythium* spp.) may result from mechanical bruising and failure to cool properly. Storages should be disinfected periodically; then the harvested product should be cured and graded carefully before storing.

Insects   Both soil and foliage insects (Figure 16.6) are frequent pests in a potato field. Several of them, including wireworms, grubs, aphids, Colorado potato beetle, and leafhoppers, either carry disease or injure the plant thereby providing an avenue for disease infection. In some production areas, other insects,

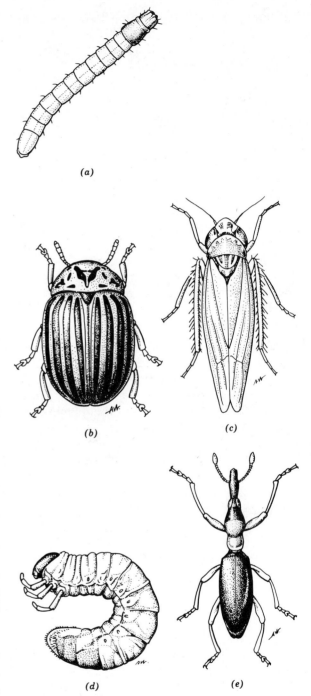

FIGURE 16.6
Potato and sweet potato insect pests: (*a*) wireworm; (*b*) Colorado potato beetle; (*c*) potato leafhopper; (*d*) white grub; (*e*) sweet potato weevil.

including the garden symphylan, seedcorn maggot, blister beetle, spider mite, thrip, and flea beetle, have caused substantial damage. Recommendations for control depend on the severity of the problem and the chemicals registered for use.

Wireworms   Wireworms (many species) are tough-bodied segmented larvae of the click beetle. They feed on roots, stolons, and tubers, providing avenues for secondary rot infection. The feeding reduces yield and defaces tubers that do not succumb to secondary rots. Soil treatment (fonofos or phorate) incorporated to a depth of 3 to 4 in. (7.5 to 10 cm) and/or seed or furrow treatment at planting time are recommended for control. If infestation persists, postemergence sidedress treatments may be necessary.

Aphids   Aphids are serious pests because of the loss of plant sap under high populations and because of virus transmission. The green peach aphid (*Myzus persicae*) is particularly serious as it is the agent for transmitting leaf roll. Both the green peach aphid and the potato aphid (*Macrosiphum solanifolii*) can be controlled with contact materials (endosulfan, parathion).

Colorado Potato Beetle   The most serious insect pest of potatoes in many production areas is the Colorado potato beetle (*Leptinotarsa decimlineata*). The variability within the species has given it resiliency in surviving pesticides, and, in many areas, no pesticide is completely successful. The adults lay eggs in clusters on the underside of leaves. These eggs hatch in about 7 days, producing reddish larvae. Both larvae and adults feed voraciously on foliage. Chemical control is most effective just after egg hatch and least effective on adult beetles. Soil-applied systemics help in control, but it is recommended that a range of different pesticide groups be used in a control program to retard development of tolerance in the beetle population. Without prompt control, the foliage can be stripped from plants in several days.

Leafhoppers   Feeding by leafhoppers (*Empoasca fabae*) causes a bronzing of the foliage, known as hopperburn. At one time, the development of hopperburn was tolerated in order to accelerate maturation. However, leafhoppers transmit virus and mycoplasma diseases and should be controlled. Both systemic (aldicarb) and contact chemicals (carbaryl and others) are effective.

White Grubs   The larval and white grub stages of the June bug (many species) feed on potato roots and tubers, as well as the roots of all grasses. The eggs are laid in the spring in the soil. The young grubs move downward as the soil cools in the fall, migrating upward again in the spring when they feed on the young developing roots. Systemics are registered for use in some areas and will control other insect pests as well.

Harvest and Market Preparation

Immature potatoes have a low specific gravity and very thin skins that rub off easily during harvest. As maturity advances, the skins become suberized, adding thickness and toughness, and the specific gravity of the flesh increases. Because of the potential for early harvest damage and the importance of specific gravity as a quality factor, growers prefer to delay harvest until tubers are fully mature.

When most tubers have reached a 2-in. (5-cm) diameter, maleic hydrazide may be applied to the foliage to retard tuber sprouting in storage. Chloro-IPC also can be used for this purpose, applied 2 to 3 weeks after harvest (aerosol or emulsifiable concentrate) via the storage ventilation system.

To prepare for harvest, irrigation is stopped 2 to 3 weeks prior to digging. Also prior to harvest, vines may be eliminated by chemical vine killers, flaming, or mechanical flails. If vines are killed chemically or mechanically, digging should be delayed for 7 to 10 days for a crop nearing physiological maturity or up to 3 weeks for immature potatoes. During this period, the tuber skin will set and thicken, the danger of disease, skin slipping, and bruising is reduced, and harvest operations will be facilitated. Application of vine-killing methods too early reduces specific gravity and may cause internal vascular discoloration in the tubers.

The digging equipment should be adjusted properly and padded to avoid excessive bruising, and the drop elevation should not exceed 4 to 6 in. (10 to 15 cm). Most large production areas utilize self-propelled harvesters, loading and moving the harvested tubers to the packing shed or storage in pallet boxes or dump trucks.

New potatoes, those harvested before skins thicken, are marketed fresh, not placed in storage. The major part of the potato crop, however, is stored, and proper storage is essential for maximum quality. Storage facilities are designed to provide uniform forced-air ventilation and control of temperature and humidity. Relative humidity is maintained at a high level (95 percent), but condensation on the tubers must be avoided.

Before the tubers are placed in storage, the area should be disinfected with quaternary ammonium compounds and rinsed. Tubers then are misted with a protectant to control *Fusarium* rot and cured to heal (suberize) cuts and bruises by placement in storage at 50 to 59°F (10 to 15°C) for 2 to 3 weeks. Cell regeneration takes place within 10 to 20 days. During this period, there is little change in content of starch and reducing sugar. For the remainder of the storage period, temperature is held at 40°F (4.4°C) for seed or for table stock and at 50°F (10°C) for chipping and processing. Tubers exposed in the field to rot pathogens normally are cooled immediately without curing.

A proper storage environment will induce rapid wound healing while reducing shrinkage from respiration and water loss. During storage, however, there will be changes in composition affecting flavor. Potatoes contain sucrose, glucose, and fructose; the latter two

reducing sugars accumulate in tubers stored at 40°F (4.4°C). These sugars react with amino acids, ascorbic acid, and other compounds under the heat of processing. The resulting dark color is associated with undesirable flavors (caramelized or burned flavor). Chip manufacturers tolerate little reducing sugar. French fries, potato puffs, and other products can be processed using potatoes slightly higher in sugar levels than is acceptable for chips.

Potatoes found to be excessively sweet can be conditioned for processing by increasing the storage temperature to 59 to 70°F (15 to 21°C). The success of this procedure depends on length of time in storage, specific gravity, and other factors (Table 16.5).

TABLE 16.5
The influence of fertility level, specific gravity, tuber portion, and storage period on loss of sugars by reconditioning

Treatment	Percentage reducing sugars[a]		
	Before reconditioning	After reconditioning	Change
Fertility			
High	5.37[b]	1.72	72[c]
Low	5.59	2.08	68
Specific gravity			
High (1.085)	5.70	1.74	73[d]
Low (1.075)	5.27	2.07	67%
Tuber portion			
Stem	7.82	3.54	55[e]
Center	4.91	1.57	69
Bud	3.72	0.59	89
Storage period			
Early	6.40	3.14	51[e]
Mid	4.50	1.54	73
Late	5.54	1.04	85

Source: Iritani and Weller (1980).

[a] Average of 2 years data.

[b] Average of eight analyses.

[c] LSD = 4, $P = .05$.

[d] LSD = 5, $P = .01$.

[e] LSD = 7, $P = .01$.

Cultivars differ with respect to reducing sugar levels, and stress conditions during growth also can affect quality (refer to Translucent End). High specific gravity normally is associated with reduced sugar levels, high dry matter, and high starch. These qualities are associated with dryness, mealiness, and, to some degree, flavor and color. The high-starch products normally absorb less fat during cooking than those of medium to low specific gravity.

When potatoes are removed from storage, temperatures may be raised to 50 to 55°F (10 to 13°C) to avoid the bruising or splitting that can occur when tubers are cold. The tubers are packaged in 5- to 10-lb (2.2- to 4.5-kg) paper, vented plastic, or net bags for consumers, or in 50- to 100-lb (22.7- to 45-kg) mesh bags or in cardboard cartons for institutional sales.

JERUSALEM ARTICHOKE

The Jerusalem artichoke (*Helianthus tuberosus*) is a tuber-bearing perennial plant native to North America. A member of the Asteraceae, it as known also as girasole and sunchoke. During the 1600s, specimens were introduced to Europe, and the crop became important in areas not suited for growing Irish potato.

Production of the Jerusalem artichoke in the United States once was promoted as a source of inulin, a sugar form that can be utilized by diabetics. Although production did not succeed economically, it did result in the distribution of the plant as a weed in several parts of the United States, particularly in the northern states.

The plant grows up to 8 ft (2.4 m) tall in cultivation, bearing rather irregular knobby tubers 3 to 4 in. (7.5 to 10 cm) in length. These tubers constitute the edible and propagative part of the plant. Seed pieces with one to several buds are cut from the tubers and planted 4 to 5 in. (10 to 15 cm) deep as early in the season as possible, since a long frost-free period is needed to ensure reasonable yields. Soils should be fertile, sandy, or well-drained loams, fertilized with 1–1–1 or 1–2–1 ratio materials, depending on soil tests. Tuberization begins as daylength increases in late summer, resulting in yields averaging 5 to 7 tons per acre (11 to 16 MT/ha), reaching 10 to 15 tons (22 to 34 MT/ha) under optimum growing conditions. Tubers may be stored for several months at 32 to 36°F (0 to 2°C) and 95 percent relative humidity and must be placed in storage soon after harvest because of their susceptibility to rapid water loss.

SWEET POTATO

Historical Perspective and Current Status

The origin of the sweet potato is obscure, but records indicate that the crop had been cultivated since prehistoric times by Polynesians in the South Pacific and by Mayan and Peruvian indians of tropical South and Central America. Although the true center of origin has been debated, evidence favors South America. How the sweet potato migrated to the South Pacific is a matter of speculation, but its appearance in widely separate geographic areas constitutes evidence of extensive exploration using very primitive modes of transportation (Baker, 1965).

Columbus mentioned sweet potato in records of his fourth voyage and is believed responsible for introducing the plant to Spain. When grown in temperate areas, it was not impressive. Spanish explorers, however, introduced the crop to the Far East, where it thrived and became an important staple food crop.

Prior to the mid-1600s, sweet potato was unknown to North American Indians. It was introduced to Virginia, probably by way of the West Indies, and quickly gained popularity in the United States among both colonists and Indians.

Sweet potato has become a valuable food, especially in the tropics where the white potato is not adapted. It is consumed as a staple in at least one-half of the temperate zone. The highly colored roots are rich in carbohydrates, β-carotene (provitamin A), and ascorbic acid (vitamin C). The content and quality of protein are average (Table 16.2). Asia leads in acreage devoted to sweet potato (over 12 million), followed by Africa and South America.

Production of sweet potato in the United States, a fraction of that in Asia, is concentrated along the southeast coast through the southwest to California. There is some acreage in New Jersey and in areas of suitable soils and climate in some of the central states. Approximately 103,000 acres (41,509 ha) is devoted to sweet potato production in the United States, of which nearly 60 percent is located in Louisiana and North Carolina. North Carolina leads in farm value, followed by California and Louisiana (Table 16.6). Over one-half of the crop produced in the United States is used fresh; the remainder is processed. Per capita consumption in the United States has declined steadily over the past 30 years and is now 5 lb (2.2 kg).

Classification, Growth, and Development

The sweet potato (*Ipomoea batatas,* family Convolvulaceae) is a warm season, tender perennial grown as an annual. Although some

FIGURE 16.7
Sprouting of sweet potato from mother root. Note dominant sprouting at proximal end. (Photo courtesy of University of New Hampshire.)

cultivars are termed "yams," the true yam is a member of the unrelated genus *Dioscorea* and is native to Africa or China.

The young sweet potato plant develops as an adventitious shoot arising from the vascular cambium of a root. The initial shoots are produced predominantly at the proximal end of the root (Figure 16.7) and develop their own root systems rapidly. After transplantation, this root system becomes very fibrous and branched, extending to a depth of 18 in. (46 cm), eventually to $2\frac{1}{2}$ ft (76 cm). Within this root system, 4 to 12 roots become enlarged and fleshy. Their structure is typical of most roots, except for the tendency to develop adventitious shoots. Enlargement of the storage roots is due to the development of secondary tissue, mostly storage parenchyma within the vascular cambium. The outer portion of the root consists of cortex, a cork cambium (phellogen), and the periderm and cork (phellem).

Foliage is semivining and vigorous with variously lobed leaves. Flowers seldom develop, especially on northern-grown sweet potato vines. Sweet potato is a short-day plant and will not flower at daylengths much in excess of 12 to 13 h. Commercial propagation is entirely asexual, for the plant is a heterozygous hexaploid and will not breed true from seed.

TABLE 16.6
Sweet potato production statistics, 1982 to 1983

	Production			
	1000 cwt		1000 MT	
Area	1982	1983	1982	1983
North Carolina	5,940	4,440	269	201
Louisiana	2,500	2,280	113	103
California	1,748	1,716	79	78
U.S. total	14,833	12,083	673	548

Source: USDA Econ. Res. Serv. Vegetable Outlook & Situation Rep. (November 1984).

Crop Establishment and Maintenance

Sweet potatoes require a growing season of 4 to 5 frost-free months. There is little growth below 59°F (15°C). Optimum temperatures are 70°F (21°C) soil and 84°F (29°C) air.

The greatest management cost is in growing transplants. Transplants are produced by bedding whole roots in a hotbed or equivalent structure, or in field seedbeds, approximately 4 to 6 weeks before outdoor planting. Seed roots often are presprouted by warming for 2 to 4 weeks at 75 to 84°F (24 to 29°C) at high humidity in well-ventilated rooms. Just prior to bedding, all roots with evidence of disease should be discarded and the remainder treated with a fungicide dip or spray to minimize rot. The use of certified seed also will reduce the chance for disease.

The seedbed must be a well-drained, light soil free of carryover disease. The treated seed roots are placed in the seed bed about 1 in. (2.5 cm) apart and covered with 2 in. (5 cm) of soil. After covering, a 1–1–1 ratio fertilizer normally is applied. Alternatively, a broadcast application may be used prior to field seedbed formation.

When hotbeds are used, the temperature should be held between 74 and 84°F (24 and 29°C) until sprouts are 6 in. (15 cm). Then, heat is required only during cold nights. If early season temperatures are below 65°F (18°C), new root growth on developing shoots is retarded. Sprouting from seed roots may be reduced at excessively high temperatures. In field seedbeds, a vented polyethylene row cover or polyester cover is applied to increase temperature and sprouting and also to minimize insect infestation.

Seed potatoes will produce up to 15 transplants per root over an extended period. Approximately five to six shoots develop at one time, and these are pulled by hand when 8 to 12 in. (20 to 30 cm) in size, taking care not to uproot the seed potato. If the shoots become too long, they may be cut in half. The basal portion is treated as a normal transplant, and the top cuttings are usable after being placed stem down in water for 72 h at 70 to 84°F (21 to 29°C) to initiate new roots. Terminal vine cuttings also can be used for propagation in the same way.

After each harvest of shoots, the seedbed is topdressed with ammonium nitrate to encourage new sprouting. Once all shoots have been harvested, the seed potatoes should be destroyed. Transplants may be stored for several days before planting, but must be kept moist by placing in moist sphagnum or wet soil or by periodic spraying with water.

Fine sandy loams or sands with a pH near or slightly below 5.5 are preferred for sweet potato production. Heavy soils or those with organic matter exceeding 2 percent are not recommended, since shape may be poor, and scurf and black rot fungi persist in such soils. The roots also become rough and cracked. If possible, fields that have not been in sweet potato production for at least 2 years should be used.

Because the soils used for production are light and generally lacking in organic matter, fertilizer must be provided at several stages of growth. A crop of 300 cwt (13.6 MT) will remove 140 lb (63 kg) of nitrogen, 20 lb (9 kg) of phosphorus, and 200 lb (91 kg) of potassium (both roots and tops). In the mid-Atlantic states, where phosphorus is fixed rapidly, the average application to a low-fertility soil is 65 lb/acre (73 kg/ha) nitrogen, 200 lb/acre (224 kg/ha) phosphate, and 300 lb/acre (336 kg/ha) potash. Approximately one-third of the nitrogen, 20 percent of the potash, and all of the phosphate are applied as a broadcast or band before planting. At "lay-by," or the last cultivation, the remaining potash and another one-third of the nitrogen are sidedressed, and a final sidedress of nitrogen is applied 5 to 6 weeks later. An adequate supply of nitrogen has been found to improve grade, total yield, and quality (protein, carotenoids, low fiber). Excessive nitrogen may increase the number of jumbo roots. Lime, if needed, is applied several months in advance of planting. However, the crop tolerates acid soils, and low pH helps to reduce *Streptomyces* soil rot.

There are occasional problems that arise because of nutrient deficiencies or other environmental stresses. Growth cracks will develop during erratic growing conditions (uneven moisture or soil fertility). Inadequate potassium has been reported to impair root shape in some production areas, causing some elongation that detracts from market grade. A physiological disorder, termed blister, results from a deficiency of boron, but the problem does not become apparent until after the roots are placed in storage. There are differences in susceptibility to this disorder, and the most effective control is addition of $\frac{1}{2}$ to 1 lb/acre (0.5 to 1.1 kg/ha) of boron to the regular fertilizer application. In those soils deficient in boron, an application may be necessary each year.

When the soil temperature approaches 64°F (18°C), shoots are planted in ridges 8 to 10 in. (20 to 25 cm) high that are spaced 32 to 42 in. (0.8 to 1.1 m) on center. Plant spacing in the row ranges from 8 in. (20 cm) (early cultivars) to 18 in. (46 cm) (late cultivars) and must be consistent to promote uniform root size. Early planting, providing soil temperatures are suitable, generally increases yield, and the ridges enhance drainage and facilitate harvest.

Although sweet potato is considered drought tolerant, an even moisture supply throughout the growing season will enhance the yield and market appearance. The amount of irrigation required depends upon soil type. Maximum yields have been reported at 20 percent available moisture on fine soils to 50 percent in coarse sands. From $\frac{3}{4}$ in. (19 mm) to 2 in. (5 cm) of water weekly may be required to maintain these levels.

Because top growth is vigorous and dense, two to three shallow cultivations in the early season will control weeds satisfactorily. Preplant and postplant herbicides are available and useful in areas or soil types in which cultivation may be difficult. Chloramben, DCPA,

and diphenamid are registered for use in sweet potato as postplant applications. Vernolate may be incorporated prior to planting.

Cultivars

Two types of cultivars are produced in the United States. Those generally termed yams, the moist flesh types derived from the original 'Porto Rico,' constitute the most accepted product. The dry lighter colored flesh types (termed "Jerseys") are well adapted to northern areas, yielding short, chunky roots, but the dry consistency is not as popular among consumers, and the storage period is limited.

Moist-flesh sweet potatoes are produced throughout the southeastern states, the southwest, and California. Dry-flesh types are grown in the "DelMarVa" area (Delaware, Maryland, Virginia) into New Jersey and in localized production sites elsewhere.

Sweet potatoes are subject to variation in spite of vegetative propagation. Some cultivars may accumulate deleterious mutations with repeated propagation and are said to "run out." To minimize this problem, growers are encouraged to use certified seed produced from carefully selected mother roots.

The characteristics of several commercial cultivars are listed in Table 16.7.

Disease and Insect Pests

Diseases   Although there are foliar diseases of sweet potato, the major disease pests are soilborne. The most common are scurf, blackrot, soil rot (pox), and *Fusarium*.

Scurf   Scurf, caused by the fungus *Monilochaetes infuscans*, has been reported extensively throughout production areas of the United States. The roots become brownish black with small coalescing spots. The injury is superficial, confined to the surface skin, but the appearance destroys root marketability. The organism can survive in soil without the sweet potato host for up to 2 years. Therefore,

TABLE 16.7
Characteristics of commercial cultivars of sweet potato

Cultivar	Type	Skin	Flesh	Resistance	Assets
				Characteristic	
Centennial	Yam	DC[a]	DO	Wireworm, *Fusarium*, internal cork	Yield, sprouting
Eureka	Yam	DC	DO	Pox, *Fusarium* wilt, root knot	—
Jewel	Yam	LC	DO	*Fusarium* wilt, root knot	Chill tolerant
Jersey Orange	Dry	C	O	—	Northern adaptation
NC Porto Rico	Yam	RP	O	—	Yield, baking
Nemagold	Dry	C	O	Nematodes	Northern adaptation
Nugget	Dry	C	O	Root knot	Northern adaptation
Travis	Yam	R	MO	Pox, cork	Very early yield
Georgia Red	Yam	R	LO	Root knot	Fresh market
				Flea beetle	Yield

[a]DC = deep copper; LC = light copper; R = rose; RP = rose-pink skin; DO = deep orange; O = orange; MO = medium orange; LO = light orange flesh.

a 2- to 3-year rotation should provide effective control as long as new inoculum is not introduced. Seed roots developed by vine cuttings or those certified free of scurf are recommended where scurf has been a problem.

Black Rot   Also a widespread disease, black rot produces black cankers on the roots, and affected plants show yellow foliage. The causal organism (*Ceratostomella fimbriata*) is most easily spread by way of seed roots. Once in the soil, it will persist for up to 2 years without a host plant. Treatment of the seedstock prior to bedding, rotation of production fields, and sanitation provide the most efficient and effective controls.

Soil Rot   A soilborne rot, caused by *Streptomyces ipomoea,* is particularly severe in areas of neutral or alkaline soils and may become a problem with use of lime. The organism is persistent and becomes increasingly active at a pH over 5.2 and in dry soils. Affected plants have a reduced number of small pale green leaves. Lesions develop on the roots, and, if young roots are affected, they often become deformed. Severely infested fields should be avoided; other fields can be improved with fumigation. Certified seed should be used for producing disease-free transplants, and some cultivars ('Eureka,' 'Travis') are reported to be resistant.

Fusarium Wilt   Wilt, or stem rot or surface rot, can be prevalent in seedbeds as well as production fields. It is caused by *Fusarium oxysporum* f. sp. *batatas* and appears as yellowing between veins, leaf drop, stunting of the stems, and discoloration of vascular tissue. Once introduced, *Fusarium* becomes a permanent resident of the soil. A number of cultivars with resistance have been developed ('Centennial,' 'Jewell,' 'Pope').

Rhizopus Soft Rot   Soft rot is the most serious storage disease. The responsible organism (*Rhizopus nigricans*) affects a wide array of crops and survives as a saprophyte on crop debris. The zygospores also persist for several months and therefore can reinfect a crop placed in contaminated storage. *Rhizopus* is a wound parasite producing a watery rot that progresses rapidly through the tissue. A coarse, whiskery fungal growth often develops as infection advances and breaks the skin of the sweet potato root. Control of *Rhizopus* is achieved by careful grading, strict sanitation, and crop rotation.

Foliar Diseases   The foliar diseases include white rust (*Albugo ipomoeae-panduratae*), leaf spots (*Alternaria, Cercospora, Septoria* spp.), and blight (*Stemphyllium botryosum*). These diseases are not uncommon, but they seldom limit production in most production areas.

Insects   The major insect and related pests, including nematodes, wireworms, weevil, grubs, banded cucumber beetle, and flea beetle, are found wherever sweet potatoes are grown.

Root Knot Nematodes   Nematode (*Meloidogyne* sp.) feeding reduces the yield, quality, and grade of sweet potatoes. Roots often show enlarged scars or "eyes," and longitudinal cracks may develop. Infestations of root knot nematode are common in warm soils and may be reduced by soil fumigation. In addition, resistance is available in several cultivars ('Jewel,' 'Eureka,' 'Pope,' 'Vardeman').

Sweet Potato Weevils   Weevils (*Cylas formicarius elegantulus*) are small [$\frac{1}{4}$ in. (6 mm)] destructive insects with dark blue head and wings and a bright orange body (Figure 16.6). All parts of the plant are eaten, but most of the damage is on the storage roots, both in the field and in storage. In addition to mechanical damage to the roots, larval feeding results in an unpleasant flavor. Control has been obtained by chemical treatment of stored potatoes and field spraying as needed.

Wireworm   The damage to sweet potato from wireworm is similar to that described for other crops (refer to White Potato). Several species of wireworm can affect the sweet potato, including the tobacco and southern wireworms and the common wireworm. All damage is confined to the roots and consists of small holes in the surface of the potato. Be-

cause the adult (click beetle) lays eggs in undisturbed grass, such areas should be avoided for planting. As protection, a granular insecticide (fonofos, diazinon) may be applied midway through the season when wireworm activity increases.

Flea Beetle   The flea beetle (*Chaetocnema confinis*) is a common pest on foliage of many vegetables. On sweet potato, however, the larvae will attack the roots, producing very small pinholes and surface tracks. Marketability of the roots is reduced, and the wounds create opportunities for disease to develop. Some resistance is available ('Georgia Red'), but control normally is provided by preplant pesticide incorporation (endosulfan).

Banded Cucumber Beetle   The cucumber beetle (*Diabrotica balteata*) damages thousands of acres of sweet potatoes in parts of the southern United States. The larvae bore small holes, characteristically in groups, through the periderm, forming irregular cavities under the surface which enlarge as the root expands. When damage occurs early in the season, the roots become unsightly. No insecticides have been particularly effective.

White Grubs   Several white grub species occasionally cause damage, particularly in soils of high organic matter. Grub feeding results in scars on the potato surface, again reducing marketability. There is no resistance among cultivars, but preplant chemical treatments (diazinon) can reduce damage.

Harvest and Market Preparation

Sweet potatoes must be harvested before freezing weather. Severe foliage loss by freezing or soil temperatures lower than 50°F (10°C) for several hours will damage the roots. Potatoes exposed to moderate chilling may be dug, cured, and sold, but they should not be placed in long-term storage. Chilling of roots below 40°F (4.4°C) will lead to internal discoloration or a "hard core" disorder (hard tissue after cooking). Potatoes exposed to light frosts should be dug immediately, since they retain moisture that otherwise would be translocated to the tops. High temperature [greater than 90°F (32°C)] also can damage harvested sweet potatoes, causing sunscald and eventually breakdown and decay.

There is no obvious indicator of maturity. Sweet potato vines continue to develop, and roots will grow as long as the temperature is suitable. Several sample diggings will indicate size and the approximate date for harvest, although size will vary considerably within a crop. The harvest date also will be affected by cultivar, plant spacing, and other factors affecting root size. A normal crop may produce approximately 65 percent Grade 1's, 5 percent jumbos, and 30 percent culls, Grade 2's and canners.

Sweet potato roots are very susceptible to skinning during harvest. Yet, hand picking has become costly and, in some areas, has been replaced by one of several mechanical handling systems. Mechanical harvesters are available that top, dig, and elevate roots to boxes. Other harvest systems involve digging with Irish-potato diggers, modified moldboard plows, or other devices, followed by hand picking or using harvest aids that combine hand sorting labor with mechanical digging and conveyer assemblies. The harvested potatoes are placed in bushel baskets or, most frequently, pallet bins for transit to curing and storage facilities. Sweet potato roots are much more susceptible to handling damage than white potato tubers. All handling systems must be designed to minimize physical damage.

The sweet potato roots then are graded and placed in pallets or bins. The Grade 1 potatoes are cured immediately and stored. Curing is necessary to heal the cuts and bruises. During this process, several layers of parenchymal cells under the damaged area desiccate and thicken (suberize). A corky wound periderm three to seven cell layers thick then develops which provides an effective barrier to pathogen penetration. Quality is improved during curing as starches are converted to sugars. At harvest, maltose is the major sugar and sucrose the predominant secondary sugar. During curing and well into storage, maltose decreases, and sucrose, fructose, and glucose increase. The optimum conditions for rapid curing are 85°F (29°C) temperature and 90 to 95 percent relative humidity in a well-ventilated curing–storage room. From 4 to 8 days normally is required, depending on soil temperature at harvest. The lower this soil temperature, the longer the time required for curing.

After curing, the temperature should be reduced gradually to 55 to 59°F (13 to 15°C) at 85 to 90 percent relative humidity for long-term (up to 6 months) storage. Temperatures below 55°F will result in the same chilling injury that occurs in the field and will also impair cooking quality. Storage temperatures 60°F (16°C) and above increase weight loss and sprouting and may increase the incidence of internal cork.

When sweet potatoes are removed from storage, they are washed, sorted for size, and graded before marketing. The roots should be firm, smooth, oval and free of insect or mechanical damage or disease, and color should be typical of the cultivar. A chemical (2,6-dichloro-4-nitroaniline) often is added during market preparation to prevent decay. Most of the crop is packed in 50-lb (23-kg) corrugated boxes and sold as bulk potatoes.

MISCELLANEOUS TUBER AND TUBEROUS ROOTED CROPS

Cassava (Manioc)

Cassava (*Manihot esculenta*) is a monoecious, dicotyledonous perennial native to Brazil and is the crop from which tapioca is derived. When harvested, the bitter cultivar forms contain poisonous cyanogenic glucosides that must be leached from the roots before use. Sweet cultivar forms may contain some bitterness in the skin, but leaching is not required.

The roots are high in carbohydrate and vitamin C, and the leaves, which may be cooked as greens, contain protein and vitamins A and C.

The crop is adapted to the long growing seasons of the tropics. Growth is favored by temperatures above 65°F (18°C) and ceases below 50°F (10°C). Propagation is vegetative, placing stems cut from the previous crop vertically in the field. As the plant develops, root thickening will occur in response to short days.

The harvested roots are highly perishable and must be used within 2 or 3 days after removal from the ground unless refrigerated. They can be held for up to 2 weeks at 42 to 45°F (5 to 7°C) and 85 to 95 percent relative humidity.

Yam

The true yam (*Dioscorea* spp.) is grown primarily in tropical and semitropical areas for its tubers. Some species of *Dioscorea* contain saponins, and several have contents sufficiently high to be grown for use in synthesizing cortisone. Other species contain diosgenin, used in manufacturing steroid drugs. The primary food value is the carbohydrate.

Yams are propagated by small tubers or tuber pieces showing sprouts, or by rooted stem cuttings. The soil must be well drained, deep, and fertile, with rainfall between 50 and 200 in. (1300 to 5000 mm) annually. Growing seasons are long, from 6 to 8 months to 12 months for some species. As the plants develop, the vines are trained, either by erecting a trellis or by using trees and shrubs left from land clearing. Tubers are susceptible to bruising and must be harvested and handled gently. They can be stored in well-ventilated, shaded houses for up to 2 months with little loss.

Other Crops

Several other vegetables are grown for tuberous roots or tubers, including arrowroot (*Maranta arundinacea*) and edible canna (*Canna edulis*), both native to South America, and oca (*Oxalis tuberosa*) and ulluca (*Ullucus tuberosa*), both cultivated by the Incas in Peru. Yam bean (*Pachyrrhizus erosus*) and potato bean (*Pachyrrhizus tuberosus*) are also native to Central and South America. Chinese artichoke (*Stachys tuberifola*), a perennial, is native to the Far East where the tubers are used as white potatoes.

SELECTED REFERENCES

Baker, H. G. (1965). *Plants and Civilization.* Wadsworth, Belmont, Calif.

Bennett, A. H., R. L. Sawyer, L. I. Boyd, and R. C. Cetas (1960). *Storage of Fall-Harvested Potatoes in the Northeastern Late Summer Crop Area,* USDA Marketing Res. Rep. 370.

Edmond, J. B., and G. R. Ammerman (1971). *Sweet Potatoes: Production, Processing, Marketing.* AVI, Westport, Conn.

Harris, P. M. (ed.) (1978). *The Potato Crop.* Chapman & Hall, London.

Heiser, C. B., Jr. (1969). *Nightshades: The Paradoxical Plants.* Freeman, San Francisco.

Iritani, W. M., and L. D. Weller (1980). *Sugar Development in Potatoes,* Washington State University Coop. Ext. Bull. 0717.

Kleinkopf, G. E. (1979). *Translucent-End of Potatoes,* University of Idaho College of Agriculture Coop. Ext. Curr. Inf. Ser. No. 488.

Onwueme, I. C. (1978). *The Tropical Tuber Crops: Yams, Cassava, Sweet Potato, and Cocoyams.* Wiley, New York.

Salaman, R. N. (1949). *The History and Social Influence of the Potato.* Cambridge Univ. Press, New York.

Smith, O. A. (1977). *Potatoes: Production, Storing, Processing.* AVI, Westport, Conn.

Thornton, R. E., and J. B. Sieczka (eds.) (1980). *Commercial Potato Production in North America,* Potato Assoc. of Amer. Handb. *American Potato Journal* **57**: Supplement.

William, R. D., N. J. Tielkemeier, and L. H. Halsey (1978). *Growing Sweet Potatoes for Profit,* University of Florida Coop. Ext. Circ. 440-11.

Wilson, L. G., *et al.* 1980. *Growing and Marketing Quality Sweet Potatoes,* North Carolina State University Agric. Ext. Bull. AG-09.

Yamaguchi, M. 1983. *World Vegetables.* AVI, Westport, Conn.

Yen, D. E. 1976. Sweet potato. In *Evolution of Crop Plants* (N. W. Simmonds, ed.). Longmans, Green, London.

STUDY QUESTIONS

1. Contrast the development of shoots on a white potato tuber and a sweet potato root.

2. What environmental factors and cultural techniques affect the specific gravity of potato.

3. Describe the internal changes that take place in potato tubers and sweet potato roots during curing and storage.

4. What cultural techniques can be applied to potato or to sweet potato to maximize their Grade 1 yield?

17

Solanaceous Crops

Tomato Eggplant
Pepper Husk tomato

Within the family Solanaceae are some of our most significant plant species, including not only several major vegetable crops but also drug species (*Atropa belladonna*), tobacco (*Nicotiana tabacum*), and several important ornamentals. Many of the solanaceous species contain highly toxic alkaloids. Those in potato and tomato, solanin and tomatine, respectively, occur in toxic quantities in foliage and in certain other green tissues. In ripe tomato fruit, tomatine changes enzymatically to a nontoxic form. Potato tubers, however, can become toxic if exposed to sufficient light to cause greening of the skin.

Tomatoes and peppers, although not among the most valuable crops in nutrients per pound, are important contributors to dietary needs because of the substantial per capita consumption of each. In the United States, tomato consumption is exceeded only by that of potato. Peppers, although a significant commodity in the United States, have a higher per capita consumption in other countries.

TOMATO

Historical Perspective and Significance

The tomato is native to the Peru–Equador region of South America, evolving from the cherry form (*Lycopersicon esculentum* var. *cerasiforme*). It is not known by what means the plant migrated north to Central America, but it was there that plants with a tremendous diversity of size, color, and other features were selected and maintained by Mexican indian tribes.

The first historical mention of tomato, by Matthiolus in 1544, placed it in Italy. However, it was first introduced to Spain by explorers returning from South and Central America. The plant received little notice in Spain and probably was introduced to Italy by way of Morocco or Turkey. In Italy, it was termed *pomi d'oro* by Matthiolus, suggesting that the first introduction may have been yellow fleshed. Subsequently, the tomato became known as *poma amoris* in Italy and *pomme d'amour* in France, both translated as "love apple." The name *tomato* was derived from the Nahuatl language of Mexico (the name *tomatl* was used by the American Indians).

For a long period, the tomato was not popular because of the widely held belief that it was poisonous. It was not until the eighteenth century that its use as a food became accepted, and this acceptance in England and France was achieved only after frequent testimonials from botanists and gourmets.

The tomato was introduced to the United States in 1710 but was not reported as a food crop until later in the eighteenth century. By 1779, catsup was produced in New Orleans, and Thomas Jefferson reportedly was consuming fresh tomatoes. Yet, even as late as 1900, George Washington Carver was attempting to convince people of the safety of the tomato by consuming the fruit in full view of doubters.

In the years since 1900, the tomato has become the second most important vegetable crop, exceeded only by potato. The total U.S. tomato crop [approximately 415,000 acres (168,000 ha)] exceeds 8 million tons (7,300,000 MT), with 7,030,000 tons (6,377,616 MT) processed into juice, soup, catsup, sauce, whole tomatoes, and prepared foods, and 1,361,850 tons (1,235,470 MT) sold as fresh tomatoes. These statistics do not include small acreages producing a very large number of tomatoes for fresh use and home processing. The leading fresh market produc-

tion areas are Florida, California, and South Carolina, whereas California (84 percent of the total processed crop), Ohio, and Indiana are the leading processing states. In the 1950s, the average yield per acre was 8 to 12 tons (18 to 27 MT/ha). By the mid-1980s, the processing average exceeded 27 tons/acre (60.5 MT/ha) in California, and yields over 40 or 50 tons/acre (89.6 or 112 MT/ha) were no longer uncommon. Fresh market average yields are less than those for processing tomatoes because of stringent market requirements, but yields of 40 tons/acre (89.6 MT/ha) or more have been reported in the field, and up to 80 tons/acre (179 MT/ha) in the greenhouse.

The tomato contains significant amounts of vitamins A and C (Table 17.1), although levels of both are affected by environment. Ascorbic acid (vitamin C) is not as high in fruit from shaded plants as in those in strong sunlight. The carotenoids are affected by temperature (light intensity), but vitamin A (β-carotene) is relatively stable.

Classification, Growth, and Development

The tomato (*Lycopersicon esculentum*) is a tender warm season perennial cultivated as an annual. It is characterized by glandular hairs (trichomes) that emit a strong aroma when

TABLE 17.1

Nutritional constituents of the major solanaceous crops[a]

Crop	Water (%)	Energy (cal)	Protein (g)	Fat (g)	Carbo-hydrate (g)	A[b] (IU)	C[c] (mg)	Thia-mine (mg)	Ribo-flavin (mg)	Niacin (mg)	Ca (mg)	P (mg)	Fe (mg)	Na (mg)	K (mg)
Tomato	94	22	1.1	0.2	4.7	900	23	0.06	0.04	0.7	13	27	0.5	3	244
Sweet pepper	93	22	1.2	0.2	4.8	420	128	0.08	0.08	0.5	9	22	0.7	13	213
Hot red pepper	80	65	2.3	0.4	15.8	21,600	369	0.10	0.20	2.9	16	49	1.4	25	564
Eggplant	92	25	1.2	0.2	5.6	10	5	0.05	0.50	0.6	12	26	0.7	2	214

Source: National Food Review (1978), USDA.

[a] Data expressed per 100 g sample.

[b] 1 IU = 0.3 μg vitamin A alcohol.

[c] Ascorbic acid.

broken. Although considered to be day neutral, the tomato is not productive in long days without a diurnal temperature variation of at least 10°F (6°C). Maximum growth is achieved at a mean temperature of 75°F (24°C) day [range 70 to 84°F (21 to 29°C)] and 65 to 68°F (18 to 20°C) night. Temperature means below 60°F (15.5°C) or above 80°F (27°C) curtail growth, and brief exposures to extreme temperatures may impair fruit set. Cultivars differ in the lowest temperature at which fruit set will occur. Many early determinates set fruit at 40 to 50°F (4.4 to 10°C). Cultivars developed for warm growing conditions often will not set fruit below 50°F (10°C). At temperatures above 90°F (32°C), pollen and stigmatic surfaces may desiccate, causing poor set in many cultivars. Often, if fruit set occurs at extreme temperatures, physiological or anatomical damage may occur that detracts from product appearance and quality.

The growth types of tomato include indeterminate (vining), semideterminate, determinate (bush), and dwarf. Indeterminate and determinate plants both produce terminal inflorescences; but, in indeterminate types, every inflorescence is forced to a lateral position when the apical bud continues vegetative development and growth. In determinate plants, the apical bud does not develop; each stem thus ends in a flower cluster (selfprunes), and axillary shoots develop rapidly, producing a compact, bush plant type. Because flowers terminate each branch of determinate plants, the fruit are exposed to the heat of direct sun and tend to ripen early. Semideterminates basically are determinate plants in form, but they produce several lateral inflorescences before terminating in a flower cluster. Dwarf plants may be determinate or indeterminate, but the internodes are very compressed.

Tomato flowers are relatively small and consist of a five-lobed corolla and calyx (Figure 17.1). The staminal cone represents a fusion of five anthers around the ovary, style, and stigma. This structure, with release of pollen on the interior of the anther, ensures a high level of self-pollination and homozygosity. Pollination is not a function of insect activity but occurs as flowers vibrate from wind currents.

The fruit, botanically a berry, has two or more cavities (locules) containing seeds imbedded in a gelatinous matrix that softens as fruit reach mature size and seeds are fully developed. Small-fruited cultivars generally have only two locules; those with large fruit have many. Ripe fruit color may be red, pink, yellow, or orange, a function of several independent genes controlling either flesh or skin color. Red fruit color, for example, is conditioned by dominant genes for red flesh and yellow skin color. Pink fruit color differs only by the recessive gene responsible for colorless skin. In addition to these basic fruit colors, other genetic factors modify flesh color intensity (crimson, or high pigment) or chlorophyll expression in unripe fruit (green shoulder, uniform green).

Ripe color is a function of two carotenoid pigments, lycopene and β-carotene. A high ratio of lycopene : β-carotene is responsible for deep red, and, as the β-carotene fraction increases, the fruit become increasingly orange. It is only the β carotenoid, however, that is associated with vitamin A content. β-Carotene is synthesized at a wider temperature range than is lycopene. At high temperatures, lycopene synthesis slows, and normally red tomatoes take on an orange hue as the β-carotene, unaffected by the heat, accumulates. This phenomenon is observed frequently in determinate plants with exposed fruit or in other plant types defoliated by disease.

Environmental fluctuations at different developmental stages can cause several physiological or morphological defects that affect productivity or marketability of tomatoes (Figure 17.2).

Fruit Cracking   There are three types of fruit cracking—radial, concentric, and burst—all occurring under erratic moisture conditions. Some cultivars crack only under very

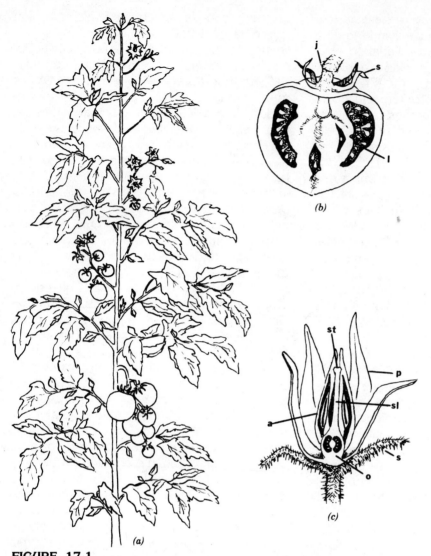

FIGURE 17.1

Morphology of (*a*) tomato plant, (*b*) fruit, and (*c*) flower. Sepal, s; locule, l, containing seeds and gel; stigma, st; petal, p; style, sl; ovary, o; anther, a; stem joint, j.

severe conditions. Resistance to radial cracking (radiating from the stem scar) is particularly effective. However, if a dry period is followed by a substantial rainfall, even resistant cultivars may show some damage. Concentric cracking, forming concentric rings on shoulders at the stem end of the fruit, occurs frequently, sometimes as a fine russet rather than as large skin ruptures, but normally before fruit are fully ripe. Concentric cracking is not as heritable as radial cracking, but many cultivars show some resistance. Burst cracking often

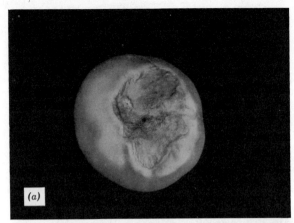

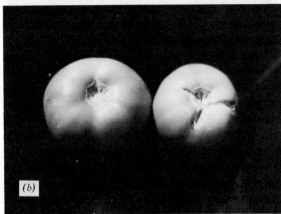

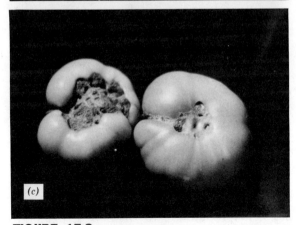

FIGURE 17.2

Tomato fruit defects: (*a*) blossom end rot; (*b*) radial cracking; (*c*) catfacing. (Photos courtesy of University of New Hampshire.)

occurs after harvest and usually is the result of handling. Such cracks generally are exacerbated by erratic moisture conditions, and affected tomatoes are unmarketable.

Catface and Blossom Scar Cold temperature during flowering and fruit set, and other stress conditions, including excessive heat, 2,4-D contamination, pruning, and erratic moisture, have been associated with catfacing or scarring of the blossom end of the fruit. This defect may be only a small scar of corky tissue, or it may occur as "whiskers" of corky tissue extending up the sides of fruit. Severe deformation is termed catfacing. Lesser damage is called blossom scar. This severe scarring usually is accompanied by a highly lobed and asymmetrical shape. Some cultivars are more susceptible than others.

Puffiness Occasionally, fruit develop locules that do not fill with gel, and such puffy fruit are very soft and unsuitable for fresh market. This condition often occurs in the fall, when cold temperature exposure is frequent but can occur when either high or low extreme temperatures cause poor fertilization in the flowers. It also may occur when nitrogen is excessive or when plants are field treated with ripening hormones.

Blotchy Ripening Blotchy ripening is characterized by failure of small to large areas of the outer fruit wall to ripen properly. The tissue remains hard with a white or yellow surface color, occasionally becoming brown in the interior (internal browning). No single agent has been implicated; improper fertilization (especially potassium), soil moisture levels, and virus disease have been associated with the condition. The frequency of occurrence also seems to relate to cultivar; some cultivars are totally immune to blotchy ripening under conditions normally favoring its appearance.

Yellow Top Yellow top affects green-shouldered cultivars and may appear after extended periods of hot or cold temperatures or after defoliation by infectious diseases. It appears as a patchy area of poor color on fruit shoulders.

Large Core, Green Gel This disorder can result from extended high or low temperature exposure, particularly with excessively high nitrogen fertilization. There are cultivar differences in tendency to show such defects.

Sunburn When the fruit are exposed directly to full sun, the exposed tissue bleaches white, becomes wrinkled, and often shows secondary infection by fruit rot organisms. Cultivars with satisfactory leaf cover seldom show sunburn.

Crop Establishment and Maintenance

In areas where soil moisture can be controlled and temperatures favor rapid germination, tomatoes are direct seeded, using precision seeders that place three to six seeds at a regular spacing. In heavy soils that tend to crust, vermiculite or phosphoric acid may be surface applied to enhance emergence. Normally one to three seedlings will develop in a clump; however, the competition within this clump does not reduce the total yield on an area basis. Gel seeders have been tested, but few are used commercially.

Most market gardeners and all tomato growers in the east and north utilize transplants. Over 800 million transplants are shipped from Georgia annually to northern growers, and other contract suppliers are located in Florida. Growers with forcing structures sow seeds for transplants 4 to 6 weeks before transplanting, sufficient time to develop a compact plant 4 to 6 in. (10 to 15 cm) tall with no flower buds. Larger transplants, showing buds or open flowers, may not develop to their full potential, since premature fruit set diverts photosynthate from vegetative growth, resulting in stunted plants with reduced productivity. The problem is especially severe in determinate cultivars characterized by restricted vine growth.

The most desirable soil for tomato production is a deep, well-drained fertile loam, but excellent crops have been grown in well-managed sandy loams and heavy clay loams free of hardpan. The tomato root system is deep and can extend to 10 ft (3 m) deep. Tomatoes are not productive on muckland, since the abundant nitrogen encourages excessive vine growth and poor fruit set. However, some organic matter added to mineral soils to improve physical characteristics will enhance tomato productivity.

Tomatoes tolerate a broad pH range, but the optimum is 6.0 to 6.5. Lime should be applied if the pH falls below 6.0. Soils exceeding 7.0 are not uncommon in the west, where deep plowing and leaching are used to provide a temporary adjustment of accompanying high surface concentrations of salts.

Fertilizer applications vary according to soil type, but ratios of 1–2–1 or 1–2–2 would be normal for an average northeastern soil, applying nitrogen at the rate of 75 to 100 lb/acre (84 to 112 kg/ha). It is recommended that one-half be applied before plowing, the remainder at planting. In California, nitrogen application rates exceed those for phosphate or potash, averaging 120 lb/acre (135 kg/ha). Phosphate and potash rates are 80 and 55 lb/acre (90 and 62 kg/ha), respectively. Florida rates are substantially higher, particularly for potash. Average rates are 210, 160, and 420 lb/acre (235, 179, and 470 kg/ha) for nitrogen, phosphate, and potash, respectively. Assuming a yield of 30 tons (27.2 MT), the plants remove 180 lb (82 kg) of nitrogen, 21 lb (9.5 kg) of phosphorus, and 280 lb (127 kg) of potassium, of which 100, 10, and 180 lb (45, 4.5, and 82 kg), respectively, are removed by the fruit. Fertilizer and lime are broadcast during field preparation or banded at planting. Sidedressings, principally of nitrogen and potassium, may be necessary for light soils. Starter solutions high in phosphorus are used frequently to stimulate rooting and establishment of transplants.

In sandy soils or those tending to be strongly acidic, deficiencies of magnesium, calcium, and, occasionally, other trace elements appear. These deficiencies, rare when fields received regular applications of barnyard manure, have increased as manure has become unavailable and as levels of nitrogen, phosphorus, and potassium have increased. Careful soil testing and tissue analyses are recommended in such areas to enable the grower to apply corrective fertilizer treatments.

The most common symptom directly or indirectly related to low pH and/or nutrient deficiency is blossom end rot. This physiological disease affects developing fruit, causing a leathery black patch at the blossom end which becomes a site for development of secondary rots. The fundamental cause of blossom end rot is lack of calcium in the fruit, caused by a lack of calcium in the soil or by soil moisture inadequate to move available calcium into the plant, and it is aggravated by excessive soil nitrogen. It also occurs when excessive soil moisture limits root respiration. Blossom end rot, frequent in sandy locations or in soils with low pH and unirrigated soils, can be alleviated by liming and regular irrigation. Dolomitic lime should not be used to correct pH where blossom end rot is a problem, and nitrogen and potassium sidedressings should be withheld.

Field spacing of tomato differs according to purpose of the crop (processing or fresh market) and the cultivar (vining or bush). Vine-type tomatoes are planted in single rows spaced 4 to 5 ft (1.2 to 1.5 m) apart, with plants generally 3 ft (0.9 m) apart in the row. Determinate plants may be grown in single or twin rows with plant density ranging from 6000 to 14,000 plants/acre (14,800 to 34,600/ha), compared with approximately 3000 (7400) for indeterminates. The high populations are typical of processing crops harvested mechanically, and the high plant density contributes to uniformity of flowering, fruit development, and concentration of fruit maturation.

Trellis tomatoes are popular in some areas, and both indeterminates and semideterminates are used. Indeterminates usually are trained on 6-ft (1.8-m) stakes, or on vertical strings stretched between two parallel horizontal wires, and are pruned to one or two leaders or main stems. Spacing between plants is approximately 20 in. (51 cm) in rows $4\frac{1}{2}$ to 5 ft. (1.4 to 1.5 m) apart. The semideterminates are becoming popular for a 3- to 4-ft (0.9 to 1.2-m) stake culture system. In this system, sucker growth is pruned only up to the first flower cluster, and the tying system contains all growth within horizontal strings pulled to opposite sides of alternate plants and secured on 1-in. (2.5-cm) square posts spaced at close intervals (Figure 17.3). Tomatoes from staked plants are more uniform in size and appearance than those harvested from prostrate plants, and there is less fruit damage from soilborne rot organisms. The consequent increase in percentage marketable grade and the ease of harvest are economic advantages that may justify the added cost of training and pruning. Similar results may be obtained using wire cages (livestock fencing) around individual plants. The support keeps developing fruit from contact with the soil, but the labor and material costs exceed those of low stake culture, and the system is recommended primarily for home gardeners.

Weeds are controlled with plastic mulch, chemical herbicides, and/or careful cultivation. In high-density plantings, cultivation may not be feasible, and most growers have adopted chemical control systems or black polyethylene. Several chemicals are available, including preplant incorporated (trifluralin), and postemergence or postplant applications (diphenamid, napromamide, and metribuzin). Metribuzin may damage certain genotypes or plants grown under certain cultural systems. Specific materials and precautions must be researched prior to use. Black plastic as a weed control is especially suitable for fresh market tomatoes, since the temperature modification often enhances early yield.

Although the tomato plant is relatively drought tolerant, it responds to a regular moisture supply. The amount of water required differs according to soil type and natural rainfall, but tomatoes should receive significant mois-

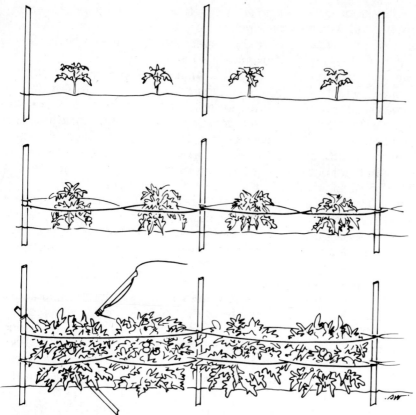

FIGURE 17.3

Short-stake (San Diego) system of trellising tomatoes. Plants are tied by weaving the twine to alternate sides of the row, securing at each stake.

ture at 5- to 10-day intervals if on loamy soils, more often if on sandy soils. Surface irrigation is preferred, since fruit and foliage remain dry, minimizing disease spread.

Cultivars

Cultivars differ in size and habit, maturity date, fruit size, color, shape, quality, disease and defect resistance, and general adaptation. Many cultivars are hybrids, generally more productive per vine than nonhybrids. Although seed is expensive, hybrids have proven to be cost effective for both market and processing-type tomato growers. A classification of fresh market, processing, and home garden tomato types is presented in Table 17.2.

In addition to the major economic factors for selecting a cultivar, a number of simple genetic traits have been introduced, some of which have contributed to quality of the harvested product. **Uniform ripening** eliminates the dark green shoulder. As the fruit ripen, they do so uniformly with no green streaks or yellow shoulders. **Jointless,** of which there are several forms, eliminates the "knee" or joint on the fruit stem (pedicel), or, if a joint appears, it does not form an abscission layer (Figure 17.4). Consequently, most fruit separate from the stem entirely when picked. At-

TABLE 17.2
Classification of tomato cultivars

I. Cultivars for processing (all red fruit)	
A. Determinate	
1. Paste type (high solids, firm, small fruit, for puree)	Roma
2. Table type (round or elongate, medium size, for whole or chunk pack)	Chico III
B. Indeterminate	
1. Pickling (small fruit; round, ovate, or pear shape)	Yellow Pear
II. Cultivars for fresh market	
A. Field grown	
1. Green wrap (indeterminate or semideterminate, red or pink color, globe or flattened globe shape)	Walter, Flora-Dade
2. Vine ripe (picked turning to full color, mostly globe)	
a. Red fruit	
i. Indeterminate	
• Early	Early Cascade
• Midseason	Better Boy
• Late	Beefsteak
ii. Determinate	
• Early	Springset Hybrid
• Midseason	Floramerica
b. Tangerine or yellow fruit	
i. Indeterminate (late)	Golden Queen
ii. Determinate (early)	Taxi
3. Cherry tomatoes (small, globe)	
a. Red fruit	
i. Indeterminate	Sweet 100, Large Red Cherry
ii. Determinate	Cherry Grande
4. Window box or patio (small determinate or dwarf)	Patio
B. Greenhouse grown (indeterminate; red or pink; globe)	Vendor

tached stems are a frequent cause of fruit punctures during bulk handling. **Color intensifiers** include dark green, high pigment, and crimson genes. Each gene changes the lycopene: β-carotene ratio, enhancing red color. The crimson gene, however, enhances red color by reducing β-carotene, thereby reducing vitamin C content.

Disease and Insect Pests

Many pests affect tomato, some of which are identical, or nearly so, to those affecting potato. The common pests include several bacterial diseases, those caused by fungi (early blight, late blight, *Rhizoctonia*, *Verticillium* and *Fusarium* wilts, and others), and sev-

eral viruses (Figure 17.5). The major insects are Colorado potato beetle and the ubiquitous aphids, leaf hoppers, flea beetles, tomato fruitworm and hornworm, and cutworms (Figure 17.6).

Diseases *Fusarium Wilt* The wilt caused by the fungus *Fusarium oxysporum* f. sp. *lycopersici* is a common vascular disease in all but the most northerly production areas. The fungus is soilborne and attacks plants through the root systems, eventually causing plugs (tyloses) to form in vascular tissue. These plugs restrict water and nutrient uptake. As a result, the leaves wilt, usually preceded by drooping and curving of oldest leaves. Yellowing also is

FIGURE 17.4

Jointed (*a*) and jointless (*b*) tomatoes. Normal separation is at the swollen joint as the abcission layer forms. Jointless stems form no abscission layer. (Photos courtesy of Peto Seed Company, Saticoy, Calif.)

typical, sometimes confined to one side of a stem. The vascular tissue becomes typically discolored. The only effective control is resistance, and cultivars are available that resist either or both races 1 and 2.

Verticillium Wilt   Similar in effect to *Fusarium*, the organism *Verticillium dahliae* (*albo-atrum*) has a broad host range. Unlike *Fusarium*, it is most severe in the northern areas. The symptoms progress from intense yellowing of old leaves to slight, then severe wilting. The yellowed leaves drop off until the center of the plant becomes totally defoliated. The vascular system shows a discoloration similar to that caused by *Fusarium*. The most effective control is resistance.

Bacterial Wilt   Bacterial wilt (*Pseudomonas solanacearum*) is predominantly a southern disease but has occurred elsewhere in some seasons. This disease causes sudden wilting with no yellowing or necrosis. The stem pith gradually decays, leaving a hollow stem. The organism is soilborne, infecting the roots and stem, then moving into the vascular system. Management through rotation, control of drainage, and strict sanitation will minimize the problem.

Bacterial Canker   Canker disease, caused by *Corynebacterium michiganense*, can be introduced by contaminated seed and contaminated transplants. The disease often affects one side of the plant, causing wilting and necrosis of lower leaves, and streaks extend down the stems and the underside of the petiole. These streaks eventually break to form cankers. In late stages, the stem interior becomes yellow, and fruit may develop characteristic dark rough spots surrounded by a light halo. Fermentation of seed pulp prior to seed cleaning helps to eliminate surface contamination, and clean seed is the best control.

Bacterial Speck   Bacterial speck (*Pseudomonas syringae*) causes lesions of fruit and stems. The specks on green fruit are small, sunken black spots surrounded with dark green halos. The disease is favored by cool, moist weather, when the bacteria are spread easily by mechanical means or by rain and wind. It is a common problem in Florida and in other areas providing prolonged periods of moist weather. Maneb and copper sprays have been recommended for control.

Bacterial Spot   An organism causing fruit lesions similar to speck (*Xanthomonas vesica-*

FIGURE 17.5

Diseases of Solanaceae: (*a*) anthracnose; (*b*) bacterial speck; (*c*) cucumber mosaic on tomato; (*d*) bacterial spot on pepper. (Photos (*a*), (*b*), and (*c*) from *Diseases of Plants*, slide set 33, Tomato Diseases. Photo (*d*) from *Diseases of Plants*, slide set 37, Diseases of Carrots, Eggplant, Peas, Pepper, Sweet Corn. Reproduced by permission of the American Phytopathological Society.)

toria) also thrives in moist climates, generally at temperatures of 75 to 86°F (24 to 30°C). All parts of the plant can be infected. The spots on leaves and stems tend to aggregate. On fruit, the spots become raised and scabby. Seed treatment with acid or chlorox will reduce inoculum, and maneb–copper sprays applied prior to infection will provide some protection. Workers should not handle plants while wet, since the bacterium is transported easily under those conditions.

Early and Late Blights Blights (caused by *Alternaria solani* and *Phytophthora infestans*, respectively) on tomato show symptoms similar to those on potato (Chapter 16). Early blight, particularly, is an inevitable problem, appearing as fruit loads become heavy. Both fungus diseases can be controlled with carbamates and copper sprays. Resistance has not been effective for late blight, but some cultivars show reduced early blight infection, enabling growers to use fungicides effectively.

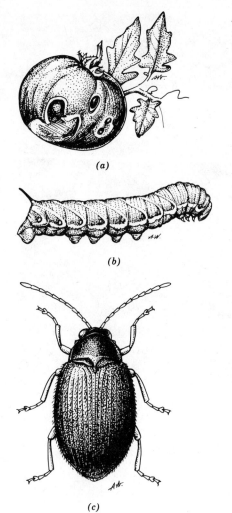

FIGURE 17.6

Insects of Solanaceae: (*a*) tomato fruitworm damage; (*b*) tomato hornworm; (*c*) flea beetle (black and covered with hairs).

Septoria Leaf Spot   *Septoria* leaf spot (caused by *Septoria lycopersici*) is particularly severe in the central United States. Leaves develop dark spots with gray, sporulating centers. The organism also attacks several solanaceous weeds. It is spread by splashing water and is seldom a problem in dry areas. Carbamates have been used successfully to control *Septoria* blight.

Anthracnose   Anthracnose (*Colletotrichum phomoides*) is common late in the harvest season, appearing as discolored sunken spots on the fruit and as dark spots on the leaves. The fungus is favored by moist weather and appears first as small spots which enlarge symmetrically, with a target board effect. The organism overwinters on infected debris, and although carbamates can provide good control, it is often difficult to spray such materials during the harvest season. A 3-year rotation is helpful in keeping inoculum levels low.

Virus Diseases   Viruses include tobacco mosaic, cucumber mosaic, curly top, spotted wilt, single virus streak, and double virus streak. Mycoplasma diseases, particularly aster yellows, also occur periodically. All of these diseases are controlled by sanitation and good weed and insect control, and resistance has been obtained for tobacco mosaic and curly top viruses. The symptoms of each virus are reasonably distinct (Table 17.3).

Other Diseases   There are several other diseases that may occur in specific production areas or under certain weather conditions. **Rhizoctonia** (*Rhizoctonia solani*), causing a soil line decay of tomato stems and subsequent wilting and death, occurs in prolonged periods of mild, moist weather and is a major problem in the southern states. Chemical control is not feasible, but the problem can be minimized by well-drained soils and adequate aeration among the plants. **Phoma rot** (*Phoma destructiva*) and **buckeye rot** (*Phytophthora* spp.) are the principal fruit decay organisms in the field. They are particularly active during moist weather and where fruit can be spattered by soil. Chemical control becomes difficult during the harvest season because of potential residual problems, and spray programs must be planned carefully.

Gray leaf spot (*Stemphylium solani*), common in the southeast, affects only foliage and is favored by warm moist conditions. Older

TABLE 17.3
Virus disease symptoms in tomato

Virus	Vector	Symptoms
Tobacco mosaic	Mechanical	Leaf mottle; deformed flower and fruit (early infection); internal browning
Cucumber mosaic	Aphid, beetle	Shoestring, straplike leaf
Curly top	Beet leafhopper	Upward leaf curl, leathery foliage, yellowing
Spotted wilt	Mechanical	Dark round dead spots on bronzed leaves, fruit spots with concentric yellow and red zones
Single virus streak	Mechanical	Mottling, occasional broad stem streaks; fruit with depressed round rings
Double virus streak	Mechanical	Combined effect of potato virus X and tobacco mosaic virus; severe spotting of leaf and fruit; streaks on stem

leaves are the first to become infected, show-ing small gray spots that become glazed and often crack. Measures used to control early blight are applicable to gray leaf spot. **Leaf mold** (*Cladosporium fulvum*) is very destruc-tive in greenhouses, developing quickly in the high humidity and limited air circulation. It ap-pears first as diffuse white spots on upper leaf surfaces, and lower surfaces become covered with an olive-brown fungus. Regular applica-tions of carbamate sprays (maneb, zineb) are recommended. **Black mold** (*Alternaria al-ternata*) is caused by a weak ubiquitous fun-gus that infects only ripe tomatoes, usually after a period of rain. Treatments nor-mally applied for buckeye rot should minimize black mold. **Nailhead spot** (*Alternaria tomato*), closely related to early blight, first appears as small, irregular dark spots on lower leaves. On fruit, the spots become slightly sunken, sur-rounded by a green halo after ripening begins. Many cultivars resist infection, and sprays for other diseases will be effective against this or-ganism.

Insects   Several insects, each with a broad host range, may infest tomatoes, including cutworms, flea beetles, aphids, leafhoppers, stink bugs, leaf miners, spider mites, and the fruit fly. Other insects, restricted to tomato or related crops or with a more narrow host range, include Colorado potato beetle, tomato

hornworm, corn borer, and tomato fruit worm. Occasionally, such insects as the pinworm and the stem borer cause damage, particularly on plants adjacent to grassy areas.

Flea Beetles   Flea beetles (*Epitrix cucu-meris*) feed on foliage early in the season, of-ten as soon as true leaves appear or when transplants have been field planted. The adults overwinter in protected areas and become ac-tive as surface soil temperatures reach 70°F (21°C). Their feeding produces a "shot hole" effect that diminishes the photosynthetic area and can, if severe, stunt early growth. Several insecticides (carbaryl, methoxyclor) are effec-tive controls.

Colorado Potato Beetle   Of all insect pests of solanaceous crops, the Colorado potato beetle (*Leptinotarsa decimlineata*) is becom-ing one of the most serious pests of solana-ceous plants and will destroy a tomato plant in a short time. This insect is described in detail in Chapter 16. Control methods must be ap-plied carefully, since the insect population rap-idly develops immunity to chemicals. Most growers utilize several materials in the course of a season to achieve economic control.

Aphids   The most common aphids on So-lanaceae, the green peach aphid (*Myzus persi-cae*) and the potato aphid (*Macrosiphum so-*

lanifolii), cause injury by direct feeding and by injecting toxins as they feed, resulting in deformation of the leaves and stems. With large aphid numbers, the "honeydew" excreted and deposited on leaves and fruit becomes visible and supports development of unsightly sooty mold fungus. Some virus diseases also may be transmitted by aphids. The green peach aphid has a broad host range, but its primary host is the peach. Infestations in tomato usually occur as a migration from peach orchards. Controls include diazinon, malathion, and other organophosphates.

Leafhoppers   There are several species of leafhopper affecting tomato. The most frequent pest is *Circulifor tenellus,* a migratory insect that overwinters in warm locations, moving northward during the growing season. These leafhoppers are sucking insects, important vectors of virus and mycoplasma diseases (curly top, aster yellows). Control is difficult because of insect mobility and the wide host range. Regular insecticide applications and elimination of weedy border areas are recommended.

Tomato Hornworm   The hornworm (*Protoparce quinquemaculata*) is a voracious feeder, the larva of a large moth. Full-grown larvae are up to 4 in. (10 cm) long, are green with diagonal white stripes on abdominal segments, and are characterized by a single horn on the tail. Serious losses seldom occur, since the insect is vulnerable to several insecticides used for other larval insects.

Tomato Fruitworm   The fruitworm (*Heliothis armigera*), alias the corn ear worm, migrates as a light brown moth from southern locations. The larvae are brown with two narrow dark dorsal stripes. The insect bores into the fruit, and the tissue normally becomes infected with rot fungi or attracts other insects, such as sap beetles and fruit flies.

Other Insects   Other than the general feeders, the damage caused by leaf miner (*Li-riomyza pusilla*), corn borer (*Pyrausta nubilalis*), spider mite (*Tetranychus bimuculatus*), and common stem borer (*Papaipema nebris*) is sporadic or confined to specific regions. The fruit fly (*Drosophila melanogaster*) principally is a postharvest problem, with the insect attracted to fermenting tissue. Tomatoes that have cracked become infested as the flies lay their eggs on exposed flesh. Processors especially are concerned with contamination of their raw product.

Harvest and Market Preparation

Fresh market tomatoes are harvested at specific stages of maturity determined by proximity to market. Those for local sale are vine ripened to full red color before harvest. Those for distant market are picked at full size, but green (mature green), and these fruit are usually ripened in 100 to 150 ppm ethylene in a closed 70°F (21°C) storage for 48 h prior to shipping or distribution. The maturity classes are listed in Table 17.4. Commercially, most fresh tomatoes are picked at breaker to mature green stages. Market gardeners normally harvest at the pink to firm ripe stages. Preripe tomatoes tend to be lower in soluble solids and reducing sugars than vine-ripened tomatoes.

Processing tomatoes are picked fully ripe. Those utilized for pureed products (soup, juice, sauce) are left on the vine until over 85 percent of the fruit are ripe. Those for whole pack must be picked while still firm, and only 65 percent of a crop may be ready to pick at a single harvest. Some growers spray the plants with ethephon to accelerate rate of ripening, thereby increasing the percentage that can be harvested at one time.

Harvest of processing tomatoes is completely mechanized (Figure 17.7), using equipment that cuts the plant at or just below ground level, conveys it to a shaker assembly to remove fruit, and sorts with the aid of several workers. Fruit are trucked to processors where they are emptied in a flume conveyer system that cleans while moving fruit to sorting lines.

TABLE 17.4
Tomato maturity classes and criteria

Maturity stage	Days to full color[a]	Criteria
Immature green	19–23	No red or yellow; skin dull green; seeds not fully developed, white in color, gel not formed; skin rubs off; harvested fruit will not ripen properly
Mature green	9–13	No red color; seeds developed and brownish, gel formed; skin bright or whitish green, glossy and waxy; blossom end whitish
Breaker	8	Trace of color, usually at blossom end, starting as a slight yellow to pink blush
Turning	7	Pink, starting from blossom end, covering 10–30% of the fruit.
Pink	6	More than 30% but less than 60% color
Light red	3	Color covering 60–90% of the fruit
Firm ripe	1	Well colored, fruit not soft
Table ripe	0	Full color, fruit softening

Source: Kader *et al.* (1977); Yamaguchi (1983).

[a] As measured at 20°C.

Harvest of mature green tomatoes is still predominantly a hand operation, although mechanization has been introduced. Mechanized harvest yields for fresh market tomatoes are reduced by 25 to 45 percent in comparison with hand harvest. All harvests for local market are by hand labor.

Full-ripe tomatoes can be stored at a temperature of 55°F (13°C) for up to several days.

FIGURE 17.7

Mechanical harvest of processing tomatoes. (Photo courtesy of University of New Hampshire.)

Tomatoes, especially those not fully ripe, exposed to temperatures below 55°F will not ripen properly, producing poor colors and off flavors.

Tomato quality factors differ according to the purpose of the raw product. Fruit for puree are evaporated to approximately 35 percent moisture for storage and transit, eventually to be reconstituted into juice, soup, and other consumer products. Thus, the desired raw product should be high in solids and viscosity, thick walled, and firm with good color and low pH. The sugar:acid ratio and solids content are influenced by the thickness of the outer wall relative to locule size. Plum types generally have high solids and sugar:acid ratio. For fresh market, appearance is important, and a tomato must be attractive in shape, color, and size, be free from defects, and have a "table quality" flavor with no textural mealiness. Soluble solids and acidity largely determine tomato flavor. Tomatoes termed "low acid" may differ little from other cultivars in pH or titratable acidity; the impressions of mild flavor apparently relate instead to a higher content of sugars.

Acidity is an important factor in processing, since it decreases the chance of contamination by the anaerobic bacterium *Clostridium botulinum.* In general, pH values below 4.5 are satisfactory. Those exceeding 4.8 may require acidification to ensure a safe pack. Acidity is affected by environment and by fruit maturity. Organic acids increase during ripening, falling as the fruit softens. Diseased plants produce fruit higher in pH than healthy plants. Acidity is maximized for any cultivar by harvesting at peak maturity from healthy vines. Likewise, the best fresh market quality is obtained by harvesting near or slightly before peak ripeness and moving the product to the consumer without delay and with no physical damage.

Mature green fruit are bulk-packed in ventilated fiberboard or wooden containers for shipment. Later, they are regraded and packed in tubes or trays covered with cellophane acetate. Fruit that are pink are marketed in lugs packed in two or three layers.

Greenhouse Tomato Production

Greenhouse tomato production has been a small but significant industry for many years. The techniques, however, have changed substantially, especially with the advent of soilless media, hydroponics, and energy conservation.

Tomatoes can be forced at temperatures of 70 to 80°F (21 to 27°C) day and 65°F (18°C) night. The most common forcing structure, largely replacing glasshouses, is the twin-layer plastic greenhouse with forced ventilation. The plants are grown in soil or, most frequently, soilless media or sand. Soilless media may be placed in fixed benches, or plastic bags of medium may serve as the plant container (Figure 17.8).

Both sand culture and soilless culture systems usually employ automatic feeding and watering systems. Such systems, using a modified Hoagland's solution, must be adjusted for pH and nutrient level periodically to ensure optimum growth. Slow-release fertilizers also have been used successfully.

Indeterminate cultivars are started in Speedling or similar containers and transplanted to the greenhouse in staggered twin rows 12 to 16 in. (30 to 40 cm) apart with

plants spaced 16 to 18 in. (41 to 46 cm) apart to maximize space for the root system. When bags of medium are used, the planting system is the same, transplanting in single or double staggered rows of holes punched through the top face. The larger the bag, the greater the yield, and bags made of dark plastic warm faster than those of white plastic. Twine is tied to a wire placed 6 to 7 ft (2.7 to 3.2 m) above the plant and to the plant base. As the plant grows, the twine is wrapped in a continual spiral around the stem to provide support. Sucker growth from leaf axils must be removed weekly before the growth of the axillary shoots exceeds 2 in. (5 cm). As flowering begins, a mechanical vibrator is applied daily to each flower cluster to pollinate.

Insect and disease control methods are the same as those applied to field tomatoes, except that the intensified environment encourages rapid development of pest problems. Regular applications of pesticides to

FIGURE 17.8

Greenhouse tomato production. Flowers require daily vibration to ensure pollination. (Photo courtesy of W. Bauerly, Ohio State University.)

control aphids, leaf mold, and white fly, in particular, are essential.

The fruit are harvested by hand when pink to firm red, leaving stem and sepals attached. The fruit then are graded for uniformity and marketed in cardboard containers, preferably separated into individual cells. Approximately 3 lb can be produced per square foot (4.5 kg/m^2) in the fall, 5 to 6 lb (7.4 to 8.9 kg) in the spring.

PEPPER

Historical Perspective and Current Status

The garden pepper is used throughout the world as a food or condiment. It originated in Mexico and neighboring areas of Central America where several species had been used by Indians from 3400 to 5200 B.C. Of these species, two have become particularly important: *Capsicum annuum* (both hot and sweet pepper) and *C. frutescens* (tabasco pepper).

Columbus discovered pungent peppers in 1493 and reportedly used them as food and spice a year later. The popularity of peppers spread quickly through Europe and Asia, and reports suggest that the uses went beyond that of food. Some cultivars, so pungent that they cause skin burns, were used in forms of torture. Red pepper juice has been used in recent times as a repellent in spray guns to deter attacking dogs and people.

For medicinal purposes, peppers were deemed once to be effective for curing dropsy, colic, toothache, and cholera, and early advertisements for tabasco sauce extolled its potential to relieve headache, neuralgia, and rheumatism. In modern medicine, extracts are used as a counterirritant for rheumatism and as a constituent of some throat gargles and lozenges. Petroleum jelly used to soothe muscle aches contains an active ingredient from pepper.

The active principle in pepper responsible for its pungency is capsaicin, a phenolic closely related to vanillin. It is not located in all parts of the fruit, and cultivars differ widely in amounts contained.

The tabasco pepper was introduced from Tabasco, Mexico, following the Mexican War of 1846 to 1848. Seeds given to Edward McIlhenny were grown at Avery Island, Louisiana, and used for a spicy sauce. After the War between the States, McIlhenny developed and marketed the sauce. It is today the only name under which tabasco sauce is manufactured, and the production and processing still are confined to Avery Island. The sauce is made by macerating the peppers, adding salt, and aging in barrels for at least 3 years. It then is mixed with vinegar, strained, and bottled.

There are several distinctive cultivars of pepper, each with a specific use. Products include paprika, a mild spice of European origin, and chili, derived from peppers grown in Mexico and the southwest United States. Peppers bearing their fruit well above the foliage have been developed for ornamental use.

Although pungent peppers predominate in Latin America and Asia, the sweet bell pepper is the most popular type in the United States. Its quality is controlled primarily by a single recessive gene that eliminates the capsaicin. These sweet peppers are harvested normally at a mature green stage but can be allowed to ripen to full red or yellow. The vitamin content of pepper (A and C) is superior to that of other solanaceous vegetables and tends to increase with maturity (Table 17.1). Red pepper food value is especially high, reflecting increased carotenoids and the low moisture content of the fruit at maturity.

China leads world production of peppers, nearly six times that of the United States. Total U.S. production of green peppers is 5,875,000 cwt (266,490 MT), with Florida and California accounting for nearly 65 percent. Approximately 71,000 acres (28,755 ha) of green peppers is harvested in the United States. In addition to the green pepper production, chilis, pimientos, paprikas, and tabasco and assorted hot peppers are grown, largely for processing and prepared foods.

Classification, Growth, and Development

Peppers are tender perennials grown in temperate regions as annuals. Their morphology is similar to that of tomato, although the appearance differs. Roots are fibrous, and top growth of shiny glabrous simple leaves generally is more compact and erect than in tomato. Flowering is solitary or in twos or threes, and flowers are greenish white with anthers not fused in a tight cone. In *Capsicum frutescens,* the calyx is cup-shaped, and the fruit, borne singly or in pairs or clusters, are linear and erect, appearing above the foliage. In *C. annuum,* the calyx is saucer-shaped, and the fruit may be erect or pendant. The fruit is a berry with large locules devoid of gel, and the seeds are tightly compressed to the central stalk. The capsaicin content is restricted to glands on internal partitions or ribs and to the placental tissue. Outer wall tissue is not pungent.

The optimum temperature for pepper growth and development is higher than that for tomato. Fruit set does not occur below or above the range 75 to 86°F (16 to 32°C), and maximum set occurs at 60 to 70°F (16 to 21°C). Although there is some evidence that pepper plants are drought resistant, fruit set tends to be depressed by any extreme environmental conditions.

Crop Establishment and Maintenance

Peppers may be direct seeded or grown from transplants. The appropriate system depends on length of the growing season and on soil temperature. Pepper seed germinates very slowly if at all in cold soils, but emergence accelerates in soils 75 to 86°F (24 to 30°C). In many production areas, warm soils do not exist in the spring, and transplants are started in greenhouses, hotbeds, or, in mild climates, in outdoor seedbeds 6 to 8 weeks prior to field planting. Prior to field planting in areas of environmental stress, pepper transplants should

be hardened, but not excessively so. Pepper growth can be delayed substantially by over-hardening.

Sandy loams or loams are preferred for field production. Soil preparation and fertilization are generally as described for tomato, although peppers utilize lower amounts of nitrogen, phosphorus, and potassium. Uptake, particularly of nitrogen, accelerates in the last two-thirds of the season. Therefore, one or more sidedressings and irrigation sufficient to move the nitrogen into the plant will enhance performance, particularly on light soils or those subject to leaching. Pepper fruit developing under erratic water conditions or low pH often show blossom end rot, rendering the fruit unmarketable.

Raised beds are used in some areas because of furrow irrigation, in others to ensure drainage. For direct seeded crops, raised beds allow improved control of surface moisture, thereby reducing the chance of infection by soilborne organisms. Single, or more often, twin rows are planted on beds spaced 40 in. (1 m) on center, with plants 14 to 18 in. (35 to 46 cm) apart in the row and 14 in. (35 cm) between the twin rows. The increased density of foliage in the twin rows improves shading of the fruit, reducing the incidence of sun scald. Direct seeded peppers established by clump planting may require some thinning, since the competition from more than two or three plants will reduce fruit size.

Pepper yields often suffer because of poor fruit set, in part related to elongation and desiccation of the style and, consequently, poor pollination. Hybrid cultivars appear to be superior in fruit set, and flower abortion also has been reduced by using irrigation and black plastic mulch.

Weeds can be controlled by cultivation, but, with increased population densities, herbicides now are used frequently. Such treatments are more effective than cultivation in maintaining weed suppression into the harvest season. Materials used for tomato generally are approved for pepper, although peppers are more sensitive than tomato to selective herbi-

TABLE 17.5
Classification of commercial peppers

Capsicum annuum	
Bell group:	Large, three- to four-lobed, blocky fruit; immature color green, turning red (some gold or yellow) at maturity
Sweet	Lady Bell, Yolo Wonder, Golden California Wonder
Pungent	Bull Nose Hot, Rumanian Hot
Pimiento group:	Large, thick-walled, green conical or heart-shaped fruit, turning red at maturity
Sweet	Pimiento, Pimiento Select
Cheese group:	Small or medium, generally flat fruit, round or irregular surface, medium to thick wall; green or yellow, maturing to red
Sweet	Sunnybrook, Cheese
Ancho group:	Large, thin-walled, heart-shaped fruit; green or dark green, ripening to reddish brown or dark brown
Mildly pungent	Mexican Chili, Mulatto
Anaheim group:	Long, thin, green tapered fruit, maturing to red
Sweet	Paprika
Slightly pungent	New Mexican Chili
Mildly pungent	Anaheim Chili
Pungent	Sandia
Cayenne group:	Long, thin, tapered green fruit, turning red and wrinkled at maturity; thin wall
Very pungent	Long Red Cayenne
Cuban group:	Yellow-green large fruit, turning red at maturity; fruit irregular and blunt with a thin wall
Sweet	Pepperoncini
Mildly pungent	Cuban, Cubanelle
Jalapeno group:	Fruit rounded, slightly conical with smooth dark green skin turning red at maturity; thick wall, occasionally russet skin
Very pungent	Jalapeno
Small hot group:	Small, green, thin-walled fruit turning red at maturity
Highly pungent	Serrano, Red Chili
Cherry group:	Small, round or slightly flattened green fruit, red at maturity
Pungent	Large Red Cherry
Wax group:	Fruit small or large, yellow turning red or orange red, conical to rounded conical
Sweet	Sweet Banana
Pungent	Hungarian Wax
Capsicum frutescens	
Tabasco group:	Thin yellow or yellow-green fruit, becoming red
Highly pungent	Tabasco

Source: Based on Erwin (1932); Greenleaf, (1986).

cides. Fresh market growers have found black plastic to be excellent for weed control.

Cultivars

In addition to the cultivars classified as bell types, there are others which have been grouped according to pungency, appearance, and use (Table 17.5, Figure 17.9).

The differences among cultivar groups, particularly among sweet peppers used mature green, have become obscured as new intermediate cultivar types have been developed. Of the groups listed, the major fresh market cultivars are within the bell group, whereas the pimiento group includes the major processing cultivars.

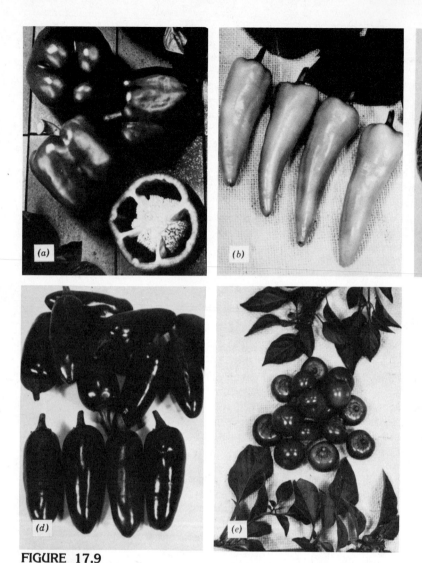

FIGURE 17.9

Cultivar types in pepper: (*a*) sweet bell; (*b*) sweet banana; (*c*) cayenne; (*d*) jalapeno; (*e*) red cherry. The latter three are hot peppers. (Photos courtesy of Peto Seed Company, Saticoy, Calif.)

In recent years, hybrids have proven superior under a wide range of environmental conditions. Fruit set, erratic in standard cultivars, generally has been reliable in hybrids.

Disease and Insect Pests

The disease problems described for tomato are the major concerns in pepper production, and the control recommendations are identical. Insect infestations also involve the same insect pests as described for tomato, except that there may be slight differences in preference. For example, the Colorado potato beetle will feed on pepper, but it generally prefers the other solanaceous vegetables, particularly potato and eggplant. Regardless, control measures must be applied as described for tomato.

Harvest and Market Preparation

Bell peppers are harvested when full size but still fully green (mature green). The fruit should be firm and crisp in texture. There is an increasing market for fully colored (red or yellow) peppers; these must be marketed promptly to avoid excessive fruit softening. Chili and paprika peppers are allowed to remain on the vine until most are fully ripe and may be left on the plant to air dry. Peppers are picked by hand for fresh market, and those for processing may be hand picked or harvested mechanically in a single destructive harvest.

Bell peppers are wiped clean, graded for size, and packed in paper cartons. Fruit can be maintained in good condition for up to 2 weeks at 45 to 50°F (7 to 10°C) and high relative humidity. Temperatures lower than 45°F will result in surface breakdown (chilling injury) shortly after exposure.

Chilis and paprikas are dehydrated artificially in most production areas, although some sun drying may be feasible. Controlled drying is preferred, since exposure to excessive heat results in a dried product with undesirable dark color.

EGGPLANT

Historical Perspective and Current Status

The eggplant is native to southeast Asia, probably India. The ancestral form was very likely a spiny plant with small, bitter fruit, but selection for improved palatability and for relative spinelessness resulted in gradual emergence of an acceptable type. There are several names by which the crop is known in India, but *brinjal* is the most familiar. The name *eggplant* is believed to derive from Gerard's description[1] of early forms with small, white fruit resembling eggs. However, in early years, eggplant also was termed *mala insana* and the Italian *melazana,* both of which translate to "mad apple." It was common belief that ingestion of eggplant could cause madness. An early authority[2] warned that eggplants

> engender melancholly, the Leprosie, Cancers, Piles, Imposumes, the Headache, and a stinking breath, breed obstructions in the Liver and Spleene, and change the complection into a foule blacke and yellow colour, unlesse they be boyled in Vinegar.

Others recommended limiting their use to ornamental purposes only.

Introduced to American gardens in 1806, it was primarily an ornamental curiosity until the present century. Although never a major crop, it has gained popularity and now is grown to some extent in most areas of the United States. Total U.S. production on 5748 acres (2328 ha) averages between 70 and 75 million lb (31,752 to 34,000 MT) annually, more than two-thirds of which is from Florida's winter and spring crops. Mexico imports almost one-third of our consumption, and the remainder is largely from New Jersey and California. Per capita consumption is only 0.5 lb (0.2 kg). Eggplant generally is limited to areas with a growing season of 4 or more months, although there are a few short season cultivars that can be grown successfully in northern states.

Classification, Growth, and Development

The eggplant (*Solanum melongena*) is a tender perennial grown as an annual and is characterized by a bush, indeterminate erect plant, 2 to 4 ft (0.6 to 1.2 m) in height. The leaves are large, lobed, hirsute on the underside, occasionally bearing sharp spines. The lavender flowers, borne singly or in small clusters, are similar to tomato, but larger. The fruit may vary in shape (oval to oblong) and color (purple, purple-black, white, red). The seeds

---

[1] John Gerard, *The Greate Herbal* (1596).

[2] John Parkinson, *Theatrum Botanicum* (1640).

are scattered through the fruit, imbedded in a firm placenta.

Of all the solanaceous vegetables, eggplant is the most sensitive to low temperature. Daily means should be in excess of 65°F (18°C), with 75 to 85°F (24 to 29°C) day and 65 to 76°F (18 to 24°C) night preferred.

Crop Establishment and Maintenance

Eggplant can be direct seeded, but most of the crop is established with transplants. The field soils should be well drained and fertile, preferably sandy loams of pH 5.5 to 6.5, with organic matter added. Soil preparation and fertilization requirements are similar to those of tomato and pepper. Because the crop requires a long season, several sidedressings of nitrogen fertilizer, at 2- to 3-week intervals, are advised.

Plants are grown 2 to $2\frac{1}{2}$ ft (58 to 76 cm) apart in single rows spaced $3\frac{1}{2}$ to 4 ft (1 to 1.2 m) apart, either on flat ground or on raised beds. Weeds are controlled by cultivation, herbicides, or black plastic mulch, the latter also enhancing early productivity of eggplant.

Cultivars

There are two basic types of eggplant, based on shape (Figure 17.10). The standard oval shape, typified by 'Black Beauty,' is characterized by an attractive, glossy black skin. The oriental types are elongate, also with purple-black skin, and seem to have improved fruit set under a range of environmental conditions. Quality of both is excellent. Fruit skin colors include not only the black typical of commercial cultivars, but also golden yellow, white, light green, and brown, most of these used in home gardens.

Disease and Insect Pests

Diseases   In addition to those foliage and vascular diseases described for tomato, eggplant is susceptible to several fruit diseases, the most important of which is caused by *Pho-*

(a)

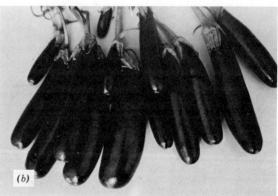

(b)

FIGURE 17.10

Oval-type (*a*) and elongate (oriental) (*b*) eggplant fruit. (Photo (*a*) courtesy of Harris–Moran Seed Company, Rochester, N.Y. Photo (*b*) courtesy of Peto Seed Company, Saticoy, Calif.)

mopsis vexans. This disease appears first as a seedling blight, but the fruit rot stage in the field and in transit is the most serious. The organism is seedborne and also overwinters in infected debris. It infects young seedlings at the soil line, girdling the stem. On fruit, it causes a watery rot that eventually mummifies and produces abundant pycnidia. Resistant cultivars provide the most effective control.

Insects   Insect pests are confined to those affecting tomato and pepper, and the control recommendations are similar.

Harvest and Market Preparation

Fruit are harvested by hand at immature and mature sizes, while the skin is still glossy, but before the skin toughens and seeds become large. Quality is superior in fruit harvested at less than full size. The fruit are clipped from the plant, leaving the calyx cup attached. The skin is very tender and easily punctured; thus, care in harvest, grading, and packing is paramount. Eggplant fruit are packed in any or several types of containers, including bushels, crates, and cartons.

The crop is not suitable for prolonged storage and will suffer serious injury if exposed to temperatures below 50°F (10°C). A high-quality fruit must be heavy in relation to size, with glossy dark uniform color, free from surface cuts or bruises and showing no decay spots.

HUSK TOMATO

The genus *Physalis* (ground cherry or husk tomato) features a calyx which enlarges or inflates to cover the berry at maturity. Its appearance and use as an ornamental is reflected in the name "Chinese Lantern Plant," although the ornamental plant is a separate species.

There are three species used for food. *P. ixocarpa* has large, rather sticky green or purplish berries that completely fill the inflated calyx. This species is an old cultivated form from Mexico and Guatemala and is still grown there as **tomatillo.** It is important in the Mexican diet, used in tacos and enchiladas and stews. *P. peruviana,* or cape gooseberry, is characterized by tall growth, hairy leaves, and large,

sweet berries, approximately $1\frac{1}{2}$ in. long and $1\frac{1}{4}$ in. wide (4 × 3 cm), which usually are eaten uncooked. This form is probably native to the Andes. *P. pruinosa* is the ground cherry most familiar in northern gardens. It is weaker growing than *P. peruviana* but otherwise similar except for flower color and a sweet, nonglutinous yellowish berry, approximately $\frac{1}{2}$ in. (12 mm) in diameter.

Ground cherry is cultivated on a wide range of soil types, primarily for home use as a preserve. It is an annual vine, generally low growing, occasionally reaching 1 to 2 ft (30 to 61 cm) in height. It can be direct seeded or transplanted in soils acceptable for tomato production. Plant spacing is 18 in. (46 cm) within rows 3 ro 4 ft (0.9 to 1.2 m) apart. The fruit are ready for picking when the husk turns brown.

SELECTED REFERENCES

Beasley, E. O., *et al.* (1975). *Growing Trellised Tomatoes in Western North Carolina,* North Carolina State University Coop. Ext. Serv. Circ. 475.

Doolittle, S. P., A. L. Taylor, and L. L. Danielson (1961). *Tomato Diseases and Their Control,* USDA Agric. Handb. 203.

Erwin, A. T. (1932). *The Peppers,* Iowa State College Res. Bull. 293.

Gould, W. A. (1974). *Tomato Production, Processing and Quality Evaluation.* AVI, Westport, Conn.

Greenleaf, W. H. (1986). Pepper breeding. In *Breeding Vegetable Crops* (M. J. Bassett, ed.), pp. 67–134. AVI Publishing, Westport, Conn.

Heiser, C. B., Jr. (1969). *Nightshades: The Paradoxical Plants.* Freeman, San Francisco.

Kader, A. A., M. A. Stevens, M. Albright-Holton, L. L. Morris, and M. Algazi (1977). Effect of fruit ripeness when picked on flavor and composition in fresh market tomatoes. *Journal of the American Society for Horticultural Science* **102**:724–731.

Rick, C. M. (1978). The tomato. *Scientific American* **239**:76–87.

Seelig, R. A., and C. Magoon (1978). *Eggplant. Fruit and Vegetable Facts and Pointers.* United Fresh Fruit & Vegetable Assoc., Alexandria, Va.

Shear, C. B. (1975). Calcium related disorders of fruits and vegetables. *HortScience* **10**:361–365.

Sims, W. L., and P. G. Smith (1976). *Growing Peppers in California,* University of California Coop. Ext. Leaflet 2676.

Sims, W. L., M. P. Zobel, D. M. May, R. J. Mullen, and R. P. Osterli (1979). *Mechanized Growing and Harvesting of Processing Tomatoes,* University of California Coop. Ext. Leaflet 2686.

Wittwer, S. H., and S. Honma (1979). *Greenhouse Tomatoes, Lettuce and Cucumbers.* Amer. Vegetable Grower, Willoughby, Ohio.

Yamaguchi, M. (1983). *World Vegetables.* AVI, Westport, Conn.

STUDY QUESTIONS

1. What are the various factors, environmental and genetic, that determine color in tomato fruit?

2. Under what conditions is fruit set poor in tomato? Are the same factors responsible for set problems in pepper and eggplant?

3. Genetic improvement has been a major factor in improving tomato yields. What specific genetic attributes have been added to cultivars to improve yield?

4. What are the symptoms of chilling injury on the fruit of the three major solanaceous crops and in what circumstances does it usually appear? Does chilling early in plant development damage the fruit?

18
Legumes

significance include garden pea, cowpea, snap and dry beans, lima bean, soybean, mung bean, and small amounts of chickpea, lentil, scarlet runner bean, and fava bean (broad bean). Both immature and mature seeds of these food legumes constitute an important dietary source of carbohydrate and protein. The immature pod contains significant amounts of vitamins A and C, whereas protein, carbohydrate, and some of the minerals are major constituents of dry seed (Table 18.1). In uncooked beans, the protein value is limited by heat-labile antitrypsin factors and hemagglutinins.

Roman farmers recognized that clover would enhance soil productivity, although the basis for this effect, the biochemical reduction of atmospheric nitrogen to NH_4 in lateral root nodules by the symbiotic bacterium *Rhizobium,* was not understood until the nineteenth century. Nitrogen fixation is considered an important agricultural process. To some degree, however, this symbiotic relationship limits economic yields of legumes, since energy (carbohydrate) must be diverted from dry matter accumulation to drive the reduction of nitrogen. For the major vegetable legumes, snap beans and peas, nitrogen fixation is not an efficient source of nitrogen for plant growth in the cultivated forms used.

Dry, snap, and shell beans	Cowpea
Lima bean	Soybean
Garden pea	Broad bean
Scarlet runner	Mung bean
Chick pea	Tepary bean
Rice bean	Adzuki bean
Winged bean	

Beans and peas are members of the Leguminoseae, a family of plants with worldwide distribution. There are nearly 30 legume species used as vegetables, most of which are important in countries other than the United States. A few are grown for tubers or tuberous roots; most are valued for seeds and/or pods. In the United States, those with commercial

PHASEOLUS

Originally, *Phaseolus* was subdivided into New World and Old World beans. The Old World beans, generally small seeded, evolved in Asia and Africa and included mung, moth, adzuki, and rice beans and the black gram, or urd. This group since has been reclassified to the genus *Vigna* (Table 9.1). New World beans,

TABLE 18.1
Nutritional constituents of some vegetable legumes[a]

Crop	Water (%)	Energy (cal)	Protein (g)	Fat (g)	Carbo-hydrate (g)	A[b] (IU)	C[c] (mg)	Thia-mine (mg)	Ribo-flavin (mg)	Niacin (mg)	Ca (mg)	P (mg)	Fe (mg)	Na (mg)	K (mg)
Bean, dry	11	340	22.3	1.6	61.3	—	—	0.65	0.22	2.4	144	425	7.8	19	1196
Bean, green	90	32	1.9	0.2	7.1	600	19	0.08	0.11	0.5	56	44	0.8	7	132
Bean, wax	91	27	1.7	0.2	6.0	250	20	0.08	0.11	0.5	56	43	0.8	7	243
Lima bean, dry	10	345	20.4	1.6	64.0	—	—	0.48	0.17	1.9	72	385	7.8	4	1529
Lima bean, fresh	68	123	8.4	0.5	22.1	290	29	0.24	0.12	1.4	52	142	2.8	2	650
Pea, dry	12	340	24.1	1.3	60.3	120	—	0.74	0.29	3.0	64	340	5.1	35	1005
Pea, green	78	84	6.3	0.4	14.4	640	27	0.35	0.14	2.9	26	116	1.9	2	316
Pea, edible pod	83	53	3.4	0.2	12.0	680	21	0.28	0.12	—	62	90	0.7	—	170

Source: National Food Review (1978) USDA.

[a] Data expressed per 100 g sample.

[b] 1 IU = 0.3 μg vitamin A alcohol.

[c] Ascorbic acid.

including the common snap and dry beans and lima bean, constitute most of the bean acreage in the United States.

Historical Perspective

Both common bean and lima bean originated in Central America, most likely Guatemala to southern Mexico. Both crops migrated north and south along primitive trade routes. One route extended through Mexico to the southwest United States, then east to the Virginia area. Two routes extended southward, one through the West Indies to South America, the other through Central America to Peru.

Evidence of common bean, dating to 5000 B.C., has been found in Peru (large-seeded types) and in the Tehuacan Valley of Mexico (small-seeded forms). The large-seeded lima beans, dating to 5000 to 6000 B.C., also have been discovered along the coast of South America toward Peru. Small-seeded limas (sieva beans), dating to 300 to 500 B.C., have been found in Mexico and Guatemala and were the types introduced to North America.

The settlers to North America found Indians interplanting common beans and maize in each hill, using both crops as staples in their diet. Explorers and traders stocked their ships with both common and lima beans, and through this trade, New World bean seed was introduced to Asia, Africa, and Europe.

COMMON BEAN

Current Status

The common bean, used as both a snap and a dry bean, ranks fifth in the United States among all vegetables in per capita consumption. The snap bean is used fresh, frozen, and canned, the last [5.7 lb (2.6 kg) consumed per capita] constituting the largest volume. Approximately 204,410 acres (82,786 ha) is devoted to processing bean production, yielding 643,860 tons (584,110 MT). The fresh market production totals nearly 3 million cwt (136,080 MT) on 83,038 acres (33,630 ha). Florida produces the largest volume for fresh market. Wisconsin leads in production of snap beans for processing (Table 18.2). Dry beans [25,563,000 cwt (1,159,538 MT)] are produced on approximately 1,924,500 acres (719,422 ha), of which 350,000 (141,750 ha) producing 4,550,000 cwt (206,388 MT) is located in Michigan.

Classification, Growth, and Development

The common bean (*Phaseolus vulgaris*) is a tender, warm season, annual dicotyledonous

TABLE 18.2

Area and production of green snap (processing) and dry beans in the major producing states, 1983

| | Area harvested | | | | Production | | | |
| | Green beans | | Dry beans | | Green beans | | Dry beans | |
State	1000 acres	1000 ha	1000 acres	1000 ha	1000 tons	1000 MT	1000 cwt	MT
Arkansas	3.3	1.3	—	—	7.3	6.6	—	—
California	—	—	143.0	57.9	—	—	2,412	109.4
Colorado	—	—	150.0	60.8	—	—	1,680	76.2
Idaho	—	—	88.0	35.6	—	—	1,452	65.9
Michigan	14.8	6.0	350.0	141.8	41.4	37.6	4,550	206.4
Nebraska	—	—	131.0	53.1	—	—	2,188	99.2
New Jersey	6.2	2.5	—	—	11.2	10.2	—	—
New York	33.8	13.7	25.0	10.1	88.2	80.0	255	11.6
North Dakota	—	—	160.0	64.8	—	—	1,648	74.8
Oregon	22.4	9.1	—	—	123.7	112.2	—	—
Pennsylvania	4.5	1.8	—	—	7.9	7.2	—	—
Tennessee	4.9	2.0	—	—	8.9	8.1	—	—
Washington	1.4	0.6	16.0	6.5	5.5	5.0	355	16.1
Wisconsin	72.4	29.3	—	—	210.7	191.1	—	—

Source: *Agricultural Statistics* (1985), USDA.

plant with an epigeal germination habit (Figure 18.1). During germination and emergence, the radicle grows rapidly, developing a primary root. Subsequently, the primary root growth subsides, and many laterals develop. Some nodulation may occur on these laterals in soils with a compatible *Rhizobium* species.

The plants are dwarf (bush or determinate), tall (pole or indeterminate), or semivining. Determinate plants develop terminal racemes of up to 12 flowers. As each vegetative stem terminates in a flower cluster, branching occurs, resulting in the typical bush appearance. Indeterminate vines produce axillary flower racemes; consequently, vegetative growth of a stem continues with little or no branching. Semivining types have terminal inflorescences, but growth is vining with some branching. The twining habit of indeterminate and semideterminate plants is stimulated by rubbing the sensory hairs along the stem.

The domestic bean plant is considered to be day neutral. Some short-day forms, particularly those adapted to tropical areas, however, flower more readily than temperate zone cultivars at daylengths less than 13 to 14 h.

Bush beans have two major yield-bearing inflorescences, the terminal and first trifoliolate mainstem nodes. The flower is typical of legumes, with a standard, two wings and a keel, the latter enclosing the ovary, style, and stigma within coiled anthers. Common beans are highly self-pollinated, and trueness of cultivar type is maintained easily. The pods are flat to round, with two valves separated by sutures, and the seeds (ovules) are attached to the pod along the ventral suture. As the pods of stringy types dry, the two valves under increasing tension will spring apart, forcibly ejecting the seeds. Stringless beans do not shatter in this way.

The bulk of a bean seed is endosperm, confined primarily to the cotyledons. Seed coat color is green when immature. By the time seeds are 50 percent of full maturity, various pigments of colored-seeded types mask

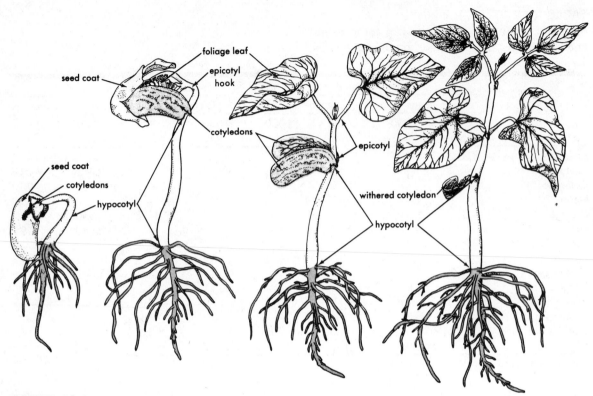

FIGURE 18.1

Germination of bean seed. Germination is epigeal, with hypocotyl lifting cotyledons above ground. (From *Botany. An Introduction to Plant Biology*, T.E. Weier, C.R. Stocking, M.G. Barbour, and T.L. Rost. Copyright © 1982 by John Wiley & Sons, Inc., New York, p. 322. Reprinted by permission.)

the green color. Seed coats can be rather hard and resistant to water penetration. The "eye" of a seed is the hilum, a scar marking the point of attachment (funiculus) to the pod. Germination occurs rapidly when soil temperature is 60 to 80°F (16 to 29°C) and air temperature is 60 to 70°F (16 to 21°C). Excessive heat or cold suppresses germination and early growth, and the plants do not tolerate even a light frost.

Crop Establishment and Maintenance

Snap and dry beans are grown on a wide range of well-drained, friable soils, including sands and mucks, although growth in the latter is generally poorer than in mineral soils. Soils high in clay that cake or crust will impede emergence and result in uneven or poor stands.

Prior to planting, soils of pH 5.2 and below must be treated with ground limestone or, in those areas deficient in magnesium, dolomitic lime. The optimum pH for bean production is 6.0 to 6.8, but satisfactory growth occurs between 5.5 and 7.0. Relatively small amounts of nitrogen are fixed by the plant. The nitrogen needs are provided by fertilizer applied primarily as a band, 2 in. (5 cm) below and to the side of seed placement. The rates used in New

York are 20 to 40 lb/acre (22 to 45 kg/ha) each of nitrogen and potash and 20 to 80 lb/acre (22 to 90 kg/ha) of phosphate. On light soils, or if early rainfall has been heavy, a nitrogen sidedressing of 30 lb/acre (34 kg/ha) is recommended. Zinc at 2 lb/acre (2.2 kg/ha) may be required on soils heavily limed. In Oregon and other western states, nitrogen rates are two to three times those of New York, and ratios average 2–2–1 ($N–P_2O_5–K_2O$). Soils in Florida require 150 lb/acre (168 kg/ha) of potash and 90 to 95 lb/acre (100 to 106 kg/ha) each of nitrogen and phosphate.

If heavy fertilization is suggested by soil test, some nitrogen and potash may be broadcast prior to seeding, but the bulk of the application would be most effective as a band. Based on a yield of 100 cwt (4.5 MT) of beans, the plant utilizes approximately 170 lb (71 kg) of nitrogen, 16 lb (7 kg) phosphorus, and 100 lb (45 kg) potassium. Maximum pod yields are obtained using the nitrate form of nitrogen, particularly applied as a sidedressing. Growth is reduced if NH_4 is the primary nitrogen source. The snap bean does not tolerate excessive boron, and rotation with crops requiring boron application may lead to toxicity.

Prior to seeding, the soil is plowed and disked. Seeding rates per acre depend on cultivar type, seed size, percentage germination, and row spacing. Dry beans are planted in rows approximately 30 in. (76 cm) apart. Spacing of bush snap beans may range from 6 to 36 in. (15 to 91 cm). Conventional single rows are spaced at 30 to 36 in. (76 to 91 cm). High-density plantings are spaced in rows 6 to 16 in. (15 to 41 cm) apart. The ideal spacing for high-density snap bean culture has been found to be 36 in.^{2} (232 cm^2) per plant, and the yield advantage is greatest when plants are in triangular arrangement, thereby maximizing root area per plant. However, because of equipment, a 12-in. (30-cm) spacing between rows is most common in those areas using high-density cultivation.

High-density plantings require increased fertilizer, but with the narrow row spacings, the rate per lineal foot can be reduced somewhat. On an acre basis, the rate may be doubled. Irrigation rates also would need to be adjusted.

Pole beans, once grown extensively in Oregon and to a lesser extent elsewhere, are seeded 6 to 9 in. (15 to 23 cm) apart in rows spaced at 3 to 4 ft (0.9 to 1.2 m). Mechanical post setters and stringers are used to establish the training system. Irrigation and fertilizer needs are increased somewhat over those for bush plantings, and development of the crop requires a longer season. Because the pole crop matures sequentially, it must be harvested by hand. As a consequence, growers have substituted high-quality bush cultivars amenable to mechanical harvest.

At a given spacing, the rate of seeding will vary widely according to seed size of the cultivar. Depth of seeding may be adjusted for soil conditions, with shallow seeding [1 in. (2.5 cm) or less] suggested for cool and moist soils, deep seeding [2 in. (5 cm)] for dry situations. Preplant herbicides, such as EPTC or trifluralin, are preferred for weed control and are effective if incorporated immediately after application; but, as an alternative, the field may be treated after seeding with a preemergence material (chloramben, dinoseb). Cultivation, if practiced, must be shallow to avoid damage to surface roots. Some growers seed in a shallow furrow, allowing the soil from light cultivation to cover the small weeds in the row as the bean plants develop.

Irrigation is required for most commercial production areas. Although the crop has an extensive root system, the plant is quite sensitive to dry soils, particularly at flowering and pod set. Flower drop is a serious limiting factor in yield and seems to occur when soil moisture is below 60 percent of field capacity and/or air temperature is high with low relative humidity. Up to 85 percent of the flowers will abort under high temperature–low moisture stress. Mauk et al. (1983) found substantial yield increases from irrigation and determined that both primary yield-bearing nodes responded to soil moisture (Table 18.3). The

TABLE 18.3

Yield per unit area of 'Oregon 1604' snap bean as affected by irrigation level and plant population for the 1978 and 1979 seasons

	Pod yield (MT/ha)	
Treatment	1978	1979
Irrigation		
High	18.7[a]	20.5[a]
Low	8.5	14.6
Plant population		
High	15.9[b]	21.3[b]
Low	11.3	13.8
Yearly mean	13.6	17.5

Source: From C. S. Mauck, P. J. Breen, and H. J. Mack (1983) *Journal of the American Society for Horticultural Science* **108**:938. Reprinted by permission of the authors.

[a,b] Significance at $P = .05$ ([a]) and $P = .01$ ([b]) for irrigation or population means within the same year.

yield gain from irrigation exceeded that from increasing population density.

Cultivars

Early cultivars of snap bean, then termed string bean, had tough fibrous strands in each suture of the pod. In the late 1800s, Calvin N. Keeney of LeRoy, New York, eliminated this dominant genetic trait, producing two important cultivars, 'Brittle Wax' and 'Stringless Green Refugee.' Subsequently, genetic improvements in quality have included development of fiberless pod walls, further improving tenderness, and white seed, which ensures a clear liquid component of canned snap beans.

Bean cultivars are categorized by growth habit (pole, vining, or bush), pod color (green or yellow) and shape (flat, oval, or round), seed color (colored or white), and use (pods or seeds). Snap beans with flat pods are sold fresh, and some are used for "French style" cut processed beans. Oval- or round-podded cultivars, however, are preferred by processors for most packs. Both bush and pole snap beans are processed, although trends toward mechanical harvest now favor bush plants. Green snap beans constitute the bulk of both fresh and processing volume.

Shell beans, those in which the mature green seeds are shelled from the pod, once were popular in New England, but have declined in recent years. Dry beans, or field beans, may be classified by seed size and color. The largest is the kidney, with the marrow, medium, and pea beans (navy beans) progressively smaller. The medium and pea bean sizes constitute close to three-fourths of the national dry bean production. 'Pinto,' a medium-size pink-buff bean, is the single most important cultivar type (Table 18.4). 'Sanilac' is the leading small white type. 'Great Northern' is a medium-size white bean important in Nebraska. Marrow beans are the least important, although some forms ('Yellow Eye,' 'Cranberry') are particularly valued in some areas for their baking quality. New York acreage is primarily planted to 'Light Red Kidney' types and 'Black Turtle Soup,' the latter for export. Other classes and their relative importance are listed in Table 18.4.

Seed quality, regardless of the bean cultivar and classification, is critical in achieving a good stand. Several diseases carried in seed can be avoided by using western-grown seed. However, seed viability also is related to seed harvest and cleaning. Threshing can damage bean seed, causing seed coat cracks, detached cotyledons, or cracking of the epicotyl

TABLE 18.4

U.S. production of classes of dry edible beans, 1981 to 1983[a]

	Production (1000 cwt)		
Class	1981	1982	1983
Pea (navy)	5,550	7,937	4,618
Pinto	14,593	7,217	4,372
Great Northern	2,690	2,896	1,940
Small White	312	257	381
Red Kidney	1,542	2,027	997
Pink	1,941	880	639
Small Red	610	529	302
Cranberry	320	420	285
Black Turtle Soup	2,244	259	48

Source: Agricultural Statistics (1985), USDA.

[a] Cleaned basis; excludes beans grown for garden seed.

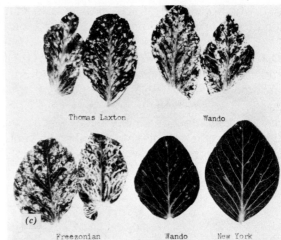

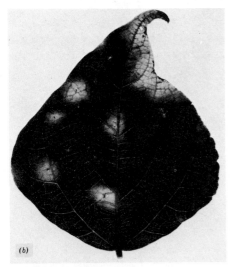

FIGURE 18.2

Diseases of beans and peas: (*a*) common bean mosaic; (*b*) halo blight; (*c*) pea enation mosaic; (*d*) gray mold (right) and white mold (left) in contrast to normal pods (center). (Photos (*a*), (*b*), and (*c*) courtesy of J. Baggett, Oregon State University. Photo (*d*) from *Diseases of Plants*, slide set 32, Bean Diseases. Reproduced by permission of the American Phytopathological Society.)

below the plumule. This damage, more severe in some cultivars than in others, results in poor viability or in seedlings that lack a growing point (baldheads). Transverse cracking also has been implicated in infection by *Pythium* spp.

Disease and Insect Pests

Damage by plant pests (Figures 18.2 and 18.3) is determined largely by relative humidity and soil moisture. Production in tropical climates is difficult because of persistent disease and insect problems, and high-density planting, which alters the microclimate, also can lead to an increase in pests.

Diseases *Root Rots* Excessive soil moisture increases activity of soilborne fungi, including *Fusarium solani* f. sp. *phaseoli, Rhizoctonia solani, Pythium* spp., and *Thelaviopsis basicola.* Infection causes decay just above the soil line, thereby girdling the plant, preventing movement of water to the leaves. Root rots often are implicated when tops die, leaves fall off or turn yellow, or plants and pods appear stunted. *Fusarium* dry rot symptoms appear early after emergence, first as a slight discoloration, becoming red and then brown as longitudinal cracks develop. The fungus can build to serious levels with continuous cropping. *Rhizoctonia* attacks the soil line during and after emergence. Reddish brown stem cankers appear at the soil line, often covered with

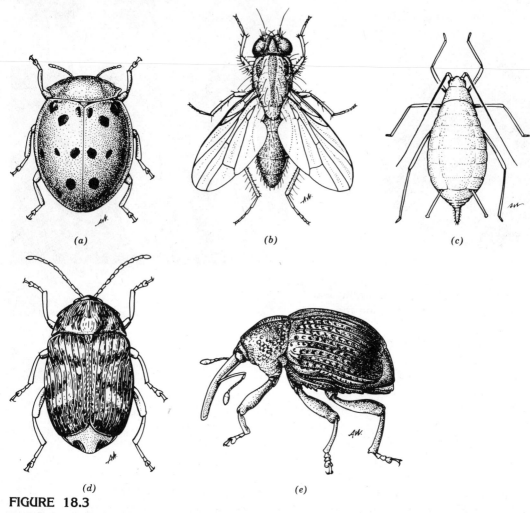

(a) (b) (c)

(d) (e)

FIGURE 18.3

Insect pests of vegetable legumes: (*a*) Mexican bean beetle; (*b*) seed corn maggot fly; (*c*) pea aphid; (*d*) pea weevil, (*e*) cowpea curculio.

a white mold. *Pythium* also affects the lower stem, most often of young seedlings, but occasionally of mature plants, their stems becoming characteristically hollow. *Thielaviopsis* infects the taproot, causing brown to black rot. Control of all root rots is difficult, since all causal organisms are long-term soil pathogens, generally with a wide host range. Rotation to avoid soil buildup of the pathogen, proper drainage and irrigation practices, and shallow seeding depth are the best preventative measures.

Bacterial Blights   Several bacterial leaf blights, carried in seed and spread easily throughout a field, reduce yields in many production areas. Four species are causal agents. **Halo blight,** caused by *Pseudomonas phaseolicola,* is recognized by many small, dead leaf spots surrounded by a yellow halo. The yellowing may be extensive at cool temperature, absent where temperatures are high. The bacteria are systemic within the plant, and thus entire leaves will be killed. Halo blight predominates in the east, where it is favored by cool moist weather. **Common blight** and **fuscous blight** are both caused by *Xanthomonas phaseoli,* but by different strains. Watersoaked spots appear on the leaves, enlarging and becoming dry and brown with a narrow yellow halo. The spots eventually coalesce, affecting the entire leaf, and pods also are affected. Both common and fuscous blights are favored by warm moist weather. **Brown spot blight** inoculum (*Pseudomonas syringae*) is carried in lilacs and wild cherry and infects lima bean more commonly than snap or dry bean. Small circular leaf spots develop, becoming brown with gray centers.

Bacterial blights generally are carried in seed, and seed becomes infected in plants grown in areas with prolonged rain and high humidity. The most effective control, therefore, is the use of western-grown seed. A 3-year rotation also is recommended to allow infected plant debris to break down completely, and plants should not be handled or cultivated when foliage is wet. The lilac and wild cherry hosts of brown spot blight should be eliminated from the production area. Resistance to halo blight has been developed in dry beans ('Redkote'), and cupric hydroxide may be used as a preventative spray.

Leaf Spots   **Anthracnose,** caused by the fungus, *Colletotrichum lindemuthianum,* occurs on leaves, pods, and stems. Black sunken lesions develop, each up to $\frac{1}{2}$ in. (1.2 cm) in diameter, often covered with salmon-colored ooze. The organism overwinters in crop debris and also in bean seed and is spread the same way as bacterial blights. Western-grown seed and crop rotation are essential control procedures, and basic copper sulfate spray may be applied when weather conditions favor the disease. Some cultivars resist certain races of the fungus.

Angular leaf spot (*Isariopsis griseola*) is another fungal disease favored by wet conditions, affecting both common and lima bean. The lesions appear first on the oldest leaves and are dark brown with sharp margins limited by veining pattern. Infected leaves senesce and drop off, and the fungus will persist for 2 years in infected plant debris. Transmission is by spores on seed and by splashing water. The control recommendations are the same as for anthracnose.

Cercospora leaf spot results from infection by *Cercospora canescens* and *C. cruenta*. *C. canescens* causes circular to slightly angular spots with gray centers and red borders. Lesions from *C. cruenta* infection are irregular in shape and size and brown to rust colored, and a dark mycelial growth occurs on the underside of the leaf. Control of both organisms is attained with clean seed, rotation, and carbamate or copper sprays.

Molds and Mildews   Powdery mildew (*Erysiphe polygoni*) and **downy mildew** (*Phytophthora phaseoli*) both occur on beans in some areas. Powdery mildew affects leaves, often late in the season, causing leaf drop and distortion of pods. Downy mildew seldom affects leaves, but does cause damage to pods. Of the two mildew diseases, downy is the more serious. The spores are carried in plant debris and in seed and are spread rapidly by wind.

Gray mold and **white mold** are caused by *Botrytis cinerea* and *Sclerotinia sclerotiorum,* respectively, two common organisms that infect a wide range of plants. Both are favored by wet weather and are spread by wind. Losses of up to 20 percent of a dry bean crop have occurred in Nebraska, and it is both a field and transit disease of snap bean, causing substantial decay. The appearance of the disease differs; gray mold is powdery and gray in color, whereas white mold is characterized by a thick cottony growth in which black sclerotia may be imbedded. Plants severely infected by white mold develop a bleached appearance. Preventative sprays (carbamate, copper) are recommended, and cultivars are being developed with some resistance to white mold. Management practices that reduce humidity within the plant canopy—irrigation system and frequency and plant spacing—may be adjusted to minimize infection.

Mosaics   Three bean virus diseases, **common bean mosaic, bean yellow mosaic,** and **peanut stunt,** occur throughout each production area. Infected plants become stunted, the leaves appear mottled, small, and puckered, the shoots have dieback, and pods and seeds may be distorted. Common bean mosaic is seedborne, and all three viruses overwinter in white clover. Bean yellow mosaic also overwinters on gladioli. These viruses are spread by aphids and by mechanical transfer. The use of clean seed helps to control common bean mosaic, and all viruses can be minimized by controlling the aphid vector. Resistance to the common bean mosaic virus and to a variant strain of the virus (NY-15) has been introduced into a number of cultivars. No resistance to bean yellow mosaic or peanut stunt virus is available.

Rust   The rust pathogen (*Uromyces phaseoli typica*) overwinters on infected plant residue. Spores developing from the red lesions travel for great distances and infect not only leaves, but also pods. The disease, however, is more serious in tropical than in temperate areas, is usually found on mature plants, and can be controlled with a maneb spray.

Insects   The predominant insect problems vary in different geographical areas. Some insects, feeding primarily on other plants, cause substantial damage to beans. Among the important pests are corn earworm, leafhopper, aphid, tarnished plant bug, Mexican bean beetle, striped or spotted cucumber beetles, flea beetle, mites, worms and slugs, and corn seed maggot.

Mexican Bean Beetle   The Mexican bean beetle (*Epilachna varivestis*) is sporadic in some areas but, when present, causes substantial damage. Both adults and larvae feed on the underside of the leaf. The adult beetle is $\frac{1}{4}$ to $\frac{1}{3}$ in. (6 to 8 mm) long and yellow to copper in color. Each wing cover has eight black spots. These adults lay eggs on the underside of the leaf, often shortly after bean emergence. Control measures include carbaryl or methomyl sprays.

Aphids   Bean aphids (*Aphis fabae*) occur on terminal bean growth. This species is similar to the pea aphid, but black in color and smaller. Infested plants become covered with a black "soot." Aphids can be controlled with organic phosphate insecticides.

Bean Weevil   The bean weevil (*Acanthoscelides obtectus*) bores holes in seeds, rendering them useless. The insect breeds continuously in dry seeds, but adults escape in the spring to feed lightly on foliage. They lay whitish eggs along the suture of developing bean pods. The emerging grubs then burrow through the pod wall into the seed, leaving only a tiny brown scab as evidence of penetration. Fumigation to control the weevil has been restricted by the carcinogenic properties of certain fumigating chemicals. Careful timing of spray materials in the field will reduce the risk that seed will become infested.

Harvest and Market Preparation

Bush beans mature within a concentrated time period, approximately 50 to 60 days after seeding, in contrast to pole beans which may be harvested over a period of 3 to 4 weeks, 60 to 70 days after seeding. A series of plantings of a bush cultivar therefore is necessary to ensure a constant supply, whether for a processor or for fresh market. Mechanical harvesters are utilized for most of the bush bean harvest; pole beans are picked by hand.

Processing bean quality is based upon uniformity of color, straightness of pods, low percentage of seeds, low fiber, and sieve size. Maturity is judged by pod and seed development. Overmature beans (large sieve size) can show "creaseback," where the pod walls expand above the suture. Pods should be smooth and crisp, with little or no seed bulge. The length of time a pod remains in prime harvest condition is short; thus, timing of harvest is very important. Overmature beans are tough and woody, and seed development is excessive.

For fresh market, beans are washed free of debris and are graded to eliminate pods with mechanical damage, disease, or discoloration. Pods have a high respiration rate and must be cooled after harvest to approximately 41°F (5°C). If pods are exposed to temperatures below 38°F (3°C), chilling injury, appearing as pitting and/or russetting of pods, may occur, and damaged beans are susceptible to storage diseases. Hydrocooling is preferred to retain turgor.

Beans are packed in wirebound crates, waxed cartons, or bushel baskets for transit. Humidity should be maintained at 95 percent at 41°F (5°C) for both transit and short-term storage. Decay can be a problem if holding temperatures exceed 45°F (7°C). Some beans are repackaged at destination in trays overwrapped with a perforated film, but many are sold from the bulk shipping containers.

Dry beans are allowed to dry in the field, preferably to 18 percent or less moisture content. Those harvested at higher moisture may be threshed and dried with circulating air at 80°F (27°C). Before harvest, chemical vine killers may be applied to facilitate cutting and threshing operations. Seeds are susceptible to cracking and checking during dry bean harvest, the extent of injury increasing with seed dryness, excessive thresher speed, and decreasing volume of beans with respect to thresher capacity.

LIMA BEAN

Current Status

Lima beans are grown primarily for processing or as a dry bean in all parts of the United States where environmental conditions are suitable. Almost all the acreage of dry limas is confined to the coastal region of California, where the season is long but temperatures are not excessively high. The leading states for processing green limas are California, Delaware, and Wisconsin, but substantial acreage also is located in Washington and New Jersey and in southern and midwestern states, extending to New York in the east and to Idaho and Oregon in the West. The total U.S. lima bean acreage (primarily processing) approximates 56,113 (22,726 ha). Lima beans for fresh market are grown only for limited local markets or direct sales.

Classification, Growth, and Development

Lima beans are predominantly annual plants, classified as *Phaseolus lunatus* and *P. limensis*. The latter species includes the large-seeded lima (potato lima or Fordhook type), a tender perennial grown as an annual. *P. lunatus,* the butter bean or sieva bean, is a climbing small-seeded annual form. The bush baby lima, also an annual, is designated as *P. luna-tus* var. *lunonnus.*

Species differences are noticeable primarily in pod characteristics. The thick-seeded bush lima, *P. limensis* (Figure 18.4), is somewhat more robust than the baby bush lima, and the pods have thicker edges with blunt tips and contain large plump seeds. *P. lunatus* pods and seeds are small and thin.

Developmentally, lima germination (epigeal) and early growth are similar to those of common bean. Bush forms, used for commercial production, become erect plants, terminating in a long raceme bearing small white or yellow-white flowers. Pole limas are vigorous twining plants.

Limas are more sensitive to environmental extremes and require a longer growing season than snap beans. Pod set, particularly of the thick-seeded limas, may be reduced substantially at temperatures above 80°F (27°C), particularly if relative humidity is less than 60 to 65 percent. The areas in which the thick-seeded cultivars are grown successfully, therefore, are restricted; the baby lima, in contrast, is widely adapted.

FIGURE 18.4

Thick-seeded lima bean pod. This pod type is noticeably thickened at the suture and is larger than the baby lima pod. (Photo courtesy of University of New Hampshire.)

Crop Establishment and Maintenance

The preferred soil types, soil preparation, and planting operations are the same as those described for common bean. Limas, however, require a full frost-free season. Emergence occurs within 1 week when soils are between 70 and 80°F (21 and 27°C), but is delayed considerably by temperatures less than 60°F (16°C); in some areas, this delay may increase the incidence of root rots. In regions of short growing season, light-textured soils therefore are recommended, and seeding should be delayed until the soil warms to 60°F (16°C).

High-density plantings are not utilized for lima bean production. Seeds of bush cultivars normally are planted 1 to 2 in. (2.5 to 5 cm) deep, spaced 4 to 6 in. (10 to 15 cm) in rows 24 to 30 in. (61 to 76 cm) apart. Pole cultivars are spaced 6 to 12 in. (15 to 30 cm) apart in trellised rows 4 ft (1.2 m) on center. Fertilization and general cultural systems are the same as for snap bean, although the lima has been reported to be somewhat more sensitive to pH than snap bean. A pH between 6.0 and 6.8 is considered optimum.

Lima bean seed is more susceptible to seed harvesting damage than is common bean seed, showing an increased incidence of cotyledon cracking. As in snap bean, such damage increases susceptibility to root rot infection. Crusted or compacted soils that increase resistance to cotyledon emergence also may cause breakage of the hypocotyl.

Cultivars

The three major categories of lima bean are the large-seeded (Fordhook), small-seeded (baby lima), and pole or sieva bean. Of these types, the large-seeded type is considered to have superior quality, but it is narrowly adapted. The small-seeded type is grown widely, and most new cultivars now have a green seed color which is more attractive than the white seed of early cultivars.

The original large-seeded dwarf plant leading to the cultivar 'Fordhook' was found in a field of pole limas in 1903, introduced by the Burpee Company in 1907, and, with minor improvements, remains as a leading cultivar today. The prototype of our modern baby limas was 'Henderson Bush,' a chance discovery in 1885. Current cultivars of the baby lima type include 'Thorogreen' and 'Nemagreen.' Several cultivars of the pole lima are grown, largely for home gardens.

Disease and Insect Pests

Lima beans are susceptible to many of the same disease and insect pests as common beans, and control recommendations are similar. Of the diseases described for common bean, downy mildew is found frequently in limas; in addition, pod blight may become serious.

Downy Mildew   Downy mildew, caused by *Phytophthora phaseoli,* overwinters in diseased plant debris and seed and is more serious on lima bean than on common bean. During wet weather, the fungus develops and spreads rapidly by wind. The pods develop a white downy fungal growth and will shrivel, blacken, and die rapidly. Young shoots and flowers also become infected. Fully developed leaves seldom develop mold but do show vein darkening. Plants may be destroyed within several days. Control is achieved through carbamate sprays.

Pod Blight   Pod blight is caused by the fungus *Diaporthe phaseolorum* and first appears as brown patches on leaves, later developing concentric rings of pycnidia. Eventually, pods may be infected, and seeds may fail to form or may be shriveled and unmarketable. The organism overwinters in diseased plant material and on seeds and is spread by wet conditions. Preventative sprays applied for other diseases should reduce the incidence of pod blight.

Harvest and Market Preparation

Most lima beans are grown commercially for processing. Harvests are mechanized, utilizing strippers that remove pods from the vine and shell the beans. Previously, beans were cut and windrowed, then picked up and shelled. At harvest, pods should be well filled, but not overmature, and seeds should be plump with tender, predominantly green seed coat. Because flowering is sequential, the pods do not mature uniformly. A once-over harvest thus will include some white seeds, an indication of overmaturity. A high-quality harvest should include less than 10 percent white seeds. For cultivars characterized by persistent green seed color, detection of overmaturity on the basis of seed color is somewhat unreliable.

Harvested beans, shelled and unshelled, are perishable and should be processed or sold immediately. For local fresh market, prepackaging or occasional misting of bulk boxes will reduce pod wilting, and with a temperature of 40°F (4.4°C) and 90 percent relative humidity, pods may be stored for up to 2 weeks.

SOUTHERN PEA

Historical Perspective and Current Status

The southern pea (cowpea) is a native of India. In prehistoric times, it was carried through India and the middle east to Africa, where wild forms still persist. Southern peas from Africa were brought to the New World by slave traders in the late 1600s, and the plant quickly adapted to the tropical climate of the West Indies. It is believed that the plant was introduced to Florida in 1700. By 1714, it was growing in North Carolina, and by 1775, in Virginia. At present, most southern pea production is confined to the south, extending from Virginia through Texas.

Southern peas are grown primarily for green-shelled seeds, although immature pods also are harvested. At the most popular edible stage, the pods and the seeds have developed fully, but the seeds have not dried. At this stage of maturity, the seeds contain approximately

24 percent protein, 55 percent carbohydrates, and 2 percent oils.

Classification, Growth, and Development

Edible cowpea is a warm season annual adapted to hot dry climates. There are two botanical forms commonly used as vegetables:

1. Southern pea, *Vigna unguiculata* (*V. sinensis*).
2. Yard-long, or asparagus bean, *V. unguiculata sesquipedalis.*

The most important form, southern pea, includes a range of cultivar types, many with an "eye" of color surrounding the seed hilum. It is a bush, twining plant, but not climbing. The pods are 3 in. (7.5 cm) or longer and solid.

Although termed a pea, it is considered a bean in its growth characteristics. Its flower shows an arched or curved keel in the corolla, as distinguished from a coiled keel in *Phaseolus.* The asparagus bean pods are distinctively long, inflated, and flabby, and the plant is taller than southern pea at maturity.

Crop Establishment and Maintenance

Slightly acidic sandy loams or sandy clay loams are preferred for southern peas. High pH levels may lead to iron chlorosis (iron deficiency). The soil should be plowed in the fall or winter preceding planting to speed decomposition of previous crop vegetation and to reduce overwintering of curculio. Fumigation may be necessary in areas with little rotation in order to reduce nematode infestation.

Southern pea does not respond to liberal application of nitrogen. Inoculated with the appropriate strain of *Rhizobium,* the plants will fix nitrogen. Therefore, fertilizer analyses generally are from 1–3–3 to 1–6–6 ($N–P_2O_5–K_2O$). Both broadcast and band applications are recommended; the latter is the most widely used system.

The crop is seeded at 12 to 25 lb/acre (13.4 to 28 kg/ha), depending on seed size, such that the plants are spaced 2 to 4 in. (5 to 10 cm) apart in 6-in. (15-cm) drilled rows. Most are grown as a dryland crop with little or no irrigation. Excessive moisture promotes vine growth, reducing yield. However, deficient moisture at flowering can cause flower abortion and sparse pod set.

Weed control is essential in a drilled planting. Trifluralin incorporated prior to seeding is common practice, and little cultivation is required if the herbicide is effective.

Cultivars

Five groups of cultivars have been described within *Vigna unguiculata,* based on seed and pod coloration and appearance.

Blackeye White seeds with black eyes around the hilum; upright to semispreading plant, pods 6 to 8 in. (15 to 20 cm) long; cultivars include 'California Blackeye No. 5,' 'Queen Anne,' and 'Princess Anne.'

Cream Seeds cream-colored when cooked; plants upright, seed shape from kidney to almost round; cultivars include several Texas lines (Nos. 40, 8, and 12), 'Elite,' and 'Conch.'

Purple Hull Red to purple pods (hulls) when mature; seeds can be any color; cultivars include 'Crimson,' 'Purple Hull,' 'Burgundy,' 'Mississippi Purple,' and others.

Crowder Seeds thickly packed (crowded) in hulls; seeds are brown when mature; predominant cultivar is 'Mississippi Silver.'

Miscellaneous Colored peas, but not of the crowder type; examples are 'Dixielee' and 'Blue Goose,' both suited to local fresh market.

A few cultivars exist within the **catjang group,** recognized by the erect small pods containing small spherical seeds crowded in the pods. These types are not of commercial significance.

Disease and Insect Pests

The diseases of southern pea resemble those of common bean, including root rots, rust, powdery mildew, *Fusarium* wilt, and several viruses. Resistance to virus infection is available in a few cultivars; otherwise, control of these and other diseases is achieved by avoiding problem soils (poorly drained or *Fusarium* infested) and by applying fungicides and insecticides as needed.

The major insect problems are cowpea curculio, corn earworm, and stinkbugs. In addition, soil insects (grubs, wireworms, cutworms) can reduce stands.

The most serious pest is the curculio (Figure 18.3). There are two curculio species, the clover root curculio (*Sitona hispidula*) and the cowpea curculio (*Chalcodermus aeneus*). The latter is an insect that pierces the pods to lay eggs in the seeds. After eggs hatch, the grubs may eat out the seed interior or, at the least, cause surface blemishes. The clover root curculio, particularly active when cowpeas follow clover that has been plowed under in the spring, destroys the seedlings as they emerge. Fall plowing is an effective control of the latter insect, and parathion will control the cowpea curculio.

Most cultivars are susceptible to nematodes, although 'Mississippi Silver' does show high resistance. Treatment with nematicides or fumigation periodically and rotation will help to reduce nematode population levels.

Harvest and Market Preparation

Commercial southern peas are harvested mechanically using equipment designed for snap beans or garden peas, depending on the stage of maturity to be harvested. The mechanically harvested crop largely is contracted for processing. For local fresh market, peas are harvested both by hand and by machine.

Most peas are harvested when pods show a slight yellow color, although immature pods and dry seeds also are marketed. Fresh peas, both immature and green shell, are perishable and must be refrigerated after harvest. Dry peas are cleaned and graded and stored for future packaging.

MISCELLANEOUS BEANS

Scarlet Runner Bean

Scarlet runner (*Phaseolus coccineus*), a perennial bean with hypogeal germination, is a native of the middle altitudes of Central America. As its name suggests, it is vining, attaining lengths of 12 ft (3.7 m) or more, and bears bright red flowers. A similar plant with white flowers is sold as 'White Runner.' Runner beans are trellised as described for pole beans, although in England, some acreage is grown unsupported following pruning to a bush form.

The scarlet runner produces tuberous starchy roots that reportedly are poisonous. In most areas, immature green pods are consumed, similar to snap bean, and shelled green and dried beans also are used in Central America.

Mung Bean

The mung bean (*Vigna radiata*), also known as green gram or golden gram, is grown for bean sprouts in many parts of the world and also is used for its edible green pods and seeds. In the southwest United States, it is used as a field crop and as a green manure. Mung bean probably originated in tropical Asia, and it is considered heat and drought tolerant.

The seeds are small, usually green or golden, and plants are hairy annuals bearing pods 2 to 4 in. (5 to 10 cm) long with 10 to 15 seeds each. In favorable environments, mung bean plants may reach a height of 3 ft (0.9 m) or more. The nutritional value of seeds is high, and caloric content relatively low.

Bean sprouts are produced by placing presoaked seeds in trays in a warm dark room, moistening the seeds periodically. In about 1

FIGURE 18.5

Chick pea (garbanzo bean). Pods are small and hairy and have large angular seeds. (Photo courtesy of University of New Hampshire.)

week, the sprouts may be harvested, placed in plastic bags, and kept refrigerated. The sprouts are served both cooked and raw. Production as a field crop is similar to chick pea.

Chick Pea (Garbanzo Bean)

The chick pea (Figure 18.5) or garbanzo bean (*Cicer arietinum*) is a large-seeded, bushy, low-growing legume with hairy pods, thought to be native to the Middle East. Dried pods, sprouts, and dried seeds are consumed, but the most popular product in the United States is the dried seed. Chick peas are adapted to cool, dry areas and have been produced for the U.S. market in coastal and interior valleys of California and in several areas of Mexico. A full growing season is required, and plant nodulation and growth are enhanced by *Rhizobium* inoculation. The crop is drilled in rows, 6 in. (15 cm) between plants in a row and 6 to 7 in. (15 to 18 cm) between rows, as soon as possible in the spring, when soil temperatures exceed 42°F (6°C). Defoliation is used to speed maturity in the fall, and the crop then is cut, windrowed, and combined. Most cultivars used in the United States are of California origin.

Soybean

The soybean (*Glycine max*), a major field crop in the United States and in other parts of the world, is used primarily for its oil and protein. The vegetable soybean, consumed in the immature bean stage, is not widely grown in the United States but has been an important commodity in China for over 5000 years. It was not until 1800 that soybeans were introduced to this country.

All soybeans are daylength sensitive. This sensitivity tempered widespread interest in this crop until the phenomenon of photoperiodicity was understood and new cultivars genetically superior in new daylength situations could be developed. Cultivars now include those categorized as Group 0 (adapted to long days of the north) to Group VIII (short days of the deep south). All cultivars require a full warm season to mature a crop.

Vegetable soybean acreage is limited to small areas, but those systems used for establishing plantings of the oil and protein soybean cultivars are applicable. In other countries, soybean is used for intercropping.

Vegetable soybeans have been selected for superior eating quality and are quite unlike field cultivars in that respect. If the beans are harvested at the mature green stage, the flavor is mild. As the beans become overmature, the flavor becomes strong and unappealing to many. Its protein, vitamin, and mineral content make it a valuable food.

Other Beans

There are several other bean genera and species important in restricted geographical areas. The large-seeded form of **broad bean** (*Vicia faba*) is popular in China and is grown as a vegetable in parts of Europe, Africa, and South America. The small-seeded type is more often used for animal feed. The green or ripe seeds provide an excellent source of protein. **Lentils** (*Lens esculenta*) are vetchlike plants bearing small pods filled with lens-shaped seeds. An excellent source of protein and used in vegetables mixes and soups, lentils are grown primarily in Washington state. **Adzuki bean** (*Vigna angularis*), a drought-resistant bean native to Japan, and **rice bean** (*Vigna umbellata*) are grown extensively in Asia, mostly for mature seeds. The adzuki bean has received attention in the United States as a potential export crop to Japan. The **winged bean** (*Psophocarpus tetragonolobus*) is a twining perennial, grown as an annual, that has attracted interest because of its adaptation to humid, hot climates and because the entire plant may be used for food. The use of **tepary bean** (*Phaseolus acutifolius* var. *latifolius*) dates back 5000 years. It is similar to snap bean in many characteristics, but is adapted to semidesert conditions. It is largely confined to areas of the southwest United States and Mexico. Prior to the extensive use of irrigation for bean production, there was interest in the tepary bean as a source of drought resistance in snap bean breeding; however, it is receiving little attention at the present time.

GARDEN PEA

Historical Perspective and Current Status

The garden pea, also termed the English pea, is a native of middle Asia, with a center of origin extending from northwest India through Afghanistan. Seeds of primitive types have been found in archaeological deposits dating to the Bronze Age and earlier. They were grown before the Christian era by the Greeks and Romans only for dried seeds. The edible-pod form did not appear until after the Norman conquest of England. By the end of the sixteenth century, many forms of garden pea had been described, including tall, dwarf, sugary, and starchy, but peas did not become a common vegetable until the eighteenth century. In 1865, experiments with the garden pea by Gregor Mendel established the fundamental laws of inheritance. The term "English pea" reflects the importance of English cultivars developed through early breeding efforts.

The nutritional value of the garden pea is related to its maturity (Table 18.1). The contents of calcium, phosphorus, iron, sodium, and potassium increase substantially with seed maturity as do those of protein and carbohydrate. Vitamins A and C decline with drying.

Most peas are consumed in the green shell stage, either fresh or processed. Per capita consumption of green peas totals 5.2 lb (2.4 kg) of which 3.6 (1.6 kg) is canned, and 1.6 (0.7 kg) is frozen. Total consumption has declined over a long period of time. Fresh consumption particularly has declined, and fresh pea marketing is considered of no commercial significance. Fresh peas are marketed, however, at roadside stands and are popular among home gardeners. In the United States, pea production for processing is estimated at 495,340 tons (449,372 MT) from 330,470 acres (133,840 ha). Wisconsin leads in tonnage, followed by Washington and Minnesota, and other significant processing acreage is located in other northern tier states into southern Canada. Dry pea production is approximately 290 million lb (131, 544 MT) annually.

Classification, Growth, and Development

The garden pea (*Pisum sativum* var. *macrocarpon*) is a hardy, cool season annual crop. Although considered daylength sensitive, the plant response changes with temperature. Flowering is accelerated by long days with low temperature. Plant growth is favored by intermediate temperature: yield is reduced as average temperature increases, and plants may die if exposed to prolonged periods above 78°F (26°C).

Germination (Figure 18.6) of the garden pea differs from that of the bean in that the cotyledons remain below ground during emergence (hypogeal). The optimum temperature for germination is 75°F (24°C), and the germination rate gradually slows as temperature declines; 40°F (4.4°C) is considered the

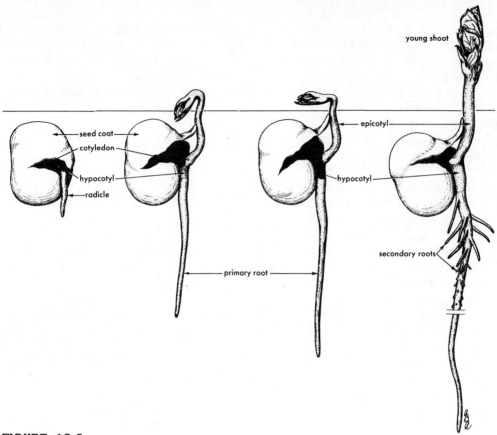

young shoot

seed coat
cotyledon
epicotyl
hypocotyl
hypocotyl
radicle

secondary roots

primary root

FIGURE 18.6

Germination of pea seed. Germination is hypogeal, with cotyledons remaining below ground. (From *Botany. An Introduction to Plant Biology*, T.E. Weier, C.R. Stocking, M.G. Barbour, and T.L. Rost. Copyright © 1982 by John Wiley & Sons, Inc., New York, p. 324. Reprinted by permission.)

minimum for germination and growth. After emergence of the radicle and the start of its downward growth, the epicotyl emerges and grows upward, retaining a curvature (plumule hook) that acts to protect the shoot apex within the soil. As the seedling emerges, it gradually straightens to show two small leaves which seldom develop further, but the rachis of subsequent compound leaves arising from the shoot apex bears one to three pairs of leaflets and one to five tendrils (modified leaflets). Each unexpanded leaf is enclosed by two sti-

pules which remain at the base of the petiole. The large stipules are considered the primary photosynthetic organs. Leafless (**afila**) plant forms, in which most leaflets are modified to tendrils, maintain the same productivity as leafy cultivars, suggesting that the leaflets indeed contribute little to ultimate yield. The leafless forms do have the normal well-developed leafy stipules (Figure 18.7).

Floral initiation can occur 20 days or more before anthesis, with the node at which the first flower appears varying with cultivar. The

FIGURE 18.7

Leafless (afila) and standard leafy garden peas. Yields of the two types are comparable, indicating that the leaflets are not responsible for most of the photosynthetic activity. (Photo courtesy of University of New Hampshire.)

flowers are a typical legume structure, with five sepals, two wings, a standard, and keel. The keel completely encloses the stigma, style, ovary, and stamens, resulting in almost complete self-pollination. Pollination usually precedes anthesis.

Peas may be determinate (dwarf) or indeterminate (tall). Determinate types produce two or more pods at each blossom node. Flowering on indeterminate plants occurs in axillary buds on the main stem. Because of the ease of harvest, most commercially important cultivars are determinate.

In early development, the pea plant develops a strong taproot which branches only in the top 6 in. (15 cm) of the soil. At blossom time, root branching becomes extensive, with an eventual lateral spread of 2 ft (61 cm).

Crop Establishment and Maintenance

Peas normally are seeded in cool to cold soils early in the spring. Early plantings will produce the best yields. The cold temperatures are not inherently harmful, although germination may be delayed. If soils are not well drained or of a good structure, emergence could be delayed sufficiently to allow soil fungi to infect seeds or emerging seedlings, reducing stand and yield. Well-drained gravelly loams or sandy loams are preferred sites for pea production.

The soil should be supplied with organic matter to enhance retention of soil nutrients and soil moisture throughout the growing season. Growth is favored by a pH between 6.2 and 6.8, and the crop is sensitive to acid soils. Soils to be used for pea production should receive either ground limestone or dolomitic limestone if pH falls below 6.0. Fertilizer is applied as a band 2 in. (5 cm) beside and below placement of the seed. Where heavy fertilization is required, some may be broadcast, the remainder banded. Based on a yield of 40 cwt (1.8 MT) of shelled peas, the total plant needs for nitrogen, phosphorus, and potassium are 170 (77), 22 (10), and 80 (36) lb/acre (kg/ha), respectively. The crop utilizes substantial amounts of nitrogen. However, plants may become viny if excessive nitrogen is used, and maturity will be delayed somewhat. In soils low in nitrogen, maturity accelerates. The quality of peas receiving adequate nitrogen and moisture is superior, and pod and seed development tend to be more uniform than on stressed plants. Nitrogen fertilization rates vary from 40 to 50 lb/acre (45 to 56 kg/ha) in eastern areas to 20 to 30 lb/acre (22 to 34 kg/ha) in western regions. Average rates for phosphate and potash are similar in both areas, from 50 to 100 lb/acre (56 to 112 kg/ha) of phosphate and 50 to 150 lb/acre (56 to 168 kg/ha) of potash.

Little nitrogen is fixed by peas, and inoculation with an appropriate *Rhizobium* strain for nitrogen fixation is not a commercial practice. Inoculation will benefit plants growing in soils lacking nitrogen but, with the existence of rhizobia in most fertilized pea-producing fields, fixation would not be enhanced by adding more inoculum.

After plowing and disking, the seeds are drilled 1 in. (2.5 cm) deep in 6- to 8-in. (15- to 20-cm) rows to achieve a final spacing of 2 in.

(5 cm) between plants. The seeding rate must be adjusted for cultivar growth habit and for growing conditions, but ranges from 120 to 180 lb/acre (134 to 202 kg/ha). Unbranched 'Alaska' types are seeded early at the high rate; late, branching cultivars at the low rate. When the soil is droughty, or other environmental conditions do not favor plant growth, seeding rate and depth should be increased somewhat to ensure uniform stands. In market gardens and home gardens, peas are seeded in single or double rows at a much lower rate. Late, tall cultivars often are trellised. Single rows are spaced 18 to 24 in. (46 to 61 cm) apart; twin rows are 8 to 10 in. (20 to 25 cm) apart in rows 24 to 30 in. (61 to 76 cm) on center.

Irrigation is utilized in some commercial pea acreage, but seldom in the eastern and north central states. Some western acreage is irrigated, but peas also are grown east of the Cascades in a dryland rotation with wheat. Soil moisture is most critical as the plants approach maturity because of the need for adequate nitrogen and other nutrients during the period of seed development.

Herbicides are used exclusively for weed control. The close-spaced drilled rows preclude effective cultivation. Preplant materials (trifluralin), incorporated into the soil, any of several preemergent materials (dinoseb, glyphosate), and a postemergence application of dinoseb have been used successfully. Selection of an herbicide is critical and must take into consideration the subsequent use of the land for a succession or rotation crop.

Cultivars

Pea cultivars are classified by seed color, growth habit (determinate or indeterminate), seed quality (starchy or sugary), and pod appearance, and certain combinations of these traits are used for specific purposes. Within each category, cultivars may be normally leafy or leafless (afila).

Dry peas Light green seed color; indeterminate; starchy with blunt pods: 'Alaska' type.

Canning peas Light green seed color; some indeterminate plants (e.g., 'Alaska'), but most cultivars are determinate; predominantly sugary with blunt pods.

Freezer peas Dark green seed color; some indeterminate cultivars, but most determinate; sugary, with either pointed or blunt pods.

Edible-pod pea Used as flat immature pod, or, in snap pea, with sugary seeds slightly enlarged; snap peas have a thickened pod wall as in snap beans but have more seed development than snap bean pods at harvest; most cultivars indeterminate.

Starchy peas can be recognized by the smooth round dry seed. These types are preferred in some restaurants and institutions because the seeds hold together during prolonged boiling. Sugary peas have wrinkled seeds and superior quality that is best maintained by minimal cooking. The afila types develop a rigid tendril system that interlocks with adjacent plants. Plants therefore remain erect, reducing soilborne pathogen activity on the pods. 'Novella' is the prototype cultivar.

Disease and Insect Pests

Diseases *Root Rot* Root rot (Figure 18.8) is the most widespread disease and the most difficult to control, since it can be caused by any of three organisms: *Pythium* spp., *Fusarium solani,* and *Aphanomyces euteiches.* Although efforts have been directed toward acquiring resistance, progress has been slow because of the several organisms involved and because different strains may occur in different production areas. Control is best achieved by proper site selection and crop management. Soils must be well drained, with good structure. Compaction will exacerbate root rot. A 4- to 5-year rotation is the primary way of maintaining a reasonably low inoculum level.

Viruses Pea enation mosaic, bean yellow mosaic, and red clover mosaic are widespread

FIGURE 18.8

Root/stem rot in peas. Foliage wilts and yellows as the disease girdles the stem base and destroys roots. (Photo courtesy of University of New Hampshire.)

virus diseases. Each can be controlled with resistance and, since transmission of these diseases is by aphids, by appropriate insect control measures. All mosaics cause mottling and plant stunting. Enation mosaic also produces abnormal growth (enations) on leaflets and pods.

Wilt   *Fusarium* wilt, caused by any of several races of *Fusarium oxysporum* f. sp. *pisi*, seldom occurs because of resistance incorporated into most popular cultivars. Its symptoms are typical of the vascular wilts, with yellowing and wilting progressing from lower to upper leaves, discoloration of the vascular system, and, finally, death of the plant.

Foliage Diseases   The major foliage diseases include bacterial blight (*Pseudomonas pisi*), *Ascochyta* blight (several species of *Ascochyta*, one with perfect stage, *Mycosphaerella*), powdery mildew (*Erysiphe polygoni*), and downy mildew (*Peronospora pisi*). Bacterial blight is characterized by lesions on leaflets, stems, and pods. The lesions are elongate linear streaks on stems, translucent spots on leaflets, and water-soaked spots on pods. *Ascochyta* blight is caused by three species: *A. pisi, A. pinodella,* and *Mycosphaerella pinodes.* Although there are some differences in symptoms, they are relatively minor. Each produces spots on leaves which become black and zonate. Similar lesions occur on stems, and, on pods, the spots are sunken with dark margins. All of the blights become severe in areas of cool moist weather. Bacterial blight and *Ascochyta* can be controlled to a great extent by using clean, western-grown seed.

Of the two mildew diseases, powdery mildew, characterized by white, dusty mold patches on leaves, can be controlled with application of sulfur. The disease is favored by humid, but not wet, conditions and is consequently a problem primarily in the southern states. Inoculum of downy mildew can be reduced by rotation, but spray applications usually are not effective.

Insects   Three insects are common pests on peas: aphid, which carries virus, pea weevil, and seed maggot.

Seed Maggot   The seed maggot (Figure 18.3) is the larva of a small fly (*Hylemya cilicrura*) that lays its eggs on or just beneath the soil surface. The small larvae bore into the seed before seedling emergence. Seed treatment in the planter box with diazinon or lindane generally is recommended.

Aphids   Pea aphids (*Macrosiphum pisi*) lay eggs on one of several host plants, and the winged forms migrate to peas and multiply rapidly under favorable conditions, sucking plant sap from leaves, flowers, and pods. In addition to virus transmission, they cause distortion of leaves and pods, plant stunting, and low yields. Insecticides, including diazinon, malathion, and parathion, must be used before infestations become severe.

Pea Weevil   The pea weevil (*Bruchus pisorum*) is a small-snout beetle overwintering in fence rows and under the bark of trees. At about the time that peas are in blossom, the adults migrate in, laying eggs on the young developing pods. The larvae enter the pod and complete their cycle, emerging in late summer. A small infestation can cause a field to be rejected for processing. Although the insect is not usually a major problem, regular inspections at blossom time should be made and appropriate sprays (parathion) applied if adults are found.

Harvest and Market Preparation

For the processing industry, plantings are scheduled to provide a steady supply of raw product at peak maturity. For this scheduling, the concept of heat units was introduced in the 1940s, based on the principle that the quanta of heat, not length of time, determine when a crop will mature. Different cultivars have different heat unit requirements, ranging from approximately 1100 to 1600. Heat units, also termed degree days, are calculated as the deviation of the daily mean temperature from the base temperature for growth, in the case of peas, 40°F (4.4°C). Thus, a 24-h day averaging 58°F (14°C) would accumulate 18 degree days (in Fahrenheit units) toward pea maturity. Similarly, no appreciable growth occurs above the maximum base of 85°F (29°C). To use the system for scheduling planting dates, the grower uses historical averages to predict the date on which the number of heat units required for a specific cultivar to mature will have accumulated. These averages also are used to determine the number of degree days

that will accumulate between harvests, and it is this number that determines the successive planting date.

Example   The grower's first planting is April 15. Based on weather averages, this planting of peas will accumulate the necessary degree days and mature on July 3. The grower is expected to deliver to the processor four fields of peas spaced 3 days apart. Based on weather records, the first 3-day interval between projected harvest dates (July 3 and 6) is computed as 54 degree days. Therefore, the second planting date also should be separated by 54 degree days. This scheduling is done by recording the number of heat units that accrue daily, starting with the April 15 seeding date. On the day on which 54 heat units have been accumulated, the second seeding is made. In a similar way, heat units for days exceeding the maximum base would be discounted in determining actual seeding dates.

Although there is some variation among methods and results in predicting harvest date by the heat unit system, it is, in general, an efficient scheduling procedure. As fields reach maturity, processing schedulers take several quality samples to determine the exact time for harvest, testing for tenderness, seed size, and weight.

Peas are harvested mechanically with a stripper as used for lima bean harvest. As harvests arrive at the processing plant, tests of quality, foreign matter (weed seed), and disease determine the acceptance or rejection and, normally, the price per ton.

Pea quality is, largely, a function of sugar content and tenderness. Shelled peas lose quality rapidly, primarily due to a reduction in sugar and increases in starch, proteins, and other insoluble substances. The seed coat becomes tough with age, reflecting increased migration of calcium. Since the changes are a function of the maturation process, and maturation accelerates with heat, it is important to process a crop immediately after harvest or, in the case of fresh market peas, hydrocool and then refrigerate the product at 32°F (0°C) until sale. Quality is evaluated by a tenderometer which assesses pericarp resistance and shear resistance of the developing cotyledons, the bulk of the young seed. This test is highly correlated with organoleptic evaluation of quality and with the starchiness of seed. As the pods and seeds develop, the yield will increase, but the tenderometer reading also will increase. Grower—processor contract prices are based on the tenderometer reading, and the higher price paid for peas with a value of 90 to 95 will more than compensate for the lower yield for peas of that developmental stage.

Off-color pods or those showing damage or diseases are removed in grading peas for fresh market. The shelled peas for processing also are graded, either by sieve size or using salt brine or both. Overmature peas sink in the salt brine and can be discarded. Off-color peas are removed from the grading conveyers.

Fresh market peas normally are sold in bulk in the pod, not prepackaged. Those shipped for any distance are top-iced to preserve turgor and sugar level.

SELECTED REFERENCES

Auld, D. L., et al. (1982). Garbanzo Beans—A Potential New Pulse Crop for Idaho, University of Idaho Agric. Exp. Sta. Bull. 615.

Hagedorn, D. J. (ed.) (1984). Compendium of Pea Diseases. Amer. Phytopathol. Soc., St. Paul Minn.

Mack, H. J., and G. W. Varseveld (1982). Response of bush snap beans (Phaseolus vulgaris L) to irrigation and plant density. Journal of the American Society for Horticultural Science 107:286–290.

Makasheva, R. Kh. (1983). The Pea. Oxonian Press, New Delhi.

Mansour, N. S., W. Anderson, and T. J. Darnell (1984). Producing Processing Peas in the Pacific Northwest, Oregon State University Regional Publ. PNW 243.

Mauk, C. S., P. J. Breen, and H. J. Mack (1983). Yield response of major pod-bearing nodes in bush snap bean to irrigation and plant population. Journal of the American Society for Horticultural Science 108:935–939.

Meicenheimer, R. D., and F. J. Muehlbauer (1982). Growth and developmental stages of Alaska peas. Experimental Agriculture 18:17–27.

Menges, T., et al. (1981). Keys to Profitable Southern Pea Production. Texas A & M University System Agric. Ext. Bull. L-1862.

Robertson, L. S., and R. D. Frazier (eds.) (1978). Dry Bean Production—Principles and Practices, Michigan State University Ext. Bull. E-1251.

Sandsted, R. F., R. Becker, and R. Ackerman (1977). Production of Processing Peas, New York State College of Agriculture and Life Sciences, Cornell University, Coop. Ext. Inf. Bull. 118.

Sandsted, R. F., R. B. How, A. A. Muka, and A. F. Sherf (1974). Growing Dry Beans in New York State, New York State College of Agriculture and Life Sciences, Cornell University, Coop. Ext. Inf. Bull. 2.

Sayre, C. B. (1952). Tenderometer grades, yields and gross returns of peas. New York Agricultural Experimental Station Farm Research 18(3):14–15.

Seelig, R. A. (1982a). Bean sprouts. Fruit and Vegetable Facts and Pointers. United Fresh Fruit & Vegetable Assoc., Alexandria, Va.

Seelig, R. A. (1982b). Lima beans. Fruit and Vegetable Facts and Pointers. United Fresh Fruit & Vegetable Assoc., Alexandria, Va.

Sims, W. L., J. F. Harrington, and K. B. Tyler (1977). Growing Bush Snap Beans for Mechanical Harvest, University of California Coop. Ext. Leaflet 2674.

Sutcliffe, J. F., and J. S. Pate. 1977. The Physiology of the Garden Pea. Academic Press, New York.

Yamaguchi, M. 1983. World Vegetables. AVI, Westport, Conn.

STUDY QUESTIONS

1. For what diseases is seed certification used in dry beans?

2. What cultural systems of techniques can be used to maximize the yield of snap beans? Would the same systems be applicable to dry beans?

3. What are the compositional changes that take place in a pea seed as it matures and how is each change affected by temperature?

4. Although the seeds of food legumes generally have high protein, the food value may be less than expected. Why?

The uses of the vine crops range from fresh desserts and salads (melon, watermelon, cucumber) to boiled or baked vegetables, utilizing immature fruit (summer squash) or hardened mature fruit (winter squash), to pies (mature squash and pumpkin), pickles (cucumber and watermelon rind), and snack food (pumpkin and squash seeds).

Breeding tests have indicated that there is little or no spontaneous hybridization between species of either *Cucumis* or *Cucurbita*. Cultivars of different types but within the same species can cross-fertilize; for example, pumpkins, summer squash, and some gourds are interfertile (all *Cucurbita pepo*). Growers saving seeds must maintain purity by isolation of one cultivar from another. Although isolation is needed for seed purity, there is no effect of cross-pollination on the flesh quality of resulting fruit.

Cucumber   Watermelon

Muskmelon   Squash

Chayote

The vegetable species of Cucurbitaceae ("vine crops") are similar in growth and culture and contribute significant amounts of vitamins and minerals (Table 19.1). They are warm season, tender annuals, thriving in hot and humid weather, and most have a spreading growth habit and bear tendrils at leaf axils. Dwarf forms are increasingly common among new cultivars of winter squash, watermelon, muskmelon, and cucumber; summer squash typically shows a bush habit. Botanically, the fruit is a pepo, a fruit type in which the ovary wall is fused with receptacle tissue to form a hard rind.

CUCUMBER

Historical Perspective and Current Status

The cucumber is native to an area of India between the Bay of Bengal and the Himalayas. Although records of its use are sparse, cucumber may be among our oldest vegetables. It is one of the plants specifically mentioned in the Bible, and, at the beginning of the Christian era, it was grown in North Africa, Italy, Greece, Asia Minor, and other countries. Emperor Tiberius reportedly developed forcing methods to ensure a steady supply of cucumbers. The crop was first introduced to England in the early 1300s, but not cultivated until it was reintroduced 250 years later. Columbus planted seeds in Haiti; by 1539, cucumbers were

TABLE 19.1
Nutritional constituents of the major vine crops[a]

Crop	Water (%)	Energy (cal)	Protein (g)	Fat (g)	Carbohydrate (g)	Vitamins A[b] (IU)	C[c] (mg)	Thiamine (mg)	Riboflavin (mg)	Niacin (mg)	Minerals Ca (mg)	P (mg)	Fe (mg)	Na (mg)	K (mg)
Cucumber	95	15	0.9	0.1	3.4	250	11	0.03	0.04	0.2	25	27	1.1	6	160
Muskmelon	91	30	0.7	0.1	7.5	3400	33	0.04	0.03	0.6	14	16	0.4	12	251
Watermelon	93	26	0.5	0.2	6.4	590	7	0.03	0.03	0.2	7	10	0.5	1	100
Winter squash															
Butternut	84	54	1.4	0.1	14.0	5700	9	0.05	0.11	0.6	32	58	0.8	1	487
Hubbard	88	39	1.4	0.3	9.4	4300	1	0.05	0.11	0.6	19	31	0.6	1	217
Summer squash															
Zucchini	95	17	1.2	0.1	3.6	320	19	0.05	0.09	1.0	28	29	0.4	1	202
Yellow	94	20	1.2	0.2	4.3	460	25	0.05	0.09	1.0	28	31	0.6	1	202

Source: National Food Review (1978), USDA.

[a] Data expressed per 100 g sample.

[b] 1 IU = 0.3 μg vitamin A alcohol.

[c] Ascorbic acid.

grown by Indians in Florida, and they reached Virginia by 1584. Cucumbers were planted in the first permanent settlements in Massachusetts in 1629.

Today, fresh cucumbers are available throughout the year. The total production is over 6,000,000 cwt (272,160 MT), grown on approximately 55,000 acres (22,275 ha). Florida, Texas, and California are the most important producing areas, primarily for the winter–spring market. Processing acreage, concentrated in Michigan, North Carolina, and Ohio, is more than double the fresh market total. An average annual total of 654,650 tons (593,898 MT) was recorded from 1975 to 1978. Per capita consumption of both fresh and processed totaled 12.7 lb (5.8 kg).

Nutritionally, cucumber has little to offer. The vitamin A content, low in comparison to many vegetables, is nil if the green rind is removed. Vitamin C content is moderate, relative to other salad crops. Cucumbers may develop noticeable bitterness of the skin due to accumulation of terpenes (cucurbitacin). This accumulation is controlled genetically, at least in part, and excessive soil nitrogen will increase bitterness. Some cultivars for fresh market are relatively free from bitterness.

Classification, Growth, and Development

Cucumber (*Cucumis sativus*) is a coarse, prostrate, annual vining plant with stiff hairs or spines on leaves and stems. Unbranched lateral tendrils develop at the leaf axil, and vining begins after two or three true leaves form. Branching also begins at this time. As soon as lateral branches develop, flower clusters appear at the leaf axils. In conventional cultivars, the first cluster always consists of male flowers, a response to photoperiods in excess of 14 h. Solitary, or occasionally pairs of, female flowers normally do not appear until the daylength begins to decline, then become numerous until fruit set is at a maximum level. As fruit are removed, new female flowers continue to develop. Eventually, however, flowering again becomes predominantly male. It is believed that environmental factors change the length of each flowering phase, but not the order of each phase.

Although flowering normally is monoecious, most cultivars developed since the mid-1960s have been gynoecious, producing one or several female blossoms at a node. Male flowers appear on gynoecious plants only if the plant is stressed or crowded, or if ethaphon or gibberellin is applied. Gibberellin and ethephon both alter the plant's endogenous auxin level. High auxin levels are associated with femaleness, low levels with maleness. Application of gibberellin increases maleness by decreasing auxin level. Ethephon increases the auxin level and appearance of female flowers.

The root system of cucumber is classified as deep, although many feeder roots spread laterally through the top 8 in. (20 cm) of soil. As the plants mature, the laterals grow downward, and the total root extension may be up to 7 ft (2.1 m).

Crop Establishment and Maintenance

Soils light in texture or otherwise well drained are required for maximum production. Sandy, silty, or clay loams enriched with organic matter, as either green manure or animal manure, are preferred. Heavy soils, or those poorly drained, are not suitable, since the root systems of cucumber and other cucurbits are very susceptible to oxygen deficiency. Cucumbers tolerate moderately acid soils, but maximum growth and fruit set occur at pH 6.0 to 6.8. Very low pH (below 5.5) can reduce fruit set.

Land previously seeded to a cover crop is prepared for cucumber planting by applying manure, lime, and, in some instances, ammonium nitrate prior to plowing. A total application of 10 to 15 tons/acre (22.4 to 33.6 MT/ha) of animal manure or green manure equivalent is recommended. A complete fertilization program includes a broadcast or band application, followed by one or more side-dressings. Based on a yield of 13,300 lb (6038 kg), the fruit remove 15 lb (6.8 kg) nitrogen, 4 lb (1.8 kg) phosphorus, and 21 lb (9.5 kg) potassium from the soil, and the vines remove approximately twice those amounts. In the northeast, the fertilizer ratios range from 1–1–1 to 1–2–2, with nitrogen supplied at 50 to 75 lb/acre (56 to 84 kg/ha), often supplemented

with 2 percent MgO. Half of the total is broadcast before plowing; the remainder is applied as a band at planting. When vines begin to spread, the rows are sidedressed with 30 to 60 lb/acre (34 to 67 kg/ha) of nitrogen. In California, the rates of nitrogen, phosphate, and potash average, respectively, 121, 60, and 24 lb/acre (135, 67, and 27 kg/ha). Cucumbers grown in Florida require an average 150 lb/acre (168 kg/ha) of nitrogen, 120 lb/acre (134 kg/ha) of phosphate, and 180 lb/acre (202 kg/ha) of potash. Similar ratios are common in the middle Atlantic states, except phosphate rates are generally higher.

Insufficient potassium will result in misshapen fruit ("bottlenecks"), and low nitrogen restricts growth, modifies the length-to-diameter ratio of fruit, and reduces fruit set and color development.

Banded applications of fertilizer with direct seeding have not been recommended in the past, perhaps because of potential damage to germinating seed or because broadcast applications seem better suited to a spreading root system. Used properly, however, banding has been superior. In Florida, fertilizer is banded in either side of a row, 4 to 6 in. (10 to 15 cm) from placement of the seed, and, in some instances, plastic mulch strips then are applied over the bands. This system reduces leaching losses and is particularly effective with seep (subsurface) irrigation. In sandland areas, some growers have direct seeded cucumber and other vine crops on soil just covering a layer of manure placed in a furrow.

Cucumber seeds do not germinate at soil temperatures below 52°F (11°C). Percentage germination increases to 70 percent or higher at 60°F (16°C), increasing further as temperatures approach 77°F (25°C). When soils have warmed to 60°F (16°C), seeds of slicing cucumbers are drilled in rows 4 to 6 ft (1.2 to 1.8 m) apart, with a final stand at 8 to 20 in. (20 to 51 cm), depending on season, soil, and other environmental considerations. Pickling cucumbers, subject to a single destructive harvest, are drilled at 2 to 3 in. (5 to 7.5 cm) spacing in rows 3 to 4 ft (0.9 to 1.2 m) apart. Beds or ridges are used to accommodate surface irrigation systems; in fields with overhead irrigation, raised beds seldom are used.

Once emerged, the plants are highly susceptible to cold soils and to wind erosion. Extended periods of cold weather followed by bright sun will cause "sudden wilt," because the transpirational water loss cannot be sustained by root absorption. Complete crop losses have occurred following chilling. The young plants also are damaged easily by the abrasive action of soil particles. Windbreaks of tall grass spaced at intervals through the field reduce soil movement substantially.

Cucumbers normally are monoecious, but many cultivars are gynoecious, requiring separate pollinator plants to ensure fertilization and fruit development. Regardless of type, pollen must be conveyed to each female flower. Honeybees are the predominant vector, and growers have found it worthwhile to establish one to three bee colonies per acre (two to six per hectare) of cucumbers as soon as the first blossoms appear. Poor fruit set or misshapen fruit may reflect poor pollination.

Weeds are controlled effectively early in the season by cultivation, but this practice is not feasible once vines of adjacent rows merge. Fresh market growers occasionally use black plastic mulch, either as a full bed mulch or as strips on each side of the row. Plastic has been particularly beneficial in northern growing areas, not only in providing weed control, but also in modifying temperature and moisture conditions within the cucumber root zone. Herbicides are preferred for full season weed control and are particularly useful in high-density plantings. Preplant (bensulide) and preemergence (chloramben, DCPA, and others) chemicals provide control of a broad spectrum of weeds.

Cultivars

There are three major cultivar types: processing (pickling), fresh market (slicing) (Fig-

ure 19.1), and greenhouse (slicing). Each is distinguished by fruit appearance and by specific quality standards for fresh market or for pickling use.

Pickling cucumbers are blunt and angular, warty, and light green in color. Conventionally, processing cultivars once were bred to include black spines, since this trait was associated with improved color after brining. Black spine color, however, also is associated with a dull orange ripe fruit color. This color develops prematurely under southern growing conditions, whereas white-spined cultivars tolerate

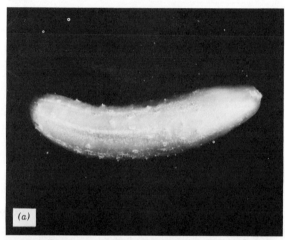

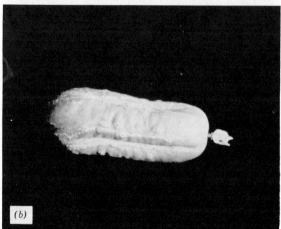

FIGURE 19.1

Slicing (*a*) and processing (*b*) cucumbers. (Photos courtesy of University of New Hampshire.)

high temperature and ripen slowly to a whitish yellow. Black-spined picklers therefore were traditionally dominant only in the northern states. However, in recent years, cultivars have been developed that combine superior brining quality with the ripening pattern associated with white-spined types. These types are now beginning to dominate the pickling cucumber market.

Slicing cucumbers are always white spined and are characterized by smooth, symmetrical fruit with glossy, dark green skin. The fruits tend to be long and tapered, with widely scattered, less pronounced warts and slow seed development.

Greenhouse cucumbers (Figure 19.2) are parthenocarpic, requiring no insect activity to develop. As a result, the fruit are seedless. Most greenhouse cultivars originated in Europe, and the fruit are very elongate, cylindrical, and smooth, with a thin tender skin and thick walls.

Most new cultivars of each type are gynoecious (female-flowered) hybrids with multiple resistance to disease. The gynoecious trait can be somewhat variable, and staminate flowers may develop under periods of growth stress.

FIGURE 19.2
Greenhouse cucumbers are seedless, elongate, and very thin-skinned. From 12 to 14 are produced per plant. (Photo courtesy of Merle Jensen, University of Arizona, Tucson.)

The gynoecious trait, however, has simplified development of productive F_1 hybrids which also have the gynoecious trait. The gynoecious hybrid, on average, is earlier than monoecious hybrids, but total yield has not been significantly superior.

The West India gherkin (*Cucumis anguria*), apparently a nonbitter cultigen of the African wild species, was transported to North America via the slave trade. It is a monoecious annual with small oval to oblong prickly fruit. It is seldom grown commercially in the United States; the gherkin pickles of commerce are small *C. sativus* fruit.

Disease and Insect Pests

Diseases   Genetic resistance has been developed against several important diseases: angular leaf spot, anthracnose (some races), cucumber mosaic, scab, downy mildew, and some races of powdery mildew. These diseases occur in most production areas, and resistant cultivars should be grown.

Although some diseases are controlled effectively by resistance, others, such as anthracnose, occur as many races which reduce effectiveness of genetic control. There are also diseases for which no resistance is available, and some of these can cause serious losses. Symptoms of several diseases are shown in Figure 19.3.

Bacterial Wilt   Bacterial wilt, caused by *Erwinia tracheiphila*, is present wherever there is feeding by the striped, banded, or spotted cucumber beetles. The bacterium is injected into seedlings as early as cotyledon emergence, but the disease is not visible until the plants have branched, flowered, and set fruit. At first, a few leaves show wilting, often on a single branch, and the plants recover at night. Gradually, however, the entire plant wilts and dies. The disease can be identified by cutting a branch, squeezing out a drop of juice from each end, touching the two cut surfaces, and slowly pulling them apart. A string of bacterial ooze connecting the two ends is a cursory but

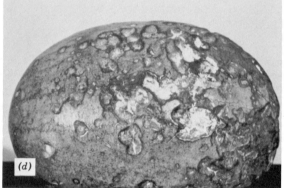

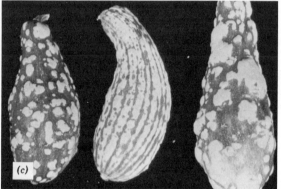

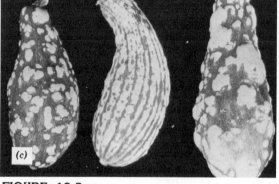

FIGURE 19.3
Diseases of the Cucurbitaceae: (*a*) bacterial wilt, showing diagnostic string of ooze from affected stem; (*b*) cucumber scab; (*c*) cucumber mosaic on 'Zucchini' squash; (*d*) anthracnose on watermelon. (From *Diseases of Plants*, slide set 36, Cucumbers, Melons, Squash Diseases. Reproduced by permission of the American Phytopathological Society.)

positive test for bacterial wilt. Control at this time can be achieved only through control of the cucumber beetle.

Angular Leaf Spot   The bacterial disease angular leaf spot (*Pseudomonas lachrymans*) has its most conspicuous symptoms on leaves, appearing as water-soaked spots that become irregular in shape, often conforming to vein patterns. Gradually, they become tan or gray "windows" and drop out. Fruit develop

small sunken water-soaked lesions that often become infected with secondary rots. The disease overwinters in seed and on plant debris. Clean seed, produced under dry conditions, and a 3-year rotation will reduce the incidence of leaf spot, and some resistance is available.

Anthracnose   The anthracnose fungus (*Colletotrichum lagenarium*) infects leaves, stems, and fruit. On leaves, infection produces rapidly developing yellowish or water-soaked areas, often progressing inward from leaf margins, which rapidly turn brown. Stem spots are elongated lesions, and leaves beyond the infection site may die. Fruit develop round sunken spots with a pink ooze in the center. The disease is spread rapidly in wet weather and is carried in the seed and on plant residue. Clean seed and a 3-year rotation will reduce inoculum. Resistant cultivars may help in some areas. Protective sprays also can be effective.

Scab Scab (*Cladosporium cucumerinum*) is recognized by dry corky lesions on the fruit. A green fungal growth may cover the spot during moist conditions. Lesions also may develop on terminal stem growth and on petioles, normally killing all tissue beyond the infection point. Disease is favored by cool nights and moist, humid weather. The scab organism overwinters on the seed and in plant residue. Resistance is effective, and clean seed and a crop rotation will reduce inoculum levels.

Cottony Leak A soilborne organism, *Pythium aphanidermatum,* invades wounds, old flower parts, and fruit touching the soil. A dark green water-soaked area develops, followed by a soft wet rot that becomes covered with a white mold during moist weather. Fruit infected with the cottony leak fungus will spread the disease rapidly in storage or transit. It is impossible to prevent, but careful grading can reduce market losses.

Gummy Stem Blight Gummy stem blight (*Didymella bryoniae*) begins as a pale brown or gray spot on leaves, petioles, or stems. The infection affects stem nodes first, then elongates into streaks from which a gummy exudate may appear. Small black pycnidia may appear on stem and leaf spots, and leaves on infected vines will yellow and die; fruit also may become infected. The fungus overwinters in seed and on plant debris. A long rotation, use of clean, treated seed, and application of early sprays have been effective control measures.

Powdery Mildew Airborne spores of the fungus *Erysiphe cichoracearum* are especially serious in very warm growing areas and in greenhouses. The disease first appears as a white powdery spot on leaves, with many spots eventually coalescing to cause infected leaves to wither and die. Although resistance to some races has been incorporated in several cultivars, the primary control is with benomyl sprays.

Downy Mildew Downy mildew infection, caused by *Pseudoperonospora cubensis,* can be extensive in pickling cucumbers because of the high population density at which they are grown. The disease is characterized by yellow-brown spots on upper surfaces of crown leaves, with a purplish mold on the underside. Eventually, the affected leaves die, and infection spreads toward new growth. Downy mildew is windborne from southern areas, and the most practical control is the use of resistant cultivars.

Cucumber Mosaic Mosaic is recognized by stunting, foliage mottling, and occasional wilting and death of the leaves. It is thought by some to be partly responsible for "crown blight," the yellowing and premature senescence of crown leaves. The fruit become distinctly mottled and unmarketable. The primary vector is the aphid, but striped and spotted cucumber beetles also carry the virus. The control measures include weed and insect control programs and the use of resistant cultivars.

Insects **Cucumber Beetles** Three beetles—banded (*Diabrotica balteata*), striped (*Acalymma vittatum*), and spotted (*D. undecimpunctata*)—each small, yellow insects with 3 distinct black bands or stripes or 12 spots, respectively, will feed on cucumbers (Figure 19.4). Although leaf feeding by adults and root damage by larvae can destroy plants, the most serious effect is the transmission of bacterial wilt. Feeding normally begins as soon as green tissue appears, and a careful spray program is essential to prevent an infestation. Carbaryl is effective but must not be used

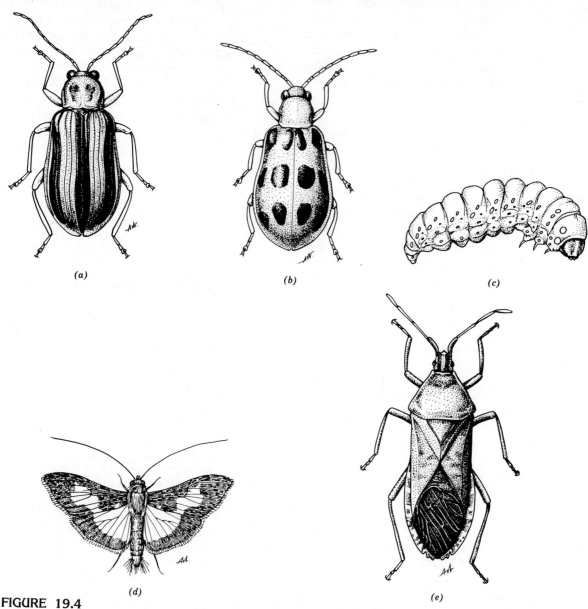

(a)

(b)

(c)

(d)

(e)

FIGURE 19.4
Insect pests of the Cucurbitaceae: (a) striped cucumber beetle; (b) spotted cucumber beetle; (c) squash vine borer; (d) pickle worm (adult moth); (e) squash bug.

once flowering has begun, since it is extremely toxic to bees. Cucurbitacin within leaves is toxic to some insects; however, cucumber beetles ingest cucurbitacin during feeding, giving them a degree of protection against bird predation. Nonbitter (low cucurbitacin) cultivars are unattractive to beetles and may provide a measure of control.

Aphids   Aphids (*Aphis gossypii*) also carry disease, primarily mosaic, and severe infestations can kill young leaves and stems. The leaves of heavily infested plants curl and wilt and eventually shrivel. Aphids appear on the plants in late spring or early summer and spread rapidly. The winged forms establish new infestations throughout the field or adjacent areas. Several organic phosphates are effective in control.

Pickle Worm   The larvae of *Diaphania nitidalis* (Figures 19.4 and 19.5) feed on leaves, flowers, and fruit, and the insect can be found from Canada to South America. The adult moth, recognized by its wing coloration, brown tips and white centers, lays its eggs on the foliage, and larvae migrate to other parts of the

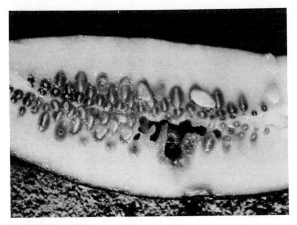

FIGURE 19.5
Damage caused by pickle worm. (Photo courtesy of Kenneth Sorensen, North Carolina State University, Raleigh.)

plant. The affected fruit become rotten or moldy. Control is through carbaryl or endosulfan sprays.

Other Insects   Other general insects, such as leaf miner, cutworms, and flea beetles, also attack cucumber, and nematode damage can be extensive in warm soils. These pests are discussed in other chapters.

Harvest and Market Preparation

Harvests for fresh market are repetitive, whereas those for processing are once-over. Repetitive harvests may continue as long as all market-size fruit are removed from the plant. If one fruit is allowed to mature seeds on the vine, additional fruit set will terminate, thereby reducing total yield.

Cucumbers are harvested several days to 7 to 10 days after pollination. Slicing cucumbers, picked several times per week, should be firm, deep green, and well developed in length [longer than 6 in. (15 cm)] and diameter [maximum $2\frac{2}{3}$ in. (6.8 cm)]. Pickling cucumbers normally are harvested mechanically and graded by size. The smallest market-size fruit, up to $1\frac{1}{16}$ in. (27 mm) in diameter, are processed whole, often as "sweet gherkins." Medium-size fruit, $1\frac{1}{16}$ to $1\frac{1}{2}$ in. (27 to 38 mm) in diameter (5 to 6 days after pollination), and large sizes, $1\frac{1}{2}$ to 2 in. (38 to 51 mm) in diameter, may be used as sliced or whole dills or as sweet or sour slices and chunks. The price paid to a grower is related to size: the small size commands the highest price; the largest marketable class brings the lowest price. Diameters over 2 in. (51 mm) are of no value.

Cucumbers are processed after long-term aerobic fermentation or as a fresh pack. Recently, "overnight" fermentation products, which must be kept refrigerated, have become popular.

Fresh market cucumbers should be graded for appearance, size, and uniformity and packed in cartons or wirebound crates for shipment. In transit or in short-term storage, the temperature and humidity must be controlled. Temperatures in excess of 60°F (16°C) will cause rapid yellowing, an indication of senescence. Most storage or transit is thus at an average of 50°F (10°C) or slightly less. Below 50°F, the fruit may suffer chilling injury, although the incidence would not be high until exposed to less than 45°F (7°C). Relative humidity must be maintained at 95 percent; otherwise, the fruit become flaccid. Fruit shipped long distances often are waxed to reduce moisture loss. Although cucumber skin is naturally waxy, an additional coating can reduce transpirational loss by 50 percent, primarily by covering the stem scar or petiole base. Fruit must be clean and free of pathogens before waxing, and the amount used must not be excessive; otherwise, the normal gas exchange may be altered, resulting in surface pitting.

Greenhouse Cucumbers

The European seedless cucumber can be produced in greenhouses as a fall or spring crop, the latter the more profitable of the two. Most growers use soilless media in greenhouses maintained at 78 to 85°F (26 to 29°C). Plants are started in peat pots and transferred to ground beds, using twin rows, with plants spaced 18 in. (46 cm) in the row arranged alternately to maximize utilization of the rooting area. The leaders are trained vertically, and the side branches are pruned to the second leaf from the main stem. Early fruit set is removed to stimulate growth and subsequent yield. When the leader reaches the top horizontal wire, two laterals are trained downward, and emerging side branches are removed completely.

A single plant should produce 18 to 20 fruit under spring conditions, fewer in the fall. Each fruit is 14 to 16 in. (35 to 41 cm) long with a very thin rind and must be cooled and overwrapped rapidly to preserve quality.

MUSKMELON

Cantaloupe Casaba
Honeydew Crenshaw
Persian

Historical Perspective and Current Status

There is indirect and incomplete evidence that muskmelon was known in Egypt in 2400 B.C. It is native to Iran and adjacent areas, but little was reported of its use until Pliny described it in the first century A.D. Galen, a Greek physician, wrote of its medicinal value in the second century, and Roman writers gave advice on growing and on preparing melons for eating. Charles VIII of France introduced the plant to central and northern Europe.

Early muskmelon lines were not introduced rapidly to new geographical regions, perhaps because quality was so variable. The crop did not become common in Spain until the fifteenth century, and through subsequent explorations, it was introduced to the New World. By the mid-1600s, muskmelon was grown by Indians from Florida to New England. The modern cantaloupe is derived from the 'Netted Gem,' a highly netted cultivar introduced by the W. Atlee Burpee Company in 1881. Subsequent improvements in quality have made the muskmelon a popular dessert vegetable.

Most commercial shipments of melons now originate from the west coast and the south, but many states produce melons for local sales and some interstate commerce. The leading producing state is California (70 percent of the total), followed by Texas and Arizona. The total U.S. production of muskmelon exclusive of small acreages, is approximately 13,340,000 cwt (605,102 MT). An additional 3,780,000 cwt (171,460 MT) of honeydew is produced annually. Other melon types represent a small total acreage relative to muskmelon production. The total harvested acreage in the United States is approximately 113,980 acres (46,162 ha).

Nutritionally, the muskmelon is much higher in vitamin A than the white-fleshed honeydew or winter melons (casaba or crenshaw). Otherwise, the differences in food value are minor.

Classification, Growth, and Development

All melons are forms of *Cucumis melo* but traditionally have been subdivided into several botanical varieties:

1. *C. melo* var. *cantaloupensis* is the true cantaloupe, not grown in the United States. Fruit are medium size with an unnetted warty tough rind.

2. *C. melo* var. *reticulatus* comprises the netted melons, both the aromatic melon of commercial importance and the Persian melon. Flesh color can range from orange to green.

3. *C. melo* var. *inodorous* lacks the aromatic flavor and includes the casabas, crenshaws, and honeydews.
    a. Casabas have a bright yellow, corrugated rind and white flesh.
    b. Crenshaws are green, ripening to a dull yellow with a pale orange flesh. The skin is mostly smooth with some wrinkling at the blossom end.
    c. Honeydew ripens to a creamy white rind and a light green flesh. The skin is very smooth.
    Melons in this group are late maturing, and some are termed "winter melons" for their adaptation to moderately long storage.

4. *C. melo* var. *flexuosus* is termed snake melon because of the slender, long shape. These usually are consumed immature, similar to cucumber.

5. *C. melo* var. *conomon,* the oriental pickling melon, is characterized by small fruit with mottled skin.

6. *C. melo* var. *chito* includes mango melon and garden lemon. The fruit are small and smooth with an acidic flavor and are used for pickling.

7. *C. melo* var. *dudaim* includes the pomegranate melon or Queen Anne's pocket melon. The small round fruit become very pubescent and have a musky odor.

Crosses among some of these groups have led to the development of new cultivars combining features of previously distinct botanical types.

The commercial melon types are warm season annuals, with branched vines extending 2 to 7 ft (0.6 to 2.1 m). The tendrils are unbranched, located in the leaf axils. Rooting characteristics are similar to those of cucumber.

The first flowers to appear are clusters of staminate flowers. As the photoperiod begins to decline, the plant branches, and the "female" blossoms develop at the first and second nodes of the branches. The female flowers are structurally perfect flowers, although in some, the stamens do not develop. The plant, however, is termed andromonoecious. As the days become short, female flowers again are inhibited, and male flowers predominate.

Most flowers have three carpels, reflected in the number of seed locules at maturity. The fruit shape ranges from elongate to round, ribbed, or smooth. The netting on *reticulatus* develops as rapidly dividing cork (callus) cells located just beneath the surface break through the epidermis and enlarge. The edible tissue of the fruit is pericarp.

At maturity, the muskmelon forms an abscission layer at the attachment point of the fruit, first appearing as a slight crack. As this crack encircles the stem, the fruit slips from the vine, a stage correlated with peak maturity and flavor. Winter melons do not slip from the vine.

Muskmelons are more sensitive to environmental variation than cucumbers. Soil temperatures below 60°F (16°C) severely restrict water absorption by the roots. Optimum growth requires 65 to 75°F (18 to 24°C). Germination and emergence, as in cucumber, increase as soil temperatures increase from 60 to 90°F (16 to 32°C). Pollination is reduced at temperatures below 70°F (21°C).

Crop Establishment and Maintenance

Well-drained, sandy or silty loam soils are preferred for melon production. Fruit developing on plants grown in peat or muck soils develop excessive size with poor sugar content and soft flesh. Light soils generally are free from compaction, and the early warming extends the effective growing season. Muskmelons are sensitive to acidic soils (below pH 6.0), showing yellowed foliage and a marked increase in abortion of the perfect flowers. However, there are cultivar differences in pH sensitivity. 'Saticoy,' for example, is tolerant; 'Harper Hybrid' is very sensitive. The optimum range for all cultivars is pH 6.0 to 6.8.

Muskmelon growth is enhanced noticeably by organic matter, regardless of soil type. Tillage therefore should be preceded by a green manure crop and/or a generous application of animal manure. Melons are especially heavy users of nitrogen and potassium, and quality is improved with reasonably high levels of potassium and calcium. The fruit remove 30, 12, and 62 lb (13.6, 5.4, and 28 kg) of nitrogen, phosphorus, and potassium, respectively, assuming a yield of 13,500 lb (6129 kg). Supporting vine growth removes 20, 4, and 28 lb (9, 1.8, and 12.7 kg), respectively. Fertilizer recommendations are similar to those recommended for cucumber for the same soil type.

Melons are established by direct seeding or by transplants. Where the crop is direct seeded, fertilizer generally is applied in split applications, broadcast or banded at planting, followed by one or more sidedressings of nitrogen until lay-by. In Texas, phosphorus is applied during field preparation and also banded with the seed.

Melons should be planted as soon in the season as soil temperatures allow. Studies have shown consistently that early planting increases vine growth and, as a consequence, total yields. Melons should be direct seeded approximately 2 weeks after the average frost-free date or transplanted when the risk of cold soils is minimal. The planting dates can be advanced by 1 or 2 weeks using plastic mulch and row covers and/or light sandy soils.

In areas with short growing seasons, transplants with two to four true leaves are field planted after a broadcast fertilization. Large transplants, those showing root growth extending well beyond the container, will suffer transplant shock. Substantial yield gains have been attained by applying starter fertilizers through the transplant water, especially in an integrated program with black plastic mulch and row covers. The plastic promotes rapid early growth, capitalizing on nutrients provided by starter fertilizer, thereby increasing early yield and total productivity.

Melons are direct seeded or transplanted in single rows or twin rows on flat ground or raised beds. The spacing varies according to melon vining habit. In Arizona and California, beds range from 4 to 7 ft (1.2 to 1.8 m) on center, with rows 18 to 24 in. (46 to 61 cm) apart on a bed, and plants thinned to 9 to 12 in. (23 to 30 cm) in the row. For a single-row hill system, 4 ft (1.2 m) between plant clumps is optimum. Close spacings in the row can be used to increase yield per acre, although fruit size will be reduced.

Plants exposed to periodic drought stress are not as productive as unstressed plants; however, excessive soil moisture, particularly when fruit are nearing maturity, has been shown to reduce soluble solids. In coarse sands of the midwest and east, $1\frac{1}{2}$ in. (3.8 cm) of water per week is required, somewhat less in soils with high-water holding capacity. Higher amounts would be required in arid climates.

Windbreaks are recommended where wind erosion is likely. Temporary windbreaks of winter wheat or rye prevent sandblasting of young seedlings and whipping of vines as plants begin to run. They also provide additional heat accumulation in the spring. The strips can be seeded the width of tillage equipment and removed before lay-by.

After lay-by, the heavy leaf cover of merging vines will suppress young weeds; thus, early season weed control is most important. Weeds can be controlled during early growth by cultivation, but many growers prefer to use preplant (bensulide) or preemergent (chloramben or dinoseb) herbicides or black plastic for full season weed suppression.

Fruit set, as in cucumber, depends on honeybee activity, and several bee colonies per acre (hectare) will improve productivity.

A large proportion of the soluble solids content of a melon fruit is acquired in the last several weeks of development prior to harvest. Maintaining soil nitrogen, avoiding water stress, and attending to other aspects of management are therefore critical.

Cultivars

The terminology applied to muskmelon is somewhat confusing. Cantaloupe may refer to all netted melons with an aromatic flavor, or the term may be used to describe only those netted melons that are relatively uniform with little or no ribbing (western shipping melons). Eastern melons for local markets, normally termed muskmelons, typically have coarse netting and prominent ribs. Other growers may refer to the latter as "Queen types" (typified by such cultivars as 'Harvest Queen'), using the name muskmelon to distinguish *Cucumis melo* from watermelon. The following is a general separation of cultivar types, exclusive of the winter melons described previously.

Shipping type (cantaloupe, 'Hales Best' type) Fruit are nearly round with dense, fine net and little or no ribbing. Flesh is orange, thick, with small seed cavity and mild flavor. Cultivars: 'Hales

FIGURE 19.6
Western shipping-type muskmelon. (Photo courtesy of Peto Seed Company, Saticoy, Calif.)

Best,' 'Top Mark,' 'Planter's Jumbo' (Figure 19.6).

Queen type Fruit are slightly longer than wide with prominent ribs and a very coarse netting, and tend to be larger than shipping melons. Flesh is deep orange, thick, smooth with a pronounced aromatic flavor and often with a very high sugar content. Cavity is often large. Cultivars: 'Burpee Hybrid,' 'Harvest Queen,' 'Gold Star'.

Mixed type Resulting from hybridization, this type includes features of honeydew, cantaloupe, and/or others, and generally is for home garden and/or local sale. Cultivar: 'Honeyloupe'.

Other melons Cultivars of honeydew, casaba, and crenshaw melons differ largely in earliness and adaptation. Most require a long growing season, although 'Earlidew' honeydew and 'Sungold' casaba are suitable for eastern areas.

Resistance to disease and to certain chemicals effective in controlling disease has been incorporated in some cultivars. In western states, sulfur-resistant melons have been valuable contributions to the management system for controlling powdery mildew.

Disease and Insect Pests

The diseases and insects of muskmelon are, to a considerable extent, the same as those described for cucumber (Figures 19.3 and 19.4). In addition, several diseases are predominantly pests of muskmelon.

Diseases *Fusarium Fruit Rot*   Fruit rot, caused by *Fusarium roseum,* a soilborne fungus, normally attacks only ripe fruit. Fruit spots, $\frac{1}{2}$ to 1 in. (13 to 25 mm), develop on the melon surface, extending $\frac{1}{2}$ in. (13 mm) deep. Affected internal flesh is white and dry, and a white mold develops on the surface. It is favored by wet conditions in the field or in storage.

Fusarium Wilt   The foliage wilt caused by *F. oxysporum* f. sp. *melonis* may be confused with bacterial wilt. However, *Fusarium* is a soilborne organism that may have several strains. Plants become stunted and turn yellow, wilt, and die. A dark streak often appears at the soil line on one side of the vine. The symptoms can be differentiated from bacterial wilt by the yellowing and lack of recovery from wilting at night and by absence of bacterial ooze. The only control is the use of resistant cultivars.

Powdery Mildew   Powdery mildew has been a serious disease problem in California for many years. Its description is provided in this chapter under Cucumber. Control in muskmelon has been achieved by breeding resistant cultivars, first by introducing cultivars resistant to sulfur, enabling growers to apply an effective chemical control, and later by adding resistance to the causal organism. Without resistance, other control measures are ineffective.

Leaf Spots   Leaf spot lesions may be due to infection by *Cercospora citrullina* or *Alternaria cucumerina. Cercospora* leaf spot is characterized by small dark brown spots with white centers. *Alternaria* spots are round, small, and water soaked, expanding and coalescing with concentric rings within the spots. Both diseases may be minimized with rotation and with protective sprays.

Insects   Melon plants are damaged by both striped and spotted cucumber beetles, aphids, pickleworm, and several of the general plant feeders (cutworms, wireworms, flea beetles). In addition, the **melon worm** (*Diaphania hyalinata*), closely related to the pickleworm, is a pest in the southern states. It is primarily a foliage feeder and, therefore, is easier to control than the pickleworm, using materials such as carbaryl.

Harvest and Market Preparation

Cantaloupes or muskmelons are harvested according to the degree of stem slip. At full slip (complete separation from the vine), the melons are at peak maturity and flavor and can be harvested for home use or on-farm sale but will soften too rapidly for marketing elsewhere. Cantaloupes that are just starting to slip (stem crack just appearing) are harvested for long-distance shipping. Although sugar content (soluble solids) in "quarter-slip" melons is lower than that at full maturity, the firmness prevents damage in transit. Preharvest ethaphon treatments have been used to advance and to concentrate the melon harvest; however, these treatments reduce soluble solids, texture, and flavor ratings in full-slip melons.

Nonslip melons—honeydews, crenshaws, and casabas—are harvested when the fruit color indicates maturity. Often, a color spot

FIGURE 19.7
Crenshaw melons packed for transit. (Photo courtesy of W. J. Lipton, USDA Market Quality & Transportation Laboratory, Fresno, Calif.)

where the fruit lies on the ground is a good indicator, and fruit size, finish color, and blossom end firmness also may be criteria. On honeydews, tiny "sugar cracks" may appear on the fruit surface. The appearance of these cracks appears to be correlated with maturity.

All melons are hand harvested, brushed free of soil, and graded according to uniformity of size and appearance. From 12 to 27 cantaloupe fruit are packed in 40-lb (18-kg) wooden crates or cardboard containers (Figure 19.7). Honeydew and other large melons are mostly packed in telescoping fiberboard containers holding 5 fruit separated by dividers.

It is important to remove field heat by hydrocooling, refrigerating, or icing, since quality changes and weight loss can occur rapidly. A melon exposed to the sun can attain interior temperatures of 78 to 90°F (26 to 32°C) and show sunburn (Figure 19.8).

The major quality factor in all melons is sugar. Although aromatic flavors certainly are important, the soluble solids content is highly correlated with a consumer's perception of quality, and it is affected by temperature and moisture during growth and maturation and by the stage of maturity at harvest. It is not unusual for a full-slip melon to approach 15 percent soluble solids. In contrast, melons picked "hard ripe" or early slip are between 8 and 12 percent. USDA standards define very good quality as a refractometer reading not less than 11 percent, good quality not less than 9 percent. Melons shipped from California must exceed 8 percent.

Some sugar increase may occur as the hard ripe melons are ripened at room temperature, but the percentage never equals that of a melon harvested at peak maturity. Peak maturity is short-lived, particularly at high temperature, and the flesh then becomes grainy, watery, with unappealing flavors.

Netting is correlated with quality, but the reasons for it are not understood. "Slickers," or melons that have not developed proper netting, invariably are poor in quality and typically develop late in the season or whenever the plants have been exposed to sufficient environmental or disease stress to interrupt normal development.

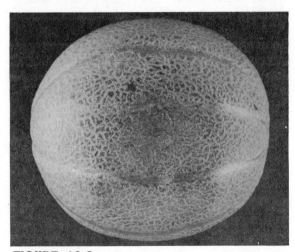

FIGURE 19.8
Solar injury on muskmelon. (Photo courtesy of W. J. Lipton, USDA Market Quality & Transportation Laboratory, Fresno, Calif.)

WATERMELON

Historical Perspective and Current Status

Watermelons have been cultivated since prehistoric times. They were grown by the ancient Egyptians and apparently cultured in areas of Asia Minor, Russia, and the Near and Middle East thousands of years ago.

The plant is native to central Africa where it has served as a source of water, a staple, and an animal feed. In some areas, it was fermented for wine. Among the wild and early cultivated watermelons were sweet and bitter forms, the latter gradually eliminated as seeds of sugary types were used to propagate the crop.

By the sixteenth and seventeenth centuries, all of the variation in color, shape, and size characterizing present cultivars was described by European botanists. Seeds were brought to America by early European colonists and planted in Massachusetts in the early 1600s. By the mid-1600s, they were being grown by Indians in Florida.

Development of cultivars for different growing requirements has expanded the areas where watermelon can grow successfully. However, commercial production, approximately 184,000 acres (74,520 ha) nationally, is concentrated in Florida, Texas, and California. Other southern states have significant acreages, and excellent melons also can be produced in certain sites in some of the northern states. The total U.S. production is 26,076,000 cwt (1,182,807 MT), over 100 million melons of normal commercial size. Some of the small acreages with roadside sales are not included in this total, nor is home garden production. Per capita consumption is 13.2 lb (6 kg).

Watermelon is considered a dessert vegetable and has relatively little nutritional value.

Classification, Growth, and Development

Watermelon (*Citrullus lanatus*) is an aggressive vining annual plant adapted to mean temperatures greater than 70°F (21°C). The root system is deep and spreading, penetrating 6 ft (1.8 m) or deeper.

The deeply lobed hairy gray-green leaves are distinctive, and the flowering is monoecious with the sequence of male and female flowering responding to daylength as described for other vine crops. Fruit may be round to oval in shape, ranging from 2 to 3 lb (0.9 to 1.4 kg) per fruit to over 25 to 30 lb (11 lb 13 kg).

Bees are required for good pollination, but female watermelon flowers will set fruit without fertilization. Normally, fertilization produces many seeds imbedded in the edible flesh. If the pollen source is, for some reason, incompatible, ovary development will be stimulated by the pollen; but, without fertilization, fully developed seeds will be lacking. Seedless watermelon cultivars, genetically triploids resulting from a cross of tetraploid females with diploid males, produce fruit normal in size and appearance, but seeds either are not formed or develop only a soft, hollow seed coat.

Crop Establishment and Maintenance

Watermelon plants require an average frost-free season of 4 months to mature a crop. Some cultivars, especially the small-fruited "ice box" types, are adapted to northern regions. The preferred soil is a light, fine blow sand, but any light soil with good drainage is suitable. Organic soils should be avoided.

Soil preparation, including incorporation of organic matter, is generally as described for muskmelon. The pH optimum is at 6.0 to 6.8, although the plant is not as sensitive to acidity as cantaloupe and will yield well in soils with a pH of 5.0 to 6.0. Fertilizer ratios appropriate for cantaloupe would be applied as a light broadcast or band at planting time, followed by side-dressings of nitrogen when plants are 3 to 5 in. (7.5 to 13 cm) high and again at lay-by in very light soils.

Most of the commercial crop is direct seeded to a final stand of one or two plants in each hill, with hills spaced 3 to 6 ft (0.9 to 1.8 m) apart in rows 8 to 10 ft (2.4 to 3 m) apart. Seeding should be delayed until the soil temperature exceeds 65 to 68°F (18 to 20°C). Germination is accelerated at temperatures of 77°F (25°C) or above, and transplants and hotcaps or row covers are used in short season areas to stimulate rapid germination and early growth. Moisture deficits, particularly late in the season, may result in poor fruit development. In California desert areas, up to 24 to 30 acre in. (24.7 to 31 ha cm) of water is needed on sandy soils, less on heavy soils.

Under normal spacing, each vine will mature no more than two full-size melons. If more than two set on a vine, size will be reduced. Some growers prune off excess fruit to ensure uniformity of size and shape.

Cultivars

Watermelon cultivars range from small (normally early maturing) to large fruited (late), and fruit may be oblong to round, with distinctive rind, flesh, and seed color. Most melons are grown for shipping, and rind toughness is an important characteristic.

"Black" melons, with dark green rind, include such cultivars as 'Klondike' and 'Peacock,' major shipping watermelons from California. Gray melons have a gray-green, sometimes mottled rind, typical of 'Charleston Gray' and 'Calhoun Gray.' Striped melons include 'Crimson Sweet,' 'Klondike Striped,' 'Jubilee,' and others. Most commercial cultivars are elongate in shape, and some are resistant to *Fusarium* wilt. Seed size in some of the new cultivars has been reduced, and the dark seed coat that contrasts with the bright red flesh color is preferred.

Ice box melons are suitable for local sales and home gardens. These cultivars are early, small fruited, quite seedy, with a thin rind. The first of this type was 'New Hampshire Midget,' but improved cultivars ('Sugar Baby' and others) have been developed, some with red flesh, others yellow.

Seedless cultivars are grown occasionally, but they are difficult to establish. Their early growth is slow, and germination requires a constant 85°F (29°C) for a good stand. As a consequence, most seedless cultivars are grown as transplants. An occasional row of a pollinator must be planted to provide pollen as a stimulus for ovary enlargement. Seedless watermelons are of exceptional quality, since softening of the flesh adjacent to the seeds is minimal compared to standard cultivars.

Disease and Insect Pests

Diseases   Among the diseases affecting muskmelons and/or cucumber, watermelon is susceptible to anthracnose, downy mildew, *Fusarium* wilt, gummy stem blight, and *Alternaria* and *Cercospora* leaf spot. Two virus diseases occasionally cause substantial losses. Watermelon virus and pimples virus stunt plant growth, and leaves become mottled and fruit deformed. These viruses are transmitted primarily by aphids. Elimination of weeds serving as alternate hosts and effective insect control are the only applicable control measures.

Insects   Aphids, cucumber beetles, leaf miners, leafhoppers, red spider mites, wireworms, and cutworms occur wherever the crop is grown. Each has been discussed previously with respect to other crops or as general feeders.

Harvest and Market Preparation

Watermelon maturity can be judged by several indicators. At prime eating quality, a pale yellow "ground spot" should be evident where the melon lies on the soil. If this spot is vivid yellow, the fruit may be overripe. In addition, there may be a slight change in bloom on the melon rind, and the tendril nearest the fruit attachment normally dies back. Thumping a melon is a reliable test for those with experience: a hollow metallic sound indicates that it is not yet ripe. As the internal temperature of a melon climbs during the day, however, this test becomes less reliable.

A crop may be sold as a field to brokers or may be sold by weight. Watermelons should be cut from the vine, not pulled, as damage to the stem end of the fruit may cause cracking. The fruit are placed in bulk pallets or trucks and marketed in bulk or packed in fiberboard containers holding four to five melons.

As in muskmelon, watermelon quality is highly correlated with content of the primary sugars, fructose, glucose, and sucrose (preferably 10 percent or higher at maturity). Proper handling of fruit picked at prime quality is essential. The melon flesh must be crisp and sweet and well colored. Tissue breakdown due to poor handling or overmaturity occurs rapidly, and texture becomes grainy and flavor insipid.

Watermelons are sensitive to chilling temperatures [lower than 50°F (10°C)], and a storage range of 50 to 60°F (10 to 16°C) is recommended. Relative humidity is not critical, since little moisture is lost from the fruit. Prime quality can be maintained in storage for up to 2 weeks.

SQUASH AND PUMPKIN

Historical Perspective and Current Status

All species of squash and pumpkin are native to America. Winter squash originated in northern Argentina near the Andes or in Andean mountain valleys and had not appeared in Central or North America prior to the Spanish conquest. Pumpkins and cushaw, butternut, and summer squashes apparently originated in Mexico and Central America. Long before the discovery of the New World, seeds of these squash types had been carried north and had become important food plants for Indians.

Squash became popular in early New England, in part because it was adapted and also because it provided an excellent food source well into the winter. Seeds of 'Hubbard' squash were listed by J. J. Gregory, a seedsman from Marblehead, Massachusetts, in 1855, although the cultivar had originated there approximately 60 years earlier. Squash seeds carried to Europe by returning sea captains became popular in several European countries in the 1500s. Summer squash, particularly, became important in Italy, and many current cultivar names ('Cocozelle,' 'Zucchini') reflect its popularity among Italian horticulturists.

Today, squash and pumpkins are grown primarily in northern and central regions of the United States. Both are grown for fresh market and processing.

Winter squash, fully matured on the vine, is superior in food value to summer squash, but there are substantial differences among winter squash cultivars. 'Butternut' is the highest in vitamin A, carbohydrate, and minerals. 'Hubbard' also is superior in nutritive value to many other winter squash cultivars.

In the United States, seeds usually are discarded, although pumpkin seeds, roasted and salted, provide a nutritious confection. A "naked-seed" mutant, in which seeds lack the normal tough seed coat, has been introduced in some cultivars of pumpkin and squash. Such cultivars, listed as edible seeded, provide an excellent source of protein (34 percent of seed weight) and fat (46 percent) in addition to the nutritional value of the fruit flesh. Squash seeds have almost the same percentages of constituents as peanuts and compare favorably in yield of seeds per acre.

Classification, Growth, and Development

There are four dominant species of squash and pumpkin (Table 19.2, Figure 19.9). *Cucurbita maxima* (winter squash, including 'Hubbard,' 'Buttercup,' and 'Banana' types and several very large pumpkins) is characterized by relatively nonlobed leaves, round soft stems, and a round fruit stem (peduncle) with an abrupt, corky attachment to the fruit. Fruit may be very large or small, but all are grown to the hard-shell mature stage, and those used as squash have very fine-textured flesh. *Cucurbita pepo* (Figure 19.10) includes summer squash (vegetable marrow), many pie, Jack-o-Lantern, and field pumpkins, the baking cultivar 'Acorn' or 'Table Queen,'

TABLE 19.2
Tabulation of several characteristics differentiating common cultivated *Cucurbita*

Species	Seta	Stem	Androecium	Peduncle	Fruit flesh	Funicular attachment of seed	Seed margin
C. pepo	Spiculate	Hard, angular	Short, thick conical	Hard, angled, ridged	Coarse	Obtuse, symmetrical	Smooth, obtuse
C. moschata	Lacking	Long, slender, columnar	Long, slender columnar	Hard, smoothly angled, flared	Fine grain	Obtuse, slightly asymmetrical	Scalloped, obtuse
C. mixta	Lacking	Hard, angular	Long, slender columnar	Hard, angular, enlarged by hard cork	Coarse	Obtuse, slightly asymmetrical	Barely scalloped, acute
C. maxima	Moderately spiculate	Soft round	Short, thick columnar	Soft, round, enlarged by soft cork	Fine grain	Acute, asymmetrical	Smooth, obtuse

Source: From T. W. Whitaker and G. W. Bohn (1950) *Economic Botany* **4**:65. Reprinted by permission of the authors.

and most ornamental gourds. These squashes are characterized by distinctly lobed leaves, hard angular stems, and an angled fruit peduncle with a strongly flared attachment to the fruit. Fruit shapes and colors and stage of harvest maturity vary widely among cultivars. In the United States, summer squash fruit are consumed at a young immature stage, before the skin becomes hardened and tough. Marrow summer squash have a smooth, thin but hard rind. Pumpkins are grown to maturity for use as Jack-o-Lanterns or as pie ingredients and have a stringy coarse flesh. *Cucurbita moschata,* used as a winter squash or pumpkin, has slightly lobed leaves and angular stems, but the fruit attachment is not flared. *Cucurbita mixta* once was included within *C. moschata,* but there are some taxonomic and quality differences that distinguish them. *C. mixta* has an angular peduncle, like *C. moschata,* but it is enlarged by corky growth, and there are differences also in seed margins and flesh texture. *C. mixta* is used primarily as a processing squash (Table 19.3). A fifth species, *C. ficifolia,* has been cultivated in Mexico and other Latin American countries.

Most pumpkins and winter squash are vining, bearing tendrils, although some bush cultivars which have no tendrils recently have been introduced. All common summer squash cultivars are bush plants. The root systems are deep and spreading, especially so for vining types.

Flowering in all squash is monoecious. Pollination requires bee visitation, and the appearance of male and female flowers responds to daylength, similar to other cucurbits.

Plant growth is favored by mean temperatures between 60 and 80°F (16 and 27°C), and the plants do not tolerate temperatures near freezing. Plant roots of pumpkin and squash, as of other cucurbits, are inefficient in cold soils, and plants may suffer permanent wilting under such conditions.

Crop Establishment and Maintenance

Squash and pumpkins can be planted on a wide variety of well-drained soils. Light-textured soils are preferred in northern areas, since they warm quickly to 59°F (15°C), the minimum temperature for germination. Hot caps, black plastic, and plastic row covers have been used to warm the soil in early spring, and the consequent acceleration of emergence has improved stand and final yield.

The optimum soil pH is 5.5 to 7.5. Soil preparation, including addition of organic matter, is as described for other vine crops.

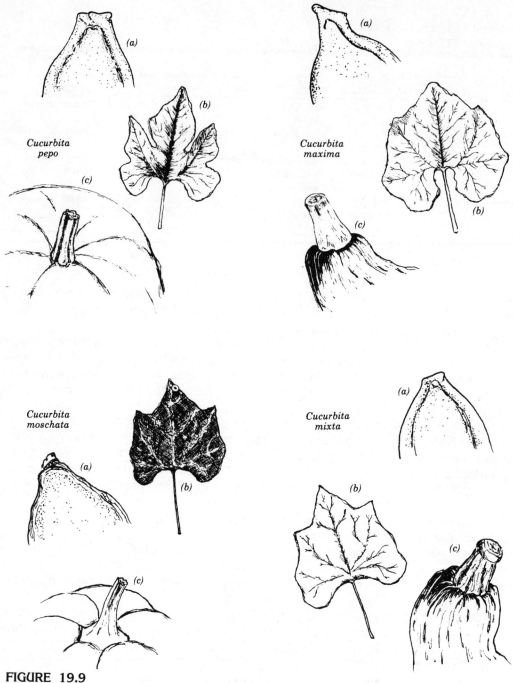

FIGURE 19.9
Characteristics separating the species of *Cucurbita*: (*a*) seed margin and attachment; (*b*) leaf margin and lobing; (*c*) shape and flaring of fruit peduncle.

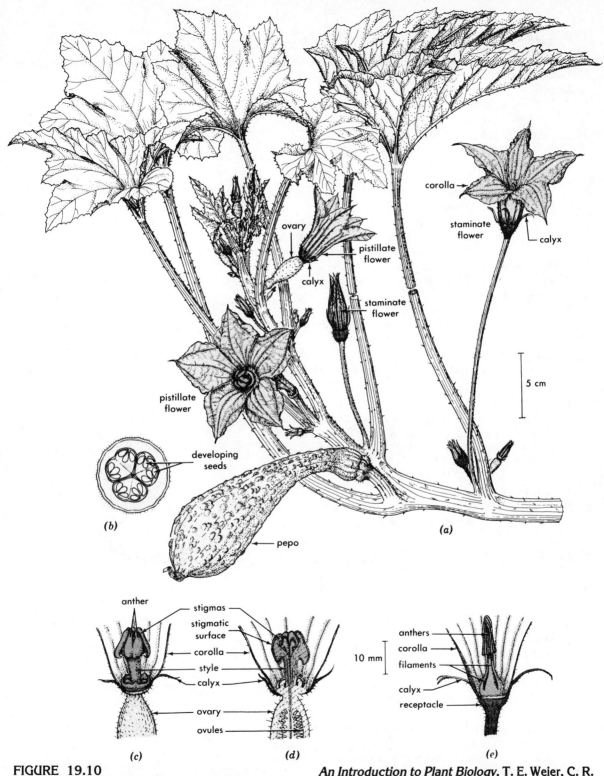

FIGURE 19.10
Typical growth habit and features of the vine crops (shown is *Cucurbita pepo*): (*a*) plant; (*b*) ovary; (*c*, *d*) female flower; (*e*) male flower. (From *Botany.*

An Introduction to Plant Biology, T. E. Weier, C. R. Stocking, M. G. Barbour, and T. L. Rost. Copyright © 1982 by John Wiley & Sons, Inc., New York, p. 637. Reprinted by permission.)

TABLE 19.3
Characteristics of the major types of cucurbits

Curcurbita pepo

I. Squash
  A. Summer squash
     1. Straightneck type: neck straight and more slender than base, surface yellow and warty — Early Prolific, Straightneck
     2. Crookneck type: neck slender and curved, surface yellow and warty — Sundance, Crookneck
     3. Marrow type
        a. Zucchini: straight, elongate fruit with thin smooth skin, whitish flesh, green or gold skin — Zucchini, Goldrush
        b. English marrow: fruit cylindrical, short and blunt, greenish flesh, skin pale green becoming white at maturity — Vegetable Marrow
        c. Italian marrow: primarily striped fruit, but size and shape similar to English marrow — Cocozelle
     4. Scallop type: flat or saucer shape, with thickened waved edges, skin green or white — Peter Pan, Scallopini
  B. Winter squash
     1. Acorn type: deeply furrowed, small fruit with pointed blossom end, dark green or orange hard rind — Table Queen, Jersey Golden
     2. Novelty types
        a. Vegetable spaghetti: flesh soft and stringy, appearing shredded — Vegetable Spaghetti

II. Pumpkin
  A. Standard types: coarse flesh, thick orange rind, shallow ribs, used in pie mixes or as Jack-o-Lantern
     1. Processing — Connecticut Field, Early Sweet Sugar, Howden's Field, Small Sugar
     2. Jack-o-Lantern — Tricky Jack, Jack-o-Lantern, Spirit
  B. Naked-seed types: variable in appearance, but seeds lack seed coat and can be roasted and eaten as confection — Lady Godiva

Cucurbita moschata

I. Squash: neck often more narrow than base, thin hard tan rind, deep orange fine-textured flesh — Butternut, Waltham
II. Pumpkin: large fruit, larger at base than neck, neck often curved — Golden Cushaw, Large Cheese

Cucurbita maxima

I. Squash: gold, gray-green to green thick hard rind, variable shape, fine-textured flesh
  A. Hubbard type: large, warted fruit, constricted at both ends, gold or blue-green rind — Blue Hubbard, Golden Hubbard
  B. Delicious type: top-shaped, warted, large fruit, gold or green rind — Golden Delicious, Green Delicious
  C. Marrow type: large lemon shape with irregular rind surface, orange color — Boston Marrow
  D. Buttercup or turban type: medium-size fruit in which the rind does not cover the ovary at the blossom end, green or gold rind — Buttercup, Golden Turban
  E. Banana type: elongate fruit with pointed ends, smooth to slightly warted, orange or gray-green skin — Banana

TABLE 19.3 (*Continued*)
Characteristics of the major types of cucurbits

Cucurbita mixta	
I. Pumpkin	
A. Cushaws: constricted and/or curved neck, rind solid green or white or striped	Green Striped Cushaw, White Cushaw
B. Pear-shaped: large pear-shaped fruit, solid rind color	Tennessee Sweet Potato

Source: Whitaker and Davis (1962).

Fertilizer applications are split, with half applied at seeding (band or broadcast) and the remainder as one or two sidedressings prior to lay-by. Nutrient uptake by squash is similar in ratio to the other cucurbits.

Vining cultivars of squash and pumpkin are direct seeded 1 in. (2.5 cm) deep in hills spaced 6 to 8 ft (1.8 to 2.4 m) in rows 10 to 12 ft (3.0 to 3.6 m) apart. Bush squash is spaced 3 to 5 feet (0.9 to 1.5 m) apart in rows 6 ft (1.8 m) on center. Some thinning may be necessary to obtain a final stand of two plants per hill. Cross-cultivation may be used to control weeds until lay-by, but the root system can be damaged if cultivation is deep. Herbicides such as chloramben (at planting) or dinoseb (preemergent) may be used, but some herbicides can cause injury to certain cultivars or under certain soil conditions. All label directions must be followed carefully.

Cultivars

There is a wide choice of cultivars of all types of squash and pumpkins, most of which are widely adapted. Many of the differences among summer squash cultivars (Figure 19.11) are cosmetic—differences in color, shape, and size rather than quality or productivity (Table 19.3). Wide differences in quality and use exist among winter squash and pumpkin cultivars.

Disease and Insect Pests

Diseases Of the diseases described for other vine crops, powdery mildew, *Alternaria* leaf spot, and viruses are the primary pests of squash and pumpkins. The virus diseases, cucumber mosaic virus and watermelon virus 2, are particularly serious, causing a vivid fruit mottle and severe deformation. In addition to these diseases common to other cucurbits, black rot and *Choanephora* wet rot may become problems.

Black Rot Black rot is caused by the gummy stem blight fungus (*Didymella bryoniae*) and affects mostly winter squash and pumpkins. It first appears as an irregular green or yellow spot that gradually turns brown, then black as the fungus penetrates the rind. A dry rot results, but the affected area often becomes infected with secondary organisms that eventually spread through the entire fruit. The organism overwinters in seed and in plant residue. Crop rotation and use of clean seed are important control measures.

Choanephora Wet Rot The organism (*Choanephora cucurbitarum*) attacks summer squash as the blossoms wilt and quickly spreads down the fruit. A black mold, resembling tiny pinheads, appears on the necrotic area, and spores are spread by insects and splashing water. This disease is common under high moisture conditions.

Insects Of the insects common on other cucurbits, aphids, cucumber beetle, pickleworms, leafhoppers, and spider mites attack squash. In addition, stem borers and squash bugs (Figure 19.4) often cause substantial losses.

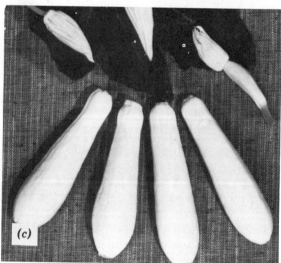

FIGURE 19.11
Summer squash fruit types: (*a*) 'Zucchini'; (*b*) scallop; (*c*) straightneck; (*d*) crookneck. (Photos courtesy of Peto Seed Company, Saticoy, Calif.)

Squash Vine Borers   The vine borer (*Melittia cucurbitae*) can be very destructive. The first symptoms of feeding are wilting of the vine or a portion of it. The stem will be hollowed out, filled with a slimy frass from borer feeding. The caterpillar is up to 1 in. (2.5 cm) long and white with a brown head. Occasionally, frass can be seen at the base of the plant, evidence of borer penetration. Such vines normally rot and die. The insect overwinters in the soil as a larva or pupa. When the vines begin to run, the pupa surfaces and splits to release the small black wasplike moth. The moth lays eggs on the basal portion of the stem, and in 1 to 2 weeks, borers emerge and penetrate the stem. Control is difficult, but sprays applied during egg laying and hatching (methoxyclor, carbaryl) are recommended.

Squash Bug   Squash bugs (*Anasa tristis*) are distributed throughout all production areas. The adult, brownish black, flat-backed, $\frac{5}{8}$ in. (15 mm) long, lays its eggs on the underside of leaves, usually in rows at a right angle. The eggs hatch, giving rise to bright-colored nymphs which, together with adults, feed on plant sap. This insect can kill small plants completely and cause leaves of large vines to wilt and die. The control recommendations are the same as for borer.

Harvest and Market Preparation

Summer squash are harvested normally 2 to 6 days after anthesis, when fruit are 6 to 8 in. (15 to 20 cm) long. The fruit are cut from the plant, leaving a short peduncle [1 to 2 in. (2.5 to 5 cm)] on the fruit. Because summer squash are thin skinned, they must be marketed quickly and maintained at high relative humidity, with storage and transit temperatures no lower than 50°F (10°C). Chilling injury will occur after several days below 50°F.

Winter squash normally remain in the field until the rind hardens and dulls in surface color. At 80 to 85°F (27 to 29°C) and 85 percent relative humidity, surface cuts will suberize, and the squash then may be placed in a relatively dry (60 percent relative humidity) storage of 50 to 59°F (10 to 15°C). Most winter squash, but especially the Hubbards, can be stored for up to 6 months under these conditions. During storage, starch converts to sugars rapidly, and total carbohydrates decline. At 6 months of storage, total carbohydrates are approximately one-half of those at harvest. Water loss also is high during storage.

Marketing differs by product. Summer squash fruit are graded for uniformity and appearance and packed in cardboard containers. Pumpkins and squash are moved by large bulk boxes or bulk trucks. Supermarkets may sell squash sections, prepeeled chunks, or whole fruit. For whole fruit, 'Buttercup,' 'Butternut,' and 'Acorn' are popular because of their convenient size.

CHAYOTE

Chayote (*Sechium edule*) is native to southern Mexico through Central America. The plant is a monoecious vining perennial reaching 50 ft (15 m) in length. Chayote is grown primarily for its pear-shaped fruit, each containing a single large seed. However, the swollen starchy roots and young leaves also are consumed.

The crop is propagated both sexually and vegetatively. Shoots can be removed from the crown and rooted in sand. The rooted cuttings then are established in pots before being moved to the field. For seed propagation, the fruit is allowed to ripen fully on the vine and is harvested before the seed begins to sprout. The entire fruit is placed in the soil.

As the plant develops, it is normally trellised. Chayote can be grown in tropical and subtropical climates and in mild temperate areas with a long season. Soils must be well drained and high in organic matter. The fruit are sensitive to chilling and should be kept at 50 to 59°F (10 to 15°C) for marketing.

SELECTED REFERENCES

Atkins, E. L., E. Mussen, and R. Thorp (1979). *Honeybee Pollination of Cantaloupe, Cucumber and Watermelon,* University of California Coop. Ext. Leaflet 2253.

Flocker, W. J., J. C. Lingle, R. M. Davis, and R. J. Miller (1965). Influence of irrigation and nitrogen fertilization on yield, quality and size of cantaloupes. *Proceedings, American Society for Horticultural Science* **86:**424–432.

Kasmire, R. F., K. B. Tyler, D. M. May, K. S. Mayberry, H. Johnson Jr., J. E. Dibble, D. G. Kontaxis, and D. E. Johnson (1981). *Muskmelon Production in California,* University of California Coop. Ext. Leaflet 2671.

Longbrake, T., J. Parsons, and R. Roberts (1980). *Keys to Profitable Watermelon Production,* Texas A & M University System Coop. Ext. Bull. B-1317.

Nitsch, J. P., E. B. Kurtz, J. L. Liverman, and F. W. Went (1952). The development of sex expression in cucurbit flowers. *American Journal of Botany* **39:**32–43.

Scholz, E. (1979). *Muskmelon Production and Marketing,* University of North Dakota Coop. Ext. Bull. 27.

Schweers, V. H., and W. L. Sims (1977). *Watermelon Production,* University of California Coop. Ext. Leaflet 2672.

Whitaker, T. W., and W. P. Bemis (1975). Origin and evolution of the cultivated *Cucurbita. Bulletin of the Torrey Botanical Club* **102:**362–368.

Whitaker, T. W., and G. W. Bohn (1950). The taxonomy, genetics, production and uses of the cultivated species of *Cucurbita. Economic Botany* **4:**52–81.

Whitaker, T. W., and G. N. Davis (1962). *Cucurbits: Botany, Cultivation, and Utilization.* Interscience, New York.

Wittwer, S. H., and S. Honma (1979). *Greenhouse Tomatoes, Lettuce and Cucumbers.* Amer. Vegetable Grower, Willoughby, Ohio.

STUDY QUESTIONS

1. Why is gynoecy important in a cucumber cultivar? How does the culture of gynoecious cultivars differ from that of standard cultivars?

2. What are the important interrelationships between insects and diseases in several of the vine crops? What are the control mechanisms?

3. Although there are similarities among the vine crops, the quality criteria differ substantially. How do they differ?

4. The vine crops thrive in soils amended with manure or other organic matter, yet they do not produce particularly well in organic soils. Why?

20

Sweet Corn

Historical Perspective and Current Status

There is no consensus on the origin of modern corn. Wild maize never has been found, but a related species, teosinte (*Zea mexicana*), appears to be either an immediate ancestor of corn or a related species that evolved from the same primitive source. It is theorized that this primitive source was carried north from the lowlands east of the Andes. Domestication of corn apparently took place in Central America, resulting from selection for adaptation and productivity. This domestication resulted in a plant ill-equipped to survive in the wild, since natural seed dispersal is restricted by the tight ear husk.

Although the ancestry of corn is a mystery, its antiquity is not. Carbon dating of pollen grains has established an age of 8000 years, and there are several archaeological records of maize seeds dating to 3000 to 5000 B.C. During its domestication, the specific pop, dent, flour, and flint corns developed, and these eventually were carried north and east through North America. Sugary forms probably had occurred throughout this evolutionary development, but they were not favored, because the grain was not suited to storage. Not until white settlers were introduced to corn, by then grown widely by the American Indians, did the sugary form become popular. The first reference date for sweet corn is 1779, when an eight-rowed, red cob type called 'Susquehanna' or 'Papoon' was introduced. However, sweet corn probably had been grown for at least 20 years before that date. By 1900, 63 cultivars were grown. The first F_1 hybrid was introduced in 1924 by the Connecticut Agricultural Experiment Station.

In its modern history, corn has been the keystone of an agricultural revolution. Grain corn has become a world staple, exceeded in use only by wheat and rice. The grain and fodder are integral to the livestock industry and to dairy production; fiber by-products have been used in construction; distillates have supplemented or, in some instances, replaced gasoline as a fuel and are used in the alcoholic beverage industry; corn oil has become important in cooking and in oleomargarine; corn sugar has partially replaced beet and cane sugar in many processed food products; and popcorn and sweet corn have become consumer favorites available throughout the year.

Sweet corn does not have the food value of matured grain. A typical cultivar of field corn will contain 72 percent starch, 10 percent protein, and 4.8 percent oil at full maturity. Sweet corn changes rapidly in composition as it approaches prime quality, but at its peak, it contains 5 to 6 percent sugar, 10 to 11 percent starch, 3 percent water-soluble polysaccharides, and 70 percent water and contributes moderate levels of protein, vitamin A, and potassium (Tables 20.1 and 20.2).

Sweet corn production is concentrated in temperate to subtropical areas, the latter providing the fresh winter crop. The total area in the United States devoted to processing and to commercial fresh sweet corn is 642,168 acres (260,078 ha). Florida (winter crop) leads in production of fresh sweet corn, with 30 percent of the total, followed by New York and California. For the processing crop, Minnesota and Wisconsin produce 50 percent of the total, followed by Washington and Oregon. Approximately 14 million cwt (635,040 MT) of corn is marketed as corn on the cob; the processing crop totals nearly 2,400,000 tons (2,177,280 MT), of which two-thirds is canned. Total per capita consumption in the United States (Table 20.3) is approximately 28 lb (12.7 kg, on-cob basis).

TABLE 20.1
Nutritional constituents of sweet corn

Constituent	Average content [a]
Water (%)	73
Energy (cal)	96
Protein (g)	3.5
Fat (g)	1.0
Carbohydrate (g)	22.1
Vitamin A (IU)[b]	400
Vitamin C (mg)[c]	2.0
Thiamine (mg)	0.15
Riboflavin (mg)	0.12
Niacin (mg)	1.7
Calcium (mg)	3.0
Phosphorus (mg)	111
Iron (mg)	0.7
Sodium (mg)	Trace
Potassium (mg)	280

Source: National Food Review (1978), USDA.

[a] Data per 100 g sample fresh wt.

[b] 1 IU = 0.3 μg vitamin A alcohol.

[c] Ascorbic acid.

TABLE 20.2
Changes in carbohydrates during the ripening of sweet corn

Age after silking (days)	Total sugars (%)	Reducing sugars (%)	Nonreducing sugars (%)	Starch (%)
5	3.81	3.07	0.74	1.38
10	4.37	2.73	1.64	1.82
15	5.31	1.49	3.82	9.12
20	3.95	1.06	2.89	16.82
25	3.02	0.75	2.27	21.76
30	2.68	0.61	2.07	24.97

Source: Culpepper and Magoon (1924).

Classification, Growth, and Development

Corn (*Zea mays*), a member of the grass family (Gramineae), has been classified into several distinct types.

Dent (var. *indentata*) Sides of kernel are flinty with a central soft floury core. Upon drying, the soft core contracts, resulting in a dented crown.

Flint (var. *indurata*) Endosperm is mostly hard. Kernel is smooth with rounded crown. Pericarp is thick.

Flour (var. *amalacea*) Endosperm is soft. Pericarp is thin.

Pop (var. *everta;* syn. *praecox*) Endosperm is hard. Pericarp is very thick. Kernel is pointed. When heated, trapped steam causes the kernel to explode, exposing the white endosperm.

Sweet (var. *rugosa;* syn. *saccharata*) Endosperm is translucent. Kernel is distinctly wrinkled on drying. Pericarp is variable in thickness.

Pod (var. *tunicata*) Kernels are enclosed in glumes or pods, entire ear in a husk. Thought by some to be involved in evolution of modern corn.

Sweet corn is a herbaceous, annual warm season monocot. It is more tolerant of cold temperatures, however, than other warm season vegetables and is the only one handled as a cool season crop to preserve harvested

TABLE 20.3
U.S. sweet corn consumption and processed pack, 1974 to 1982

	Canned sweet corn			Frozen							Fresh per capita consumption (on-cob basis)	
				Cut		Corn on cob (cut basis)						
Year	Cases 303s (1000)	Per capita consumption		1000 lb	1000 kg	1000 cwt	1000 MT	Per capita				
		lb	kg					lb	kg	lb	kg	
1974	46,431	6.0	2.7	2,982	135.3	1.52	68.9	1.6	0.7	7.6	3.5	
1976	54,694	5.7	2.6	2,824	128.1	1.88	85.3	1.5	0.7	8.1	3.7	
1978	57,907	6.0	2.7	3,031	137.5	3.07	139.3	1.8	0.8	7.0	3.2	
1980	50,574	5.8	2.6	2,705	122.7	2.59	117.5	1.9	0.9	7.4	3.4	
1982	60,522	5.4	2.5	4,557	206.7	3.79	171.9	2.0	0.9	7.0	3.2	

Source: Agricultural Statistics (1984), USDA.

product quality. The morphology, growth, and development features of all corn types are quite similar. Of the features distinguishing sweet corn from grain (dent) corn, the sugary gene (*su*) is the most obvious, but there are other growth and quality differences as well, including pericarp characteristics and extent of plant tillering.

Corn is monoecious (Figure 20.1): the tassel, the terminal inflorescence, comprises the male flowers, and the lateral ear shoots include female florets borne on a central axis. The tassel consists of a central spike with several lateral branches arranged in a loose panicle. The flowers occur as paired spikelets, and, within each, there are two florets, one usually maturing slightly earlier than the other. Each floret contains three anthers, and the pollen is vectored entirely by wind. The female flowers also occur as paired spikelets, but on a central axis (cob), each flower with two florets. In most corn, the lower floret aborts, and the subsequent caryopses, or grains, one per spikelet, therefore develop as paired rows. In zig-zag cultivars ('Country Gentleman' types), no florets abort, and the kernels become crowded and develop with no rowing. The long silk, comprising the style and stigma, protrudes beyond the ear and intercepts the airborne pollen. At the base of each kernel, near the cob, are the glumes, lemma, and palea (chaff), all of which are, morphologically, modified leaves.

Because the basal nodes of an ear shoot are concentrated, with little internode space, they form a tight envelop or husk around the developing ear.

Each individual kernel is a single-seeded, dry indehiscent fruit (caryopsis) composed of a relatively small embryo and a large area of endosperm (Figure 20.2). The outer layer of the endosperm is termed aleurone, and it is enclosed by the pericarp, a maternal tissue.

Leaf structure in corn is typical of the grasses. As leaves develop, the central stem becomes enveloped by leaf sheaths which extend from the node to the ligule, then angle outward in a broad, flattened leaf blade (Figure 20.1). The stem is a rigid and tall, with prominent nodes. At the basal nodes, adventitious roots (brace roots) develop, many serving to anchor or brace the plant. Most of the mature root system in corn is adventitious, arising from these basal nodes. The primary roots, developing with early seedling growth, are important only in early establishment and support.

The threshold temperature for corn growth is 50°F (10°C). The optimum range is 75 to 86°F (24 to 30°C), but maximum dry matter is accumulated where the days exceed 65°F (19°C) and night temperatures average

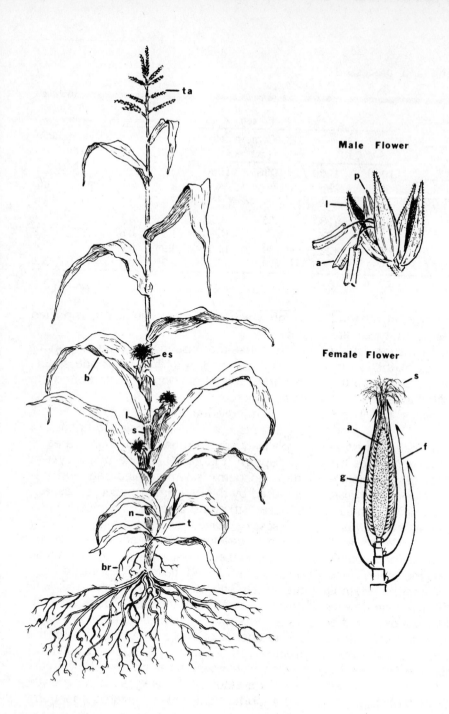

Male Flower

p

l

a

Female Flower

s

a

f

g

FIGURE 20.1

Characteristics of a corn plant: (*a*) *Plant*—tassel, ta; earshoot,es; leaf blade, b; ligule, l; sheath, s; node, n; tiller, t; brace root, br. (*b*) *Male flower*—anthers, a; lemma, l; palea, p. (*c*) *Female flower*—silk, s; flag leaves, f; axis or cob, a; grain, g. (Portions redrawn from W. C. Galinat, Botany and Origin Of Maize. In *Maize* (E. Hefliger, ed.). Copyright © 1979. CIBA–GEIGY, Basle, Switzerland.)

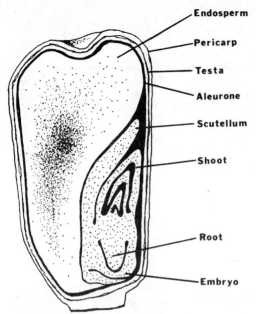

FIGURE 20.2

Anatomy of a corn kernel. At the proper stage for harvest, sweet corn endosperm is largely liquid, and pericarp is tender. As the ear matures, the endosperm becomes solid and starchy, and pericarp toughens.

55°F (13°C). Cool nights are particularly important at harvest time, slowing maturity and extending the time at which sugar content and other quality factors in sweet corn are maximum.

Daylength does affect time of flowering in corn but is not a determinant for flowering in most cultivars. Some cultivars developed for a tropical ecology may not flower under temperate conditions until short days. Others have shown delayed flowering in short days.

Crop Establishment and Maintenance

Sweet corn can be grown successfully in a wide range of soil types. Fine sandy loams are used for early market production, and heavy loams and mucks are well suited for late crops or those for processing. In acidic soils, corn growth and quality are improved by liming (Table 20.4) to an optimum pH of 6.0 to 6.8; however, productivity can be maintained at acidic or alkaline levels with suitable fertility management.

If heavy applications of lime are required, approximately half should be added prior to plowing to distribute it throughout the plow depth. Light applications, or the remainder of the split application, would be disked in after plowing. Organic matter, as a green manure cover crop, is important, particularly for those soils not rotated periodically to a sod.

Sweet corn thrives on liberal amounts of soil nitrogen, although high fertilization rates are not cost effective. For a yield of 130 cwt (5.9 MT) the harvested ears remove 55, 8, and 30 lb (25, 3.6, and 14 kg), respectively, of nitrogen, phosphorus, and potassium. The vegetative material removes 100, 12, and 75 lb (45, 5.4, and 34 kg), respectively. The fertilizers used to supply nutrients depend on initial soil tests. On a soil high in potassium, ammonium phosphate (monoammonium or diammonium) is applied, usually as a band 2 in. (5 cm) to the side and below the seed. Placement and amount of fertilizer are important considerations, since fertilizers can reduce germination or emergence if placed too close to the seed. New York fertilizer ratios vary around a basic 2–1–2 or 2–1–1, with 40 to 80 lb/acre (45 to 90 kg/ha) of nitrogen. A sidedressing of 30 to 60 lb/acre (34 to 68 kg/ha) of ammonium nitrate may be justified on light soils or when rainfall has been heavy early in the season. Other northeastern and mid-Atlantic states have similar requirements for loam soils. Gravelly soils will require greater amounts. In western regions, the nitrogen rates are somewhat higher than those in the east, particularly if following a crop of grain. For winter production in Florida, 150 lb/acre (168 kg/ha) each of nitrogen and potash and 120 lb/acre (134 kg/ha) of phosphate are recommended.

Where heavy fertilizer applications are suggested by soil test, split applications are used, one-half broadcast prior to field preparation and one-half banded at planting. A single band

TABLE 20.4

Main effects of lime treatments on vigor, yield, maturity, and quality of combined sweet corn hybrids[a,b]

Lime type	Yield (MT/ha)	Market-able (%)	Ear length (cm)	Ear weight husked (g)	Flag leaf (rating)
Control	7.2 c	67.2 b	12.0 c	139 d	28.1 c
Calcitic	8.8 b	78.5 a	12.7 b	149 c	30.5 b
Calcitic, 3% Mg	9.3 a	78.6 a	12.9 a	154 b	30.6 b
Dolomitic	9.4 a	80.8 a	13.0 a	156 a	31.4 a

Source: From C. B. Smith (1984) *Journal of the American Society for Horticultural Science* **109:**573. Reprinted by permission of the author.

[a] Mean separation within each variable by Duncan's modified (Bayesian) LSD test with $K = 100$.

[b] Original soil: pH 5.2; low in P, Ca, and Mg.

application should not exceed 80 to 100 lb/acre (90 to 112 kg/ha) nitrogen, phosphorus, or potassium, singly or in combination.

Trace elements, including boron, zinc, magnesium, and manganese, may be inadequate for normal growth at certain pH levels, particularly in light soils or in mucks. Sweet corn does not have a high requirement for trace elements, but light sandy soils or occasionally mucks may be deficient. Excessive potassium also can create a nutrient imbalance affecting availability of several other elements. Corrective treatments with borax, dolomitic limestone (Table 20.4), manganese sulfate, zinc chelate, or sulfate can be added as a supplement or formulated in a bulk mix.

Sweet corn is planted mechanically, using seed graded for a specific planter or planter plate. Seeds not matched in size and shape with a specific planter will clog the seed delivery mechanism resulting in skips in plantings and reduced yield per acre. Corn is planted in single rows on raised beds 40 in. (1 m) on center, or as two rows per wide bed where furrow irrigation is used. In nonirrigated areas, or where supplemental water is applied by an overhead system, rows are spaced at 30 to 36 in. (76 to 91 cm) without beds or ridges. In the row, seeds of early cultivars are spaced at 8 to 10 in. (20 to 25 cm), late cultivars at 10 to 12

in. (25 to 30 cm). Depth of seeding should be only enough to place the seed in contact with moisture. Deep seeding may result in variable rates of emergence.

Early seeding is important in obtaining early harvest. Although soil temperatures in excess of 50°F (10°C) are preferred, seedings 1 to 2 weeks earlier than the optimum date generally do not suffer from light frosts or cold soils. Prolonged cold and wet soil conditions, however, will increase incidence of seed decay, resulting in a poor stand.

The date of maturity of hybrid cultivars is very uniform. To ensure a continuous supply of corn in prime condition, the expected maturity dates must be spaced at regular intervals. Processors and large market growers grow relatively few cultivars, using the heat unit system to separate seeding and expected maturity dates. Growers for local or direct sales usually plant a range of cultivars differing in maturity class to maintain supply.

Corn is considered relatively drought resistant, but yields are enhanced by regular applications of water (Figure 20.3). The most critical periods are at silking and pollination and when ears are filling. If irrigation is not available, initial plant density should be reduced to minimize interplant competition. From 12 to 16 in. (30 to 41 cm) of water has

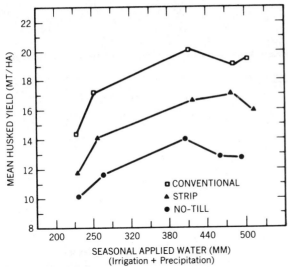

FIGURE 20.3

Yield versus applied water for sweet corn grown under three tillage methods and five irrigation levels in 1983. (From K. L. Petersen, H. J. Mack, and R. H. Cuenca (1985) *HortScience* 20:901–903. Reprinted by permission of the authors.)

been reported as necessary for early cultivars, 18 to 20 in. (46 to 51 cm) for late. Since some of the absorbing root system is shallow, regularity of supply is as important as total amount.

Weed control is achieved primarily by cultivation and by herbicides. While plants are young, cultivation is effective, but early weed growth can become a serious competitor if, for some reason, cultivation is not possible. Chemical herbicides are used quite generally in sweet corn production, but growers involved with a series of crops in rotation must be careful to select herbicides of which the residuals will not carry over to subsequent crops (Table 20.5). Atrazine, for example, should not be used if corn is to be followed by a different crop, in either the same or the following year. Black plastic mulch has been used by a few growers, seeding through polyethylene already laid.

No-till culture, utilizing herbicides to kill an established cover crop (often stubble from a small grain or sod rotation crop) and seeding with a chisel planter in undisturbed soil, thus far has been researched more in corn than in any other vegetable crop. For sweet corn, research results have shown yields to be less than those of standard tillage (Figure 20.3). Under some soil conditions, increases in yield have been obtained, largely ascribed to an increase in soil moisture. In northern areas, however, soil temperature under stubble is measurably less than that in conventionally tilled soils, often delaying emergence. The sod stubble, especially in a cool wet spring, also increases damage from slugs or disease and reduces stand.

Although pollen release is abundant during hot dry conditions, fertilization of egg cells may be erratic. Poor fertilization is recognized on harvested ears by skips along the kernel row or by blank ear tips. In either case, market appearance and yield are impaired. Fertilization from nearby field corn pollen also has an immediate effect on ear quality, since it carries the dominant gene *Su* (starchy), affecting constituents of the kernel endosperm (xenia). In a drying ear, such kernels will stand out as smooth or nonwrinkled. A few scattered starchy kernels in a fresh ear of corn normally will not be noticed by the consumer, but excessive stray pollen will reduce quality.

Cultivars

Early sweet corn introductions were open-pollinated and somewhat variable in growth and maturity. Several cultivars are still available today, including 'Stowell's Evergreen' (1853) and 'Country Gentleman' (1890). Modern sweet corn cultivars are F_1 hybrids, providing greater yield and uniformity, important for mechanical harvest and processing, than open-pollinated types. Hybrids are developed by crossing two inbred lines, one of which carries a cytoplasmic factor for male sterility and both of which carry complementary genetic systems to restore fertility in the F_1. Most hybrids, however, still are produced by detasseling normal fertile corn, allowing it to be fertilized by an adjacent inbred. At one time, all corn was produced by detasseling, but male sterility offered a system for producing seed at

TABLE 20.5

Activity of several herbicides against grass and broadleaf weeds in corn

Weed	Herbicide[a]					
	Atrazine	Butylate	EPTC	Dicamba	Alachlor	Metolachlor
Grasses						
Quack grass	G[a]	0	F	0	0	P
Crabgrass	P	G	G	0	G	G
Yellow nutsedge	P	G	G	0	G	G
Barnyard grass	F	G	G	0	G	G
Panicum	P	G	G	0	G	G
Giant foxtail	F	G	G	0	G	G
Yellow foxtail	G	G	G	0	G	G
Broadleaf						
Velvetleaf	F	F	F	G	P	P
Pigweed	G	F	F	G	G	G
Ragweed	G	P	F	G	P	P
Milkweed	P	0	0	F	0	0
Mustard	G	P	P	F	P	P
Lamb's-quarter	G	P	F	G	F	F
Canada thistle	P	0	0	G	0	0
Field bindweed	0	0	0	G	0	0
Smartweed	G	P	P	G	P	F

Source: From R. Behrens, Weed Control in the United States. In *Maize* (E. Hefliger, ed.). Copyright © 1979 by CIBA–GEIGY, Basle, Switzerland.

[a] Atrazine may be preplant, preemergence, or postemergence; alachlor and metolachlor may be preplant or preemergence; butylate and EPTC are preplant; and dicamba is postemergence.

[b] G = good; F = fair, P = poor; 0 = none.

minimal cost. Although the male sterile system is effective and efficient, it has not been without problems. In developing usable male sterility, a cytoplasmic factor conditioning susceptibility to southern corn leaf blight inadvertently was bred into the male sterile line. Its presence was not detected until conditions favoring disease development occurred. Because all field and sweet corn hybrids had been developed using virtually the same source of male sterility, the disease became a severe threat, and a significant acreage was affected. New sources of sterile cytoplasm have been found that carry resistance, but the incident illustrates the problem of genetic vulnerability in highly bred cultivars.

Sweet corn cultivars (Table 20.6) are classified by kernel color (yellow, white, or bicolor), by earliness, and by use (market, freezing, canning, or shipping). Market types must be high quality in both flavor and appearance, have good tip fill, light silks, and complete ear cover. The ear size is generally correlated with maturity. The earliest cultivars normally are small plants with small ears, whereas late cultivars are tall, with increased ear length and kernel row number. White and bicolor types are primarily for local sales. Cultivars for long-distance shipping should have uniform ears of yellow kernels and be adapted to machine harvest. Ears must be covered with tight husks and have three dark green flag leaves. Re-

TABLE 20.6
Characteristics of some commercial cultivars of sweet corn

Cultivar	Season[a]	Color[b]	Use[c]	Row	Type
Golden Miniature	EE	Y	FM	12	Hybrid
Spring Gold	E	Y	FM	14	Hybrid
Golden Beauty	E	Y	FM	12–14	Hybrid
Gold & Silver	E	B	FM	12–14	Hybrid
Aztec	E	Y	FM,S	14–16	Hybrid
White Sunglow	E	W	H	12	Hybrid
Extra Early SS	E	Y	FM	12–14	Hybrid, supersweet
Sundance	E	Y	FM	14–16	Hybrid
Harmony	M	B	FM	16	Hybrid
Florida Sweet	M–L	Y	FM,S	16	Hybrid, supersweet
Seneca Chief	M	Y	FM,H	12–16	Hybrid
Iochief	M–L	Y	P	16	Hybrid
Golden Jubilee	M	Y	FM,P	16	Hybrid
Symphony	L	B	FM	16	Hybrid, supersweet
Silver Queen	L	W	FM	18	Hybrid

[a] EE = extra early; E = early; M = midseason; L = late.

[b] Y = yellow; W = white; B = bicolor.

[c] FM = fresh market; S = shipping; H = home garden; P = processing.

quirements for corn for processing differ according to the kind of pack (Table 20.7).

Supersweet corns are any of several distinct genetic types that contain up to twice the normal sugar content at edible maturity. Standard sweet corns contain the recessive gene, *su*. Field corn is dominant for this gene and therefore starchy. The original supersweet cultivar was 'Illini Xtra Sweet,' an F_1 cultivar with the recessive gene sh_2 incorporated into each inbred parent. This gene enhances the sugar content whether the plant is normally sugary (*su*) or starchy (*Su*). Supersweets of this type therefore require isolation from normal sugary or starchy corns to ensure the self-pollination required to preserve the high level of sweetness. The sh_2 gene also reduces the starch content and produces highly shrunken, chaffy seed, often with poor germinability. Recent supersweet introductions are substantially improved in germinability, and a high percentage of the winter corn acreage in Florida now is planted to 'Florida Sweet,' a supersweet cultivar with sh_2 donated by one inbred.

The gene bt_2 also increases sugar content and has been used for several Hawaiian cultivar improvements. A single gene *se* (designated as SE or ES or EH in cultivar descriptions), slowing the conversion of sugar to starch, has been incorporated into several cultivars. This factor extends the time during which the corn will maintain its maximum sugar content. ADX cultivars also have been released, combining three genes to enhance sugar while retaining high starch and germinability of the seed. The highly shrunken seeds of high-sugar corns require special handling, since they are easily damaged and are susceptible to poor soil environment. These seeds should be planted at a higher soil temperature than required for normal sweet corn.

In addition to high sugar and pericarp tenderness, other plant traits important in a cultivar are wide leaves, medium placement of

TABLE 20.7
Relative importance[a] of traits of sweet corn hybrids in the fresh market and processing trades

Trait[b]	Fresh market	Processing		
		Whole kernel	Cream style	Frozen corn on cob
Yield				
Green weight	1	3	3	2
Weight of cut corn	1	3	3	1
Usable ears/acre	3	2	2	3
Ear characteristics				
Husk cover	3	2	2	2
Flag leaves	3	2	2	2
Ear length	2	2	2	3
Light silk color	3	2	3	3
Tip fill	3	2	2	3
Ease of husk removal	1	3	3	3
Husked ear appearance	3	2	2	3
Color of cob (light)	3	3	3	3
Kernel characteristics				
Size	2	3	1	3
Depth	1	3	2	2
Color	2	3	3	3
Tenderness	2	3	2	2
Flavor	3	3	2	3
Silk attachment color	2	2	2	2
Black layer	1	3	2	1

Source: From K. Kaukis and D. W. Davis, Sweet Corn Breeding. In *Breeding Vegetable Crops* (M. J. Bassett, ed.). Copyright © 1986 by the AVI Publishing Company, Westport, Conn. 06881.

[a] Rated 1, relatively unimportant, to 3, very important.

[b] Seedling vigor, uniformity, resistance to insects and diseases, lodging resistance, and stress tolerance are examples of traits that are equally important in fresh market and processing sweet corn.

ears on stalks, good husk cover to impede earworm and bird damage, green silks, and even rows of deep kernels.

Disease and Insect Pests

Diseases   The diseases of sweet corn include seed rots and seedling diseases, stalk rots, leaf spots and blights, rust and smut, mildews, and viruses (Figure 20.4).

Seed Rots and Seedling Diseases   These diseases are caused by any of several soil- and seedborne fungi. *Pythium* species constitute the most important group, infecting through cracked seed coats and through root and stem tissue. These organisms, found in every soil, are favored by cold, wet conditions. Several organisms normally associated with mature plants (*Helminthosporium maydis, Fusarium moniliforme,* and others) also cause seed or seedling damage. The best control is the use of high-quality seed treated with a protectant before sowing.

Stalk Rots   Losses from stalk rots, under conditions favorable for pathogen activity, may

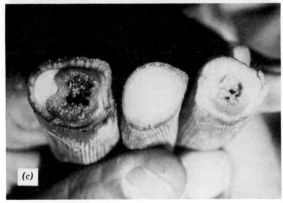

FIGURE 20.4
Diseases of corn: (*a*) smut; (*b*) southern corn leaf blight; (*c*) Stewart's bacterial wilt. (Photo (*a*) courtesy of D. N. Maynard, Florida Agricultural Research and Education Center, Bradenton. Photos (*b*) and (*c*) from *Diseases of Plants*, slide set 37, Diseases of Carrots, Eggplant, Peas, Peppers, Sweet Corn. Reproduced by permission of the American Phytopathological Society.)

exceed 50 percent. The cause is usually a complex of several fungi and bacteria (*F. moniliforme, Diplodia maydis, Gibberella zeae, Cephalosporium maydis, Macrophomina phaseolina, Pythium aphanidermatum, Erwinia carotovora*). These organisms overwinter on seed and plant debris and become active in wet weather or in fields receiving heavy nitrogen applications. Proper cultural practices will minimize the problem.

Leaf Spots   Included among the leaf spots are **northern corn leaf blight,** caused by *Helminthosporium turcicum*, **southern corn leaf blight,** caused by *H. maydis,* **yellow leaf blight** (*Phyllosticta maydis*), **anthracnose** (*Colletotrichum graminicola*), and **bacterial leaf spot** (*Pseudomonas precipitans*). These diseases, distinguished by lesion shape, size, and color, have different geographical distributions. Primary infection by *Helminthosporium* occurs from infected crop debris or seeds. Damp weather and moderate temperatures favor spread, with southern corn leaf blight favored by slightly higher temperatures than northern blight. Both anthracnose and bacterial spot develop at high temperature. Severe infection may require protective sprays, but normally the diseases are sporadic.

Stewart's Bacterial Wilt   Bacterial wilt, caused by *Xanthomonas stewartii,* is particularly severe on young plants, occasionally causing death. Leaves show the earliest symptoms, yellowish steaks becoming brown. In young plants, cavities may form in the center of the stem. The organism is transmitted by flea beetles. Warm winter temperatures that favor survival of the insect vector increase prevalence of the disease in the following season. When the sum of December, January, and February mean temperatures exceeds 90°F (32°C), the flea beetles could be abundant with consequent increased infection by Stewart's wilt. Cultivar resistance and eradication of the flea beetle are recommended control measures.

Rust   Rust (*Puccinia sorghi*) usually appears at tasselling as elongate reddish brown pustules on either surface of the leaf. The pustules break, and the red dusty spores are disseminated widely by wind. Black spores, the overwintering stage, are produced later, and these germinate and infect the alternate host, *Oxalis.* In the south, the red spore stage persists from year to year. In the north, the disease will develop only if reintroduced from the south or if spores develop on the alternate host. Infection is enhanced in 60 to 70°F (16 to 21°C), moist environments. Under heavy inoculum, control is difficult. The related southern corn rust (*P. polysora*) has small round lesions with light-colored spores that do not have an alternate host.

Smut   Appearance of the smut fungus (*Ustilago maydis*) is exacerbated by drought stress, wide plant spacing, very high nitrogen levels, and plant injury caused by insects (particularly corn borer), cultivation, or hail. The disease is recognized by the large puffy galls on various plant parts. These galls are white when immature, becoming dark as the powdery spores disseminate. The fungus spores overwinter in the soil and are favored by dry, warm weather. A number of cultivars are tolerant to smut, and careful management and pest control will reduce the number of infected plants in all cultivars. In the west and northwest, head smut, caused by *Sphacelotheca reiliana,* occasionally becomes an economic problem. The disease is seedborne, and seed treatment is an effective control. Wide differences in resistance have been observed among cultivars.

Downy Mildew   Downy mildew (*Peronosclerospora sorghi*) occurs in some states, causing considerable damage. Young plants are very susceptible, and the organism apparently becomes systemic. The leaves develop distinct yellow streaks on which a mold eventually may appear. Tassels become deformed on plants infected when young. Once plants have become well established, 4 or more weeks after sowing, the plants are virtually immune. There are several strains of the fungus with different adaptations, and all have increased since 1960. A different species (*Sclerophthora macrospora*) causes a disease called "crazy top." Infection causes abnormal proliferation of the tassel, increased earshoots, an increase in internodes above the ears, excessive tillering, and stunted, straplike leaves. The fungus is systemic and infects young plants, particularly in waterlogged soils. The crazy top fungus has a wide host range, including several other grain crops. The threat posed by the mildew diseases has increased efforts in both genetic and chemical control approaches.

Virus Diseases   Viruses include maize dwarf mosaic, maize chlorotic dwarf, and wheat streak, but maize dwarf mosaic is the most important. The disease is transmitted by any of several species of aphid, and infected plants develop a fine, stippled streaking of light and dark green. The upper internodes become compressed, and leaves eventually become yellow and stunted. The plants tiller profusely and also develop multiple ear shoots. Several strains of the virus exist in overwintering weed hosts. Aphid control and tolerant cultivars provide reasonably effective control.

Insects   The insect pests may be grouped as soil insects (rootworms, cutworms, wireworms, and grubs), foliar insects (aphid, corn earworm, armyworms, and leaf and flea beetles), and stem borers (European corn borer and stalk borer) (Figure 20.5).

Corn Rootworms   Rootworms include three species: northern (*Diabrotica longicornis*), southern (*D. undecimpunctata howardi*), and western (*D. virgifera*). The southern rootworm is found in all production areas, but it rarely damages maize. The northern and western rootworms are similar in life cycle, and both are exclusively corn insects. Of the two,

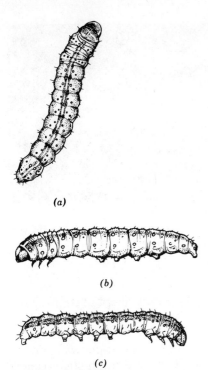

FIGURE 20.5
Sweet corn insects: (*a*) corn earworm; (*b*) European corn borer; (*c*) fall armyworm.

the northern corn rootworm has the widest geographical distribution. Both overwinter in the soil as eggs which hatch in late spring, and the larvae feed on developing maize roots. In heavy infestations, the damage may cause lodging or crooked growth. Adults feed on ears, silks, and pollen. Crop rotation reduces insect buildup, and organophosphates and carbamates have been applied in the row during planting to minimize early damage.

Worms   **Wireworms** (species of *Conoderus* and *Melanotus*) cause sporadic damage, feeding on roots and occasionally boring into the stem. Several soil grubs and cutworms also may cause similar isolated damage. Soil treatments applied for rootworm will help control other soil insects.

Of the insects feeding on leaf or ear tissue, **corn earworm** (*Heliothis zea*) is the most destructive. It is a migratory insect, moving northward from overwintering sites each year. Control must begin as soon as the ear silk appears. The moths lay eggs in the silk, and larvae feed in the developing whorl or in the ear. Although only a portion of an ear may be damaged, such damage is unacceptable on the market. It is estimated that 15 percent of the yield may be lost from earworm damage. Several pesticides (carbaryl, diazinon) may be applied to the developing ear shoot and silk, and treatments must be repeated until silks are no longer green. Cultivars with tight husks completely covering the ear tip seem to show less damage than those with incomplete tip cover.

The **armyworm** (*Pseudaletia unipuncta*) and **fall armyworm** (*Spodoptera frugiperda*) also are migratory insects, and their damage is more noticeable in the south than in the north. Both feed on leaves and in the whorl, occasionally causing complete leaf loss. Granular insecticides applied over the whorl or complete plant sprays have been recommended controls.

Aphids   Aphids (*Rhopalosiphum maidis*) do not cause widespread damage but do carry virus diseases. Infestations usually are sporadic, but, where heavy, can render plants unproductive. Organophosphates are effective controls.

Stem and Stalk Borers   Of the stem borers, the most significant is the **European corn borer** (*Ostrinia nubilalis*). This insect is widely distributed and causes extensive losses. The insect has over 200 plant species as alternate hosts, but prefers corn. Normally, there are two broods per year, occasionally three. It overwinters in the larval stage on field debris. In the spring, the larvae pupate in the debris, and by early summer, adult moths lay eggs on leaf surfaces. The young larvae feed on leaves and eventually bore into the stem, completing larval development there. The damage weakens the plant, interrupts the transport of nutrients, and provides opportunities for disease. Because sweet corn is harvested when the stalk is still green, the most serious damage is caused by direct feeding in the ear. The larvae may enter through the shank, husk, or ear tip. Progress has been made toward resistance; however, practical control is limited to chemical sprays applied to foliage (diazinon, carbaryl). Insect traps are used to indicate the proper time for spray applications.

Stalk borers seldom cause economic damage, but these insects are widespread. Often, they are associated with grassy or weedy borders, and corn at the perimeter of the field may be affected. Control of perimeter weed growth and occasional insecticide applications may be necessary if infestation is severe.

Spider Mites Mites are a problem primarily in hot and dry production areas, causing severe leaf damage. Several species may infest corn, and once established, they are difficult to eradicate. Organophosphate miticides are used as needed, and the spray must cover both upper and lower surfaces of the leaf.

Harvest and Market Preparation

Sweet corn must be picked at peak maturity and moved quickly to the processor or to the consumer. At harvest, the kernels should be pale yellow and plump, and ears should show miniature kernels at the tip $\frac{1}{2}$ in. (1.3 cm) of the ear (Figure 20.6). Ears with full-size kernels at the tip are likely to be overmature. As the kernels mature, they increase in sugar content (through premilk to milk stage), then begin to lose sugar and increase in starch (early dough to dough stage). The moisture percentage of the kernels at maturity should be 68 to 78 percent.

During maturation, the pericarp remains tender through the milk stage, then becomes tough. Before harvest, high temperature does not affect the percentage sugar at any time, but it does hasten maturity. After harvest, however, temperature does accelerate the rate of sugar loss. Ears kept at 68°F (20°C) lose sugar

FIGURE 20.6
Ear of sweet corn showing proper tip fill. (Photo courtesy of Harris—Moran Seed Company, Rochester, N.Y.)

six times faster than those kept at 32°F (0°C). Corn should be harvested in early morning, then hydrocooled or vacuum cooled to 32°F. An advantage offered by the supersweet corns is a prolonged shelf life. Although sugar is lost during storage and transit, the initially high content ensures superior sweetness at the marketplace.

Corn is not husked prior to marketing. The husks are protection against abrasion; however, if the ears are not kept misted or top-iced to maintain humidity and a temperature below 40°F (4.4°C), the husks will draw moisture from the kernels, resulting in kernel denting and poor quality.

Harvest of processing acreage is programmed using a heat unit system as described for garden peas, but with 50°F (10°C) as a base temperature. As a given crop nears maturity, it is sampled for tenderness and sugar to determine appropriate harvest time. Contract price is based on quality, freedom from insect or disease damage, and percentage cut (depth of kernel and ear fill).

Although most sweet corn for fresh market and processing is harvested at the milk stage for peak quality, cream-style corn requires a starchier kernel, and harvest is delayed until the early dough stage.

Increasing acreages of sweet corn for

fresh use are machine harvested, requiring uniform stands of plants. All processing acreage is machine harvested. Hand harvest is preferred by most market gardeners, since it ensures the best quality product. Others argue that machines enable the grower to time the harvest properly, taking advantage of cool morning temperatures to preserve quality while harvesting a large acreage. Investment in a machine is not justified for limited acreage or where ample labor is available. On many large commercial farms, hand harvest is accomplished using "mule trains," mobile conveyers that allow field sorting and packing.

For shipping, the ears are packed in cardboard cartons or wirebound crates, then placed on pallet loads in vacuum coolers or in hydrocoolers. The temperature of corn premoistened and vacuum cooled can be reduced from 85°F (29°C) to 40°F (4.4°C) in a half hour. The crates must be iced thoroughly for transit.

SELECTED REFERENCES

Allemann, D. V. (1979). Maize pests in the USA. In *Maize* (E. Hefliger, ed.). CIBA–GEIGY Ltd., Basle.

Behrens, R. (1979). Weed control in the U.S. In *Maize* (E. Hefliger, ed.). CIBA–GEIGY Ltd., Basle.

Bonnett, O. T. (1953). *Developmental Morphology of the Vegetative and Floral Shoots of Maize,* University of Illinois Res. Bull. 568.

Cassini, R., and T. Cotti (1979). Maize diseases. In *Maize* (E. Hefliger, ed.). CIBA–GEIGY Ltd., Basle.

Culpepper, C. W., and C. A. Magoon (1924). Studies upon the relative merits of sweet corn varieties for canning purposes and the relation of maturity of corn to the quality of the canned product. *Journal of Agricultural Research* **28:**403–443.

Fletcher, R. F. (1975). *Commercial Sweet Corn Production,* Pennsylvania State University Coop. Ext. Spec. Circ. 208.

Galinat, W. C. (1979). Botany and origin of maize. In *Maize* (E. Hefliger, ed.). CIBA–GEIGY Ltd., Basle.

Goodman, M. M. (1976). Maize. In *Evolution of Crop Plants* (N. W. Simmonds, ed.). Longmans, London.

Huelson, W. A. (1954). *Sweet Corn.* Interscience, New York.

Kaukis, K., and D. W. Davis (1986). Sweet corn breeding. In *Breeding Vegetable Crops.* M. Bassett (ed.) AVI, Westport, Conn.

Petersen, K. L., H. J. Mack, and R. H. Cuenca (1985). Effect of tillage on the crop-water production function of sweet corn in western Oregon. *HortScience* **20:**901–903.

Shurtleff, M. C., Q. Holdeman, C. W. Horne, T. Kommedahl, C. A. Martinson, R. R. Nelson, G. C. Schiefle, J. L. Weihing, D. R. Wilkinson, G. L. Worf, D. S. Wysong, H. E. Smith, G. J. Muller (1973). *A Compendium of Corn Diseases.* Amer. Phytopathol. Soc., St. Paul, Minn.

STUDY QUESTIONS

1. Outline the specific changes that take place in a sweet corn kernel as it progresses to hard seed stage.

2. Outline the differences in cultivar attributes necessary for sweet corn for shipping, early local market, pick-your-own, canning, and freezing.

3. What are the advantages of high-density plantings of sweet corn? Disadvantages?

4. Under what conditions is no-till production of sweet corn most appropriate?

21
Miscellaneous Vegetables

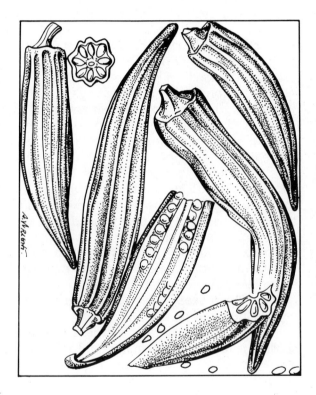

Okra   Mushroom   Taro

Of the many plants used as vegetables throughout the world, relatively few are significant crops in North American agriculture. Yet, production of adapted minor commodities can be profitable for specialized growers. Among these commodities are okra, mushroom, and taro, each described in this chapter, and several oriental vegetables. Production of okra, mushroom, and taro is restricted to only a few states, either by geographical adaptation or by consumer demand.

OKRA

Historical Perspective and Current Status

Okra (gumbo) is thought to be native to an area extending from Ethiopia to the Sudan. Its early history and distribution are not known, but it apparently was introduced to Egypt in the seventh century. Okra then was carried through north Africa and areas bordering the Mediterranean and eastward. There is no evidence that this crop was grown in Asia before the Christian era.

Okra was introduced to the United States in the early 1700s, either by French settlers in Louisiana or by slaves from Africa. By the mid-1700s, it was grown as far north as Philadelphia, but it is most adapted to the southeast and other warm growing areas. Most current acreage is located in Texas, Georgia, Florida, and South Carolina. A total of 6389 acres (2588 ha) was harvested in 1978 in the United States, and this volume changes little each year. A major portion of the harvest is processed as canned, frozen, or dehydrated products.

Okra has moderate levels of vitamins A and C and of calcium, phosphorus, and potassium, and is higher than many vegetables in thiamine, riboflavin, and niacin (Table 21.1).

Classification, Growth, and Development

Okra (*Abelmoschus esculentus*), a member of the mallow family (Malvaceae), is a vigorous, herbaceous warm season annual dicot reaching 3 to 6 ft (0.9 to 1.8 m) in height. The

TABLE 21.1

Nutritional constituents of okra

Constituent	Amount contained[a]
Water (%)	89
Energy (cal)	36
Protein (g)	2.4
Fat (g)	0.3
Carbohydrate (g)	7.6
Vitamin A (IU)[b]	520
Vitamin C (mg)[c]	31
Thiamine (mg)	0.17
Riboflavin (mg)	0.21
Niacin (mg)	1.0
Calcium (mg)	92
Phosphorus (mg)	51
Iron (mg)	0.6
Sodium (mg)	3.0
Potassium (mg)	249

Source: National Food Review (1978), USDA.

[a] Data per 100 g fresh wt.

[b] 1 IU = 0.3 μg vitamin A alcohol.

[c] Ascorbic acid.

leaves are broad palmate. Large flowers, with five petals, yellow toward the tip and purple at the base, develop in each leaf axil above the sixth or eighth lower leaves. Only one flower opens on a single stem in a single day. The fruit are green or creamy white, smooth or ridged pods, 6 to 8 in. (15 to 20 cm) long at maturity.

Okra is predominantly a tropical crop but can be grown in warm sections of the temperate zone, preferably where the average temperature range is 65 to 95°F (18 to 35°C). Flowering of most cultivars responds to short days; however, 'Clemson Spineless' is not sensitive to photoperiod and can be grown in areas of long days where others would show flower abortion.

Crop Production and Maintenance

Okra should be grown in frost-free areas on sandy or sandy loam soils, but also can be produced on muck. The pH optimum is 5.8 to 6.5, and organic matter, preferably barnyard

manure, should be plowed in before seeding. Fertilizer placement in a band 2 to 3 in. (5 to 7.5 cm) below and to the side of seed is preferred. Where soluble salts are a problem, however, fertilizers are broadcast and disked in. A high-yielding crop of okra uses approximately 21, 10, and 62 lb (9.5, 4.5, and 28 kg), respectively, of nitrogen, phosphorus, and potassium. A continuous supply of nitrogen, some of which may be added as a sidedress several weeks after seeding, is necessary for prolonged pod set. Phosphorus and potassium needs depend on the soil type and pH. On irrigated mineral soils in Florida, about 100 lb/acre (112 kg/ha) of nitrogen and 140 lb/acre (157 kg/ha) each of phosphate and potash, followed by one to five sidedressings of 30 lb/acre (34 kg/ha) each of nitrogen and potash, are applied, smaller amounts on nonirrigated mineral soils and on muck soils. Trace minerals may be needed periodically.

Seeds are sown in moist, 70°F (21°C) or above soil, $\frac{1}{2}$ to 1 in. (1.2 to 2.5 cm) deep in rows 3 to 4 ft (0.9 to 1.2 m) apart. The seeding rate is 4 seeds/ft (13/m), later thinned to an in-row spacing of 6 in. (15 cm) for dwarf types, 12 to 15 in. (30 to 38 cm) for standard cultivars. Germination time is not uniform because of variation in seed coat thickness. The percentage "hard seed," which is slow to germinate, is indicated on the seed tag.

Early cultivation to remove weeds must be shallow to avoid root pruning, and trifluralin applied preplant or diphenamid applied preemergence may be used in controlling both grass and broadleaf weeds.

Irrigation ensures continual growth and superior pod quality, but light applications are recommended; otherwise, soils are cooled, reducing growth rate.

Cultivars

Cultivars may be tall [6 to 7 ft (1.8 to 2.1 m)] or dwarf [3 ft (0.9 m)], bearing smooth or ridged pods. Some are suited primarily to fresh market, others for processing. The most popular cultivars include 'Clemson Spineless' (tall, ridged pods, used for fresh market and some processing), 'Emerald' (tall, smooth pods, processing type), 'Dwarf Green Long Pod' (dwarf, ridged pods, fresh market cultivar), and 'Perkins Spinelsss' (dwarf, ridged pods, for fresh and processing use). 'White Velvet' is a cultivar with white, ridgeless pods.

Disease and Insect Pests

Diseases   The most common diseases affecting okra are *Fusarium* **wilt, cotton root rot,** and *Verticillium* **wilt,** each a soilborne disease. *Fusarium* wilt is not serious in many areas, but infected plants wilt and die rapidly. It is caused by *Fusarium oxysporum* f. sp. *vasinfectum,* and no cultivars are resistant. A long (6- to 10-year) rotation is suggested. *Verticillium* wilt, caused by *Verticillium dahliae,* produces similar wilt symptoms as *Fusarium,* and the same control is recommended. Cotton root rot is caused by *Pythium* species, and the girdling of the basal stem or disintegration of the roots produces wilt and death, similar to that caused by *F. oxysporum.* Soil drainage and rotation are the most effective controls. Any rotations should avoid cotton, since the disease organisms affecting each crop are the same.

Insects   Many insects found on okra are predominantly pests of other plant species. These pests include flea beetle, Japanese beetle, cucumber beetle, aphids, leaf miner, and corn earworm and are described in detail elsewhere. In addition, many of the insect pests of cotton also feed on okra, including the **boll weevil** (*Anthonomus grandis*) and **pink bollworm** (*Pectinophora gossypiella*). These insects can be controlled with mevinphos and carbaryl.

Nematodes in okra are constant problems, and treatment of fields periodically with nematicides will be necessary. Fumigants are available that will reduce not only nematode populations but also activity of soil pathogens,

but these chemicals are more expensive than nematicides.

Harvest and Market Preparation
For the best market quality, pods are harvested immature, 4 to 6 days after anthesis [approximately 3 to 3½ in. (7.5 to 9 cm) long]. At this stage, they are free of fiber and are tender; as pods mature, quality deteriorates rapidly. Plants normally must be harvested every other day, perhaps every day in hot weather. If pods are allowed to mature on a plant, further fruit set is erratic. Total yields of well-managed fields will approach 4 tons/acre (9 MT/ha), and 6 tons/acre (13 MT/ha) has been recorded.

The harvested pods should be cooled rapidly to preserve quality and handled carefully to avoid bruising. They are graded according to length, then packed and shipped under refrigeration. Because of their perishability, okra pods should be marketed within 36 h of harvest.

MUSHROOM

Historical Perspective and Current Status
The history of mushroom use and subsequent cultivation is scant. Wild races of the common mushroom occur throughout the northern hemisphere outside the tropics and the arctic. Mushrooms were cultivated in France and England long before commercial production began in the United States. The first mushroom enterprise developed near New York City in the latter part of the nineteenth century and was followed by similar efforts in other eastern states. Currently, over 66 million tons (60 million MT) of mushroom (all species) are grown for commerce throughout the world. Of this volume, the United States produces about one-fourth, 65 percent of it in the northeast.

The mushroom is grown primarily for its flavor. However, it does rank rather high in nu-

TABLE 21.2
Approximate levels of nutritional constituents in cultivated mushroom

Constituent	Level
H_2O (% of fresh wt)	88.0
Fat (% of dry wt)	2.0
Carbohydrate (% of dry wt)	59.5
Calories (kcal/100 g)	340.0
Thiamine (mg/100 g)	1.0– 8.9
Riboflavin (mg/100 g)	3.7– 5.0
Niacin (mg/100 g)	42.5–57.0
Ascorbic acid (mg/100 g)	26.5–81.9

Source: From S. T. Chang and W. A. Hayes, *The Biology and Cultivation of Edible Mushrooms.* Copyright © 1978. Academic Press, Inc., New York.

tritive value, although the analyses of its constituents vary widely. Its amino acid content ranks just below pork, chicken, beef, and milk, and overall nutrition is, by some estimates, just below chicken, beef, pork, and soybean. Some of the nutritional constituents are listed in Table 21.2. Mineral content, particularly of phosphorus, sodium, and potassium, is substantial.

Fungi are known to contain antibacterial and other medicinal substances. The common mushroom has been reported to show some antibacterial activity and to contain an antitumor substance (quinoid).

Classification, Growth, and Development

Of the wild mushrooms, some are edible, some extremely poisonous. The cultivated mushroom, *Agaricus bisporus* (*A. campestris*), is a heterotrophic fungus, a member of the Basidiomycetes, order Agaricales, family Agaricaceae.

The mushroom mycelium grows vigorously on an organic substrate. When conditions are suitable, a small "button" breaks through the substrate surface, enlarging rapidly to become the edible fruiting body. The stem or **stipe** of this fruiting body is thick and smooth, bearing a rounded white to light tan thickened cap, or **pileus.** The underside of the pileus ruptures at maturity to show radiating gills from which spores are released.

Mushroom **spawn** is the mixture of the released spores and organic matter or grain substrate used to seed the growing medium.

Crop Establishment and Maintenance

Commercial production of mushrooms requires extreme care in maintaining a clean environment. Natural caves have been used successfully as growing chambers, but most mushrooms now are produced in environmentally controlled buildings. A typical structure would be 65 ft long and 20 ft wide (19.8 × 6 m), containing two tiers of five or six shelves each, for a total of 3600 ft^2 (118.8 m^2) of growing space. The mushrooms are cultured in trays placed on the shelves. The growing chambers and trays are sterilized regularly to minimize contamination, and the growing medium must be conditioned properly before seeding the spawn.

To develop a growing medium, several ingredients, including horse manure and other manures, peat, chopped hay (grass and clover), corn cobs, and dried brewer's grain, usually are blended and stacked in rows 6 ft wide × 6 ft high (1.8 × 1.8 m) on a concrete slab. The mixture is moistened and aerated to encourage aerobic breakdown. At 2 to 4-day intervals, it again is mixed and restacked. After the second restacking, gypsum, NH_3OH, and muriate of potash are added. Following a total composting period of 2 weeks, the conditioned medium is placed 6 in. (15 cm) deep in trays or beds in a closed room and exposed for about 3 days to 135 to 140°F (57 to 60°C). Supplemental heat may be required to raise the temperature. Once the pasteurization has been completed (no ammonia aroma), the room is aerated, and the temperature of the beds is dropped to 75 to 78°F (24 to 26°C).

Spawn then is seeded at 1 lb/50 ft^2 (10.7 kg/m^2) on the surface of the beds, and the environment is maintained at 72 to 75°F (22 to 24°C) at high humidity. The mycelium must be allowed to grow throughout the compost for a period of 2 to 4 weeks, the duration depending on the spawn mix.

After mycelial growth is established, a "casing soil" (mixture of loam and peat, pH 7.0, steam sterilized) is spread over the surface of the bed. The beds then are kept moist by light watering. The mushrooms will not appear without applying the casing soil, and the reasons for this stimulation are not known. The mushrooms will appear 3 to 4 weeks after casing at which time the room temperature must be reduced to 60°F (15.5°C) to discourage insects and diseases.

Disease and Insect Pests

Diseases *White Mold* White mold (*Mycogone perniciosa*), also termed "wet bubble," results in a malformed stipe and pileus, and the entire sporophore may not develop. The surface becomes covered with white mycelium, and in the final stages, a foul-smelling brown liquid is evident.

Verticillium *Verticillium malthousei* is considered to be the most severe disease organism other than virus. The conidia are spread rapidly by phorid flies, but a major source of infection is dust and debris within the growing chamber. The organism causes brown "dry bubbles" on the sporophore, and a mold also appears on the pileus.

Pseudomonas Several problems are caused by *Pseudomonas tolaasii.* This bacterium occurs in soil and occasionally in water. At high temperature and humidity, it develops rapidly and infects the sporophore. It may cause brown blotch—sunken surface lesions on a maturing pileus—or, if early infection, it may result in little or no development of the stipe and pileus. Secondary infection is frequent.

Virus Viral diseases are difficult to diagnose. The most important symptom is yield reduction. Six viruses have been isolated, and these may occur singly or in complexes. Phorid fly larvae and mites are known to transmit virus.

Disease control must be practiced throughout the development of the growing medium and in the production house. Zineb is used to reduce fungal problems, and chlorine is effective as an antibacterial. Proper sanitation and pasteurization are essential.

Insects *Phorid Fly* The phorid fly (*Megaselia nigra*) is a small dark-colored humpbacked fly common near decaying vegetation. Its larvae tunnel into the sporophore tissues and act as vectors of *Verticillium* and virus. The lack of light in production rooms inhibits oviposition and thereby helps to minimize infestations.

Sciarids The sciarids (*Lycoriella solani, Bradysia brunipes*), or fungus gnats, can become a problem when the crop is dense. The insects are found in moist shady areas and are attracted to composts containing nitrogen supplements. These insects also are thought to play a role in virus transmission.

Mites Several species of mites can be introduced into culture through poor composting. These insects will destroy mycelial strands and can transmit mushroom virus.

For all insect pests, proper composting and sanitation will reduce the chance of infestation, and diazinon and malathion may be used routinely in compost or casing soil to maintain an insect-free environment.

Nematode Nematode infestations develop because of contaminated casing soil or poor pasteurization of the initial medium. Two species (*Ditylenchus mycelophagus* and *Aphelenchoides composticola*) are the primary pathogenic nematodes and both are common in commercial mushroom production.

Harvest and Market Preparation

Mushroom growth occurs in flushes, initially at about 7-day intervals, gradually slowing to several weeks. In a period of 2 to 3 months, from five to seven flushes will be harvested, totaling approximately 2.7 lb/ft² (29 kg/m²).

The mushrooms are harvested by hand before the caps open to release spores, either cutting at the soil line or, more commonly, twisting gently to remove the pileus and stipe intact. The product should be white or light buff without bruises or dark marks on either the stipe or the pileus. The cap should be well rounded and the stipe thick.

Following harvest, the mushrooms are cooled rapidly to 32°F (0°C). If washed, they must be air-dried rapidly, since surface moisture causes discoloration. Mushrooms are packed in rigid containers to avoid abrasion, and they may be overwrapped with a film permeable to water vapor.

TARO

Historical Perspective and Current Status

Taro (dasheen) is believed to be native to China and was introduced to the West Indies by explorers. A wild form originating in India and southeast Asia had migrated previously to several parts of Africa and Asia and had been carried by Polynesians to several islands of the South Pacific.

The crop is primarily of value for its starch content, although the tops also may be used as a nutritious green. Some of the wild forms contain calcium oxalate crystals, which give an acrid flavor and may cause itching or burning in the mouth. Dasheen is a cultivar of taro lacking the calcium oxalate crystals. Cooking normally releases the crystals, making taro an acceptable product. Taro is grown in Hawaii and other areas with Polynesian heritage, where it is fermented to form poi. However, it has been far more important as a staple in Africa and Asia, although it is being replaced gradually by other aroids that yield a superior product.

Classification, Growth, and Development

Taro (*Colocasia esculenta*) is a herbaceous monocot grown for its storage corms. It is adapted to tropical rainforests but has been grown successfully in warm temperate zone areas. The top growth consists of a whorl of long petioles bearing large, flat heart-shaped leaves with a slightly wavy margin (Figure 21.1). The corms are cylindrical, up to 8 in. (20

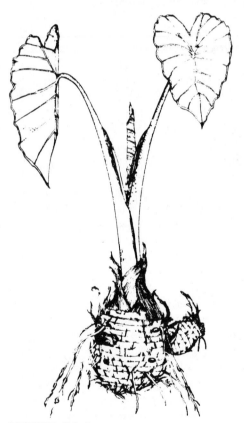

FIGURE 21.1
Taro (*Colocasia esculenta*) plant, showing corm and cormel. (From Mas Yamaguchi, *World Vegetables.* Copyright © 1983 by the AVI Publishing Company, Westport, Conn. 06881.)

cm) long and 5 in. (13 cm) in diameter. Cormels 2 to 3 in. (5 to 7.5 cm) in size develop adventitiously.

Crop Establishment and Maintenance

Taro may be grown as a paddy crop (lowland culture) or as a dryland crop (upland culture). The lowland culture is similar in some respects to rice production. The field is fertilized and flooded, then plowed and puddled before the presprouted corms or cormel pieces (corm apex plus leaf sheath enclosing growing point) are planted. Planting is in rows 3 to 4 ft (0.9 to 1.2 m) apart, with $1\frac{1}{2}$ to 2 in. (2.5 to 5 cm) between corms. After planting, the water level can be maintained at up to 4 ft (1.2 m) but should be continuously flowing.

Upland crops are established after fertilizing, plowing, and disking. The corms are set in furrows 12 in. (30 cm) deep and $2\frac{1}{2}$ to 3 ft (0.8 to 0.9 m) apart, with 16 in. (41 cm) between corms. Following planting, the field must be irrigated.

The fertilization rates and ratios differ in each production area. However, a complete fertilizer providing nitrogen, phosphorus, and potassium at close to a 1–1–1 ratio is common. The pH should be maintained between 6.0 and 7.0.

The growing season is long, with 3 to 5 months of growth before corm development,

TABLE 21.3
Exotic vegetables grown in the United States

Common name	Botanical name	Edible part
Water chestnut	*Eleocharis dulcis*	Corm
Rakkyo	*Allium chinense*	Bulb
Chinese chive	*Allium tuberosum*	Blanched and green leaves
Japanese yam	*Dioscorea batatas*	Tubers
Chinese yam	*Dioscorea alata*	Tubers
Ginger root	*Zingiber officinale*	Rhizomes
Taro	*Colocasia esculenta*	Corms
Chinese arrowhead	*Saggitaria sagittifolia*	Corms
Chinese spinach	*Amaranthus tricolor*	Leaves, shoots
Malabar spinach	*Basella alba*	Leaves
Lotus root	*Nelumbo nucifera*	Rhizomes
Mizuna	*Brassica juncea* var. *japonica*	Leaf
Daikon	*Rhaphanus sativus*, Gr. *Longipinnatus*	Root
Yambean	*Pachyrrhizus erosus*	Tuberous roots
Chinese parsley	*Coriandrum sativum*	Immature plants
Water spinach	*Ipomoea aquatica*	Shoots
Japanese pumpkin	*Cucurbita mixta*	Fruit
Chinese okra	*Luffa acutangula* and *L. cylindrica*	Immature fruit
Chinese squash, or ash gourd	*Benincasa hispida*	Fruit
Chinese squash	*Lagenaria siceraria*	Immature fruit
Japanese pickling melon	*Cucumis melo* var. *conomon*	Immature fruit
Bitter melon	*Momordica charantia*	Immature fruit, leaves, shoots
Edible burdock	*Acctium lappa*	Roots
Garland chrysanthemum	*Chrysanthemum coronariom*	Shoots, flower buds
Butterbur	*Petasites japonicus*	Petioles

Source: From M. Yamaguchi (1973) *HortScience* **8**:362–370.
Reprinted by permission of the author.

and 7 to 11 months from planting before corms are harvestable.

There are few disease and insect pests of taro. Leaf blight (*Phytophthora colocasiae*) is more frequent in lowland than upland culture. *Pythium* occasionally can cause soft rot roots and corms. Nematode infestations will cause damage to feeder roots, with consequent poor corm development.

Harvest and Market Preparation

Harvest of taro normally is by hand. The corms are trimmed of roots and tops and washed. The tops, cut below the growing point, are used for propagation. In Hawaii, yields per acre range from 15 to 20 tons (33.6 to 45 MT/ha).

EXOTIC VEGETABLES

Several oriental vegetables have been reviewed in previous chapters, and these and other such crops (Table 21.3) may, in the future, offer enhanced potential for growers as the popularity of exotic cuisine increases. Although production of most oriental vegetables currently is concentrated in Hawaii, California, and Florida, market gardens in other states have developed in recent years to meet a growing demand for these crops in large metropolitan areas.

SELECTED REFERENCES

Chang, S. T., and W. A. Hayes (1978). *The Biology and Cultivation of Edible Mushrooms.* Academic Press, New York.

Coursey, D. G. (1968). The edible aroids. *World Crops* 20:25–30.

Lambert, E. B. (1977). *Mushroom Growing in the United States,* USDA Farmer's Bull. 1875.

Schweers, V. H., and W. L. Sims (1976). *Okra Production,* University of California Coop. Ext. Serv. Leaflet 2679.

Sims, W. L., and F. D. Howard (1979). *Growing Mushrooms,* University of California Coop. Ext. Serv. Leaflet 2640.

Wang, J.-K. (1983). *Taro.* Univ. of Hawaii Press, Honolulu.

Yamaguchi, M. (1983). *World Vegetables.* AVI, Westport, Conn.

STUDY QUESTIONS

1. What are the major okra production problems and how are they alleviated?
2. Mushrooms grow wild in areas of decomposing organic matter. Since no aseptic measures are necessary for these wild mushrooms, why is it so important to maintain cleanliness in commercial mushroom facilities?

Glossary

Abscission Natural separation of leaf or fruit from stem due to death of a layer of cells (abscission layer).

Absorption Imbibition or drawing in of water.

Achene One-seeded fruit developing from one or two carpels. Seed coat can be removed from the mature seed.

Adjuvant Material added to spray formulations to extend the spreadability or adherence of the active ingredient.

Adsorption Attraction and retention of a vapor or liquid to a surface.

Adventitious Buds, roots, or shoots arising from parts of the plant not considered as the primary source of these structures.

Aggregation A property of soil whereby soil particles clump together, thereby improving aeration and water movement.

Aleurone Outer layer of cells enclosing the endosperm of a grain.

Alkaloid Organic bases occurring in plants and containing C, N, H, and O.

Allelopathy The chemical interaction between root systems of adjacent plants. Normally considered a deleterious or protective effect.

Amino acid Basic unit of proteins. There are 20 found in living organisms, with the basic formula $NH_2–CHR–COOH$.

Ammonification Initial breakdown of proteinaceous nitrogen by soil microorganisms to form ammonia.

Androecious All flowers staminate.

Andromonoecious Flowering includes perfect and staminate blossoms.

Angiosperm Flowering plant with seed in enclosed ovary.

Annual Plant that completes its life cycle within 1 year.

Anther Staminal structure within which microspores are produced.

Anthesis Flowering stage of plants marked by rupture of anthers and release of pollen.

Anthocyanins Water-soluble pigments located within the vacuole; generally red and blue colors.

Apical dominance Predominant growth at the apical meristem due to concentration of auxins; lateral growth suppressed.

Asexual Reproduction by vegetative means, without fertilization.

Autotoxicity Allelopathic phenomenon in which a plant exudate inhibits its own growth.

Autotrophic Organisms that produce food from nonfood materials.

Auxin Group of plant growth regulators with roles in bud development and cell elongation.

Bacterium Morphologically, the simplest of the plant pathogens, characterized by rapid reproduction.

Bar Measurement of pressure, 1 bar = 0.987 atmosphere (atm); current terminology in Pascals, 1 bar = 100 kiloPascals (kPa).

Berry Fleshy fruit derived from the ovary wall producing fleshy pericarp as it develops; generally derived from several carpels.

Biennial Completing a life cycle within 2 years, or having a vernalization requirement before flowering can occur to complete the life cycle.

Blade Broad, thin portion of a leaf above the sheath or petiole.

Blanching In plant production, the withholding of light to maintain white tissue (as in cauliflower or asparagus); in processing, the heat treatment applied to kill enzymes prior to packing.

Bolting Sudden seedstalk emergence.

Botanical variety Taxonomic group of plants within a subspecies.

Brace root Adventitious root developing above ground, serving to anchor the plant; common in corn.

Bud Region of differentiated and meristematic cells from which leaves, flowers, and stems develop.

Bulb Underground storage organ consisting of fleshy, enlarged leaf bases attached to an insignificant stem plate.

Bulbil Small aerial bulb arising from axillary buds and capable of regenerating a new plant.

Bulk density Soil property measured as the weight of oven-dried soil per unit bulk volume, including air space.

Calorie Energy required to raise the temperature of 1 g of water 1°C at constant pressure.

Calyx Sepals.

Cambium Meristematic cells giving rise to conductive cells.

Cane Lignified, matured stem.

Capsaicin Phenolic responsible for the pungency of pepper.

Capsule Dehiscent fruit developing from one or more carpels.

Carbohydrate Molecule composed of C, H, and O in the ratio of $1:2:1$.

Carotenoids Plant pigments responsible for yellow or orange color, some of which have vitamin A value.

Carpel Leaflike structure bearing ovules along its margins.

Caryopsis Indehiscent fruit developing from a single carpel and containing one seed. The seed coat is indistinguishable from the integuments.

Casing soil In mushroom culture, a sterilized soil or mix applied to induce production of fruiting bodies.

Cation exchange capacity Total amount of exchangeable cations that can be held in the soil, the result of charged surfaces of organic and clay particles.

Cell membrane Membrane separating the cell wall and cytoplasm, regulating flow of material to and from cell cytoplasm.

Cellulose Carbohydrate that is the main constituent of cell walls.

Cell wall Structural limit of the cell, largely composed of cellulose.

Certified seed Seed inspected for purity, freedom from disease, and germinability by an authorized agency.

Chelate Chemical compound in which a metallic ion is held by multiple chemical bonds.

Chicon Head of witloof (forced) chicory.

Chimera Genetic mosaic, in which a specific segment of tissue is genetically distinct from the remainder of the plant.

Chlorophyll Green pigment in chloroplasts; an organic molecule that captures light energy for photosynthesis.

Chlorosis Destruction of chlorophyll due to disease or insufficient nutrient uptake.

Chroma Intensity of color of an object due to absorption of specific wavelength(s) of light.

Chromosome Rodlike structure within the nucleus composed of DNA and protein; responsible for inherited traits.

Cladophyll Photosynthetic modified leaf (the fern) of asparagus.

Climacteric Physiological stage within some ripening fruit, characterized by rapid increase in respiration.

Clone Group of plants obtained by vegetative propagation and therefore genetically identical.

Clove Section of a bulb (as in garlic) capable of regenerating a new plant.

C:N ratio Soil property related to the amount and maturity of organic matter and to the activity of microorganisms. A high C:N ratio decreases nitrogen available for plant growth, since nitrogen is an energy source for microorganisms required to degrade the carbon source.

Cold frame Forcing structure without heat, used for cool season crops and for hardening-off.

Coleoptile Leaf at the second node of the embryonic stem within a grain. It serves to protect the shoot during emergence.

Common storage Storage structure without artificial refrigeration, using a system of flues and vents to draw in cool outside air.

Companion crop Plant that enhances productivity of an adjacent different plant by repelling insects.

Compensation point Period at which the rates of photosynthesis and respiration are equal.

Competition Reduction of adjacent plant performance due to foraging within a common pool of nutrients and water.

Complete flower Flower having petals, sepals, stamens, and pistil.

Composting Process of aerobic decomposition of organic matter, generally following stacking in a pile that allows buildup of heat.

Controlled atmosphere Storage in which the atmospheric contents of oxygen, carbon dioxide, nitrogen, and ethylene are regulated.

Cooperative Federation of farmers, formed for the purpose of purchasing production inputs or marketing produce.

Corm Short, thick upright underground stem in which carbohydrates are stored.

Cormel Adventitious corm arising from the main corm.

Corolla Petals of the flower.

Cortex Tissue in the root lying between the vascular elements and the epidermis.

Cotyledon Seed leaf or leaves at the first node of the primary stem.

Cover crop Crop seeded to provide herbage growth on a soil, preventing leaching and erosion and suppressing weeds.

Cross-pollination Vectoring of pollen from anthers of one plant to the stigma of another.

Crown Mass of short rhizomes, roots, and buds providing stored carbohydrate for subsequent growth.

Cultivar Group of plants of a kind with similar economic traits.

Curd Enlarged mass of undifferentiated flower buds in cauliflower.

Curing Process by which produce is placed in an environment fostering chemical changes and/or healing of wounds.

Cutin Clear waxy material on plant surfaces.

Cytoplasm Protein matrix containing cellular organelles, but excluding the nucleus.

Damping-off Soilborne seedling disease characterized by collapse of the stem at the soil line.

Day neutral Flowering not affected by length of day.

Dehiscent Dry fruits splitting open when ripe.

Denitrification Process by which soil nitrates or nitrites are reduced to ammonia or free nitrogen by bacterial action.

Determinate Stem in which growth is terminated by a flower cluster; sometimes referred to as self-pruning.

Dicotyledonae Subclass of flowering plants characterized by two cotyledons at the first node of the primary stem.

Digestion Breakdown of complex foods into simpler ones more easily respired.

Dioecious Male and female flowers occur on separate plants.

Diploid Each sporophytic cell nucleus contains homologous pairs of each chromosome.

Dominant In genetics, a trait expressed equally as homozygote and heterozygote.

Dormancy (see *Rest period.*) State of minimum metabolic activity as a result of environmental conditions.

Double cross Cross of two F_1 hybrids.

Double fertilization Sexual fertilization in flowering plants in which one nucleus from the male gametophyte unites with the egg to form the embryo, while the other unites with two polar nuclei to form the endosperm.

Drupe Dry-fleshy simple fruit with a single seed within a hard endocarp.

Edaphic Pertaining to soil conditions.

Embryo Rudimentary plant within the seed, produced by cell division following fertilization.

Embryo sac Cell developed by mitosis in the megaspore, containing egg, polar, synergid, and antipodal nuclei.

Emergence Appearance of the plant from the soil during germination and/or early growth.

Endocarp Inner layer of cells of the pericarp.

Endodermis Single layer of cells at the inner edge of the root cortex, separating cortex from pericycle.

Endogenous Developed internally.

Endosperm Triploid tissue derived from union of a generative nucleus with two polar nuclei; the food storage tissue of seed.

Enzyme Protein functioning as a biological catalyst, regulating cellular function.

Epicotyl Portion of the embryo above the cotyledons, consisting of stem tips and embryonic leaves.

Epidermis Covering or outer layer of cells of plant structures.

Epigeal germination Cotyledons pushed above the soil surface during germination (as in snap beans).

Etiolation Spindly growth, the result of little or no light, characterized by long internodes, light green color, and soft tissue.

Evapotranspiration Total loss of water from a given leaf area through evaporation and transpiration.

Exocarp Outer layer of cells of the pericarp.

Exogenous Produced externally.

Eye Apical and lateral buds of potato tubers.

Eyebrow Vestigial abscission layer subtending each potato eye.

F₁ Cross of two inbred lines.

Fallow Maintenance of vegetation-free land for a period to minimize weed growth and to conserve moisture.

Family Taxonomic group of plants above genus and below order.

Fasciation Twisted, abnormal growth.

Fats Organic compounds of C, H, and O, in which O is in much lower proportion than in carbohydrates.

Fermentation Anaerobic respiration with end product of ethyl alcohol, lactic acid, or other substances.

Fertilization Union of sperm nucleus with egg nucleus.

Field capacity Soil water status after free water has drained under the influence of gravity.

Filament Stalk bearing the anther.

Flavonoids Group of aromatic compounds including anthocyanin (reds and blues) and flavone (yellow) plant pigments.

Floret Individual floral unit of a spikelet.

Follicle Dry, dehiscent fruit developed from a single carpel. A single line of dehiscence occurs along the adaxial suture.

Forma speciales Biotypes of fungi with specificity toward distinct genera.

Fruit Mature ovary or group of ovaries with associated parts.

Fumigation Chemical treatment applied to control pests, in which the active ingredient volatilizes within the area treated.

Fungi Organisms characterized by lack of chlorophyll and deriving energy from living or dead organic matter.

Funiculus Stalk of an ovule or seed.

Gamete Either male or female haploid cell involved in fertilization.

Gametophyte Spore that ultimately produces the male or female gametes.

Gel seeding System of precision seeding in which the seed is placed into a furrow by means of a gel carrier.

Gene Basic unit of inheritance.

Generative nucleus Nucleus responsible for fertilizing the egg; formed by a division of the pollen nucleus, which also forms the tube nucleus.

Genome Set of chromosomes corresponding to the haploid number.

Genotype Genetic constitution of an organism.

Genus Taxonomic group below family and above species.

Germination Activation of growth of the embryo, beginning with imbibition and ending with external appearance of radicle.

Glume Bractlike structures at the base of a grass spikelet.

Grain Caryopsis, or dry indehiscent fruit developing from a single carpel and containing one seed.

Green manure Crop planted for its organic matter, generally a grass and/or a legume.

Group (See *Botanical variety.*)

Growth regulator Endogenous substance affecting growth and development.

Gynoecious All flowers pistillate.

Gynomonoecious Flowers on a plant include pistillate and complete.

Haploid Represented by the basic set of chromosomes, one-half the somatic chromosome number.

Hardening-off Process by which the photosynthates produced within a young plant are not utilized in supporting rapid growth. Hardening normally is achieved by withholding water or by reducing temperature while providing full sunlight.

Hardpan Dense, cemented layer of soil providing high impedance and poor water drainage.

Hard seed Seed having a hard seed coat that prevents normal germination by restricting imbibition of water.

Heat of fusion Heat released during freezing of a liquid.

Herbaceous Soft top growth; no woody tissue.

Herbicide Chemical for eliminating weed growth.

Hermaphrodite Flowers containing functional male and female sexual parts.

Heterogeneous Variability of phenotype within a population.

Heterozygous Unlike alleles at one or more chromosome loci.

Hilum Scar on a dicot seed marking the attachment to the fruit.

Homozygous Like alleles at each locus of a chromosome.

Hormone (See *Growth regulator.*)

Hotbed Forcing structure constructed as a coldframe, but including a heat source.

Humus Well-decomposed, relatively stable organic fraction of mineral soils.

Hybrid Cross of two genotypically distinct parents.

Hydrocooling Removal of field heat from harvested produce by using cold, flowing water.

Hydrolysis Chemical reaction in which water is a reactant.

Hydroponics System of soilless culture in which the plants receive nutrients through an aqueous medium.

Hypocotyl Stem of an embryo between the root and the cotyledons.

Hypogeal germination Cotyledons remain below ground during germination.

Imperfect flower Flower including only male or female reproductive parts.

Inbred line Group of plants developed by repeated self-fertilization or otherwise homozygous at most loci.

Incipient wilting Wilting that occurs during periods of high transpiration. Plants regain turgor when conditions have eased.

Incompatibility Pollen not functional on stigma. Self-incompatibility is common in the Brassicaceae.

Incomplete flower Flower that is missing some or all of its parts.

Indehiscent fruit Fruit that does not split open at maturity to disperse seeds.

Indeterminate Stem growth terminated by vegetative bud. Flowering occurs on axillary stems.

Indigenous Native to an area or region.

Inferior ovary Ovary located completely within receptacle.

Inflorescence Structurally organized group of flowers.

Inoculum Propagules of disease infection.

Integuments Maternal tissues in the ovary that envelop the embryo sac and mature to become the seed coat, or testa.

Intercropping Growing two crops simultaneously within the same area.

Internode Stem area between leaf nodes.

Kilocalorie Amount of energy required to raise the temperature of 1 kg of water 1°C at constant pressure.

Larva Stage of complete metamorphosis in insects; the most destructive stage for many insects.

Latex Milky exudate from several plants.

Laticifers Small ducts carrying latex.

Leaching Removal of soluble materials with water.

Legume Dry dehiscent fruit developed from a single carpel. The fruit splits along two sutures at maturity.

Lemma Large bractlike structures of a single floret.

Lenticel Stomatal pores of a stem.

Lignin Complex, organic substance located in secondary cell walls; responsible for strength and also woody texture.

Ligule Structure at the junction of the leaf sheath and blade.

Limestone Calcium carbonate (dolomitic contains both calcium and magnesium carbonates).

Lipid Organic compounds composed of fats and oils.

Locule Seed cavities of tomato and pepper.

Long day Plants in which flowering occurs when the length of day exceeds a threshold.

Macronutrient Nutrient required in relatively large amounts (including C, H, O, P, K, N, S, Ca, Fe, and Mg).

Male sterility Failure of the male gametes to form or to be functional.

Matric potential Pressure at which water is held in the soil by capillary force.

Megaspore Product of meiosis in the pistil which gives rise to the egg.

Meiosis Reduction division producing megaspores and microspores, each of which carries the haploid chromosome number.

Meristem Region of active cell division.

Mesocarp Middle cell layers of the pericarp.

Micronutrient Nutrient required by the plant in relatively small amounts (B, Mn, Cu, Mo, Zn, and Cl).

Micropropagation Propagation by culture of small tissue segments on artificial nutrient medium in the laboratory.

Microspore Male gametophyte resulting from meiosis in the anther.

Middle lamella Pectic layer between adjoining primary cell walls.

Mitosis Cell division that maintains chromosome number by duplication of chromosomes.

Monocotyledonae Plants with only a single cotyledon at the first node of the primary stem.

Monoculture Cultivation of a single crop over a large area.

Monoecious Male and female flowers occur on the same plant.

Mosaic Disease symptoms often caused by virus or mycoplasmas, characterized by mottled light and dark green color.

Mulch Material applied around plants to suppress weeds and to modify the soil environment.

Mycelium Fine filamentous growth of fungi.

Mycoplasma Viruslike pathogen, normally transmitted by insects, considered as the smallest independent organism.

Necrosis Dead tissue resulting from disease, insect damage, or environmental problems.

Nitrification Change of ammonia form of nitrogen to nitrate by soil microorganisms.

Nitrogen fixation Conversion of atmospheric (free) nitrogen to a form available to plants, most often through the symbiotic relationship between rhizobia and nodules of plant roots.

Node Stem region, occasionally swollen, including an apical meristem (bud) usually subtended by a leaf or petiole.

Nodule Swollen area on legume roots within which symbiotic bacteria fix nitrogen.

No-till Plant culture without tillage, seeding through stubble left by a cover crop or previous crop treated with herbicides.

Nucleus Cellular organelle containing chromosomes; surrounded by two membranes.

Nut Dry indehiscent fruit resembling an achene, but the pericarp becomes thick and hard. It is one-seeded and may be enclosed within a fleshy involucre.

Organ Morphologically distinct part of the plant (fruit, leaf, etc.).

Organelle Functional structure within the cell (mitochondria, chloroplast, etc.).

Osmosis Movement of fluids through a semipermeable membrane until concentrations on both sides have equilibrated.

Ovary Enlarged basal portion of the pistil, containing the seeds.

Ovule Rudimentary seed containing embryo sac and unfertilized egg.

Ovum Female reproductive cell, or egg.

Palea Innermost bractlike structure enclosing a grass floret.

Panicle Inflorescence that has a branched central axis and is common among grasses.

Parasite Organism that derives its energy from a living host.

Parenchyma Tissue consisting of thin-walled unspecialized cells.

Parthenocarpy Fruit development without fertilization.

Parthenogenesis Form of reproduction in which an unfertilized egg develops into a new organism.

Pathogen Organism capable of infecting and subsequently causing aberrant growth or death of the infected plant.

Pedicel Flower or inflorescence stem or stalk.

Peduncle Stem or stalk of single flower or one arising from main stem of inflorescence.

Pepo Fleshy fruit in which the receptacle tissue fuses with the ovary wall, forming a hard outer rind.

Perennial Plant that continues to live and reproduce yearly.

Perianth Term including both calyx and corolla.

Pericarp Walls of a mature ovary or fruit.

Pericycle Layer of cells immediately inside the endodermis which give rise to branch roots.

Periderm Skin of roots or tubers or other storage organs, or cortical tissue from cork cambium.

Permanent wilting point That stage of soil moisture at which plants do not recover from transpirational wilting.

Petal Single unit of the corolla.

Petiole Stalk attaching the leaf blade to the stem.

pH Hydrogen ion concentration, a measure of soil acidity or alkalinity, expressed as the negative logarithm (base 10).

Phenotype External appearance of an organism; the result of interaction of genotype with environment.

Pheromones Chemical attractants derived from insects. Some are based on sex attractants and usually are highly species specific.

Phloem Conductive tissue responsible for transport of carbohydrates and other materials synthesized within the plant.

Photoperiod Length of day; normally used to classify plants by flowering response.

Photorespiration Phenomenon occurring in chloroplasts of certain plants by which respiration of photosynthate is stimulated by light. C_4 plants do not show photorespiration and therefore have a higher net photosynthetic rate.

Photosynthesis Conversion of water and carbon dioxide to sugar in chloroplasts exposed to light.

Phototropism Growth response to directional light. Normally plants grow toward light.

Pigment Organic compounds of variable solubility that absorb light at a specific wavelength and therefore assume a specific color.

Pistil Complete female structure of a plant.

Pistillate Female.

Pith Central region of stems consisting of thin-walled cells.

Plumule Portion of the embryo above the cotyledons.

Pod (See *Legume.*)

Pollen (See *Microspore.*)

Pollination Transferral of pollen from the anther to the stigma.

Polyploid Plant in which the chromosome is a multiple of three or more of the monoploid number.

Polysaccharide Long-chain molecules of sugars.

Postemergence Referring to herbicide application, after emergence of the crop.

Preemergence Referring to herbicide application, before the crop has emerged, but after seeding.

Prepackage In marketing, packaging the produce, often in consumer-size packages, prior to retailing or wholesaling.

Protein Structural or functional groups of linked amino acids with precise order and arrangement.

Proximal dominance Predominant growth oriented toward the mother plant, as in sprouting of sweet potato roots.

Pubescence Surface hairs on leaves, stems, or fruits.

Pupa Stage between larva and adult in insects with complete metamorphosis.

Raceme Inflorescence in which flowers on pedicels arise from a single unbranched axis.

Rachis Central axis of a spike.

Radicle Portion of the embryo that becomes the primary root.

Ray flower Characteristic of the Asteraceae, flowers at the edge of the inflorescence, often with a showy petal.

Receptacle Enlarged end of the pedicel or peduncle to which flower parts are attached.

Recessive Genetic condition in which the trait is expressed only when homozygous.

Reduction division (See *Meiosis.*)

Relative humidity Ratio of actual water vapor of air to the maximum amount that can be held at the same temperature, $\times 100$.

Resistance Genetic capacity of a plant to avoid disease symptoms, insect damage, or other effects of environmental agents.

Resistometer Device by which soil water content is measured as the electrical resistance across porous gypsum blocks.

Respiration Oxidation of food, producing energy for cell activities.

Rest period Natural arrest of metabolic activity regardless of environmental conditions.

Rhizobium Genus of bacteria that live symbiotically in root nodules of legumes, fixing free nitrogen.

Rhizome Horizontal underground stem.

Rhizosphere Area of root growth in the soil.

Rogue Inferior plant, or action of removing inferior plants.

Root cap Protective hard cells at the tip of a root.

Root hair Elongated epidermal cell of a root, active in absorption.

Rotation Successive alteration of crops on the same field.

Russet Fine surface corkiness resulting from minute growth cracks and subsequent healing.

Salinity Degree of concentration of salts in the soil solution, measured as electrical conductivity (millimhos).

Samara Dry indehiscent simple winged fruit.

Savory Crinkled or blistered leaf form.

Scape Flower stalk arising from the ground.

Schizocarp Dry indehiscent fruit consisting of two or more carpels, each of which is a mericarp containing a single seed.

Scutellum Single cotyledon of a monocot.

Secondary cell wall Cellulose matrix with other materials imbedded in it, lying on the cytoplasmic side of the primary cell wall.

Secondary phloem Phloem cells formed from the vascular cambium inside the primary phloem; found in carrot and beet.

Secondary xylem Xylem cells formed from the vascular cambium outside of the primary xylem; found in carrot and beet.

Seed Mature ovule capable of regenerating a new plant; also refers to other propagative plant parts (seed pieces, sets, etc.).

Seed coat Testa, or tissue surrounding the embryo, developed from the integuments.

Self-fertilization Union of a sperm nucleus with an egg derived from the same plant.

Self-pollination Pollen transferred to the stigma of the same flower or plant.

Senescence Process of aging and dying.

Sepals Outer floral leaflike structures at the base of the flower, collectively termed the calyx.

Sheath Supportive part of a leaf that envelops the stem, below the blade.

Short day Plant that will flower when the daylength is less than a threshold.

Silique Dry dehiscent fruit in which two carpels form a bilocular ovary with a longitudinal septum.

Single cross Cross of two genetically distinct, normally inbred lines.

Soil profile Vertical section of soil extending to the parent material, showing each horizon.

Soil solution Soil water and dissociated ions available to plants.

Soil structure Manner in which the soil particles aggregate to preserve aeration and tilth.

Soil texture Composition of a soil in terms of proportions of sand, silt, and clay.

Solarization Artificial heating of soil with solar radiation through clear plastic for the purpose of soil sterilization.

Somatic tissue Nonreproductive, vegetative tissue developing through mitosis.

Spawn Inoculum used to seed sterile medium for mushroom production.

Spear Short, unopened stem, as in asparagus.

Species Group of interbreeding individuals with certain common features; taxonomically below genus.

Specific gravity Weight of an object relative to its weight in water.

Sperm nucleus Haploid nucleus formed by division within the pollen grain that effects fertilization.

Spike Inflorescence with a central axis and sessile spikelets.

Spikelet Flowering unit of a grass, composed of one or more florets subtended by glumes.

Spore Haploid cell produced in meiosis; also a reproductive cell of fungi.

Spore mother cell Diploid cells that undergo meiosis in either the ovary or the anther.

Sporophyte Phase of plant development characterized by an unreduced chromosome number.

Stamen Male structure of the flower, including anther and filament.

Staminate Male.

Starch Complex polysaccharide of glucose.

Stem Axis of the plant supporting leaves, flowers,

and fruit, through which nutrients and plant substances flow.

Stem plate Small, flattened stem characteristic of onion.

Stigma Structure of the pistil that receives pollen.

Stipules Basal appendages of a petiole, usually two in number.

Stolon Slender above-ground horizontal stem, often termed a runner.

Stoma Opening or pore bordered by guard cells that regulate its size; site of gas exchange between leaves and atmosphere.

Strip tillage Cultural system in which the crop is planted in paths tilled in an established growing or killed sod.

Stubble mulch Cultural system in which the stubble of the previous crop is not tilled in prior to planting. (See *No-till.*)

Style Portion of the pistil between the ovary and stigma.

Suberin Waxy substance found within cork tissue.

Suberize To develop a protective corky layer, containing suberin, on cut surfaces, as in potato.

Sucker Shoot or tiller arising from an axillary node.

Superior ovary Ovary completely subtended by the receptacle.

Systemic Involving the entire metabolic system.

Tendril Modified leaves or leaflets or stems, coiled to provide support for climbing plants.

Tensiometer Instrument for measuring matric potential of soil and determining irrigation needs.

Testa (See *Seed coat.*)

Tetraploid Plant in which each somatic cell contains four basic genomes; may be autotetraploid, in which the chromosomes all develop from the same basic set, or allotetraploid, in which two asynaptic sets both double (also termed amphidiploid).

Texture Quality attribute reflecting hand or mouth feel, often determined objectively by cutting or shearing a tissue.

Threshold Level at which plant response to a stimulus changes.

Tiller Stem, usually of grass plants.

Tilth Overall physical condition of the soil for plant growth.

Tissue culture (See *Micropropagation.*)

Tolerance Reaction of a plant to disease or insect pests or to environmental factors in which the plant shows symptoms, but not sufficient to cause economic damage.

Translocate To move from one place to another, as in plants, movement of sugars from leaves to roots and stems.

Transpiration Loss of water vapor by plant tissues.

Trap crop Crop planted to attract insects, thereby protecting a crop to be harvested.

Trichome Bristle or hair from an epidermis.

Trifoliolate Compound leaf divided into three leaflets from one node.

Triploid Cell with three genomes, often derived from a cross of diploid with tetraploid.

Tuber Modified, enlarged stem, developing at the tip of a rhizome or stolon.

Umbel Inflorescence in which the pedicels of individual flowers arise from the apex of a peduncle.

Vacuole Area of the cell, bounded by a membrane, within which various substances are stored.

Vacuum cooling Cooling system based upon evaporative cooling in a vacuum.

Variety (See *Cultivar.*)

Vascular Pertaining to the plant conductive tissue.

Vector Agent of transfer, as the vector of a disease.

Vernalization Process by which reproductive growth is initiated in response to cold temperature.

Virus Minute pathogen, transmitted mechanically or by insects; a filterable organic substance.

416 / **Glossary**

Vitamin Organic substance required for animal growth, often obtained from plants.

Wax Fat derived from fatty acids and alcohols other than glycerol.

Weed Plant growing where it is not desired, often competing with established crops.

Whorl Arrangement of three or more leaves or buds in a circle at one node.

Windrow Harvested product moved into a row to facilitate drying or pickup.

Xanthophylls Yellow pigments, including carotenoids.

Xenia Immediate effect of fertilization on appearance of seed, due to dominant gene in sperm nucleus uniting with recessive polar nuclei.

Xylem Cells through which water and minerals move upward through the stem.

Zygote Cells resulting from the fusion of male and female gametes.

Appendix

Conversion Factors for Metric and U.S. Units

To convert column 1 into column 2, multiply by	Column 1	Column 2	To convert column 2 into column 1, multiply by
	Length		
0.621	kilometer (km)	mile (mi)	1.609
3.281	meter (m)	foot (ft)	0.305
0.394	centimeter (cm)	inch (in.)	2.54
	Area		
0.386	square kilometer	square mile	2.59
247.1	square kilometer	acre	0.00405
2.471	hectare (ha)	acre	0.405
	Volume		
0.00973	cubic meter	acre-inch	102.8
35.32	cubic meter	cubic foot	0.0283
0.0353	liter	cubic foot	28.32
0.0284	liter	bushel (bu)	35.24
1.057	liter	quart (qt)	0.946
	Mass or weight		
1.102	metric ton (MT)	ton	0.9072
2.205	quintal (q)	hundredweight (cwt)	0.454
2.205	kilogram (kg)	pound (lb)	0.454
0.035	gram (g)	ounce (oz)	28.35
	Pressure		
14.22	kilogram per square centimeter	pound per square inch (psi)	0.0703
14.50	bar	pound per square inch	0.06895
0.9869	bar	atmosphere (atm)	1.013
0.9678	kilogram per square centimeter	atmosphere	1.033
	Rate		
0.446	metric ton per hectare	ton per acre	2.24
0.892	kilogram per hectare	pound per acre	1.12
0.892	quintal per hectare	hundredweight per acre	1.12
0.107	liter per hectare	gallon per acre	9.354
	Water measurement		
8.108	hectare meter	acre-foot	0.1233
97.29	hectare meter	acre-inch	0.01028
0.08108	hectare centimeter	acre-foot	12.33
0.973	hectare centimeter	acre-inch	1.028
0.00973	cubic meter	acre-inch	102.8

Conversion Factors for Metric and U.S. Units
(*Continued*)

To convert column 1 into column 2, multiply by	Column 1	Column 2	To convert column 2 into column 1, multiply by
0.981	hectare centimeter per hour	cubic foot per second	1.0194
440.3	hectare centimeter per hour	gallon per minute	0.00227
0.00981	cubic meter per hour	cubic foot per second	101.94
4.403	cubic meter per hour	gallon per minute	0.227
		Temperature	
$(9/5)(°C) + 32$	Celsius	Fahrenheit	$(5/9)(°F - 32)$

Index